FOOD SCIENCE AND TECHNOLOGY

PRACTICAL FOOD AND RESEARCH

FOOD SCIENCE AND TECHNOLOGY

Additional books in this series can be found on Nova's website under the Series tab.

Additional E-books in this series can be found on Nova's website under the E-books tab.

FOOD SCIENCE AND TECHNOLOGY

PRACTICAL FOOD AND RESEARCH

RUI M. S. CRUZ
EDITOR

Nova Science Publishers, Inc.
New York

For permission to use material from this book please contact us:
Telephone 631-231-7269; Fax 631-231-8175
Web Site: http://www.novapublishers.com

LIBRARY OF CONGRESS CATALOGING-IN-PUBLICATION DATA

Practical food and research / editor: Rui M.S. Cruz.
p. cm.
Includes bibliographical references and index.
ISBN 978-1-61728-506-6 (hardcover)
1. Food--Analysis. 2. Food--Composition. I. Cruz, Rui M. S.
TX541.P695 2010
664'.07--dc22

2010027143

Published by Nova Science Publishers, Inc. ✝ New York

CONTENTS

PREFACE

The field of food science deals with numerous processes and phenomenons. The development of processing treatments in order to stabilize foods and/or develop new food products is common in research. This book presents a realistic approach about the most concerning food issues in food processing, particularly in processing by heat, such as: enzymes, microorganisms, vitamins and physical properties. Focusing in their action mechanisms, research analyses, statistical analysis and the application of innovative technologies to reduce or replace the severity of heat treatment processing.

This book is intended as a reference book for food engineers and researchers. Students at undergraduate and postgraduate levels in food science will also benefit greatly from the contents of this book.

This book is the combined effort of several contributors from different countries, with expertise in food processing, that greatly improved the quality of the manuscript. The book has 16 chapters and contains 3 main parts, each dealing with important quality parameters in the processing of Fruits/Vegetables, Meat/Fish and Milk/Dairy Products. The 4th part is dedicated to application of Statistical Analysis in Food Science.

Rui M. S. Cruz

ACKNOWLEDGMENTS

I would like to recognize and thank each one of the experts and researchers for their effort and making this book a reality with their contributions. I also wish to express my gratitude to the editorial staff at Nova Science for the opportunity and support.

I would like to dedicate this work to my dear Mother Deolinda Cruz, my dear Brothers Jaime Cruz and Cândido Cruz and to my dear Niece Laurinha Cruz and Nephews Jaiminho Cruz and Martim Cruz. A final and special dedicatory goes to my dear Father Jaime Mariano da Cruz, whose love and care will always perpetuate in my memory.

Rui M. S. Cruz

PART I

FRUITS AND VEGETABLES

In: Practical Food and Research
Editor: Rui M. S. Cruz, pp. 3-50

ISBN: 978-1-61728-506-6

Chapter I

COLOUR AND PIGMENTS

P. J. Cullen[1], B. K. Tiwari[2], K. Muthukumarappan[3] and Colm P. O'Donnell[2]

[1]School of Food Science and Environmental Health, Dublin Institute of Technology, Dublin 2, Ireland, pjcullen@dit.ie

[2]UCD School of Agriculture, Food Science and Veterinary Medicine, University College Dublin, Belfield, Dublin 4, Ireland, brijesh.tiwari@ucd.ie and colm.odonnell@ucd.ie

[3]Department of Agricultural and Biosystems Engineering, South Dakota State University, SAE 225/Box 2120, Brookings, SD 57007, USA, muthukum@sdstate.edu

ABSTRACT

Plant pigments including anthocyanins, carotenoids, betalains and chlorophyll are principally responsible for the characteristic colour of fruits and vegetables and their products. These pigments have attracted attention due to their biological and health related functions. Apart from their role in human health these pigment contribute to colour which is one of the most important parameters of fruit and vegetable quality, greatly influencing consumer acceptance. This chapter provides fundamental information about these colour pigments, reviews the effects of various food processing operations on pigment stability and discusses proposed degradation mechanisms.

1.1 INTRODUCTION

Pigments present in fruits and vegetables are chemical compounds that absorb light in the visible wavelength range of the electromagnetic spectrum (380-750 nm). Natural pigments that produce colour are specific to chromophores [1, 2]. Plant compounds that are perceived by humans to have colour are generally referred to as 'pigments' [3]. Epidemiological studies suggest that consumption of foods containing natural colour pigments such as anthocyanins, beta-carotene, lycopene and other carotenoids may be effective against certain types of cancer [4]. These plant pigments may contribute to the prevention of degenerative processes, particularly lowering the incidence and mortality rate of cancer and cardio- and cerebrovascular diseases [5]. Interest in biological effects of natural pigments has increased substantially due to their reported therapeutic effects including antioxidant capacity, inhibition of cancerous cells growth, inflammation and anti-obesity affects etc [6-8]. The dietary role of plant pigments along with their preventative role in various diseases is well documented [9-14]. The major classes of plant pigments such as chlorophylls, anthocyanins, carotenoids and betalains found in fruits and vegetables are influenced by various food processing operations. These may degrade due to various factors including; pH, light, oxygen, enzymes, ascorbic acid and thermal treatment [15-17]. Natural plant pigments imparting characteristic colour is a primary factor considered by the consumer in determining quality and can be correlated with both sensorial and nutritional quality attributes [18]. This chapter deals with the changes in fruits and vegetables pigments due to processing, degradation mechanisms and evaluation techniques.

1.1.1 Food colour

Pigments present in fruits and vegetables are natural substances in cells and tissues that impart colour. Natural pigments present in plants can be classified based by (a) the chemical structure of the chromophore; (b) their origin; (c) the structural characteristics of the natural pigments or (d) food additives [2]. Natural pigments of plant origin can be divided into four major groups; carotenoids, chlorophylls, anthocyanins and betalains. However, based on the chemical structure, plant pigments can be classed into five families i.e. tetrapyrroles (e.g. chlorophyll), carotenoids (e.g. β-carotene), flavonoids (e.g. anthocyanins), phenolic compounds (e.g. catechin) and N-heterocyclic compounds (e.g. betalains) [19].

1.1.1.1 Anthocyanins

Anthocyanins (in Greek *anthos* means flower, and *kyanos* means blue) are the more important plant pigments visible to the human eye [20, 21]. Anthocyanins are also the largest family of coloured compounds. They are responsible for colour ranges from pink and salmon, through scarlet, violet to purple, and blue of a large variety of flowers, petals, leaves, fruits and vegetables and are a sub-group within the flavonoids characterized by a C-6–C-3–C-6-skeleton [22, 23]. Anthocyanins are bioactive compounds and s are believed to provide a broad variety of health benefits such as prevention of heart disease, inhibition of carcinogenesis, and anti-inflammatory activity in the brain [13]. Anthocyanins occur ubiquitously in the plant kingdom and confer bright red or blue colouration on berries and other fruits and vegetables. Apart from imparting colour to plants, anthocyanins also have an array of health-promoting benefits, as they can protect against a variety of oxidants through various mechanisms.

Anthocyanins are found in various fruits and vegetables e.g. fresh red grapes, strawberries, blackberries, raspberriesy and their products. To date about 550 anthocyanins have been identified in nature [24]. Table 1.1 shows some examples of fruits containing anthocyanins. Since anthocyanins impart characteristic colour to fruits and vegetables they influence a key quality parameter in influencing consumer sensory acceptance [25]. Berry fruits are often processed into juice and juice concentrate or purees in order to reach a more widespread market. Anthocyanin content can be greatly influenced by various genetic (cultivar), environmental and agronomic factors. Anthocyanins are highly unstable and very susceptible to degradation [26]. Their stability is greatly affected by several factors such as pH, storage temperature, chemical structure, concentration, light, oxygen, solvents, presence of enzymes, flavonoids, proteins and metallic ions [27]. Because of their possible beneficial effects, particular attention has to be given to the changes that anthocyanin pigments undergo during various food processing operations. Anthocyanins, as well as other phenolic compounds, are easily oxidized and, thus, susceptible to degradation reactions during processing (section 1.3.3).

Table 1.1. Presence of anthocyanins in some fruits and vegetables.

Fruits and vegetables	Major anthocyanins	Minor anthocyanins
Strawberry	Pelargonidin-3-glucoside	Cyanidin- 3-glucoside, pelargonidin-3-rutinoside
Blackberry	Cyanidin-3-glucoside	cyanidin-3-rutinoside, malvidin-3-glucoside
Raspberry	Cyanidin-3-glucoside	Pelargonidin- 3-glucoside, Pelargonindin-3-rutinoside
Sweet cherry	Cyanidin-3-rutinoside	Cyanidin-3-glucoside, Peonidin-3-rutinoside
Blackcurrant	Cyanidin-3-rutinoside	Cyanidin 3-glucoside, Delphinidin-3-glucoside
Bilberry	Delphinidin-3-galactoside	Peonindin-3-glucoside, Peonindin-3-galactoside
Red onion	Cyanidin-3-glucoside	Delphinidin 3-glucoside, Petunidin glucoside
Blood orange	Malvidin-3-glucoside	Malvidin-3-acetylglucoside

1.1.1.2 Betalains

Betalains are water-soluble nitrogen-containing pigments [28], which comprise the red-violet betacyanins (from Latin *beta*, red beet and *kyanos*, blue colour) and the yellow betaxanthins (from Latin *beta*, red beet and Greek *xanthos*, yellow). Betacyanin structures show variations in their sugar (e.g., 5-O-D-Glucose) and acyl groups (e.g., feruloyl), whereas betaxanthins show conjugation with a wide range of amines (e.g., glutamine) and amino acids (e.g., tyrosine) in their structures [1]. Compared to other pigment classes such as the carotenoids, chlorophylls and anthocyanins, the betalains have been studied to a lesser degree [29]. To date, the betalains comprise of about 55 structures including the red-violet betacyanins and the yellow-orange betaxanthins [29]. The betalains in red beet (*Beta vulgaris* L.) consist of betanin, isobetanin, betanidin, isobetanidin, betaxanthins and some other yellow pigments [30, 31]. Table 1.2 lists some of the health related functions of betalains.

Betalains are the red and yellow plant pigments obtained from members of the order Centrospermae, the most food-significant of these being beetroot (*Beta vulgaris* L.). Betalains are divided into two groups, the betacyanins which are purplish-red in colour and the less common yellow-coloured betaxanthins. Betacyanins differ from other naturally occurring water-soluble plant pigments, especially anthocyanins, in that their colour is not significantly affected by pH changes in the range normally encountered in foodstuffs [37]. Betacyanins are relatively stable under food-processing conditions, although heating in the presence of air at neutral pH causes breakdown to brown compounds. Davidek et al. [38] reported that the most important betalain is betanin (or phytolaccanin), the β-D-glucopyranoside of betanidin, which may be enzymically hydrolysed to the corresponding aglycon. In the presence of acids, it is transformed into its isomer, and further, to yellow betalamic acid products, containing an open ring system, and finally to brown products [39]. In alkaline medium, the red-violet pigment is decomposed into colourless products [40]. Vulgaxanthin I, the most important betaxanthin, and its amide vulgaxanthin II, are hydrolysed in acid medium to amines or amino acids bound to the dihydropyridine moiety. Like many other natural pigments, betalains are very sensitive to heat, light and oxidation, especially caused by peroxidases which is one of the major causes of discolouration of the pigment [41-44]. The stability of betalains during food processing is influenced by many factors, the most significant of which are temperature, pH, a_W, M^{n+}, O_2, and hυ [45].

Table 1.2. Some functions of betalains.

Function	References
Antioxidant	[32]
Radical scavenging activities	[33, 34]
Oxidative stress-related disorders	[32]
Chemoprevention against lung and skin cancers	[35]
Inhibition of cell proliferation	[36]

1.1.1.3 Chlorophyll

Chlorophyll is a fat soluble tetrapyrrole pigment, occurring in chloroplasts of green plants, photosynthetic bacteria and algae [46, 47]. Plants are predominantly characterized by the presence of chlorophylls (*chloros* "green" and *phyllon* "leaf") that are crucial for photosynthetic activity. Chlorophyll absorbs light most strongly in the blue and red regions but poorly in the green region of the electromagnetic spectrum, hence the green colour of chlorophyll-containing tissues like plant leaves.

Chlorophylls, are susceptible to degradation during processing, resulting in colour changes in food [48]. The major chlorophylls in plants include chlorophyll a and chlorophyll b, which occur in an approximate ratio of 3:1 [49]. Chlorophyll a has a methyl group at the C-3 carbon, while a formyl group is bonded to the same carbon atom in chlorophyll b. In addition to structural differences between chlorophyll a and b, their thermal stabilities are also different. Chlorophyll a was reported to be thermally less stable than chlorophyll b [48, 50-54]. Chlorophyll retention has been used as a measure of quality in green vegetables [55].

Chlorophyll destruction can proceed as an acid-, base- or enzyme catalysed reaction. Weak acids liberate the Mg atom bound to the porphyrin ring to form pheophytins by substitution with hydrogen [19]. This results in a colour change from green to dull brown. Magnesium may be replaced by copper (as in copper chlorophyll additives, i.e. chemically modified chlorophylls) and by tin and zinc. Alkaline salts may also be produced, but these are very unstable [56]. Chlorophylls may also undergo photo-oxidation accompanied by the loss of colour. The rate of oxidation has been shown to be dependent upon water activity and the temperature and duration of blanching in dehydrated products. Lipoprotein-bound chlorophylls are somewhat protected against acids, but can be affected by cooking and processing. In alkaline media, the decomposition of chlorophylls is very rapid and they are not stable against the action of free-radicals, e.g. during lipoxygenase- catalysed oxidation of lipids, probably due to the effect of hydroperoxides. The allomerization reaction of chlorophylls takes place spontaneously in a polar medium and is metal-ion catalysed; allomeric 10-hydroxychlorophylls and 10-methoxylactones have been detected [57]. Chlorophylls also form Schiff bases, the colour maximum of which is pH dependent [58]. For example the major organic acids involved in the degradation of chlorophylls are acetic acid and 5-oxopyrrolidinecarboxylic acid [59]. Chlorophyll destruction occurs most readily at acidic pH, with little destruction occurring above pH 8 [56].

1.1.1.4 Carotenoids

Carotenoids form one of the most important classes of plant pigments and play a crucial role in defining the quality parameters of fruit and vegetables. Carotenoids are fat soluble colour compounds that are associated with the lipidic fractions. This class of natural pigment occurs widely in nature. Like other plant pigments these also have some health related functions as shown in Table 3. Carotenoids are synthesized by plants and many microorganisms with more than 600 carotenoids isolated from natural sources [60-62]. Carotenoids are natural pigments present in chromoplasts of plants and some other photosynthetic organisms (e.g. algae). Chemically carotenoids are polyisoprenoid compounds and can be divided into two main groups: (a) carotenes and (b) xanthophylls [62]. Table 1.3 shows some of the functions for carotenoids. Apart from nutritional benefits, carotenoids also have photooxidative protection and photosynthesis functions [63]. Carotenoid content in fruits and vegetables depends on several factors such as, genetic variety, maturity, post-harvest storage, and processing and preparation parameters.

Table 1.3. Some functions of carotenoids*.

Carotenoids	Function	References
β-Carotene, α-carotene, β-cryptoxanthin	Provitamin A activity	[63]
All carotenoids	Antioxidant	[64]
β-Carotene, canthaxanthin, cryptoxanthin	Cell communication	[65]
β-Carotene	Immune function enhancers	[66]
β-Carotene		[64]
β-Carotene, lycopene	UV skin protectant	[67]
Lutein, zeaxanthin	Macula protection	[68]

*adapted from [70]

1.2 Evaluation Techniques

Colour is a fundamental property of food products, playing a major role and is consequently one of the most common quality control test carried out in industry. Fruits and vegetables and their products exhibit a wide spectrum of colour. Presently, a number of approaches are available for colour measurement but three methods seem to be widely adopted [18]. The first is a series of visual systems in which the object under consideration is compared with a series of visual standards. The second is a physical system in which a reflection or transmission spectrum is obtained and either used directly or converted into a tristimulus system. The third is a system of tristimulus colorimetry in which the signals from a sample by reflection or transmission are calculated directly into units related to human vision hence, restricted to the visible light region.

The colour of fruits and vegetables can be described using several colour coordinate systems [71-75] namely RGB (red, green and blue), Hunter *L a b* and CIE (Commission Internationale de l'Eclairage) L^* a^* b^*, CIE *XYZ*, CIE L^* u^* v^*, CIE *Yxy*, and CIE *LCH*. These colour systems differs in the symmetry of the colour space and in the coordinate system used to define points within that space [71]. The tristimulus methods based on CIE and Hunter colour system are of greatest importance to instrumental colour measurement of fruits and vegetables. According to the CIE concept, the human eye has three colour receptors—red, green and blue—and all colours are combinations of these. These instrumental measurements provide a consistent measurement of the true surface colour of the fruit and vegetables. However, current laboratory practices for instrumentally measuring colour changes of fruit have a limited viewing area (2-5 cm^2), and consequently are unable to capture and describe the entire fruit in a single measurement. Descriptions of heterogeneous fruit colour are only possible through measurement at multiple locations. Also errors are introduced due to choice of location, when tracking the quality changes of the same produce as they mature [76].

Colour can be expressed in variables that correspond to the colour perception of the average person. One system using this approach is the CIE-Lab system, with the *L, a* and *b* values expressing the 'brightness', the 'green–red' and the 'blue–yellow' axis, respectively. The CIE-Lab system is frequently used as a versatile and reliable method to assess the colour of fruit and vegetables and any changes during storage and processing [77-81]. Kidmose and Hansen [82] reported a good relationship between instrumental colour, sensory yellowness and chlorophyll content in cooked and stored broccoli florets. These chlorophyll pigments degrade during storage, acid and heat treatments, resulting in green colour changes. Colour can also be rapidly analyzed by image analysis techniques, also known as computer vision systems (CVS). CVS can be employed for measurement of uneven colouration and also the other attributes of total appearance [83]. CVS is a computerized image analysis technique, which overcome the deficiencies of visual and instrumental techniques and offer an objective measure for colour and other physical factors [84, 85]. CVS finds applications in classification and quality evaluation of various fruits and vegetables such as apples [86-88], chicory [89], banana [90], pomegranate [91] and mango [76]. Mendoza and Aguilera [90] employed CVS for colour measurements and other appearance parameters and observed that they closely correlate to parameters obtained from visual assessment and colourimeters. They employed CVS to identify the seven ripening stages of bananas and calibrated the computer vision system to quantify colour changes during ripening using the L^*, a*, b^* colour space. Kang et al. [76] employed CVS to characterise the colour change of a bicoloured mango fruit during storage (Figure 1.1).

Various dedicated commercial CVSs are available for a variety of industrial applications. These are especially recommended for measurement of colour assessments of samples with curved and irregular shapes such as whole fruits and vegetables. The knowledge of these effects, such as the variations of L^*, a^*, and b^* for a particular shape of the sample, could be useful for developing image processing correction algorithms which can permit a better correlation between product quality by CVS and human vision evaluations [92].

1.2.1 Computer vision system (CVS)

A CVS system typically consists of the following elements: standard illuminants, a digital or video camera for image acquisition, and computer software for image analysis.

1.2.1.1 Lighting

Samples are illuminated using lamps with a standard light source, D_{65} is commonly used in food research. Usually, light diffusers cover each lamp and electronic ballast assures a uniform illumination. Consistent illumination is critical for reproducible imaging [85].

1.2.1.2 Digital camera and image acquisition

A colour digital camera is generally employed to capture images. The angle between the camera lens and the lighting source axis should be approximately 45°, since the diffuse reflections responsible for the colour occurs mainly at this angle from the incident light [93].

1.2.1.3 Image processing

After image acquisition standard algorithms for pre-processing of full images, segmentation from the background, and colour analysis are done using standard image analysis statistical packages. Colour space transformation is the most common pixel pre-processing method for food quality evaluation [94]. Where colour space transformations are used for image pre-processing, hue, saturation, and intensity (HSI) colour spaces are typically employed [95, 96]. However, $L^*a^*b^*$ colour space has also been used to perform image pre-processing [97]. HSI is an effective tool for differentiating colour images. Usually, colour images are taken by a digital device and saved in a 3-dimensional RGB (red, green, and blue) colour space.

1.2.2 Juice colour

Juice colour measurement is primarily done by employing colorimetric methods. However several techniques and instruments have been developed over the years. Juice colour is mainly due to the presence of various colour pigments such as carotenoids and anthocyanins. They provide the natural yellow, orange or red colours of many fruits and vegetable juices including tomato, carrot, orange, apple and grape. Citrus juice colour is of great importance in quality control for the evaluation of the degree of fruit maturity or for differentiation between varieties [98]. The United States Department of Agriculture (USDA) assigned 40 points out of a scale of 100 points for the commercial classification of orange juices according to their colour number [99]. For example, a grade A orange juice must have a colour number between 36 and 40 points, whereas grade B orange juices are those with colour numbers ranging from 32 to 35 points [100, 101]. Similar standards are also available for grapefruit juices [102]. However, such legislation for juice colour does not exist in the

European Union (EU). EU legislation requires that all fruit juices must have the typical colour, aroma and taste of the source fruit [99, 103].

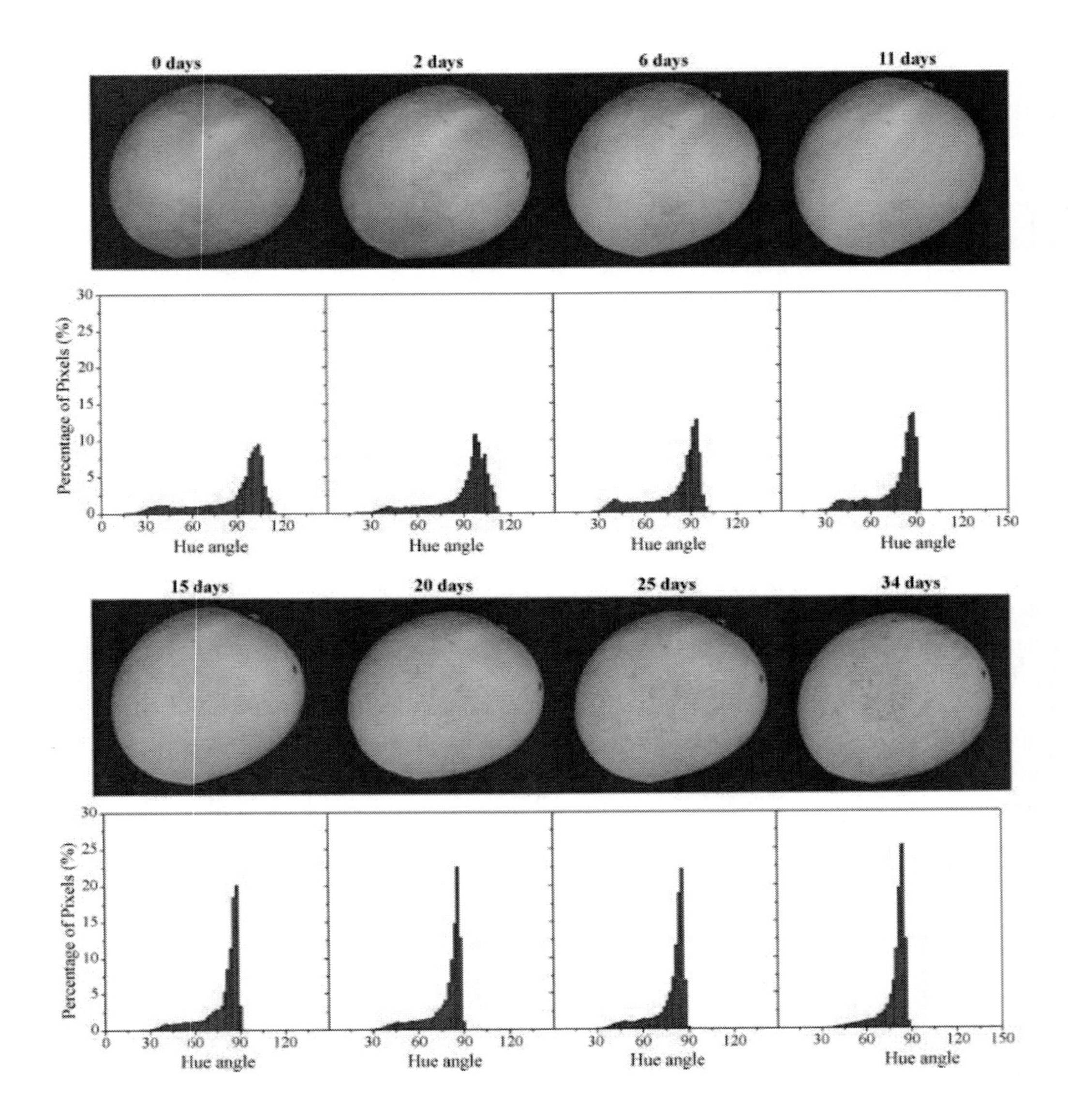

Figure 1.1. Colour change (hue angle) of an individual mango stored at 17 °C. Histograms below the images show the proportional range of hue angles observed at each time frame. Captured images and histograms of the colour change of a single fruit stored at 17 °C. The captured images clearly demonstrate the shift in background colour from green to yellow as the fruit ripens. Analysis of the images quantifies this colour change with the largest proportion of colours (pixels) shifting from a hue angle range of 95-120 (shades of green) to the 60-95 range (orange to yellow). Also noticeable is a shift in variation in colour of the fruit as it matured. At the beginning of storage (day 0), background colour spanned a large range of hue angles, while at the completion of maturation, background hue angle was compressed to a small range, as demonstrated by larger peaks [76].

As with whole fruits and vegetables, colour measurements in juice samples can be carried out either by visual evaluation or instrumental analysis. Visual analyses of juice colour include sensorial analysis. Visual assessment of colour can be also carried out by means of colour scales or atlases containing comparative standards, which are cheap in comparison to the instrumentation methods employed for objective colour measurement. In the case of instrumental measurement, juice colour is expressed by means of the colour coordinates as previously discussed. There are 3 types of instruments used for the objective measurement of colour: colorimeters, spectrophotometers and spectroradiometers [104]. Colorimeters measure the colour of primary radiation sources, which emit light, and secondary radiation sources, which are those that reflect or transmit external light. Spectrophotometers measure the spectral distribution of transmittance or reflectance of the sample. From these measurements colour is calculated under different conditions. The X, Y, Z values obtained depend on the illuminant, the measurement geometry and the observer [105]. Both transmittance and reflectance are inherent and relative properties of the objects, which do not depend on either the illumination or the observer, whereas, as has been mentioned before, colour depends on both. Transmittance measurements by means of a spectrophotometer, is the ratio between the response when the sample is in the optical pathway of the instrument and the response when the sample is not present. Spectroradiometers were designed for the measurement of radiometric quantities as a function of wavelength [104]. Tristimulus values are also mathematically obtained. This apparatus has the same components as a spectrophotometer, with the exception of the light source, which is external. Spectroradiometers can be also used for transmittance or reflectance measurements of an object.

1.3 Colour Stability During Food Processing

Fruits and vegetables are subjected to processing prior to consumption to increase their edibility and palatability. Fruits and vegetables processing also aims to prolong shelf life within microbial safety constraints [106]. The majority of fruits and vegetables are processed in the form of processed products such as juice, paste, purée, jam, marmalade, jellies and also minimally processed fruits and vegetables. Heat processing coupled with hermetic packaging is used to preserve a wide variety of products such as fruits, vegetables and various meat products. During thermal processing foods are generally heated to temperatures in the range of 50-90 °C or 100-150 °C, depending upon the pH of the product and desired shelf life. Thermal processes are established based on the heat resistance of the target microorganisms for each specific product formulation and composition, and the heating rate of the specific product [107]. Heat applied during any given stage of processing induces physical and chemical changes. Changes brought about as a result of thermal processing can be either desirable or undesirable and are influenced by the time and temperature of the process [108]. Crucially the application of heat can influence the bio-active properties of colour bearing pigments such as carotenoids and anthocyanins. These pigments influence both the quality (via their influence on colour) and nutritional properties (via their bio-active potential) of many fruits and vegetables.

1.3.1 Blanching

Blanching is one of the most common unit operations in fruit and vegetable processing and is frequently carried out prior to freezing. It is employed to inactivate enzymes and to remove occluded oxygen that causes undesirable changes during the processing and subsequent storage of the products. It is generally employed to decrease the microbial load and improve colour stability and texture [109]. Blanching inactivates enzymes responsible for the oxidative degradation of colour pigments including chlorophyllase, polyphenols oxidase and peroxidases [110, 111]. Ndiaye et al. [112] studied the heat stability of peroxidase and polyphenoloxidase in mango (*Mangifera indica* L.) slices, and observed complete inactivation after 5 min for POD and 7 min for PPO. Tijskens et al. [113] studied the effect of pH on the colour stability of broccoli and reported that the colour of blanched and thawed broccoli strongly depends on the pH of the surrounding environment. Although bacterial safety increases through the application of low pH values, the green colour of vegetable products rapidly turns to an undesirable olive green.

Thermal treatment to inactivate most heat resistant peroxidases is generally sufficient to destroy undesirable enzymes, since peroxidases activity has not been shown to be directly responsible for quality deterioration during frozen storage of vegetables. However the use of lipoxygenase as an indicator of blanching has been recommended for determining the storage stability of frozen vegetables [114, 115]. Lipoxygenase has been associated with quality deterioration because of its involvement in off-flavour and odour production, loss of pigments such as carotenes and chlorophylls, and destruction of essential fatty acids [116]. Lipoxygenases can be thermally inactivated above 60 °C with a resulting improvement in the shelf life of foods. However, heating also increases non-enzymatic oxidation and thus may exceed the oxidation due to lipoxygenase [117]. During blanching an initial brightening of colour is observed due to the removal of gases and air around fine hairs on the surface of the leaves along with occluded air from extracellular space [118]. With further heating, an undesirable fading of the colour from bright green to olive brown appears, attributed to the loss of the central magnesium atom in the tetrapyrrole ring [48]. The latter component is very labile, especially at low pH values. During the blanching treatment, organic acids are released from plant tissues, thus causing a pH decrease and consequential conversion of chlorophylls to pheophytins [119]. Heaton et al. [121, 122] monitored the degradation of chlorophyll to pheophorbide in coleslaw. The concentration of pheophorbide in this system remained unchanged upon further storage, suggesting that pheophorbide was the end product of chlorophyll degradation. Similar observations were noticed by Minguez-Mosquera et al. [122] and White et al. [123]. This is quite different from the mechanism outlined in Figure 1.3, where pheophorbide is only an intermediate catabolite of chlorophyll. Heaton et al. [121] did, however, notice that the concentrations of chlorophyll, pheophytin, chlorophyllide and pheophorbide decreased in whole, unprocessed cabbage over a period of 150 days during cold (5 °C) storage. Heaton et al. [121] observed that the colour change was from green to a lighter white colour, thus suggesting that the chlorophyll in whole cabbage could have been degraded to fluorescent compounds (colourless) or even rusty pigments, which would account for the loss of green colour and lack of accumulation of the other pigments. They suggested that the colour change (discolouration) in coleslaw is due to pheophorbide accumulation, which in turn is due to chlorophyllase activity; therefore any treatment (i.e., blanching and/or acidification) designed toward the inhibition of this activity would successfully prevent the colour change from occurring. Weemaes et al. [124] modelled the thermal degradation of chlorophyll in broccoli, based on separate but concomitant degradation of chlorophyll a and b. The physical colour, on the other hand, was modelled by a consecutive reaction mechanism, based on the conversion of chlorophyll to pheophytin and further to pyropheophytin. Minguez-Mosquera et al. [125] studied the change in green colour of green table olives based on chemical measurements of chlorophyll compounds.

1.3.2 Degradation mechanism

During blanching and other processing steps, tissue is damaged, initiating chlorophyll degradation [126]. Chlorophyll, the pigment responsible for the characteristic green colour of several fruit and vegetables, can degrade to undesirable grey-brown compounds such as pheophorbide and pheophytin. This degradation is mediated by acid and the enzyme chlorophyllase. Pheophorbide can be further metabolized to colourless compounds in metabolically active tissue. Lower pH leads to faster degradation. Due to the high concentration of hydrogen ions, the magnesium in the centre of the chlorophyllic group is rapidly replaced by hydrogen ions to form pheophytin. This constitutes one of the steps in the well-known pathway of chlorophyll degradation [126-128]. The characteristic green colour of vegetables such as broccoli (*Brassica oleracea*), spinach, green peas, green table olives etc degrades by a reaction mechanism based on the conversion of chlorophyll to pheophytin and further to pyropheophytin [125]. The replacement of Mg^{2+} in the chlorophyll compounds by action of H^{+} ions leads to the formation of pheophytin and can directly be attributed to some catalytic action of hydrogen ions indicating the importance of pH in chlorophyll degradation [129]. Changes in chlorophyll content have been monitored by the ratio of a to b value (−a/b) for canned green peas [130], blanched and frozen broccoli [81] and for canned green beans [131].

Thermal blanching is widely adopted to reduce the microbial load and inactivate deleterious enzymes to enhance shelf-life and stability during freezing. Apart from the traditional blanching treatments using boiling water and steam to inactivate enzymes, the use of microwaves to reduce PPO activity in strawberry [132] and banana slices [111]. The combination of steam and microwave for blanching and subsequent osmotic dehydration of strawberries has been reported [132].

Application of ohmic heating is also reported for blanching of fruits and vegetables. Mizrahi (1996) reported that the blanching of vegetables by ohmic heating considerably reduces the extent of solid leaching as compared to hot water processing and has a short blanching time regardless of the shape and size of the product. Ohmic heating is of advantage especially for large vegetables, having a relatively small surface to volume ratio, allowing blanching in a short time without cutting. The energy level dissipated by the electric current passing through the product, is capable of heating it uniformly and quickly regardless of its shape or size. In contrast, dicing is required in order to maintain a reasonably short water-blanching time. A reduction of one order of magnitude in solute losses may be achieved when blanching by ohmic heating due to a favourable combination of low surface area to volume ratio and short process time. Icier et al. [134] studied the application of ohmic blanching for pea purée and observed that the peroxidase enzyme inactivation occurs at a lower processing time than conventional water blanching.

1.3.3 Effect of Heat on Anthocyanins

Food processing operations involving heat such as blanching and pasteurisation can markedly affect the anthocyanin content of fruits and vegetables. A study on the stability of anthocyanins in red cabbage has shown that all processes involving heat leads to a significant loss in anthocyanins of 59%, 41% and 29% during blanching, boiling and steaming, respectively [135]. Similar results were reported by Lee et al. [136]; Srivastava et al. [137] in blueberry products. Conversely, Rossi et al. [138] suggested that the addition of a blanching step in juice processing may be beneficial, when processing fruit products for their health effects as blanching inactivates polyphenol oxidase. Kirca et al. [139] reported that anthocyanins from black carrots were reasonably stable during heating at 70-80 °C, which is in accordance with the kinetic data reported by Rhim [140] on the thermal stability of black carrot anthocyanins between 70 and 90 °C.

Hager et al. [141] found that processing of canned berries in water or syrup resulted in total anthocyanin losses of 42% and 51%, respectively. The authors suggested that approximately 16% of the original concentration of anthocyanins was lost during juice pressing of black raspberry, and 1.3% of the original concentration of anthocyanins was removed as sediment in the juice clarification step. Consistent with the above findings, significant losses (up to 65%) of anthocyanins during the primary steps of juice processing (thawing, crushing, depectinization, and pressing) were previously reported for blueberries [137, 138, 142], and strawberries [143].

Skrede et al. [142] demonstrated that addition of a blanched blueberry-pulp extract to blueberry juice resulted in no degradation of anthocyanins, whereas addition of an unblanched extract caused a 50% loss of anthocyanins, suggesting an enzymatic role in anthocyanin degradation. Similarily, Lee et al. [136] reported that less than 20% of anthocyanins were retained in clarified blueberry juices and found that only 8% of the anthocyanins were lost during the clarification step, as opposed to 25% loss in a study carried out by Brownmiller et al. [144] during clarification. In a study conducted by Dyrby et al. [145] greater anthocyanin thermal stability was reported for red cabbage compared to anthocyanins from blackcurrant, grape skin and elderberry in soft drink model systems (Table 1.4) due to the protection of flavylium system through copigmentation. Kirca et al. [146] reported that stability of monomeric anthocyanins in black carrot juice and concentrates depended on temperature, solid content and pH. Anthocyanins from black carrot were relatively stable towards heat and pH change compared to anthocyanins from other sources due to diacylation of the anthocyanin structure. Acylation of the molecule is believed to improve anthocyanin stability by preventing hydration [147, 148]. Guisti and Wrolstad [149] reported high colour and pigment stability when colouring secondary bleached cherries with radish anthocyanin extract (RAE). Similarly, Sadilova et al. [150] reported that methoxylation of the acyl moiety improves the structural integrity toward heat. The degradation rates of anthocyanins increased with increasing solid content during heating. This could be due to closeness of reacting molecules in juice with higher soluble solid content [151]. In strawberries anthocyanin degradation occurs as soon as strawberries are processed into juice or concentrate and continues during storage. This degradation of anthocyanins is greater in concentrates compared to juices [152]. Similar trends were reported for anthocyanins in sour cherry [17] and blood orange [153]. Anthocyanin stability is highly influenced by heat which can be significantly reduced by increasing temperatures during thermal processing. For instance, Sadilova et al. [150] observed that elderberry anthocyanins were very sensitive to thermal treatment. After 3 h of heating, only 50% of elderberry pigments were retained at a 95 °C.

Processing berries into purees resulted in a 43% loss in total monomeric anthocyanins [144]. Similar losses in raspberry purees were reported by Ochoa et al. [154]. During strawberry jam manufacture at atmospheric pressure, anthocyanin losses in the final product varied from 10% to 80% with boiling times from 10 to 15 mins [155]. Another study by García-Viguera and Zafrilla [156] showed that under vacuum pressure conditions, the loss incurred was approximately 40% during a 15 min process. Besides heat, various other processing factors related to the structure and pH value of anthocyanins may play an important role in its degradation. Figure 1.2 shows the structural transformation upon pH change. In summary, inter and intramolecular copigmentation with other moieties, polyglycosilated and polyacylated anthocyanins provide greater stability towards change in pH, heat and light [157, 158].

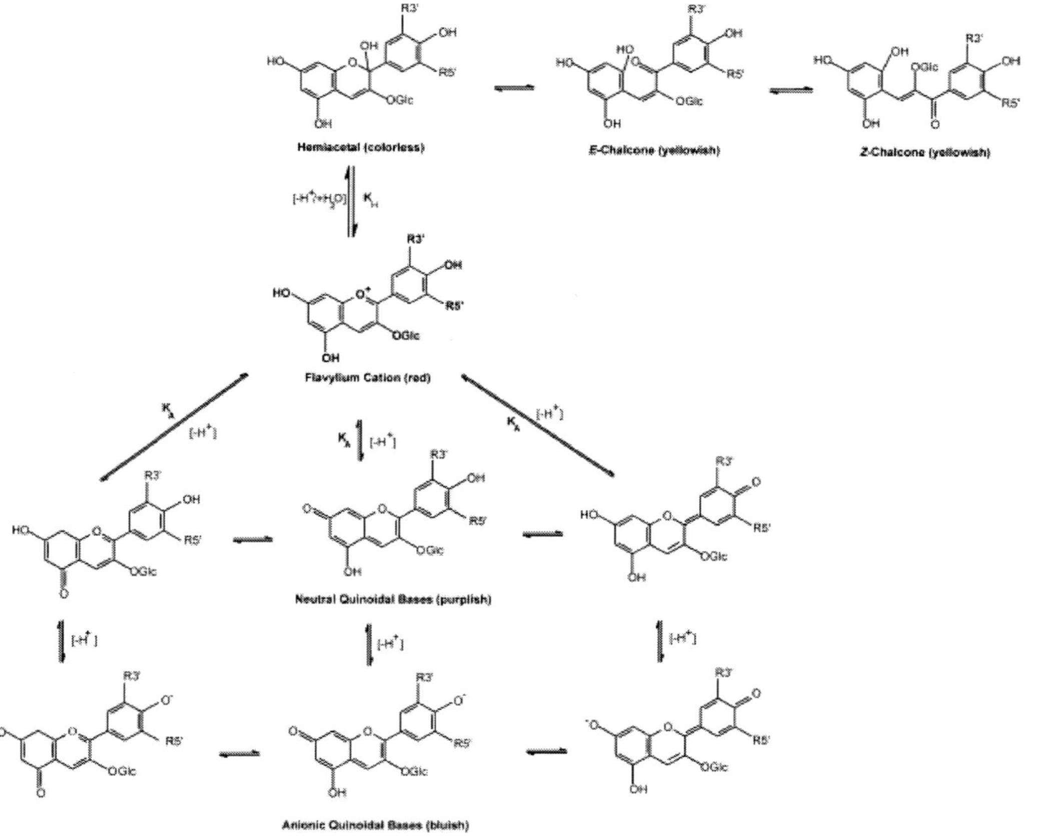

Hemiacetal (colorless)
E-Chalcone (yellowish)
Z-Chalcone (yellowish)
Flavylium Cation (red)
Neutral Quinoidal Bases (purplish)
Anionic Quinoidal Bases (bluish)

Figure 1.2. Structural transformations of anthocyanins upon pH changes [221].

1.3.4 Anthocyanin Stability During Storage

Anthocyanins, as well as other phenolics, are easily oxidized and, thus are susceptible to degradative reactions during various steps of processing and storage. For example, Skrede et al. [142] showed substantial losses of anthocyanins and polyphenolics when blueberries were processed into juice and concentrate and that different classes of compounds had varying susceptibility to degradation with different processing operations. The levels of anthocyanins continued to decline in a linear fashion during storage, with losses of 76% and 75% observed in berries when stored for 6 months. Ngo et al. [159] also reported that total anthocyanins in strawberries canned in 20 ^{o}Brix syrup declined by 69% over 60 a day period at room temperature. Polymeric colour values were reported to increase from 7.2% to 27.4% and 33.3% in syrup and fruit, respectively.

Anthocyanins present in clarified and non-clarified blueberry juices behave differently, when pasteurized. For example, Hager et al. [141] observed that total monomeric anthocyanins in nonclarified juices decreased linearly during storage with losses of 20%, 52%, and 68% observed over 1, 3, and 6 month storage, respectively. Similar losses of 9%, 48%, and 64% were observed for clarified juices over 1, 3, and 6 month storage, respectively compared to pasteurize juice. Consistent with the above results, Srivastava et al. [137] reported that only 50% of the anthocyanins were retained in blueberry juices stored for 60 day at 23 °C. Piljac-Žegarac et al. [160] suggested that fruit juices should be consumed within the initial 48 h after opening, to obtain maximum benefits from polyphenolic antioxidants. Gössinger et al. [161] suggested a significant positive effect of pre-freezing strawberries on colour stability of nectars from purée, which can be stored for more than 12 months. They reported that cold storage temperature of the nectars at 4 °C is also a suitable way to stabilize the colour of strawberry nectars (redness) which is a direct reflection of the level of anthocyanins. Patras et al. [162] reported a significant correlation between redness and anthocyanins of strawberry and blackberry purée. Kırca et al. [163] also reported that storage temperature had a significant effect on the stability of black carrot anthocyanins. They reported a fast degradation of anthocyanins in coloured juices and nectars stored at 37 °C, whereas refrigerated storage resulted in much lower degradation of anthocyanins. Rubinskiene et al. [164] demonstrated that cyanidin-3-rutinoside showed the highest stability to the effect of thermal treatment at 95 °C in black currant. Cyanidin and delphinidin-rutinosides were the most stable anthocyanins during storage for 12 months at 8 °C. Transformations at low temperature (4 °C) in an inert atmosphere may induce a slow degradation process of anthocyanins. Under these conditions, it is also probable that degradation compounds of sugars and ascorbic acid may be the prevalent cause of the transformation of anthocyanins into brown compounds [165].

1.3.5 Degradation Kinetics

Colour degradation kinetics of food products is a complex phenomenon. Kinetic modelling may also be employed to predict the influence of processing on critical quality parameters. Kinetics of pigment and colour degradation of fruits and vegetables during thermal processing has been studied by numerous researchers [124, 166-169].

Pigment degradation under isothermal heating are reported to follow first order kinetics (Equation 1) for juice and concentrate of sour cherry [17] strawberries [152] and blackberries [16]. Several other kinetic studies have been made to evaluate the colour changes in fruits and vegetable products such as garlic paste/purée [170], spinach, mustard leaves, mixed purée [171], green chilli purée [172, 173], red chilli purée and paste [174], papaya purée [175], coriander leaf purée [169], tomato paste [176], peach purée [177, 178], apple purée [179], pea purée [167, 169], cupuaçu [180], tomato purée [181] and pear purée [182]. The kinetic parameters namely, rate constant and activation energy provide useful information on the quality changes which occur during thermal processing.

Table 1.4 shows some of the examples for degradation kinetics and parameters of food and food products along with model juices containing anthocyanins. Degradation kinetics of anthocyanins or other quality parameters during thermal processing are obtained by first determining the rate constants at a given temperature against time. The key parameters of thermal degradation kinetics i.e. half life ($T_{½}$) and activation energy are calculated using the following equations.

$$C_t = C_0 \times \exp(-k \times t) \tag{1}$$

$$T_{1/2} = \frac{\text{Log}_e 2}{k} \text{ or } T_{1/2} = \frac{2.303}{k} \tag{2}$$

$$\text{Log}\left(\frac{k_T}{k_0}\right) = -\frac{E_a}{2.303 \times R}\left[\frac{1}{T_1} - \frac{1}{T_2}\right] \tag{3}$$

Where, C_t is anthocyanin concentration (mg/100 ml) at time t (min), C_0 is initial concentration (t=0), E_a is activation energy (kJ mol^{-1}), R is universal gas constant (8.314 kJ $mol^{-1}\,^{o}C^{-1}$) and k (min^{-1}) is first order degradation rate constant.

The majority of studies on the degradation kinetics of anthocyanins have been carried out under isothermal conditions at temperatures up to 100 °C. However, anthocyanin degradation in solid or semi solid foods such as fruit or berry pommace, grains, vegetables is not isothermal i.e. therefore kinetic modelling should include time-temperature history [184]. Recently Mishra et al. [184] and Harbourne et al. [189] studied thermal degradation of grape pomace anthocyanins and blackcurrant anthocyanins in a model juice under non-isothermal conditions (Table 1.4). Previous to this study Dolan (2003) proposed a one step kinetic model for determining kinetic parameters for non isothermal food process (Equations 4 and 5). This model was employed by Harbourne et al. [189] for non isothermal processing of model blackcurrant juice.

$$C_t = C_0 \times \exp(-k_t \times \beta) \tag{4}$$

Where, β is the thermal history which can be determined by Equation 5, k_t is rate constant.

$$\beta = \int_0^t \exp\left[\frac{-E_a}{R}\left(\frac{1}{T_t} - \frac{1}{T_{ref}}\right)\right] \quad (5)$$

Where T_{ref} is the arbitrary reference temperature and $T_{(t)}$ is the temperature at time (t).

Kinetic parameters for non isothermal heating can be obtained by employing non linear regression analysis techniques [191]. The degradation rate constant and activation energy during isothermal and non isothermal heating systems depends upon the stability of the anthocyanin in question which in turn is dependent on composition, structure, physicochemical properties and presence of other flavanols or organic acids in fruit juices.

1.3.6 Factors Influencing Colour

Maintenance of naturally occurring colour pigments in processed fruit and vegetable products and stored foods has been a major challenge in food processing [192]. Fruits and vegetable processing carried out at commercial or domestic level involving heat may result in the degradation of pigments influencing colour. This is due to the fact that most of these pigments are relatively unstable [193, 194]. Many studies have been reported on the thermal degradation of pigments as a consequence of blanching, cooking, pasteurization, sterilization, dehydration, and freezing [195]. There are many factors that govern the degradation of colour and pigment during thermal and processing of food products. Special care must be taken to produce food that retains a bright, attractive colour during food processing [196, 197]. Stability of plant pigments responsible for colour can be affected by isomerization and oxidation reactions during processing and storage. Factors such as temperature, light, presence of oxygen, metals, enzymes, unsaturated lipids, pro-oxidants and type of packaging material can produce oxidation while heating; acids and light promote *trans–cis* isomerization reactions. In both cases, β-carotene concentration could diminish resulting in discolouration, loss of nutritional value and formation of volatile compounds that would impart desirable or undesirable flavour [198, 199]. The interaction of colour pigments with other organic compounds such as ascorbic acid can reduce oxidative degradation of carotenoids [200]. Ascorbic acid is usually added in a variety of food products as an acidulant or fortificant; for example, in fruit salads, blood orange juices, vegetable purees [201]. In spite of the antioxidant effect of ascorbic acid [202], it has a negative effect on carotenoid pigments such as anthocyanins resulting in both degradation and an increase in browning index, decreasing the colour and nutritional quality of products [201]. Anthocyanin stability is also influenced by the presence of various other compounds such as ascorbic acid and other polyphenols. Interactions of ascorbic acid with anthocyanins leading to mutual degradation have been reported in various fruit juice model systems including cranberry juice [203, 206], strawberry juice [205, 206] and high hydrostatic pressure processed blackcurrant juice [207]. This occurs either by hydrogen peroxide formation through oxidation or condensation of ascorbic acid directly with anthocyanin pigments [208]. Kouniaki et al. [207] reported a protective effect of ascorbic acid on phenolic compounds in HHP processed blackcurrant juice.

Table 1.4. Thermal processing of selected fruits and vegetables.

Fruit/Vegetable juice	Anthocyanins	Processing conditions	Kinetic parameters	References
Roselle (*Hibiscus sabdariffa* L. cv. 'Criollo')	Delphinidin-3-xylosylglucoside Cyanidin-3-xylosylglucoside	60 to 100 °C and 20 to 120 min	Ea=15.83 kcal mol^{-1} (66.22 kJ mol^{-1}) Q_{10} =1.01	[183]
Grape pomace	Delphinidin cyanidin petunidin peonidin malvidin	Nonisothermal heating (retort)	$k_{110\,°C}$= 0.0607 min^{-1} Ea = 65.32 kJ mol^{-1}	[184]
Purple-flesh potato	Anthocyanin extract	Blanching (Boiling water) pH=3	Z=28.4 °C Q_{10}=2.25 Ea=72.49 kJ mol^{-1}	[185]
Red-flesh potato	Anthocyanin extract	Blanching (Boiling water) pH=3	Z=31.5 °C Q_{10}=2.08 Ea=66.7 kJ mol^{-1}	[185]
Grape	Anthocyanin extract	Blanching (Boiling water) pH=3	Z=28.0 °C Q_{10}=2.28 Ea=75.03 kJ mol^{-1}	[185]
Purple carrot	Anthocyanin extract	Blanching (Boiling water) pH=3	Z=26.0 °C Q_{10}=2.44 Ea=81.34 kJ mol^{-1}	[185]
Black carrots	Monomeric anthocyanin (439 mg/l)	70, 80 and 90 °C pH= 2.5, 3.0, 4.0, 5.0, 6.0 and 7.0 11, 30, 45, 64 °Brix	Ea=78.1, 72.4, 56.8, 42.0 and 47.4 kJ mol^{-1} $Q_{10(70\text{-}80\,°C)}$=(1.7-2.8) $Q_{10(80\text{-}90\,°C)}$ =(1.9-2.2)	[186]
Blond orange juice with anthocyanin-rich black carrot	Total anthocyanin	70-90 °C	Ea=72.0 kJ mol^{-1}	[187]

Table 1.4. (Continued).

Fruit/Vegetable juice	Anthocyanins	Processing conditions	Kinetic parameters	References
Purple corn cob	Total anthocyanin	70 , 80 and 90 °C at pH 4.0	Ea= 18.3 kJ/mol	[188]
Model juice (blackcurrant anthocyanin)	Total monomeric anthocyanins	4-100 °C: isothermal method 110-140 °C : non isothermal method	73 ± 2 kJ/mol (calculated over the range 21-100 °C) 91.09 ± 0.03 kJ/mol (non isothermal)	[189]
Plum purée	Total anthocyanins	50, 60, 70, 80 and 90 °C 0-20 min	Ea= 37.48 kJ mol^{-1}	[190]
Grape skin	Anthocyanin extract in McIlvaine buffer (MB) and non carbonated soft drink (SD) medium	25, 40, 60 and 80 °C. 15 min, 30 min, 60 min, 120 min, 240 min and 360 min	Ea= 69.0 kJ mol^{-1} (MB) Ea= 50.0 kJ mol^{-1} (SD)	[145]
Blackcurrant pomace	Anthocyanin extract in MB and SD		Ea= 58.0 kJ mol^{-1} (MB) Ea= 77.0 kJ mol^{-1} (SD)	[145]
Elderberry juice concentrate	Anthocyanin extract in MB and SD		Ea= 89.0 kJ mol^{-1} (MB) Ea= 56.0 kJ mol^{-1} (SD)	[145]

Talcott et al. [209] reported that the exogenous addition of ascorbic acid in Dense phase carbondioxide (DPCD) processed grape juice exerts a protective effect against anthocyanin destruction caused by enzymatic activity of PPO. The exogenous addition of cofactors such as caffeic, catechin [210], rosemary extract [209], isoflavonoid extracts of red clover or other commercial and/or purified botanical extracts of flavonoids can improve the stability of anthocyanins [211]. For example a study by Del Pozo-Insfran et al. [212] on the stability of HHP processed grape juice fortified with water soluble polyphenolic cofactors from thyme (*Thymus vulgaris L.*) indicated the efficacy of such additions on the stability of anthocyanins in the presence of residual enzymatic activity of PPO. In another study by Del Pozo-Insfran et al. [212] on DPCD processed grape juice, the importance of copigmentation with thyme extract in the retention of anthocyanins during storage was reported. The interaction of ascorbic acid with anthocyanin pigments may result in the degradation of both and could decrease both the colour and nutritional quality of the products.

1.3.7 Degradation Mechanisms

1.3.7.1 Chlorophyll degradation

Chlorophylls are susceptible to many chemical or enzymatic degradation reactions. The simultaneous actions of enzymes, weak acids, oxygen, light and heat can lead to the formation of a large number of degradation products. Major chemical degradation routes are associated with pheophytinization, epimerization, and pyrolysis, and also with hydroxylation, oxidation or photo-oxidation, if light is implicated [213]. There is general agreement that the main cause of green vegetable discolouration during processing is the conversion of chlorophylls to pheophytins.

Several mechanisms can cause chlorophyll degradation these includes loss of magnesium due to heat and/or acid substitution of Mg^{2+} by H^{+} [128] and possibly enzymatically [126] or loss of the phytol group through the action of the enzyme chlorophyllase (EC 3.1.1.14). Also loss of the carbomethoxy group may occur, leading to formation of pyropheophytin and pyropheophorbide. Figure 1.3 shows the schematic pathway of chlorophyll degradation. Van Boekel [128] reported that the degradation of chlorophyll in olives during fermentation is due a combination of enzymatic activity and chemical changes induced by higher pH (pH 8-9, 5-7 h) during fermentation. It has been reported that the enzyme (chlorophyllase) responsible for degradation of chlorophyll has maximum activity in alkaline environment [125] converting chlorophyll into chlorophyllide. Other mechanism for degradation of green colour during fermentation could be due to lowering of pH by production of lactic acid during fermentation as a result of which Mg^{2+} is replaced by H^{+}, leading to formation of pheophytin and pheophorbide.

Chlorophyll degradation in processed foods follows two distinctive pathways (Figure 1.3). The first pathway represents the primary loss of the magnesium moiety to form pheophytin, followed by the cleavage of the phytol chain to form pheophorbide, whereas the second pathway involves primary cleavage of the phytol chain to form chlorophyllide and subsequently pheophorbide. Removal of the central magnesium ion from the porphyrin ring is usually the result of acidic substitution and/or heat [122, 123, 214]. However, Langmeier et al. [215] postulated that there might be a magnesium dechelatase that could perform this function. Cleavage of the phytol chain from chlorophyll can either result from chemical hydrolysis [216] or enzymatic cleavage by chlorophyllase [217].

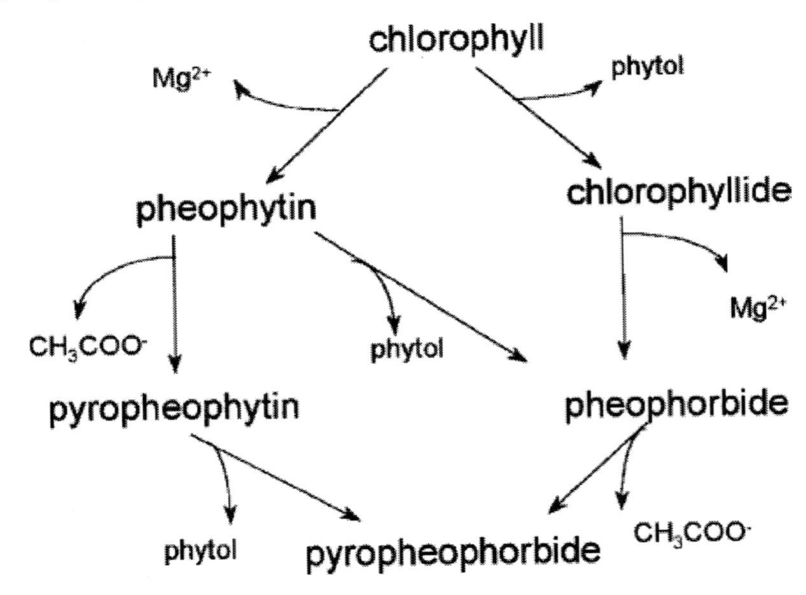

Figure 1.3. Schematic representation of pathways of chlorophyll degradation [126].

1.3.7.2 Anthocyanins degradation

Anthocyanin pigments can be destroyed during the processing of fruits and vegetables. Numerous examined the properties of anthocyanins to facilitate their retention. High temperature, increased sugar level, pH, and ascorbic acid can affect the rate of destruction [218, 219]. Temperature has been reported to induce a logarithmic destruction with time. Bleaching by effect of heat occurs because of the above-described equilibrium is changed toward the uncoloured forms. It has been suggested that flavonoid structure is opened to form chalcone, which is degraded further to brown products [218]. However, interestingly, it has been observed that optimal conditions permit the regaining of colour on cooling if there is sufficient time (several hours) for the reconversion.

Anthocyanins can be degraded by a number of enzymes found in fruits and vegetables such as glycosidases, polyphenoloxidases, and peroxidases. Glycosidases produce anthocyanidins and sugars, and anthocyanidins are very unstable and rapidly de-grade. Polyphenoloxidase catalyzes the oxidation of o-dihydrophenols to o-quinones that further react to brown polymers; however, this reaction is more favorable for other phenols. The mechanisms by which anthocyanins are degraded due to adverse processing conditions are summarized in Figure 1.4. Typical hydroxybenzoic acid derivatives are p-hydroxybenzoic acid from pelargonidin, protocatechuic acid from cyanidin, vanillic acid from peonidin, 5-methoxy-protocatechuic acid from petunidin and syringic acid from malvidin, respectively [221].

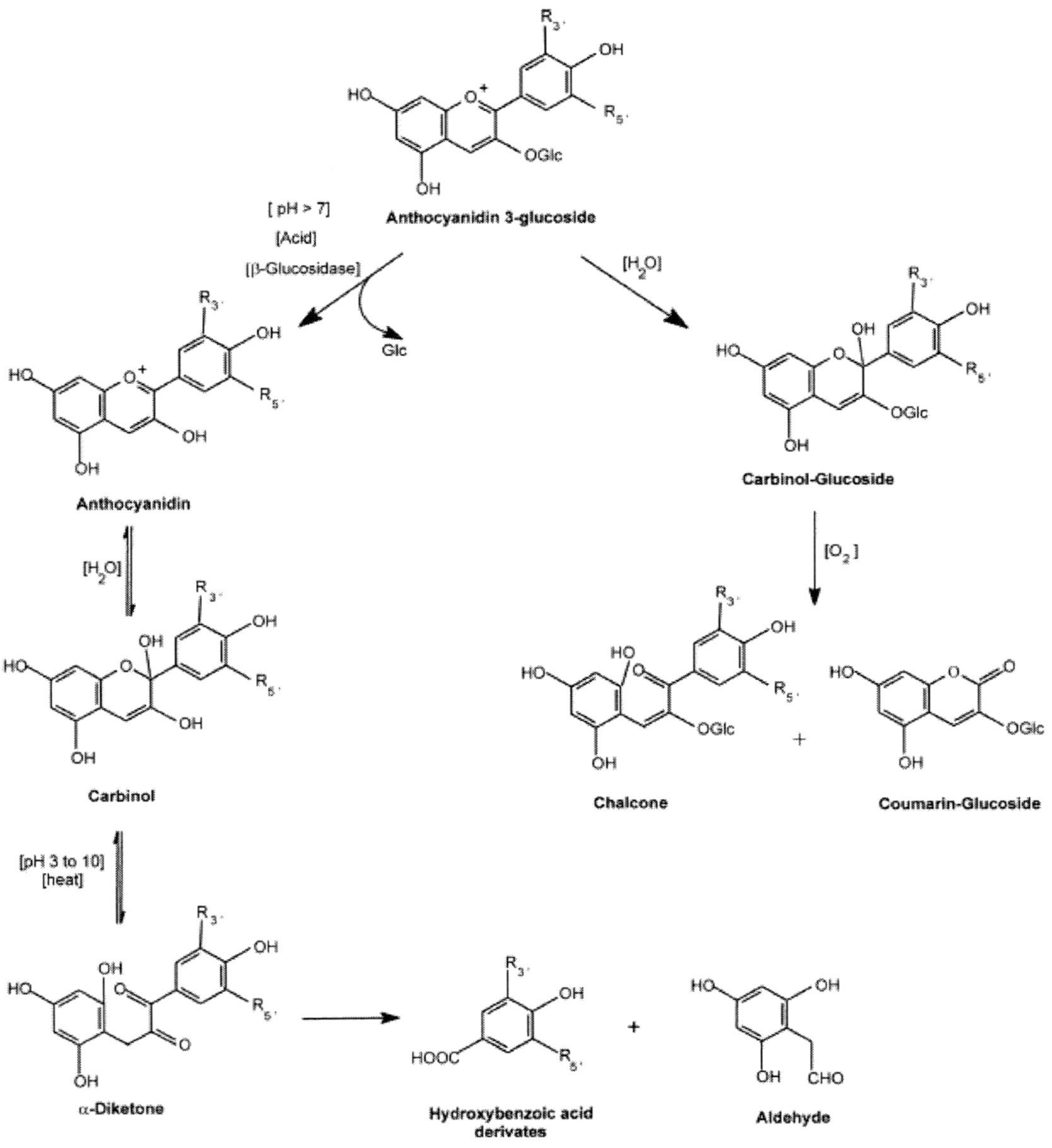

Figure 1.4. Pathways of anthocyanidin 3-glucoside degradation [208, 220].

Anthocyanins are glycosylated anthocyanidins; sugars are attached to the 3-hydroxyl position of the anthocyanidin (sometimes to the 5 or 7 position of flavynium ion) [23]. Variation in chemical structure are mainly due to differences in the number of hydroxyl groups in the molecule, degree of methylation of these OH groups, nature and number of sugar moiety attached to the phenolic molecule and to some extent the nature and number of aliphatic or aromatic acids attached to it [22, 222]. Sugar moieties are attached as 3-monosides, 3-biosides, 3-triosides, 3, 5-diglycosides, 3, 7-diglycosides consisting mostly of glucose, galactose, rhamnose and xylose [222]. Degradation is primarily caused by oxidation, cleavage of covalent bonds or enhanced oxidation reactions due to thermal processing.

Thermal degradation of anthocyanins can result in a variety of species depending upon the severity and nature of heating. Understanding the degradation mechanism is a prerequisite for minimizing nutritional and visual quality. Relatively little is known about degradation mechanisms of anthocyanins but chemical structure and presence of other organic acids are known to have a strong influence. For example, the degradation rate of anthocyanins increases during processing and storage as the temperature rises [223, 224] suggested opening of the pyrylium ring and chalcone formation as a first degradation step for anthocyanins. Adams [225] proposed hydrolysis of sugar moiety and aglycone formation as the initial degradation step possibly due to the formation of cyclic-adducts which can lead to a loss of an active OH-group in the meta-position of the A ring from the flavylium cation molecule mainly responsible for the loss, hence affecting anthocyanin molecule [226]. Figure 1.5 shows the degradation of anthocyanins due to heat. Seeram et al. [220] demonstrated that a high temperature in combination with high pH causes degradation of cherry anthocyanins resulting in three different benzoic acid derivatives. However, a separate study conducted by von Elbe and Schwartz [227] suggested that Coumarin 3, 5-diglycosides are also common thermal degradation products of anthocyanin 3, 5-diglycosides.

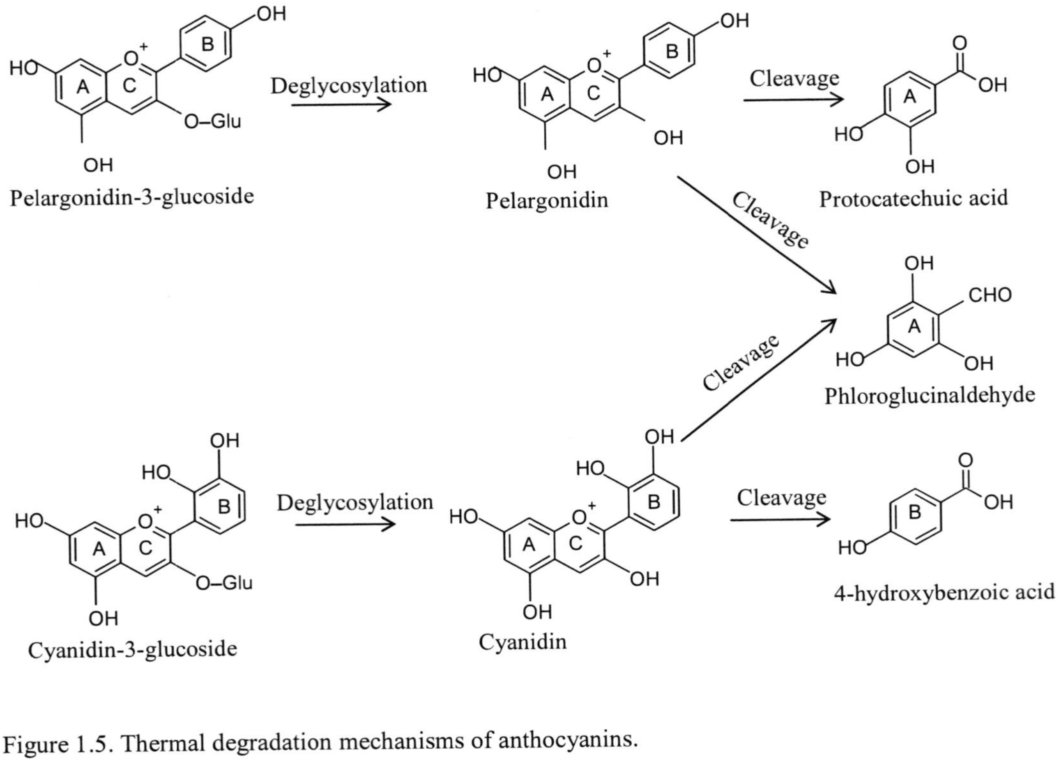

Figure 1.5. Thermal degradation mechanisms of anthocyanins.

Oxygen also plays a vital role in the anthocyanin degradation process. The presence of oxygen can accelerate the degradation of anthocyanins either through a direct oxidative mechanism and/or through the action of oxidising enzymes. In the presence of oxygen, enzymes such as PPO catalyse the oxidation of CG into the corresponding o-quinone (chlorogenoquinone, CGQ). This quinone reacts with anthocyanins to form brown condensation products [228]. Kader et al. [229] working with model solutions of purified substrates also proposed that cyanidin 3-glucoside [*ortho*-diphenolic anthocyanin, (Cy 3-glc)] is degraded by a mechanism of coupled oxidation involving the enzymatically generated *o*-quinone with partial regeneration of the *o*-diphenolic co-substrate (CG). These observations confirm that PPO plays a vital role in anthocyanin degradation. Furthermore in a subsequent study the authors suggested that regardless of the reaction conditions used (±CG), no degradation of anthocyanins was observed in the H_2O_2-free control solutions, meaning that blueberry POD is involved in the process of anthocyanin degradation [229]. In a separate study Kader et al. [230] demonstrated that anthocyanins, such as pelargonidin-3-glucoside (Pg 3-glc), are degraded by a mechanism involving a reaction between the *o*-quinone and/or secondary products of oxidation formed from the quinone and the anthocyanin pigment.

Sarni et al. [231] reported that the degradation products of Cy-3-glc and Mv 3-glc contained both caffeoyltartaric acid and anthocyanin moieties. Moreover, these degradation products were gradually replaced by colourless products as a consequence of further oxidative degradation. The authors pointed out that the reaction between Mv 3-glc and caftaric acid *o*-quinone leads to the formation of adducts [231]. This indicated that the hemiacetal form of the pigment is more reactive than the flavylium form.

1.4 Non Thermal Technologies

Conventional thermal processing of fruit and vegetables remains the most widely adopted technology for shelf-life extension and preservation of fruit juice. However, consumer demand for nutritious foods, which are minimally and naturally processed, has led to increased interest in non-thermal technologies. These preservation technologies may inactivate microorganisms and enzymes by physical and/or chemical principles [232]. Non-thermal technologies are preservation treatments that are effective at ambient or sub-lethal temperatures, thereby minimising negative thermal effects on food nutritional and quality parameters.

1.4.1 The Effect Of High Hydrostatic Pressure Processing (hhp) on Colour

HHP treatment (at low and moderate temperatures) has minimal effect on pigments (e.g. chlorophyll, carotenoids, anthocyanins, etc.) responsible for the colour of fruits and vegetables [106]. The colour compounds of HHP processed fruits and vegetables can, however, change during storage due to incomplete inactivation of enzymes and microorganisms, which can result in undesired chemical reactions (both enzymatic and non-enzymatic) in the food matrix [106]. In fruit-based food products, no visual colour differences (based on L^*, a^* and b^* values) are observed immediately after HHP treatments, for example in white grape juice after 10 min HHP treatment at 400 MPa/2 °C, 500 MPa/2 °C or 400 MPa/40 °C [233] or in mango pulps after HHP treatments at 100-400 MPa/20 °C for 15 or 30 min [234]. Ahmed et al. [234] observed that colour parameters such as (a/b), C and h values of mango pulps remained constant after HHP treatment indicating pigment stability, while increasing pressure intensity decreased the value of ΔE. Similar results for carrot and tomato puree were reported by Dede et al. [235]. They observed that high pressure treatment at 250 MPa/35 °C/15 min produced a lower colour difference compared to thermally processed juices. In the case of vegetables, the green colour of vegetables becomes even more intense (decrease in *L**, *a** and *b** values), for example green beans after HHP treatment of 500 MPa/ambient temperature/1 min [236]. This might be caused by cell disruption during HHP treatment resulting in the leakage of chlorophyll into the intercellular space yielding a more intense bright green colour on the vegetable surface [237]. Colour changes in HHP-treated fruits and vegetables can also be related to changes in textural properties. This phenomenon was observed in tomato based products. HHP treatment (400 MPa/25 °C/15 min) resulted in an increase in the *L** value of tomato purée indicating a lightening of the purée surface colour. The CIE Lab parameters were significantly higher both in the untreated and in the HHP-treated tomato purée compared to the thermally treated purees [238].

HHP treatment at ambient and moderate temperatures results in limited colour change of green vegetables. For example Krebbers et al. [237] reported a decrease in *L**, *a** and *b** values of green beans after HHP treatment (500 MPa) at ambient temperature. However, at higher temperatures they observed a slight change in green colour towards olive green. Similar results were also observed for basils after HP treatment [238].

Van Loey et al. [240] reported an extreme pressure stability of both chlorophylls *a* and *b* in broccoli juice at room temperature and at higher temperature (>50 °C) and studied the kinetics of degradation. The temperature dependency of the degradation rate constant of chlorophyll a was higher than that of chlorophyll b. However, at a constant pressure level, the values of the degradation rate constants of chlorophylls increase with increasing temperature [240] whereas at constant elevated temperatures, pressure increase accelerates the degradation of chlorophyll a and b. The pressure dependency of the degradation rate constant of chlorophyll b at 70 °C is higher than that of chlorophyll a. For example, elevating pressure from 200 to 800 MPa accelerates the degradation of chlorophyll a and chlorophyll b of broccoli by 19.4% and 68.4%, respectively [240]. Similarly, Matser et al. [241] also reported chlorophyll degradation of green beans and spinach due to HP processing at elevated temperatures. Butz et al. [242] studied the effect of high pressure on the stability of lycopene and β-carotene in tomatoes and reported no significant changes in concentration. Lycopene and β-carotene were found to be stable during pressurisation of tomato [243]. Stability of pigments may be explained by matrix effects: within tissues, the pigments are often compartmentalised and thus protected from adverse influences.

Carotenoid pigments are reported to be fairly stable towards pressure. HP treatment increases the extraction yields of carotenes from the plant matrix. Numerous studies have shown an increase in extraction yield of carotenoids due to high pressure processing, e.g., more than 40% increase in carotenoid content of pressurized carrot homogenate (600 MPa/25 °C/10 min) [244]; 20-43% increase in the carotenoid content of orange juice at 360 MPa in the temperature range between 30 and 60 °C for 2.5, 5 and 15 min [245]; 53% increase in the carotenoid content of orange juice after treatment of 400 MPa/40 °C/1 min [246] and a significant increase in the carotenoid content of pressurized (400 MPa/25 °C/15 min) tomato purée compared to either thermal treated or untreated purée [247]. Due to this benefit, HHP technology has also been studied to extract lycopene from tomato paste waste [106, 248].

HHP (at low and moderate temperatures) also has a limited effect on anthocyanin pigments. Anthocyanins are stable during HHP treatment at moderate temperature, for example, pelargonidin-3-glucoside and pelargonidin-3-rutinoside in red raspberry (*Rubus idaeus*) and strawberry (*Fragaria x ananassa*) during HHP treatment at 800 MPa (18-22 °C/15 min) [249]. Anthocyanins in pressure-treated vegetables and fruits were not stable during storage. A shelf-life study (7 days at 5, 20 and 30 °C) on pressurized (200, 400, 600 and 800 MPa/15 min/20-22.5 °C) blackcurrants (*Ribes nigrum*) during a 7 days at selected temperatures showed that cyanidin-3-rutinoside and delphinidin-3- rutinoside had different stabilities. Anthocyanins in pressurized blackcurrants remained unchanged during storage at 4 °C [207]. Enzymatic activity plays an important role in stability of anthocyanins subjected to pressure treatments. For example; Suthanthangjai et al. [250] showed that cyanidin-3-glucoside and cyanidin-3-sophorosides (the major pigments in raspberry) had the highest stability during 9 days of storage at 4 °C after pressurization at 200 or 800 MPa (18-22 °C/15 min) compared with pressure treatment at 400 or 600 MPa. A high loss of both pigments after HHP treatment at 400 and 600 MPa is probably due to a lower degree of inactivation of β-glucosidase, peroxidase and polyphenoloxidase.

Corrales et al. [251] reported an insignificant reduction in cyanidin-3-glucoside in a model solution at processing conditions of 600 MPa, 20 °C and 30 min. They also reported a 25% degradation at 600 MPa, 70 °C for 30 min compared to a 5% degradation at 70 °C for 30 min, indicating that HHP aaccelerates anthocyanin degradation at elevated temperatures. This is probably due to condensation reactions involving covalent association of anthocyanins with other flavanols present in fruit juices leading to the formation of a new pyran ring by cycloaddition [252]. Chemical compounds derived from these condensation reactions are responsible for changes in the colour of red wine towards brown or orange [253, 254]. The storage stability of anthocyanins content in raspberry [250], strawberry [255], and blackcurrant [207] were most stable when a pressure of 800 MPa for 15 min was applied. Garcia-Palazon et al. [249] reported that the stability of strawberry and red raspberry anthocyanins, namely pelargonidin-3-glucoside and pelargonidin-3-rutinoside, at 800 MPa for 15 min at moderate temperature (18-22 °C) is mainly due to complete inactivation of polyphenoloxidase. Enzymes such as polyphenoloxidase, peroxidase and β-glucosidase have been implicated in the degradation of anthocyanins. However it must be noted that the effect of HHP processing parameters such as pressure, temperature and time along with physicochemical properties of fruit such as total soluble solids and pH have varying effects on the enzymes responsible for anthocyanins stability in HHP processed fruits and fruit products [256].

Enzymatic degradation of anthocyanins by β-glucosidase (EC 3.2.1.21) is mainly by the hydrolysis of terminal, non-reducing β-D-glucose residues with release of β-D-glucose. Due to the loss of the terminal β-D-glucose moiety, the aglycon (anthocyanidin) can become extremely unstable with a consequent adverse effect on juice colour, it also represents a good substrate for other enzymes [257] such as polyphenoloxidases and peroxidases. Specificity of β-glucosidase is another cause for selective degradation of anthocyanins present in fruit. For example a higher reduction in pelargonidin-3-glucoside, compared to pelargonidin-3-rutinoside in HHP processed strawberries during storage under similar processing conditions was reported by Zabetakis et al. [255] because β-glucosidase has a greater affinity towards glucose compared to rutinose. Similarly, Gimenez et al. [258] reported greater loss of pelargonidin-3-glucoside, compared to pelargonidin-3-rutinoside in HHP processed strawberry jam. Anthocyanin degradation in HHP processed juice combined with heat could be due to condensation reactions involving the covalent association of anthocyanins with other flavanols or organic acids present in fruit juices such as reaction of malvidin 3-O-glucoside with epicatechin and acetaldehyde [259]; anthocyanins and tannins [260]. Corrales et al. [251] reported a significant loss (53%) of cyanidin-3-O-glucoside (Cy3gl) in a model solution subjected to HHP (600 MPa) and a temperature of 70 °C during longer holding times (6 h) due to condensation reaction of Cy3gl with pyruvic acid by formation of a new pyran ring by cycloaddition.

1.4.2 Pulsed Electric Field

PEF has been demonstrated to be effective against various pathogenic and spoilage microorganisms and enzymes without appreciable loss of flavour and colour pigments such as anthocyanins [261-264]. PEF treatment is reported to have minimal changes in colour pigments such as carotenoids in orange juice [265]. Cortés et al. [266] reported a reduction of 12.6% in the carotenoid concentration of pasteurised orange juice with respect to untreated fresh orange juice compared to 9.6%, 6.3% and 7.8% for PEF (25, 30 or 40 kV/cm) processed samples respectively.. PEF processing is reported to enhance some carotenoids (lycopene, β-carotene and phytofluene), the red colour of tomato juice [266, 267] and the lycopene content in melon juice [268]. PEF is aslo reported to minimize non-enzymatic browning by the application of pulse widths beyond 4.5 μs for strawberry juice and up to 2.5 μs for tomato and watermelon juices [269]. A PEF pre-treatment of is reported to increase the anthocyanin concentration in grape juice [270]. Corrales et al. [251] demonstrated that PEF treatment enhances the extraction of anthocyanins by 17% compared to conventional methods and by 10% compared to HHP. Higher extraction rates of pigments from pulp during treatment may be due to electroporation. Electroporation causes loss of membrane integrity, inactivation of proteins, and leakage of cellular contents of microorganisms, leading to microbial inactivation [271]. Studies have shown the degradation of anthocyanins during PEF treatment. Zhang et al. [272] studied the degradation kinetics of cyanidin-3-glucoside, a major anthocyanin present in blackberry and blood orange during PEF treatment. They reported that the degradation rate constant of Cy-3-glu exposed to PEF compares favourably to that for thermal treatment.

Anthocyanin retention during PEF processing is influenced by polarity, treatment time and frequency employed. Recently, Odriozola-Serrano et al. [266] reported greater anthocyanin retention in strawberry juice during PEF treatment in bipolar pulse mode compared to monopolar mode. They observed ~4% difference in the anthocyanin degradation between monopolar and bipolar modes. Storage studies on PEF processed juice indicates PEF does not have a post-treatment effect on the stability of Cy-3-glu, which is mainly governed by storage conditions.

The exact mechanism of degradation during PEF processing is difficult to establish due to the complexity of chemical reactions taking place in natural systems such as juice or concentrate. Hence, literature suggests that it is difficult to explain changes of anthocyanins in isolation due to the complexity of reactions which are often influenced by the presence of ascorbic acid and other organic acids participating in condensation reactions at higher temperatures during processing. Therefore, the evaluation of a particular parameter requires model systems in order to closely control the composition and factors influencing pigment stability. A study conducted by Zhang et al. [272] on purified cyanindin-3-glucoside from red raspberry by PEF (1.2-3.0 kV/cm) demonstrated the formation of chalcone. Opening of the pyrylium ring and chalcone formation is regarded as the first step in the degradation of anthocyanins [224]. Partial inactivation of enzymes such as β-glucosidase, polyphenoloxidases and peroxidase can also contribute to the degradation of anthocyanins during storage of PEF processed juices. Recently, Aguilo-Aguayo et al. [269] reported an enhanced activity of 113% for β-glucosidase in strawberry juice during PEF treatment (50 Hz, pulse width 1-μs at 35 kV/cm for 1000 μs).

1.4.3 Ozone Treatment

Ozone was declared as Generally Recognized as Safe (GRAS) for use in food processing by the US Food and Drug Administration (FDA) in 1997 [273]. Ozone subsequently gained approved as a direct food additive for the treatment, storage and processing of foods in the gaseous and aqueous phases in 2001 [274]. Ozone is used for washing of fruits and vegetables [275]. Ozone treatments were reported to have minor effects on anthocyanin contents in strawberries [276] and blackberries [277]. The anthocyanin content in blackberries stored in air and at 0.1 ppm ozone, remained stable while it fluctuated in the 0.3 ppm ozone treated samples during storage. The red colour of intact, whole berry fruit was optimum at 0.3 ppm ozone treated samples during storage [277]. It was also reported that the undesirable colour change from green to yellow in broccoli was significantly less pronounced for ozone treated samples [278]. However, ozone was reported to change the surface colour of some products like peaches [279] and carrots [280].

Until recently, ozone processing has been carried on solid foods however with the FDA approval of ozone as a direct additive to food, the potential of ozonation in liquid food applications has emerged. A number of commercial fruit juice processors in the USA have started to employ ozone to meet the recent USFDA mandatory 5 log reduction resulting in industry guidelines being issued by the USFDA for ozonation of apple juice [281]. Ozone has been reported for processing of various fruit juices including; apple juice (cider) [282-284], orange juice [284-286], blackberry juice [287] and strawberry juice [288].

Ozone has a high oxidation potential (2.07 V) resulting in the degradation of most organic compounds. The oxidizing ability of ozone is derived from the nascent oxygen atom. However it has been reported that ozonation of organic dyes results in loss of colour due to the oxidative cleavage of the chromophores [289] due to attack on conjugated double bonds. Similarly the chromophore of conjugated double bonds of carotenoids is responsible for orange juice colour. Carotenoids pigments which contribute to yellow, orange or red colour in orange juice contain one or more aromatic rings [290]. The ozone and hydroxyl (OH^-) radicals generated in the aqueous solution may open these aromatic rings and lead to partial oxidation of products such as organic acids, aldehydes, and ketones.

Ozonation of fruit juices rich in anthocyanins such as strawberry and blackberry juice causes a significant reduction in these colour pigments. A significant reduction of 98.2% in the pelargonidin-3-glucoside content of strawberry juice was reported at an ozone concentration of 7.8% w/w processed for 10 min [288]. Reductions of > 90% in the cyanidin-3-glucoside content of blackberry juice was reported under similar treatment conditions [287].

The degradation of anthocyanins during ozone treatment could be due to direct reaction with ozone or indirect reactions of secondary oxidators such as •OH, $HO^{2\bullet}$, $\bullet O_2^-$, and $\bullet O_3^-$. Such secondary oxidators may lead to electrophilic and nucleophilic reactions occurring with aromatic compounds that are substituted with an electron donor (e.g. OH^-) having high electron density on the carbon compounds in ortho and para positions. Direct reaction is described by the Criegee mechanism [291] where ozone molecules undergo 1-3 dipolar cyclo addition with double bonds present, leading to the formation of ozonides (1,2,4-trioxolanes) from alkenes and ozone with aldehyde or ketone oxides as decisive intermediates, all of which have finite lifetimes [291]. This leads to the oxidative disintegration of ozonide and formation of carbonyl compounds, while oxidative work-up leads to carboxylic acids or ketones. Ozone attacks OH radicals, preferentially to the double bonds in organic compounds leading to the formation of unstable ozonide which subsequently disintegrates. The degradation mechanism for anthocyanin in strawberry juice was proposed by Tiwari et al. [288].

1.4.4 Ultrasound

Ultrasound processing of juices is reported to have minimal effects on the degradation of key quality parameters such as colour in orange juice during storage at 10 °C [292]. This positive effect of ultrasound is assumed to be due to the effective removal of occluded oxygen from the juice [293]. Cruz et al. [47] reported enhancement of green colour in thermosonicated watercress samples. The effect of ultrasound on anthocyanins in strawberry juice was reported by Tiwari et al. [294]. They reported a slight increase (1-2%) in the pelargonidin-3-glucoside content of the juice at lower amplitude levels and treatment times which may be due to the extraction of bound anthocyanins from the suspended pulp. Similarly, weak ultrasonic irradiation is reported to promote an increase in the amount of phenolic compounds found in red wine [295]. Literature indicates that ultrasound processing enhances the extraction of phenolic and other bioactive compounds from grape must or wine [296]. The application of ultrasound assisted extraction improves the extraction yield of bioactive compounds by between 6 to 35% [297]. Zhao et al. [298] reported degradation of (all-E)-astaxanthin into unidentified colourless molecule(s) during extraction with sonication at an increased power level and treatment time. Similarly in a study by Tiwari et al. [294] the anthocyanin content of the juice was found to degrade when higher amplitude levels were employed, however the maximum observed degradation was less than 5%. The observed degradation of anthocyanins is mainly due to cavitation, which involves the formation, growth, and collapse of microscopic bubbles. Degradation of quality and nutritional parameters results from the extreme physical conditions which occur within the bubbles during cavitational collapse at micro-scale [299] and several sonochemical reactions occurring simultaneously or in isolation. The chemical effects produced by cavitation generate high local temperature, pressure and mechanical action between solid and liquid interfaces [300]. Cavities formed by sonication may be filled with water vapour and gases dissolved in the juice, such as O_2 and N_2 [301]. The anthocyanin degradation may also be due to the presence of other organic acid or ascorbic acid and can be related to oxidation reactions, promoted by the interaction of free radicals formed during sonication [302]. Hydroxyl radicals produced by cavitation can be involved in the degradation of anthocyanins by opening of rings and formation of chalcone [231] mainly due to the temperature rise which occurs during sonication.

1.4.5 Dense Phase CO_2

Dense phase CO_2 processing (DP-CO2) is a continuous, nonthermal processing system for liquid foods that utilises pressure (<90 MPa) in combination with carbon dioxide (CO_2) to destroy microorganisms as a means of food preservation [303]. Recently Garcia-Gonzalez et al. [304] reviewed the effects of high pressure CO_2 on microbial inactivation. The use of DPCD for inactivation of organisms continues to attract attention due to its unique properties that make it an appealing medium for food preservation. Dense phase carbon dioxide or supercritical CO_2 denotes phases of matter that remain fluid, yet are dense with respect to gaseous CO_2. Moreover, in the supercritical state, CO_2 has low viscosity ($3\text{-}7\times10^{-5}$ Pa.s) and zero surface tension, so it can quickly penetrate complex structures and porous materials. Finally, CO_2 is inexpensive and readily available [305]. DPCD is reported to have significant lethal effects on microorganisms in liquid foods [306] and to inactivate enzymes such as polyphenols oxidases [303] and peroxidases [307] which influence the stability of anthocyanins during storage. There are a limited number of studies available in the literature regarding the effect of DPCD on product quality. Application of DPCD has been reported for various fruit juices such as apple cider [308]; orange juice [309]; grape juice [310] and mandarin juice [311]. These studies indicated minimal changes in key quality parameters.

In a study conducted by Del Pozo-Insfran et al. [306] no significant changes in total anthocyanin content was reported for DPCD processed muscadine grape juice compared to a 16% loss observed in thermally processed juice. Enhanced anthocyanin stability was also observed in DPCD processed juice during storage for 10 weeks at 4 $^{\circ}$C. The greater stability of DPCD processed juice could be due to the prevention of oxidation by removal of dissolved oxygen. The exact mechanism for anthocyanin stability is difficult to establish. However Del Pozo-Insfran et al. [212] demonstrated that anthocyanin stability is also dependent on the PPO inactivation potential of DPCD treatment and governed by extrinsic control parameters of pressure and CO_2 concentration gradient.

1.5 Conclusion and Future Trends

Current knowledge indicates that in general high temperature treatments can affect levels of anthocyanins in fruit and vegetable food products. However, only limited information is available on the temperature stability of anthocyanins. Since loss of bioactivity through anthocyanin degradation is not compensated for by the respective colourless phenolics generated upon heating, further studies in this area would generate the potential for consumers to gain even more positive health benefits from foods with high anthocyanin content such as fresh or processed berry fruits. Based on current knowledge it is not possible to predict the effect of thermal treatment on anthocyanin retention. The need to optimize processes in terms of quality and operating costs, the demand for more research to streamline processes by combinations of technologies, particularly with respect to optimisation of practical applications. Process optimisation of thermal processes in combination with non thermal technologies such as high pressure and ultrasound food has been the focus of research studies in recent years. The food industry is poised to adopt new concepts and technologies that offer advantages over conventional systems. Extensive validation and verification, accuracy and cost effectiveness, controls and monitoring capabilities, are some of the key elements that will justify the adoption of developed systems. Thermal degradation should therefore be assessed on a case by case basis until a consensus can be reached. Since the degradation mechanism of anthocyanins is rather complex, it is possible that thermal processing particularly above 50 °C or higher could induce some (un)expected and (un)desired chemical reactions which (in) directly influence food quality.

REFERENCES

[1] Delgado-Vargas, F., Jiménez, A.R., & Paredes-López, O. (2000). Natural Pigments: Carotenoids, Anthocyanins, and Betalains- Characteristics, Biosynthesis, Processing, and Stability. *Critical Reviews in Food Science and Nutrition*, *40*, 173-289.

[2] Hari, R.K., Patel, T.R., & Martin, A.M. (1994). An overview of pigment production in biological systems: functions, biosynthesis, and applications in food industry. *Food Reviews International*, *10*, 49-70.

[3] Tanaka, Y., Sasaki, N., & Ohmiya, A. (2008). Biosynthesis of plant pigments: anthocyanins, betalains and carotenoids. *The Plant Journal*, *54*, 733-749.

[4] Peto, R., Doll, R., Buckley, J.D., & Sporn, M.B. (1981). Can dietary beta-carotene materially reduce human cancer rates? *Nature*, *290*, 201-208.

[5] Hertog, M.G.L., Kromhout, D., Aravanis, C., Blackburn, H., Buzina, R., Fidanza, F., Giampaoli, S., Jansen, A., Menotti, A., Nedeljkovic, S., Pekkarinen, M., Simic, B.S., Toshima, H., Feskens, E.J.M., Hollman, P.C.H., & Katan, M.B. (1995). Flavonoid intake and long-term risk of coronary heart disease and cancer in the seven countries study. *Archives of Internal Medicine*, *155*, 381-386.

[6] Rice-Evans, C.A., Sampson, J., Bramley, P.M., & Holloway, D.E. (1997). Why do we expect carotenoids to be antioxidants in vivo? *Free Radical Research*, *26*, 381-398.

[7] Wang, H., Cao, G., & Prior, R.L. (1997). Oxygen radical absorbing capacity of anthocyanins. *Journal of Agricultural and Food Chemistry*, *45*, 304-309.

[8] Hou, D.X., Ose, T., Lin, S., Imamura, I., Kubo, M., Uto, T., Terahara, N., Yoshimoto, M., & Fujii, M. (2003). Anthocyanidins induce apoptosis in human promyelocytic leukemia cells: structure–activity relationship and mechanisms involved. *International Journal of Oncology*, *23*, 705-712.

[9] Netzel, M., Netzel, G., Kammerer, D.R., Schieber, A., Carle, R., Simons, L., Bitsch, I., Bitsch, R., & Konczak, I. (2007). Cancer cell antiproliferation activity and metabolism of black carrot anthocyanins. *Innovative Food Science & Emerging Technologies*, *8*(3), 365-372.

[10] Zafra-Stone, S., Yasmin, T., Bagchi, M., Chatterjee, A., Vinson, J.A., & Bagchi, D. (2007). Berry anthocyanins as novel antioxidants in human health and disease prevention. *Molecular Nutrition & Food Research*, *51*, 675-683.

[11] Clifford, M.N. (2000). Anthocyanins nature, occurrence and dietary burden. *Journal of the Science of Food and Agriculture*, 80, 1063-1072.

[12] Rentzsch, M., Schwarz, M., & Winterhalter, P. (2007). Pyranoanthocyanins-an overview on structures, occurrence, and pathways of formation. *Trends in Food Science and Technology*, *18*(10), 526-534.

[13] Bagchi, D., Sen, C.K., Bagchi, M., & Atalay, M. (2004). Anti-angiogenic, antioxidant, and anti-carcinogenic properties of a novel anthocyanin-rich berry extract formula. *Biochemistry*, *69*, 75-80.

[14] Duthie, G.G., Duthie, S.J., & Kyle, J.A.M. (2000). Plant polyphenols in cancer and heart disease: Implications as nutritional antioxidants. *Nutrition Research Reviews*, *13*, 79-106.

[15] Jackman, R.L., Yada, R.Y., & Tung, M.A. (1987). A review: separation and chemical properties of anthocyanins used for their qualitative and quantitative analysis. *Journal of Food Biochemistry*, *11*, 279-308.

[16] Wang, W.D., & Xu, S.Y. (2007). Degradation kinetics of anthocyanins in blackberry juice and concentrate. *Journal of Food Engineering*, *82*, 271-275.

[17] Cemeroğlu, B., Velioğlu, S., & Işik, S. (1994). Degradation kinetics of anthocyanins in sour cherry juice and concentrate. *Journal of Food Science*, *59*(6), 1216-1218.

[18] Francis, F.J. (1995). Carotenoids as colorants, The World of Ingredients, Sept.-Oct., 34-38.

[19] Schoefs, B., Bertrand, M., & Lemoine, Y. (1995). Separation of photosynthetic pigments and their precursors by reversed-phase high-performance liquid chromatography using a photodiode-array detector. *Journal of Chromatography* A, *692*, 239-245.

[20] Harborne, J.B., & Grayer, R.J. (1988). The anthocyanins In: Harborne, JB, Ed. The Flavonoids London: Chapman and Hall Ltd, 1-20.

[21] Kong, J.-M., Chia, L.-S., Goh, N.-K, Chia, T.-F., & Brouillard, R. (2003). Analysis and biological activities of anthocyanins. *Phytochemistry*, *64*, 923-933.

[22] Mazza, G., & Miniati, E. (1993). Anthocyanins in fruits, vegetables and grains Boca Raton, FL: CRC Press, 362.

[23] Harborne, J.B. (1988). The flavonoids: recent advances In: Goodwin, T W, Ed. Plant Pigments Academic Press, London, 298-343.

[24] Andersen, O.M., & Jordheim, M. (2006). The anthocyanins In: OM Andersen and KR Markham (Editors), Flavonoids: Chemistry, biochemistry and application, Taylor & Francis Group, Boca Raton/London/New York, 471-489.

[25] Tiwari, B.K., O'Donnell, C.P., Patras, A., Brunton, N.P, & Cullen, P.J. (2009). Effect of ozone processing on anthocyanins and ascorbic acid degradation of strawberry juice. *Food Chemistry*, *113*(4),1119-1126

[26] Giusti, M.M., & Wrolstad, R.E. (2003). Acylated anthocyanins from edible sources and their applications in food systems. *Biochemical Engineering Journal*, *14*, 217-225.

[27] Rein, M. (2005). Copigmentation reactions and color stability of berry anthocyanins. Helsinki: University of Helsinki, 10-14.

[28] Stintzing, F.C., Trichterborn, J., & Carle, R. (2006). Characterisation of anthocyanin–betalain mixtures for food colouring by chromatic and HPLC–DAD–MS analyses, *Food Chemistry*, *94*, 296-309.

[29] Stintzing, F.C., & Carle, R. (2007). Betalains In: C Socaciu (Ed), Food colorants: Chemical and functional properties Taylor & Francis/CRC Press.

[30] Schwartz, S.J., & von Elbe, J.H. (1980). Quantitative determination of individual betacyanin pigments by high-performance liquid chromatography. *Journal of Agricultural and Food Chemistry*, *28*(3), 540-543.

[31] Vincent, K.R., & Scholz, R.G. (1978). Separation and quantification of red beet betacyanins and betaxanthins by high-performance liquid chromatography. *Journal of Agricultural and Food Chemistry*, *26*(4), 812-816.

[32] Kanner, J. Harel, S., & Granit, R. (2001). Betalains-a new class of dietary cationized antioxidants. *Journal of Agricultural and Food Chemistry*, *49*, 5178-5185.

[33] Escribano, J., Pedreno, M.A., Garcia-Carmona, & F., Munoz, R. (1998). Characterization of the antiradical activity of betalains from *Beta vulgaris* L roots. *Phytochemical Anal*, *9*, 124-127.

[34] Pedreno, M.A., & Escribano, J. (2000). Studying the oxidation and the antiradical activity of betalain from beetroot. *Journal of Biological Education*, *35*, 49-51.

[35] Kapadia, G.J., Tokuda, H., Konoshima, T., & Nishino, H. (1996). Chemoprevention of lung and skin cancer by Beta vulgaris (beet) root extract. *Cancer Letters*, *100*, 211-214.

[36] Reddy, K.M., Ruby, L., Lindo, A., & Nair, G.M. (2005). Relative inhibition of lipid peroxidation, cyclooxygenase enzymes and human tumor cell proliferation by natural food colors. *Journal of Agricultural and Food Chemistry*, 53, 9268-9273.

[37] Stintzing, F.C., & Carle, R. (2007). Betalains – emerging prospects for food scientists. *Trends in Food Science and Technology*, *18*, 514-525.

[38] Davidek, J., Velisek, J., & Pokorny, J. (1990). Sensorically active compounds. In: Development in Food Science 21, Chemical Changes in Food Processing. Elsevier Science, Amsterdam, pp. 302-78.

[39] Saguy, I., Kopelman, I.J., & Mizrahi, S. (1978). Computer-aided determination of beet pigments. *Journal of Food Science*, *43*,124-127.

[40] Mabry, T.J. (1980). Betalains. In E.A., Bell, & B.V., Charlwood (Eds.), *Encyclopedia of Plant Physiology*. Vol. 8, Secondary Plant Products. Springer-Verlag, Berlin, 513-533.

[41] Lashley, D., Wiley, R.C. (1979). A betacyanine decolorizing enzyme found in red beet tissue. *Journal of Food Science*, *44*, 1568-1569.

[42] Martinez-Parra, J., & Munoz, R. (2001). Characterization of betacyanin oxidation catalyzed by a peroxidase from *Beta vulgaris* L roots. *Journal of Agricultural and Food Chemistry*, *49*(8), 4064-4068.

[43] Shih, C.C., & Wiley, R.C. (1981). Betacyanine and betaxanthine decolorizing enzymes in the beet (*Beta vulgaris* L) root. *Journal of Food Science*, *47*, 164-166 (172).

[44] Wasserman, B.P., & Guilfoy, P. (1983). Peroxidative properties of betanin decolorization by cell walls of red beet. *Phytochemistry*, *22*, 2653-2656.

[45] Singer, J.W., & von Elbe, J.H. (1980). Degradation rates of vulgaxanthine-I. *Journal of Food Science*, *45*, 489-491.

[46] Priestley, R.J. (1979). Effects of heating on foodstuffs, Applied Science Publishers Ltd, London. 307-314.

[47] Cruz, R.M.S., Vieira, M.C., & Silva, C.L.M. (2007). Modelling kinetics of watercress (*Nasturtium officinale*) colour changes due to heat and thermosonication treatments. *Innovative Food Science and Emerging Technologies*, *8*, 244-252.

[48] Schwartz, S.J., & von Elbe, J.H. (1983). Kinetics of chlorophyll degradation to pyropheophytin in vegetables. *Journal of Food Science*, *48*, 1303-1306.

[49] von Elbe, J.H., & Schwartz, S.J. (1996). Colorants In: OR Fennema, Editor, *Food Chemistry*, Marcel Dekker, New York, 651-722.

[50] Buckle, K.A., & Edwards, R.A. (1970). Chlorophyll colour and pH changes in HTST processed green pea purée. *Journal of Food Technology*, *5*, 173-178.

[51] Canjura, F.L., & Schwartz, S.J. (1991). Separation of chlorophyll compounds and their polar derivatives by high-performance liquid chromatography. *Journal of Agricultural and Food Chemistry*, *39*, 1102-1105.

[52] Lajollo, F.M., Tannenbaum, S.R., & Labuza, T.P. (1971). Reaction at limited water concentrations. Chlorophyll degradation. *Journal of Food Science*, *36*, 850-853.

[53] Schwartz, S.J., & Lorenzo, T.V. Chlorophyll stability during continuous aseptic processing and storage. *Journal of Food Science*, *56*, 1059-1062.

[54] Tan, C.T., & Francis, F.J. (1962). Effect of processing temperature on pigments and colour of spinach. *Journal of Food Science*, *27*, 232-240.

[55] Sweeney, J.P., & Martin, M.E. (1961). Stability of chlorophyll in vegetables as affected by pH. *Food Technology*, *15*, 263-266.

[56] Scotter, M.J., & Castle, L. (2004). Chemical interactions between additives in foodstuffs: A review. *Food Additives and Contaminants*, *21*, 93-124.

[57] Schaber, P.M., Hunt, J.E., Fries, R., & Katz, J.J. (1984). High-performance liquid-chromatographic study of the chlorophyll allomerization reaction. *Journal of Chromatography*, *316*, 25-41.

[58] Maggiora, L.L., Petke, J.D., Gopal, D., Iwamoto, R.T., & Maggiora, G.M. (1985). Experimental and theoretical-studies of Schiff-base chlorophylls. *Photochemistry and Photobiology*, *42*, 69-75.

[59] Lin, Y.D., Clydesdale, F.M., & Francis, F.J. (1971). Organic acid profiles of thermally processed, stored spinach purée. *Journal of Food Science*, *36*, 240-242.

[60] Pfander, H. (1987). Key to Carotenoids. Birkha¨ user Verlag, Basel.

[61] Carotenature (The carotenoids page) (2000). Retrieved September 4, 2002 from the World Wide Web: http://www.carotenature.com.

[62] de Quiros, A.R., & Costa, H.S. (2006). Analysis of carotenoids in vegetable and plasma samples: A review. *Journal of Food Composition and Analysis*, *19*, 97-111.

[63] Armstrong, G.A., & Hearst, J.E. (1996). Genetics and molecular biology of carotenoid pigment biosynthesis. *FASEB Journal*, *10*, 228-237.

[64] Van Vliet, T., van Schaik, F., Schreurs, W.H.P., & van den Berg, H. (1996). In vitro measurement of β-carotene cleavage activity: methodological considerations and the effect of other carotenoids on β-carotene cleavage. *International Journal for Vitamin and Nutrition Research*, *66*, 77-85.

[65] Palozza, P., & Krinsky, N.I. (1992). Antioxidant effects of carotenoids in vivo and in vitro: an overview. *Methods in Enzymology*, *213*, 403-420.

[66] Stahl, W., & Siess, H. (1998). The role of carotenoids and retinoids in gap junctional communication. *International Journal for Vitamin and Nutrition Research*, *68*, 354-359.

[67] Lindley, M.G. (1998). The impact of food processing on antioxidants in vegetable oils, fruits and vegetables. *Trends in Food Science and Technology*, *9*, 336-340.

[68] Solomons, N.W., & Bulux, J. (1997). Identification and production of local carotene-rich foods to combat vitamin A malnutrition. *European Journal of Clinical Nutrition*, *51*, S39-S45.

[69] Seddon, J.M., Ajani, U.A., Speruto, R.D., Hiller, R., Blair, N., Burton, T.C., Farber, M.D., Gragoudas, E.S., Haller, J., Miller, D.T., Yannuzzi, L.A., & Willett, W. (1994).

Dietary carotenoids, vitamin A, vitamin C and vitamin E and advanced age-related macular degeneration. *Journal of the American Medical Association*, *272*, 1413-1420.

[70] van den Berg, H., Faulks, R., Granado, H.F., Hirschberg, J., Olmedilla, B., Sandmann, G., Southon, S., Stahl, W. (2000). The potential for the improvement of carotenoid levels in foods and the likely systemic effects. *Journal of the Science of Food and Agriculture*, *80*, 880-912.

[71] Abbott, J.A. (1999). Quality measurement of fruits and vegetables. *Postharvest Biology and Technology*, *15*(3), 207-225.

[72] Clydesdale, F.M. (1978). Colorimetry-methodology and applications. *Critical Reviews in Food Science and Nutrition*, *10*, 243-301.

[73] Francis, F.J. (1980). Color quality evaluation of horticultural crops. *HortScience*, *15*, 14-15.

[74] Hunter, R.S., & Harold, R.W. (1987). The Measurement of Appearance. Wiley-Interscience, New York.

[75] Minolta. (1994). Precise Color Communication Minolta Co, Ramsey, NJ.

[76] Kang, S.P., East, A.R., Trujillo, F.J. (2008). Colour vision system evaluation of bicolour fruit: A case study with 'B74' mango. *Postharvest Biology and Technology*, *49*(1), 77-85.

[77] Gnanasekharan, V., Shewfelt, R.L., & Chinnan, M.S. (1992). Detection of color changes in green vegetables. *Journal of Food Science*, *57*, 149-154.

[78] Tian, M.S., Davies, L., Downs, C.G., Liu, X.F., & Lill, R.E. (1995). Effects of floret maturity, cytokinin and ethylene on broccoli yellowing after harvest. *Postharvest Biology and Technology*, *6*, 29-40.

[79] Tian, M.S., Woolf, A.B., Bowen, J.H., & Ferguson, I.B. (1996). Changes in color and chlorophyll fluorescence of broccoli florets following hot water treatment. *Journal of the American Society for Horticultural Science*, *121*, 310-313.

[80] Barrett, D.M., Garcia, E.L., Russell, G.F., Ramirez, E., & Shirazi, A. (2000). Blanch time and cultivar effects on quality of frozen and stored corn and broccoli. *Journal of Food Science*, *65*, 534-540.

[81] Gunawan, M.I., & Barringer, S.A. (2000). Green color degradation of blanched broccoli (*Brassica oleracea*) due to acid and microbial growth. *Journal of Food Processing and Preservation*, *24*, 253-263.

[82] Kidmose, U., & Hansen, M. (1999). The influence of post-harvest storage, temperature and duration on quality of cooked broccoli florets. *Journal of Food Quality*, *22*, 135-146.

[83] Hutchings, J.B. (1999). Food Color and Appearance, Aspen Publishers, Inc, Gaithersburg, MD, 593.

[84] Paulus, I., & Schrevens, E. (1999). Shape characterization of new apple cultivars by Fourier expansion of digitalized images. *Journal of Agricultural Engineering Research*, *72*, 113-118.

[85] Shahin, M.A., & Symons, S.J. (2001). A machine vision system for grading lentils. *Canadian Biosystems Engineering*, *43*, 77-14

[86] Leemans, V., Magein, H., & Destain, M.F. (2002). On-line fruit grading according to their external quality using machine vision. *Biosystems Engineering*, *83*, 397-404.

[87] Li, Q., Wang, M., & Gu, W. (2002). Computer vision based system for apple surface defect detection. *Computers and Electronics in Agriculture*, *36*, 215-223.

[88] De Belie, N., Tu, K., Jancsok, P., & De Baerdemaeker, J. (1999). Preliminary study on the influence of turgor pressure on body reflectance of red laser light as a ripeness indicator for apples. *Postharvest Biology and Technology*, *16*(3), 279-284.

[89] Zhang, M., De Baerdemaeker, J., & Schrevens, E. (2003). Effects of different varieties and shelf storage conditions of chicory on deteriorative color changes using digital image processing and analysis. *Food Research International*, *36*, 669-676.

[90] Mendoza, F., & Aguilera, J.M. (2004). Application of image analysis for classification of ripening bananas. *Journal of Food Sci*ence, *69*, 471-477.

[91] Blasco, J., Cubero, S., Gomez-Sanchis, J., Mira, P., & Moltó, E. (2009). Development of a machine for the automatic sorting of pomegranate (*Punica granatum*) arils based on computer vision. *Journal of Food Engineering*, *90*(1), 27-34.

[92] Mendoza, F., Dejmek, P., & Aguilera, J.M. (2006). Calibrated color measurements of agricultural foods using image analysis. *Postharvest Biology and Technology*, *41*, 285-295.

[93] Francis, F.J., & Clydesdale, F.M. (1975). Food colorimetry: Theory and applications. Westport: AVI Publishing Company.

[94] Du, C.J., & Sun, D.W. (2004). Recent developments in the applications of image processing techniques for food quality evaluation. *Trends in Food Science and Technology*, *15*, 230-249.

[95] Li, J., Tan, J., & Shatadal, P. (2001). Classification of tough and tender beef by image texture analysis. *Meat Science*, *57*, 341-346.

[96] Tao, Y., & Ibarra, J.G. (2000). Thickness-compensated X-ray imaging detection of bone fragments in deboned poultry-model analysis. *Transactions of the ASAE*, *43*, 453-459.

[97] Vizhanyo, T., & Felfoldi, J. (2000). Enhancing colour differences in images of diseased mushrooms. *Computers and Electronics in Agriculture*, *26*, 187-198.

[98] Lee, H.S. (2000). Objective measurement of red grapefruit juice color. *Journal of Agricultural and Food Chemistry*, *48*, 1507-1511.

[99] Meléndez-Martínez, A.J., Vicario, I.M., & Heredia, F.J. (2005). Instrumental measurement of orange juice colour: a review. *Journal of the Science of Food and Agriculture*, *85*(6), 894-901.

[100] Lee, H.S. (2001). Characterization of carotenoids in juice of red navel orange (*Cara cara*). *Journal of Agricultural and Food Chemistry*, *49*, 2563-2568.

[101] Stewart, I. (1977). Provitamin A and carotenoid content of citrus juices. *Journal of Agricultural and Food Chemistry*, *25*, 1132-1137.

[102] Huggart, R.L., & Petrus, D.R. (1976). Color in canned commercial grapefruit juice. *Proceedings of the Florida State Horticultural Society*, *89*, 194-196.

[103] Consejo de la Unión Europea (2002). Directiva relativa a los zumos de frutas y otros productos similares destinados a la alimentación humana Diario Oficial de las Comunidades Europeas, 2001/112/CE.

[104] Wyszecki, G., & Stiles, W.S. (1982). Color Science Concepts and Methods Quantitative Data and Formulae, Wiley, New York.

[105] Hutchings, J.B. (1994). Food colour and appearance, London, UK: Blackie Academic and Professional Publication, 220-256.

[106] Oey, I., Plancken, I.V., Loey, A.V., & Hendrickx, M. (2008). Does high pressure processing influence nutritional aspects of plant based food systems? *Trends in Food Science and Technology*, *19*(6), 300-308.

[107] Awuah, G.B., & Ramaswamy, H.S. (2007). Economides, A. Thermal processing and quality: Principles and overview. *Chemical Engineering and Processing*, *46*, 584-602.

[108] Leadley, C., Tucker, G., & Fryer, P. (2008). A comparative study of high pressure sterilisation and conventional thermal sterilisation: Quality effects in green beans. *Innovative Food Science and Emerging Technologies*, *9*, 70-79.

[109] Murcia, M.A., Lopez-Ayerra, B., & Garcia-Carmona, F. (1999). Effect of processing methods and different blanching times on broccoli: proximate composition and fatty acids. *Lebensmittel-Wissenschaft und Technologie*, *32*, 238-243.

[110] Bahçeci, S.K., Serpen, A., Gökmen, V., & Acar, J. (2004). Study of lipoxygenase and peroxidase as indicator enzymes in green beans: Change of enzyme activity, ascorbic acid and chlorophylls during frozen storage. *Journal of Food Engineering*, *66*, 187-192.

[111] Cano, M.P., Marín, M.A., & Fuster, C. (1990). Freezing of banana slices Influence of maturity level and thermal treatment prior to freezing. *Journal of Food Science*, *55*, 1070-1072.

[112] Ndiaye, C., Shi-Ying, X., & Zhang, W. (2009). Steam blanching effect on polyphenoloxidase, peroxidase and colour of mango (*Mangifera indica* L) slices. *Food Chemistry*, *113*, 92-95.

[113] Tijskens, L.M.M., Barringer, S.A., & Biekman, E.S.A. (2001). Modelling the effect of pH on the colour degradation of blanched broccoli. *Innovative Food Science & Emerging Technologies*, 315-322.

[114] Williams, D.C., Lim, M.H., Chen, A.O., Pangborn, R.M., & Whitaker J.R. (1986). Blanching of vegetables for freezing-which indicator enzyme to choose, *Food Technology*, *40*, 130-140.

[115] Sheu, S.C., & Chen, A.O. (1991). Lipoxygenase as blanching index for frozen vegetable soybeans, *Journal of Food Science*, *56*(2), 448-451.

[116] King, D.L., Klein, B.P. (1987). Effect of flavonoids and related compounds on soybean lipoxygenase-1 activity. *Journal of Food Science*, *52*(1), 220-221.

[117] Wang, Y.J., Miller, L.A., & Addis, P.B. (1991). Effect of heat inactivation of lipoxygenase on lipid oxidation in lake herring (*Coregonus artedii*). *Journal of American Oil Chemist Society*, *68*, 752-755.

[118] Priestley, R.J. (1979). Effects of heating on foodstuffs (pp. 307-314). London: Applied Science Publishers Ltd.

[119] Murcia, M.A., López-Ayerra, B., Martnez-Tomé, M., & García-Carmona, F. (2000). Effect of industrial processing on chlorophyll content of broccoli. *Journal of the Science of Food and Agriculture*, *80*, 1447-1451.

[120] Heaton, J.W., Lencki, R.W., & Marangoni, A.G. (1996). Kinetic model for chlorophyll degradation in green tissue. *Journal of Agricultural and Food Chemistry*, *44*, 399-402.

[121] Heaton, J.W., Yada, R.Y., & Marangoni, A.G. (1996). Discoloration of coleslaw tissue caused by chlorophyll degradation. *Journal of Agricultural and Food Chemistry*, *44*, 395-398.

[122] Minguez-Mosquera, M.I., Garrido-Fernandez, I., & Gandul-Rojas, B. (1989). Pigment changes in olives during fermentation and brine storage. *Journal of Agricultural and Food Chemistry*, *37*, 8-11.

[123] White, R., White, R.C., Gibbs, E. (1963). Determination of chlorophylls, chlorophyllides, pheophytins and pheophorbides in plant material. *Journal of Food Science*, *28*, 431-435.

[124] Weemaes, C.A., Ooms, V., van Loey, A.M., & Hendrickx, M.E. (1999). Kinetics of chlorophyll degradation and color loss in heated broccoli juice. *Journal of Agricultural and Food Chemistry*, *47*, 2404-2409.

[125] Minguez-Mosquera, M.I., Gnadul-Rojas, B., & Gallardo-Guerrero, L. (1994). Mechanism and kinetics of the degradation of chlorophylls during the processing of green table olives. *Journal of Agricultural and Food and Chemistry*, *42*, 1089-1095.

[126] Heaton, J.W., & Marangoni, A.G. (1996). Chlorophyll degradation in processed foods and senescent plant tissues. *Trends in Food Science and Technology*, *7*, 8-15.

[127] Van Boekel, M.A.J.S. (1999). Testing of kinetic models: usefulness of the multiresponse approach as applied to chlorophyll degradation in foods. *Food Research International*, *32*, 261-269.

[128] Van Boekel, M.A.J.S. (2000). Kinetic modelling in food science: a case study on chlorophyll degradation in olives. *Journal of the Science of Food and Agriculture*, *80*, 3-9.

[129] Tijskens, L.M.M, Greiner, R., Biekman, E.S.A., & Konietzny, U. (2001). Modelling the effect of temperature and pH on the activity of enzymes: the case of phytases. *Biotechnology and Bioengineering*, *72*, 323-330.

[130] Gold, H.J., & Weckel, K.G. (1959). Degradation of chlorophyll to pheophytin during sterilization of canned peas by heat. *Food Technology*, *13*, 281-286.

[131] Hayakawa, K., & Timbers, G. (1977). Influence of heat treatment on the quality of vegetables; changes in visual green color. *Journal of Food Science*, *42*, 778-781.

[132] Moreno, J., Chiralt, A., Escriche, I., & Serra, J.A. (1998). Efecto estabilizador de tratamientos combinados de inactivación enzimática y deshidratación osmótica en fresas mínimamente procesadas Alimentaria, *295*, 49-56.

[133] Mizrahi, S. (1996). Leaching of soluble solids during blanching of vegetables by ohmic heating. *Journal of Food Engineering*, *29*, 153-166.

[134] Icier, F., Yildiz, H., & Baysal, T. (2007). Peroxidase inactivation and colour changes during ohmic blanching of pea purée. *Journal of Food Engineering*, *74*, 424-429.

[135] Volden, J., Grethe, I., Borge, A., Gunnar, B., Magnor, B., Ingrid, H., Thygese, E., & Wicklund, T. (2008). Effect of thermal treatment on glucosinolates and antioxidant-related parameters in red cabbage (*Brassica oleracea* L ssp *capitata* f *rubra*). *Food Chemistry*, *109*(3), 595-605.

[136] Lee, J., Durst, R.W., Wrolstad, R.E. (2002). Impact of juice processing on blueberry anthocyanins and polyphenolics: comparison of two pretreatments. *Journal of Food Science*, *67*, 1660-1666.

[137] Srivastava, A., Akoh, C.C., Yi, W., Fischer, J., & Krewer, G. (2007). Effect of storage conditions on the biological activity of phenolic compounds of blueberry extract packed in glass bottles. *Journal of Agricultural and Food Chemistry*, *55*, 2705-2713.

[138] Rossi, M., Giussani, E., Morelli, R., Lo Scalzo, R., Nani, R.C., Torreggiani, D. (2003). Effect of fruit blanching on phenolics and radical scavenging activity of highbush blueberry juice. *Food Research International*, *36*, 999-1005.

[139] Kirca, A., Özkan, M., & Cemeroglu, B. (2006). Stability of black carrot anthocyanins in various fruit juices and nectars. *Food Chemistry*, *97*, 598-605.

[140] Rhim, J.W. (2002). Kinetics of thermal degradation of anthocyanin pigment solutions driven from red flower cabbage. *Food Science and Biotechnology*, *11*, 361-364.

[141] Hager, A., Howard, L.R., Prior, R.L., & Brownmiller, C. (2008). Processing and storage effects on monomeric anthocyanins, percent polymeric color, and antioxidant capacity of processed black raspberry products. *Journal of Food Science*, *73*, H143-H139.

[142] Skrede, G., Wrolstad, R.E., & Durst, R.W. (2000). Changes in anthocyanins and polyphenolics during juice processing of highbush blueberries (*Vaccinium corymbosum* L). *Journal of Food Science*, *65*, 357-364.

[143] Klopotek, Y., Otto, K., & Böhm, V. (2005). Processing strawberries to different products alters contents of vitamin C, total phenolics, total anthocyanins, and antioxidant capacity. *Journal of Agricultural and Food Chemistry*, *53*, 5640-5646.

[144] Brownmiller, C., Howard, L.R., & Prior, R.L. (2008). Processing and storage effects on monomeric anthocyanins, percent polymeric color, and antioxidant capacity of processed blueberry products. *Journal of Food Science*, *5*, 72-79.

[145] Dyrby, M., Westergaard, N., & Stapelfeldt, H. (2001). Light and heat sensitivity of red cabbage extract in soft drink model systems. *Food Chemistry*, *72*(4), 431-437.

[146] Kirca, A., Ozkan, M., & Cemeroglu, B. (2007). Effects of temperature, solid content and pH on the stability of black carrot anthocyanins. *Food Chemistry*, *101*(1), 212-218.

[147] Kong, J-M., Chia, L-S., Goh, N-K., Chia, T-F., & Brouillard, R. (2003). Analysis and biological activities of anthocyanins. *Phytochemistry*, *64*, 923-933.

[148] Goto, T. (1987). Structure, stability and color variation of natural anthocyanins. *Progress in the Chemistry of Organic Natural Products*, *52*, 113-158.

[149] Giusti, M.M., & Wrolstad, R.E. (1996). Characterization of red radish anthocyanins. *Journal of Food Science*, *61*, 322-326.

[150] Sadilova, E., Stintzing, F.C., & Carle, R. (2006). Thermal degradation of acylated and nonacylated anthocyanins. *Journal of Food Science*, *71*, C504-C512.

[151] Nielsen, S.S., Marcy, J.E., & Sadler, G.D. (1993). Chemistry of aseptically processed foods. In J.V., Chambers, & P.E., Nelson (Eds.), *Principles of Aseptic Processing and Packaging* (pp 87-114) Washington, DC: Food Processors Institute.

[152] Garzon, G.A., & Wrolstad, R.E. (2002). Comparison of the stability of pelargonidin-based anthocyanins in strawberry juice and concentrate. *Journal of Food Science*, *67*(4), 1288-1299.

[153] Kirca, A., Ozkan, M., & Cemeroglu, B. (2003) Thermal stability of black carrot anthocyanins in blond orange juice. *Journal of Food Quality*, *26*(5), 361-366.

[154] Ochoa, M.R., Kesseler, A.G., Vullioud, M.B., & Lozano, J.E. (1999). Physical and chemical characteristics of raspberry pulp: storage effect on composition and color. *LWT Food Science & Technology*, *149*, 149-153.

[155] García-Viguera, C., & Zafrilla, P. (2001). Changes in anthocyanins during food processing: influence on color ACSSymposium Series775 (chemistry and physiology of selected food colorants). *American Chemical Society*, 56-65.

[156] García-Viguera, C., Zafrilla, P., Romero, F., Abellán, P., Artés, F., & Tomás-Barberán, F.A. (1999). Color stability of strawberry jam as affected by cultivar and storage temperature. *Journal of Food Science, 64*(2), 243-247.

[157] Escribano-Bailón, M.T., Santos-Buelga, C., & Rivas-Gonzalo, J.C. (2004). Anthocyanins in cereals. *Journal of Chromatography A*, *1054*, 129-141.

[158] Francis, F.J. (1992). A new group of food colorants. *Trends in Food Science and Technology*, *3*, 27-30

[159] Ngo, T., Wrolstad, R.E., & Zhao, Y. (2007). Color quality of Oregon strawberries—impact of genotype, composition, and processing. *Journal of Food Science*, *72*, 25-32.

[160] Piljac-Žegarac, J., Valek, L., Martinez, S., & Belšcak, A. (2009). Fluctuations in the phenolic content and antioxidant capacity of dark fruit juices in refrigerated storage. *Food Chemistry*, *113*, 394-400.

[161] Gössinger, M., Moritz, S., Hermes, M., Wendelin, S., Scherbichler, H., Halbwirth, H., Stich, K., & Berghofer, E. (2009). Effects of processing parameters on colour stability of strawberry nectar from purée. *Journal of Food Engineering*, *90*, 171-178.

[162] Patras, A., Brunton, N.P., & Butler, F. (2008). Impact of high pressure processing on antioxidant activity, ascorbic acid, anthocyanins and instrumental colour of blackberry and strawberry purée. *Innovative Food Science and Emerging* Technologies, doi: 101016/jifset200812004.

[163] Kırca, A., Oskar, M., & Cemeroglu, B. (2006). Stability of black carrot anthocyanins in various fruit juices and nectars. *Food Chemistry*, *97*, 598-605.

[164] Rubinskiene, M., Viskelis, P., Jasutiene, I. Viskeliene, R.C., & Bobinas, C. (2005). Impact of various factors on the composition and stability of black currant anthocyanins. *Food Research International*, *38*, 867-871.

[165] Krifi, B., & Maurice, M. (2000). Degradation of anthocyanins from blood orange juices. *International Journal of Food Science and Technology*, *35*, 275-283.

[166] Kajuna, S.T.A.R., Bilanski, W.K., & Mittal, G.S. (1998). Color changes in bananas and plantains during storage. *Journal of Food Processing and Preservation*, *22*, 27-40.

[167] Shin, S., & Bhowmik, S.R. (1995). Thermal kinetics of color changes in pea purée. *Journal of Food Engineering*, *24*(1), 77-86.

[168] Steet, J.A., & Tong, C.H. (1996). Degradation kinetics of green colour and chlorophyll in peas by colorimetry and HPLC. *Journal of Food Science*, *61*, 924-927, 931.

[169] Ahmed, J., Shivhare, U.S., & Singh, P. (2004). Colour kinetics and rheology of coriander leaf purée and storage characteristics of the purée. *Food Chemistry*, *84*, 605-611.

[170] Ahmed, J., & Shivhare, U.S. (2001). Thermal kinetics of color change, rheology, and storage characteristics of garlic purée/paste. *Journal of Food Science*, *66*(5), 754-757.

[171] Ahmed, J., Kaur, A., & Shivhare, U.S. (2002). Color degradation kinetics of spinach, mustard leaves, and mixed purée. *Journal of Food Science*, *67*(3), 1088-1091.

[172] Ahmed, J., Shivhare, U.S., & Debnath, S. (2002). Colour degradation and rheology of green chilli purée during thermal processing. *International Journal of Food Science and Technology*, *37*, 57-63.

[173] Ahmed, J., Shivhare, U.S., & Raghavan, G.S.V. (2000). Rheological characteristics and kinetics of colour degradation of green chilli purée. *Journal of Food Engineering*, *44*, 239-244.

[174] Ahmed, J., Shivhare, U.S., & Ramaswamy, H.S. (2002). A fraction conversion kinetic model for thermal degradation of color in red chilli purée and paste. *Lebensmittel Wissenchaft und Technology*, *35*, 497-503.

[175] Ahmed, J., Shivhare, U.S., & Sandhu, K.S. (2002). Thermal degradation kinetics of carotenoids and visual color of papaya purée. *Journal of Food Science*, *67*, 2692-2695.

[176] Barreiro, J.A., Milano, M., & Sandoval, A.J. (1997). Kinetics of colour change of double concentrated tomato paste during thermal treatment. *Journal of Food Engineering*, *33*, 359-371.

[177] Garza, S., Ibarz, A., Pagan, J., & Giner, J. (1999). Non-enzymatic browning in peach purée during heating. *Food Research International*, *32*, 335-343.

[178] Ibarz, A., Pagan, J., & Garza, S. (1999). Kinetic models for colour changes in pear purée during heating at relatively high temperatures. *Journal of Food Engineering*, *39*, 415-422.

[179] Ibarz, A., Pagan, J., & Garza, S. (2000). Kinetic models of non-enzymatic browning in apple purée. *Journal of the Science of Food and Agriculture*, *80*(8), 1162-1168.

[180] Silva, F.M., & Silva, C.L.M. (1999). Colour changes in thermally processed cupuaçu (*Theobroma grandiflorum*) purée: critical times and kinetics modelling. *International Journal of Food Science and Technology*, *34*, 87-94.

[181] Zanoni, B., Pagliarini, E., Giovanelli, G., & Lavelli, V. (2003). Modelling the effects of thermal sterilization on the quality of tomato purée. *Journal of Food Engineering*, *56*, 203-206.

[182] Avila, I.M.L.B., & Silva, C.L.M. (1999). Modeling kinetics of thermal degradation of colour in peach purée. *Journal of Food Engineering*, *39*, 161-166.

[183] Aurelio, D.L., Edgardo, R.G., & Navarro-Galindo, S. (2008). Thermal kinetic degradation of anthocyanins in a roselle (*Hibiscus sabdariffa* L cv 'Criollo') infusion. *International Journal of Food Science and Technology*, *43*, 322-325.

[184] Mishra, D.K., Dolan, K.D., & Yang, L. (2008) Confidence intervals for modeling anthocyanin retention in grape pomace during nonisothermal heating. *Journal of Food Science*, *73*, E9-E15.

[185] Reyes, L.F., & Cisneros-Zevallos, L. (2007). Degradation kinetics and colour of anthocyanins in aqueous extracts of purple- and red-flesh potatoes (*Solanum tuberosum* L). *Food Chemistry*, *100*(3), 885-894.

[186] Kirca, A., Özkan, M., & Cemeroglu, B. (2007). Effects of temperature, solid content and pH on the stability of black carrot anthocyanins. *Food Chemistry*, *101*(1), 212-218.

[187] Kirca, A., Özkan, M., & Cemeroglu, B. (2003). Thermal stability of black carrot anthocyanins in blond orange juice. *Journal of Food Quality*, *26*(5), 361-366.

[188] Yang, Z., Han, Y., Gu, Z., Fan, G., & Chen, Z. (2008). Thermal degradation kinetics of aqueous anthocyanins and visual color of purple corn (*Zea mays* L) cob. *Innovative Food Science and Emerging Technologies*, *9*(3), 341-347.

[189] Harbourne, N., Jacquier, J.C., Morgan, D.J., & Lyng, J.G. (2008). Determination of the degradation kinetics of anthocyanins in a model juice system using isothermal and non-isothermal methods. *Food Chemistry*, *111*(1), 204-208.

[190] Ahmed, J., Shivhare, U.S., & Raghavan, G.S.V. (2004). Thermal degradation kinetics of anthocyanin and visual colour of plum purée. *European Food Research and Technology*, *218*(6), 525-528.

[191] Dolan, K.D., Yang, L., & Trampel, C.P. (2007). Nonlinear regression technique to estimate kinetic parameters and confidence intervals in unsteady-state conduction-heated foods. *Journal of Food Engineering*, *80*(2), 581-93.

[192] Clydesdale, F.M., Fleischman, D.L., & Francis, F.J. (1970). Maintenance of color in processed green vegetables. *Journal of Food Product Development*, *4*, 127-130.

[193] Lovric, T., Sablek, Z., & Boskovic, M. (1970). Cis-trans isomerization of lycopene and colour stability of foam-mat dried tomato powder during storage. *Journal of the Science of Food and Agriculture*, *21*, 641-647.
[194] Tiwari, B.K., O'Donnell, C.P., Muthukumarappan, K., & Cullen, P.J. (2008). Effect of ultrasound processing on quality of fruit juices. *Stewart-Post Harvest Review*, *4*(5), 1-6.
[195] Hurt, H.D. (1979). Effect of Canning on the Nutritive Value of Vegetables. *Food Technology*, 62-65.
[196] Meyer, L. (1987). *Food Chemistry* (CBS Publication, New Delhi, India).
[197] Ahmed, J., & Ramaswamy, H.S. (2005). Effect of temperature on dynamic rheology and colour degradation kinetics of date paste. *Food and Bioproducts Processing*, *83*(C3), 198-202.
[198] Gliemmo, M.F., Latorre, M.E., Gerschenson, L.N., & Campos, C.A. (2009). Color stability of pumpkin (*Cucurbita moschata, Duchesne* ex Poiret) purée during storage at room temperature: Effect of pH, potassium sorbate, ascorbic acid and packaging material. *LWT - Food Science and Technology*, *42*, 196-201.
[199] Schieber, A., & Carle, R. (2005). Occurrence of carotenoid cis-isomers in food: technological, analytical, and nutritional implications. *Trends in Food Science and Technology*, *16*, 416-422.
[200] Young, A.J., & Lowe, G.M. (2001). Antioxidant and prooxidant properties of carotenoids. *Archives of Biochemistry and Biophysics*, *385*, 20-27.
[201] Choi, M.H., Kim, G.H., & Lee, H.S. (2002). Effects of ascorbic acid retention on juice color and pigment stability in blood orange (*Citrus sinensis*) juice during refrigerated storage. *Food Research International*, *35*, 753-759.
[202] Mahoney, J.R., & Graf, E. (1986). Role of alpha-tocopherol, ascorbic acid, citric acid and EDTA as oxidants in model systems. *Journal of Food Science*, *51*(5), 1293-1296.
[203] Shrikhande, A.J., & Francis, F.J. (1974). Effect of flavonols on ascorbic acid and anthocyanin stability in model systems. *Journal of Food Science*, *39*(5), 904-906.
[204] Starr, M.S., & Francis, F.J. (1968). Oxygen and ascorbic acid effect on the relative stability of four anthocyanin pigments in cranberry juice. *Food Technology*, *22*(10), 91-93.
[205] Golaszewski, R., Sims, C.A., O'keefe, S.F., Braddock, R.J., & Littell, R.C. (1998). Sensory attributes and volatile components of stored strawberry juice *Journal of Food Science*, *63*(4), 734-738.
[206] Skrede, G., Wrolstad, R.E., Lea, P., & Enersen, G. (1992). Color stability of strawberry and blackcurrant syrups. *Journal of Food Science*, *57*, 172-177.
[207] Kouniaki, S., Kajda, P., & Zabetakis, I. (2004). The effect of high hydrostatic pressure on anthocyanins and ascorbic acid in blackcurrants (*Ribes nigrum*). *Flavour and Fragrance Journal*, *19*, 281-286.
[208] Markakis, P. (1974). Anthocyanins and their stability in foods. *Critical Reviews in Food Technology*, *4*, 437-456.
[209] Talcott, S.T., Brenes, C.H., Pires, D.M., & Del Pozo-Insfran, D. (2003). Phytochemical stability and color retention of copigmented and processed muscadine grape juice. *Journal of Agricultural and Food Chemistry*, *51*, 957-963.
[210] Darias-Martin, J., Carrillo, M., Dias, E., & Boulton, R.B. (2001). Enhancement of red wine colour by pre-fermentation addition of copigments. *Food Chemistry*, *73*, 217-220.

[211] Talcott, S.T., Peele, J.E., & Brenes, C.H. (2005). Red clover isoflavonoids as anthocyanin color enhancing agents in muscadine wine and juice. *Food Research International*, *10*, 1205-1212.

[212] Del Pozo-Insfran, D., Del Follo-Martinez, A., Talcott, S.T., & Brenes, C.H. (2007). Stability of copigmented anthocyanins and ascorbic acid in muscadine grape juice processed by high hydrostatic pressure. *Journal of Food Science*, *72*(4), 247-253.

[213] Mangos, T.J., & Berger, R.G. (1997). Determination of major chlorophyll degradation products. Zeitschrift für Lebensmittel Untersuchung und-Forschung, *204*, 345-350.

[214] Mahanta, P.K., & Hazarika, M. (1985). Chlorophyll and degradation products in orthodox and CTC black teas and their influence on shade of colour and sensory quality in relation to thearubigins. *Journal of the Science of Food and Agriculture*, *36*, 1122-1139.

[215] Langmeier, M., Ginsburg, S., & Matile, P. (1993). Chlorophyll breakdown in senescent leaves: demonstration of Mg-dechelatase activity. *Physiologia Plantarum*, *189*, 347-353.

[216] Schwartz, S.J., & Lorenzo, T.V. (1990). Chlorophylls in foods, In: Critical Reviews in Food Science and Nutrition; Clydesdale, F M, Ed; CRC: Boca Raton, FL.

[217] Amir-Shapira, D., Goldschmidt, E.E., & Altman, A. (1987). Chlorophyll catabolism in senescing plant tissue: in vivo breakdown intermediates suggest different degradative pathways for citrus fruits and parsley leaves. *The Proceedings of the National Academy of Sciences*, *84*, 1901-1905.

[218] Francis, F. J. (1989). Food colorants: anthocyanins. *Critical Reviews in Food Science and Nutrition*, *28*(4), 273-314.

[219] Wong, D.W.S. (1989). Colorants. In: Mechanism and Theory in Food Chemistry, AVI, Westport, CT, 147-187.

[220] Seeram, N.P., Bourquin, L.D., & Nair, M.G. (2001). Degradation products of cyanidin glycosides from tart cherries and their bioactivities. *Journal of Agricultural and Food Chemistry*, *49*, 4924-4929.

[221] Stintzing, F.C., & Carle, R. (2004). Functional properties of anthocyanins and betalains in plants, food, and in human nutrition. *Trends in Food Science & Technology*, *15*, 19-38.

[222] Mcghie, T.K., & Walton, M.C. (2007). The bioavailabity and absorption of anthocyanins: Towards a better understanding. *Molecular Nutrition Food Research*, *51*, 702-713.

[223] Palamidis, N., & Markakis, P. (1978). Stability of grape anthocyanin in a carbonated beverage. *Industrie delle Bevande*, *7*, 106-109.

[224] Hrazdina, G. (1971). Reactions of the anthocyanidin-3,5-diglucosides: Formation of 3,5-di-(O-β-D-glucosyl)-7-hydroxy coumarin. *Phytochemistry*, *10*, 1125-1130.

[225] Adams, J.B. (1973). Thermal degradation of anthocyanin with particular reference on 3 glucosides of cyanidin in acidified aqueous solution at 100 °C. *Journal of the Science of Food and Agriculture*, *24*, 747-762.

[226] Rice-Evans, C.A., Miller, N.J., & Paganga, G. (1996). Structure-antioxidant activity relationships of flavonoids and phenolic acids. *Free Radical Biology & Medicine*, *20*(7), 933-956.

[227] Von Elbe, J.H., & Schwartz, S.J. (1996) Colorants In: O.R. Fennema (Ed), Food chemistry (3rd ed., pp 651-722) New York, NY: Marcel Dekker, Inc.

[228] Volden, J., Grethe, I., Borge, A., Gunnar, B., Magnor, B., Ingrid, H., Thygese, E., & Wicklund, T. (2008). Effect of thermal treatment on glucosinolates and antioxidant-related parameters in red cabbage (*Brassica oleracea* L ssp *capitata* f *rubra*). *Food Chemistry*, *109*(3), 595-605.

[229] Kader, F., Irmouli, M., Nicolas, P., & Metche, M. (2002). Involvement of blueberry peroxidase in the mechanisms of anthocyanin degradation in blueberry juice. *Journal of Food Science*, *67*(3), 910-915.

[230] Kader, F., Irmouli, M., Nicolas, J.P., & Metche, M. (2001). Proposed mechanism for the degradation of pelargonidin 3-glucoside by caffeic acid o-quinone. *Food Chemistry*, *75*(2), 139-144.

[231] Sarni, P., Fulcrand, H., Souillol, V., Souquet, J.M., & Cheynier, V. (1995). Mechanisms of anthocyanin degradation in grape must-like model systems. *Journal of the Science of Food and Agriculture*, *69*(3), 385-391.

[232] Sadilova, E., Carle, R., & Stintzing, F.C. (2007). Thermal degradation of anthocyanins and its impact on color and in vitro antioxidant capacity. *Molecular Nutrition & Food Research*, *51*, 1461-1471.

[233] Raso, J., & Barbosa-Cánovas, G.V. (2003). Nonthermal preservation of foods using combined processing techniques. *Critical Reviews in Food Science and Nutrition*, *43*(3), 265-285.

[234] Daoudi, L., Quevedo, J.M., Trujillo, A.J., Capdevila, F., Bartra, E., Mínguez, S., & Guamis, B. (2002). Effects of high-pressure treatment on the sensory quality of white grape juice. *High Pressure Research*, *22*, 705-709.

[235] Ahmed, J., Ramaswamy, H.S., & Hiremath, N. (2005). The effect of high pressure treatment on rheological characteristics and colour of mango pulp. *International Journal of Food Science and Technology*, *40*, 885-895.

[236] Dede, S., Alpas, H., & Bayindirli, A. (2007). High hydrostatic pressure treatment and storage of carrots and juices: Antioxidant activity and microbial safety. *Journal of the Science of Food and Agriculture*, *87*, 773-872.

[237] Krebbers, B., Matser, A.M., Koets, M., & Van den Berg, R.W. (2002). Quality and storage-stability of high-pressure preserved green beans. *Journal of Food Engineering*, *54*, 27-33.

[238] Krebbers, B., Matser, A., Koets, M., Bartels, P., & Van den Berg, R. (2002). High pressure–temperature processing as an alternative for preserving basil. *High Pressure Research*, *22*, 711-714.

[239] Sánchez-Moreno, C., Plaza, L., Elez-Martínez, P., De Ancos, B., Martín-Belloso, O., & Cano, M.P. (2005). Impact of high-pressure and pulsed electric fields on bioactive compounds and antioxidant activity of orange juice and comparison with traditional thermal processing. *Journal of Agricultural and Food Chemistry*, *53*(11), 4403-4409.

[240] Van Loey, A., Ooms, V., Weemaes, C., Van den Broeck, I., Ludikhuyze, L., Indrawati, S., Denys, S., & Hendrickx, M. (1998). Thermal and pressure-temperature degradation of chlorophyll in broccoli (*Brassica oleracea* L *italica*) juice: a kinetic study. *Journal of Agricultural and Food Chemistry*, *46*, 5289-5294.

[241] Matser, A.M., Krebbers, B., Van den Berg, R.W., & Bartels, P.V. (2004). Advantages of high pressure sterilization on quality of food products. *Trends in Food Science and Technology*, *15*, 79-85.

[242] Butz, P., Edenharder, R., Garcia, A.F., Fister, H., & Merkel, C. (2002). Changes in functional properties of vegetables induced by high pressure treatment. *Food Research International*, *35*, 295-300.

[243] Tauscher, B. (1995). Pasteurisation of food by hydrostatic high pressure: chemical aspects. *Zeitschrift Lebensmittel-Untersuchung und-Forschung*, *200*, 3-13.

[244] De Ancos, B., Gonzalez, E., & Pilar Cano, M. (2000). Effect of high pressure treatment on the carotenoid composition and the radical scavenging activity of persimmon fruit purees. *Journal of Food Chemistry*, *48*, 3542-3548.

[245] De Ancos, B., Sgroppo, S., Plaza, L., & Cano, M.P. (2002). Possible nutritional and health-related value promotion in orange juice preserved by high-pressure treatment. *Journal of the Science of Food and Agriculture*, *82*(8), 790−796.

[246] Sánchez-Moreno, C., Plaza, L., Elez-Martínez, P., De Ancos, B., Martín-Belloso, O., & Cano, M.P. (2005). Impact of high-pressure and pulsed electric fields on bioactive compounds and antioxidant activity of orange juice and comparison with traditional thermal processing. *Journal of Agricultural and Food Chemistry*, *53*(11), 4403-4409.

[247] Sánchez-Moreno, C., Plaza, L., De Ancos, B., & Cano, M.P. (2006). Impact of high-pressure and traditional thermal processing of tomato purée on carotenoids, vitamin C and antioxidant activity. *Journal of Science and Food Agriculture*, *86*(2), 171-179.

[248] Jun, X. (2006). Application of high hydrostatic pressure processing of food to extracting lycopene from tomato paste waste. *High Pressure Research*, *26*(1), 33-41.

[249] Gárcia-Palazon, A., Suthanthangjai, W., Kajda, P., & Zabetakis, I. (2004). The effects of high hydrostatic pressure on β-glucosidase, peroxidase and polyphenoloxidase in red raspberry (*Rubus idaeus*) and strawberry (*Fragaria × ananassa*). *Food Chemistry*, *88*, 7-10.

[250] Suthanthangjai, W., Kajda, P., & Zabetakis, I. (2005). The effect of high hydrostatic pressure on the anthocyanins of raspberry (*Rubus idaeus*). *Food Chemistry*, *90*, 193-197.

[251] Corrales, M., Butz, P., & Tauscher, B. (2008) Anthocyanin condensation reactions under high hydrostatic pressure. *Food Chemistry*, *110*(3), 627-635.

[252] Rivas-Gonzalo, J.C., Bravo-Haro, S., & Santos-Buelga, C. (1995). Detection of compounds formed through the reaction of malvidin-3-monoglucoside and catechin in the presence of acetaldehyde. *Journal of Agricultural and Food Chemistry*, *43*, 1444-1449.

[253] Schwartz, M., Quast, P., von Baer, D., & Winterhalter, P. (2003). Vitisin A content of Chilean wines from *vitis vinifera* cv Cabernet Sauvignon and contribution to the colour of aged red wines. *Journal of Agricultural and Food Chemistry*, *51*, 6261-6267.

[254] Hayasaka, Y., & Asenstorfer, R.E. (2002). Screening for potential pigments derived from anthocyanins in red wine using nanoelectrospray tandem mass spectrometry. *Journal of Agricultural and Food Chemistry*, *50*(4), 756-761.

[255] Zabetakis, I., Leclerc, D., & Kajda, P. (2000). The effect of high hydrostatic pressure on the strawberry anthocyanins. *Journal of Agricultural and Food Chemistry*, *48*, 2749-2754.

[256] Cano, M.P., Hernandez, A., & De Ancos, B. (1997). High pressure and temperature effects on enzyme inactivation in strawberry and orange products. *Journal of Food Science*, *62*, 85-88.

[257] Barbagallo, R.N., Palmeri, R., Fabiano, S., Rapisarda., P., & Spagna, G. (2007). Characteristic of β-glucosidase from Sicilian blood oranges in relation to anthocyanin degradation. *Enzyme and Microbial Technology*, *41*(5), 570-575.

[258] Gimenez, J., Kajda, P., Margomenou, L., Piggott, J.R., & Zabetakis, I. (2001). A study on the colour and sensory attributes of high hydrostatic-pressure jams as compared with traditional jams. *Journal of the Science of Food and Agriculture*, *81*, 1228-1234.

[259] Es-Safi, N.E., Fulcrand, H., Cheynier, V., & Moutounet, M. (1999). Studies on the acetaldehyde-induced condensation of epicatechin and malvidin-3-O-glucoside in a model solution system. *Journal of Agricultural and Food Chemistry*, *47*, 2096-2102.

[260] Remy, S., Fulcrand, H., Labarbe, B., Cheynier, V., & Moutounet, M. (2000). First confirmation in red wine of products resulting from direct anthocyanin–tannin reactions. *Journal of Science of Food and Agriculture*, *80*, 745-751.

[261] Yeom, H.W., Streaker, C.B., Zhang, Q.H., & Min, D.B. (2000). Effects of pulsed electric fields on the quality of orange juice and comparison with heat pasteurization. *Journal of Agricultural and Food Chemistry*, *48*, 4597-4605.

[262] Hodgins, A.M., Mittal, G.S., & Griffiths, M.W. (2002). Pasteurization of fresh orange juice using low-energy pulsed electrical field. *Journal of Food Science*, *67*, 2294-2299.

[263] Cserhalmi, Zs., Sass-Kiss, A., Tóth-Markus, M., & Lechner, N. (2006). Study of pulsed electric field treated citrus juices. *Innovative Food Science and Emerging Technologies*, *7*, 49-54.

[264] Elez-Martínez, P., Soliva-Fortuny, R., & Martín-Belloso, O. (2006). Comparative study on shelf life of orange juice processed by high-intensity pulsed electric fields or heat treatment. *European Food Research and Technology*, *222*, 321-329.

[265] Cortés, C., Esteve, M.J., Rodrigo, D., Torregrosa, F., & Frígola, A. (2006). Changes of colour and carotenoids contents during high intensity pulsed electric field treatment in orange juices. *Food and Chemical Toxicology*, *44*, 1932-1939.

[266] Odriozola-Serrano, I., Soliva-Fortuny, R., Hernández-Jover, T., & Martín-Belloso, O. (2009). Carotenoid and phenolic profile of tomato juices processed by high intensity pulsed electric fields compared with conventional thermal treatments. *Food Chemistry*, *112*(1), 258-266.

[267] Odriozola-Serrano, I., Soliva-Fortuny, R., & Martín-Belloso, O. (2008). Changes of health-related compounds throughout cold storage of tomato juice stabilized by thermal or high intensity pulsed electric field treatments. *Innovative Food Science and Emerging Technologies*, *9*(3), 272-279.

[268] Oms-Oliu, G., Odriozola-Serrano, I., Soliva-Fortuny, R., & Martín-Belloso, O. (2009). Effects of high intensity pulsed electric field processing conditions on lycopene, vitamin C and antioxidant capacity of watermelon juice. *Food Chemistry*, *115*, 1312-1319.

[269] Aguiló-Aguayo, I., Soliva-Fortuny, R., & Martín-Belloso, O. (2009). Avoiding non-enzymatic browning by high-intensity pulsed electric fields in strawberry, tomato and watermelon juices. *Journal of Food Engineering*, *92*, 37-43.

[270] Knorr, D. (2003). Impact of non-thermal processing on plant metabolites. *Journal of Food Engineering*, *56*, 131-134.

[271] Jeyamkondan, S., Jayas, D.S., & Holley, R.A. (1999). Pulsed electric field processing of foods: a review. *Journal* of *Food Protection*, *62*, 1088-1096.

[272] Zhang, Y., Hu, X.S., Chen, F., Wu, J.H., Liao, X.J., & Wang, Z.F. (2008). Stability and colour characteristics of PEF-treated cyanidin-3-glucoside during storage. *Food Chemistry*, *106*, 669-676.

[273] Graham, D.M. (1997). Use of ozone for food processing. *Food Technology*, *51*, 121-137.

[274] Khadre, M.A., Yuoseft, A.E., & Kim, J. (2001). Microbiological aspects of ozone applications in food: a review. *Journal of Food Science*, *66*(9), 1242-1252.

[275] Karaca, H., & Velioglu, Y.S. (2007). Ozone applications in fruit and vegetable processing. *Food Reviews International*, *23*(1), 91-106.

[276] Perez, A.G., Sanz, C., Rios, J.J., Olias, R., & Olias, J.M. (1999). Effects of ozone treatment on postharvest strawberry quality. *Journal of Agricultural and Food Chemistry*, *47*, 1652-1656.

[277] Barth, M.M., Zhou, C., Mercier, J., & Payne, F.A. (1995). Ozone storage effects on antocyanin content and fungal growth in blackberries. *Journal of Food Science*, *60*(6), 1286-1288.

[278] Skog, L.J., & Chu, C.L. (2001). Effect of ozone on qualities of fruits and vegetables in cold storage. *Canadian Journal of Plant Science*, *81*(4), 773-778.

[279] Badiani, M., Fuhrer, J., Paolacci, A.R., & Giovannozzi, S.G. (1996). Deriving critical levels for ozone effects on peach trees (*Prunus persica* L *Batsch*) grown in open-top chambers in central Italy Fresenius. *Environmental Bulletin*, *5*, 592-597.

[280] Liew, C.L., & Prange, R.K. (1994). Effect of ozone and storage temperature on postharvest diseases and physiology of carrots (*Caucus carota* L). *Journal of the American Society for Horticultural Science*, *119*, 563-567.

[281] FDA. (2004). FDA Guidance to Industry: Recommendations to processors of apple juice or cider on the use of ozone for pathogen reduction purposes. http://wwwcfsanfdagov/~dms/juicgu13html.

[282] Choi, L.H., & Nielsen, S.S. (2005). The effects of thermal and nonthermal processing methods on apple cider quality and consumer acceptability. *Journal of Food Quality*, *28*, 13-29.

[283] Steenstrup, L.D., & Floros, J.D. (2004). Inactivation of *E. coli* 0157:H7 in apple cider by ozone at various temperatures and concentrations. *Journal of Food Processing Preservation*, *28*, 103-116.

[284] Willams R.C., Sumner S.S., & Golden, D.A. (2005). Inactivation of *Escherichia coli* O157:H7 and *Salmonella* in apple cider and orange juice treated with combinations of ozone dimethyl dicarbonate and hydrogen peroxide. *Journal of Food Science*, *70*,197-201.

[285] Tiwari, B.K., Muthukumarappan, K., O'Donnell, C.P., & Cullen, P.J. (2008). Modelling colour degradation of orange juice by ozone treatment using response surface methodology. *Journal of Food Engineering*, *88*, 553-560.

[286] Angelino, P.D., Golden, A., & Mount, J.R. (2003). Effect of ozone treatment on quality of orange juice. In: IFT Annual Meeting Book of Abstracts, Abstract No. 76C-2. Institute of Food Technologists, Chicago, IL.

[287] Tiwari, B.K., O'Donnell, C.P., Muthukumarappan, K., & Cullen, P.J. (2009). Anthocyanin and colour degradation in ozone treated blackberry juice. *Innovative Food Science and Emerging Technologies*, *10*, 70-75.

[288] Tiwari, B.K., O'Donnell, C.P., Patras, A., Brunton, N.P., & Cullen, P.J. (2009). Effect of ozone processing on anthocyanins and ascorbic acid degradation of strawberry juice. *Food Chemistry*, *113*(4), 1119-1126.

[289] Nebel, C. (1975). Ozone decolorization of secondary dye laden effluents In Second Symposium on Ozone Technology Montreal May, 11-14, 336-358.

[290] Melendez-Martínez, A.J., Vicario, I.M., & Heredia, F.J. (2007). Review: Analysis of carotenoids in orange juice. *Journal of Food Composition and Analysis*, *20*, 638-649.

[291] Criegee, R. (1975). Mechanism of Ozonolysis. *Angewandte Chemie International Edition*, *14*(11), 745-752.

[292] Tiwari, B.K., O'Donnell, C.P., Muthukumarappan, K., & Cullen, P.J. (2009). Effect of sonication on orange juice quality parameters during storage. *International Journal of Food Science and Technology*, *44*, 586-595.

[293] Knorr, D., Zenker, M., Heinz, V., & Lee, D.U. (2004). Applications and potential of ultrasonics in food processing. *Trends in Food Science and Technology*, *15*, 261-266.

[294] Tiwari, B.K., O'Donnell, C.P., Patras, A., Brunton, N.P., & Cullen, P.J. (2009). Stability of anthocyanins and ascorbic acid in sonicated strawberry juice during storage, *European Food Research and Technology*, *228*, 717-724.

[295] Masuzawa, N., Ohdaira, E., & Ide, M. (2000). Effects of ultrasonic irradiation on phenolic compounds in wine. *Japanese Journal of Applied Physics*, *39*, 2978-2979.

[296] Cocito, C., Gaetano, G., & Delfini, C. (1995). Rapid extraction of aroma compounds in must and wine by means of ultrasound. *Food Chemistry*, *52*, 311-320.

[297] Vilkhu, K., Mawson, R., Simons, L., & Bates, D. (2008). Applications and opportunities for ultrasound assisted extraction in the food industry - A review. *Innovative Food Science and Emerging Technologies*, *9*,161-169.

[298] Zhao, L., Zhao, G., Chen, F., Wang, Z., Wu, J., & Hu, X. (2006). Different effects of microwave and ultrasound on the stability of (all-E)-Astaxanthin. *Journal of Agricultural and Food Chemistry*, *54*, 8346-8351.

[299] Suslick, K.S. (1988). The sonochemical hot spot, In Ultrasounds: Its Chemical, Physical and Biological Effects, VHC Publishers, New York.

[300] Suslick, K.S., Hammerton, D.A., & Cline, R.E. (1986). The sonochemical hot spot. *Journal of American Chemical Society*, *108*, 5641-5642.

[301] Korn, M., Primo, P.M., & DeSousa, C.S. (2002). Influence of ultrasonic waves on phosphate determination by the molybdenum blue method. *Microchemical Journal*, *73*(3), 273-277.

[302] Portenlänger, G., & Heusinger, H. (1992). Chemical reactions induced by ultrasound and γ-rays in aqueous solutions of L-ascorbic acid. *Carbohydrate Research*, *232*(2), 291-301.

[303] Del Pozo-Insfran, D., Balaban, M.O., & Talcott, S.T. (2006). Enhancing the retention of phytochemicals and organoleptic attributes in muscadine grape juice through a combined approach between dense phase CO_2 processing and copigmentation. *Journal of Agricultural and Food Chemistry*, *54*, 6705-6712.

[304] Garcia-Gonzalez, L., Geeraerd, A.H., Spilimbergo, S., Elst, K., Van Ginneken, L., Debevere, J., Van Impe, J.F., & Devlieghere, F. (2007). High pressure carbon dioxide inactivation of microorganisms in foods: the past, the present and the future. *International Journal of Food Microbiology*, *117*(1), 1-28.

[305] Zhang, J., Davis, T.A., Matthews, M.A., Drews, M.J., LaBerge, M., & An, Y.H. (2006). Sterilization using high-pressure carbon dioxide. *The Journal of Supercritical Fluids*, *38*(3), 354-372.

[306] Park, S.J., Lee, J.I., & Park, J. (2002). Effects of a combined process of high-pressure carbon dioxide and high hydrostatic pressure on the quality of carrot juice. *Journal of Food Science*, *67*, 1827-1834.

[307] Gui, F., Chen, F., Wu, J., Wang, Z., Liao, X., & Hu, X. (2006). Inactivation and structural change of horseradish peroxidase treated with supercritical carbon dioxide. *Food Chemistry*, *97*, 480-489.

[308] Gunes, G., Blum, L.K., & Hotchkiss, J.H. (2006). Inactivation of *Escherichia coli* (ATCC 4157) in diluted apple cider by dense-phase carbon dioxide. *Journal of Food Protection*, *69*(1), 12-16.

[309] Balaban, M.O. (2003). Effect of dense phase CO_2 on orange juice enzymes and microorganisms in a batch system In: IFT Annual Meeting Book of Abstracts; No: 50-2, July 15-20, Chicago.

[310] Gunes, G., Blum, L.K., & Hotchkiss, J.H. (2005). Inactivation of yeasts in grape juice using a continuous dense phase carbon dioxide processing system. *Journal of the Science of Food Agriculture*, *85*(14), 2362-2368.

[311] Yagiz, Y., Lim, S.L., & Balaban, M.O. (2005). Continuous high pressure CO_2 processing of mandarin juice In: IFT annual meeting book of abstracts; 54F-16, 2005 July 15-20, New Orleans.

In: Practical Food and Research
Editor: Rui M. S. Cruz, pp. 51-65
ISBN: 978-1-61728-506-6

Chapter II

ENZYMES

Jaime M. C. Aníbal[1,2] and Rui M. S. Cruz[1,3]

[1]Departamento de Engenharia Alimentar, Instituto Superior de Engenharia, Universidade do Algarve, Campus da Penha, 8005-139 Faro, Portugal,
janibal@ualg.pt and rcruz@ualg.pt
[2]CIMA- Centro de Investigação Marinha e Ambiental, Universidade do Algarve, Campus de Gambelas, 8005-139 Faro, Portugal
[3]CIQA- Centro de Investigação em Química do Algarve, Universidade do Algarve, Campus de Gambelas, 8005-139 Faro, Portugal

ABSTRACT

This chapter will address some standard enzymes extraction and purification methods, and the way to put them to work with little experience and laboratorial materials. The focus will be on tricks and tips that should be followed when working with the most important food related enzymes to achieve good results, which are not usually mentioned.

In order to determine the adequacy of several food processing treatments, many enzymes have been suggested as heat treatment indicators. Among various enzymes, peroxidase (POD) is one of the most heat resistant enzymes and its inactivation increases the shelf life of various fruits and vegetables. Polyphenoloxidase (PPO) and pectinmethylesterase (PME) are other two concerning enzymes present in fruits and vegetables that need to be inactivated. The use of heat treatments and other alternative treatments in fruits and vegetables processing will also be discussed.

2.1 INTRODUCTION

One of the most important functions of proteins is their role as catalysts, allowing that the biochemical reactions can occur fast enough to sustain life. Proteins that have this function are known as enzymes [1]. The word "enzyme" is derived from the Greek meaning "in yeast" and was first used by Kühne in 1878 [2].

Enzymes are very important for the processing and preservation of foods. This is confirmed by the frequent appearance of enzymological papers in food sciences research journals. This chapter is not a review of the existing bibliography, because there are many books addressing that issue (*e.g.* [3-5]).

In living systems, biochemical reactions have two main limitations: temperature and concentration of reactants. Without enzymes, biochemical reactions would only occur at very high temperatures, harming delicate biological structures, and the concentrations of reactants would have to be very high, in order to allow sufficient collisions between molecules to result in effective chemical reactions [1]. This is valid not only for biochemical reactions inside cells, but also for food processing, where the objective is to alter molecules properties through chemical reactions, in the most energy saving way.

When working with enzymes, the main concern is to maintain intact their structures and consequently their functions. Changes in the protein conformation will diminish or stop the catalytic capacity of an enzyme. This is the main issue to be address when working, directly or indirectly, with enzymes. Generally this issue is related to three important factors: time (speed), temperature and pH.

Be quick when you work with enzymes, especially if you are extracting and isolating from raw materials. When cells are destroyed to extract enzymes, a series of chemical reactions occur, which accelerate the degradation of the enzymes. Isolation and purification is not only important to obtain a pure enzyme to work with, but also to impede its degradation. The issue here is time and speed of action!

Control the temperature at which all the work is done. The higher the temperature, the higher the catalytic rate of chemical reactions, including those that degrade the enzymes with which we are trying to work with and also higher the heat-denaturation conditions.

Try to control all the chemical factors disturbing the enzyme stability, especially those related to pH, through the use of adequate buffer solutions with specific additives to produce accurate effects (*e.g.* the use of β-mercaptoethanol for breaking the disulfide bridges between cysteine aminoacids). The aim is to create an optimal medium that preserves the enzyme with its function intact. The correct pH value of the buffer should be previously assessed, and complemented with other molecules that will stabilize the enzyme structure in order to maintain it as active as possible.

Normally, a study of a particular food enzyme includes several steps. The first is the enzyme purification. This should be carried out to the point at which there is only one enzyme present and specific inhibitors or activators are removed or neutralized. The second step is the characterization of the enzymatic activity. This step deals not only with the determination of specific activity, but also with the investigation of different enzyme properties, such as specificity for substrate, pH-dependence for activity and the rate of heat-denaturation under various conditions [6]. The third and last step is the determination of kinetic parameters that characterize an enzyme performance and will be discussed in chapter VII.

Different researchers will undoubtedly have different objectives when working with enzymes. A technician for a company that produces enzymes will dedicate a lot of effort in the first stages of purification, and a fair amount of time in understanding the factors that lead to stability of the enzyme during use. A researcher using an enzyme in food processing is perhaps more interested in the ways that pH and temperatures affect the enzyme. A graduate student might be concerned with developing a workable assay for an enzyme in a food product, most of the time, depending upon the interests and funding available in the department [6].

When starting a work with a food enzyme, there are two important questions that need to be answered:"Which is (are) the enzyme(s) that I should work with?" and "What do I really need to know about this (these) enzyme(s) and its (their) characteristics?"

The answer to the first question written above lead us to the following issue:

"From all the enzymes directly and indirectly related with food processing, which are the most important to quantify and/or control?"

2.2 Extraction And Isolation of Enzymes

If we hope to gain a detailed understanding of the behavior of an enzyme in a complex system, we must first try to understand its properties in a simple system. This simple system is normally a solution of enzyme in a medium containing for instance small ions, buffer molecules and cofactors. However, in some cases the isolated enzyme may be inactive in the absence of some molecules (e.g. phospholipid or detergent), which should be added to the simplified system. When studying an isolated enzyme we can learn about its specificity for substrates, kinetic parameters for the reaction, and possible means of regulation. This information is very important to understand the enzyme role in its natural environment.

The aim of an enzymatic purification procedure is to isolate a given enzyme with the maximum possible yield, based on the percentage recovered activity compared with the total activity in the original extract. In addition, the preparation should possess the maximum catalytic activity, and should be of the maximum possible purity. There are several sequential steps involved in the purification of an enzyme. The procedure to be adopted for a given case will involve choices of the source of enzyme, methods of homogenization, and methods of separation [2].

In the field of food enzymes, there are usually two kinds of situations: the study of enzymes related to the processing of the raw materials, or study the way processing affects the enzymes remaining active, or not, in the processed product. This means that a researcher is limited to the food product itself, even if there is not much quantity of enzyme to work with. The obvious consequence is to work with larger samples in order to isolate enough quantity of workable active enzymes.

Enzymes are generally inside cells, and food products are no exception. Consequently the methods of homogenization and extraction of enzymes depend on the kind of cells that compose the food product (animal tissue or plant, fungal, or bacterial material) and the place where those enzyme are located (membrane-bound or cytoplasm free enzymes).

Animal tissues lack a rigid cell wall, making homogenization relatively easy. The tissue is often cut up into small pieces and homogenized using high-speed blender. Extraction is performed using isotonic or low ionic strength solutions, depending on the need of breaking cellular membranes or structures.

Plant, fungal and bacterial materials have rigid cell walls, which need the use of stronger methods of extraction. Some alternatives are grinding with abrasives such as alumina or sand, freezing and thawing, long periods of blending, or using appropriate hydrolytic enzymes [2]. Particular problems can arise with plant tissues, because the release of the contents of the vacuoles during homogenization can cause damage to enzymes. This can be avoided by using buffer solutions. In addition, plant tissues often contain certain phenolic compounds that are readily oxidized to form dark pigments that can be harmful to enzymes. The pigments can be removed by adsorption on polymers such as poly(vinylpyrrolidone) [7].

The study of membrane-bound enzymes presents special problems. The best way to maintain or acquire functional enzymes is to create an environment as close as possible to the one naturally provided by the membrane. The use of detergents is usually the correct procedure, but choosing from the large range of cationic, anionic, and neutral detergents available is only possible through an intensive empirical approach of trial and error. Independently of the choice is the fact that some detergents can limit the purification methods to be used consequently. For instance, the use of cholate solution will rule out the use of ion-exchangers such as DEAE-cellulose, since they would adsorb the detergent [8].

Separating one the different molecules of a mixture can be made by using specific properties for each molecule. The main properties of enzymes that can be exploited in separation methods are size, charge, solubility, and the possession of specific binding sites. From all the different methods, this chapter will focus on the most common and simple ones, giving some tricks on how to do them with minimal laboratory resources.

Centrifugation is a very common separation method to remove precipitated or insoluble material during an isolation, for instance to remove cell debris after homogenization. Large molecules such as enzymes can be sedimented by the high centrifugal fields generated by an ultracentrifuge. Although the rate at which any particular enzyme will sediment depends on a variety of factors including the size and shape of the molecule and the viscosity of the solution. In general the higher the relative molecular mass value the greater the rate of sedimentation [9].

Another example of centrifugation is the collection of enzyme that has been precipitated by the addition of ammonium sulphate. This method is usually referred to as ammonium sulphate fractionation, and is based on the fact that large charged molecules are generally to some extent soluble in pure water. Adding ions promotes the solubility by helping to disperse the charge carried by these large molecules. This phenomenon is known as "salting in". But if the ionic strength is increased beyond a certain point, the charged molecules will be precipitated ("salting out"). Each enzyme will generally begin to precipitate at a certain concentration of ammonium sulphate and this forms the basis of an initial fractionation procedure. This method is often used to reduce the volume of solution in which an enzyme is dissolved because the precipitated enzyme can be redissolved in a smaller volume of buffer [7].

After an ammonium sulphate fractionation, the enzyme is largely bound to salts that need to be eliminated. One simple procedure to achieve this elimination is dialysis, which is based on the properties of a membrane that acts as a sieve with pores large enough to permit the passage of salts, organic solvents, low weight peptides or free aromatic amino acids, but not larger molecules such as enzymes. The dialysis is a slow diffusion method, where the mixture solution inside the membrane is purged from the small molecules to the outside solution [7]. A way to speed this process is by stirring the solution containing the membrane, which turns diffusion into an advective process.

The methods described above are often considered "coarse", being their use confined to the initial stages of enzyme purification. More refined methods are usually based on chromatographic principles, specially the ion-exchange and affinity chromatography.

Ion-exchange chromatography is based on the electrostatic relation between species of opposite charge. Ion exchangers are usually modified derivates of some support material such as cellulose or Sephadex. The enzyme is usually applied to an ion exchanger in a solution of low ionic strength and at a pH at which, for instance, the enzyme and the ion exchanger have opposite charges. The bound species can be desorpted by increasing the ionic strength of the solution, making it compete with the enzyme for the binding sites on the ion exchanger. Use of a gradient or a succession of discrete solutions, with increasing ionic strength permits the separation of proteins in a mixture on the basis of their ability to bind to the ion exchanger [8]. Ion-exchange chromatography can be performed on a large or a small scale. The costs of the materials involved make working on a large scale very expensive. On a small scale, both adsorption and desorption are performed in a column, which can be created using low budget materials (Figure 2.1).

Figure 2.1. Ion-exchange chromatography apparatus (originally developed by Gil Fraqueza, 1999).

In the apparatus illustrated in Figure 2.1, the column is built with a syringe, a plastic cork and a capillary tube. An adequate ion-exchanger is inside the syringe and all the solutions are added to the column via the capillary tube. When the small metal tap is opened, the exit of solution creates a negative pressure inside the column, which pulls the adding solution from the upper essay tube. This upper position will contain a sequence of tubes with the sample solution, the equilibrium buffer, and all the discrete solution with increasing ionic strength. The changing of the upper tubes is a critical point of the method because the ion-exchanger material is not supposed to be exposed to air, and should always be emerged on a buffer solution in order to maintain its properties. The solutions exiting the column are collected in different fractions, so that the purified enzyme is separated from the other molecules.

Affinity chromatography is based on the property of a molecule such as a substrate or competitive inhibitor to interact specifically with the enzyme of interest, which is linked covalently to an inert matrix. When a mixture is passed down a column containing the affinity matrix, only this enzyme is retained and other enzymes and proteins are washed away. The bound enzyme can be desorbed by a pulse of substrate, which will compete for the binding sites on the enzyme, or by changing the pH or ionic strength of the solution in such a way as to weaken the bonding of the enzyme to the column. However, there are some problems associated with the use of affinity chromatography in enzyme purification. It can be difficult to attach a suitable substrate or inhibitor to the matrix; linking of the ligand to the matrix may interfere with the binding to the enzyme; achieving the correct strength of the interaction between matrix-bound ligand and the enzyme is not always easy to do; and special problems are posed by enzymes that catalyse reactions involving more than one substrate. In spite of these problems, affinity chromatography has made a very significant contribution to the purification of enzymes but also of other biological molecules such as antibodies that possess specific binding sites. The technique is most likely to be successful when full use is made of information from solution studies concerning the interaction between the ligand and the enzyme [8]. This kind of chromatography can also be performed using a preparation similar to ion-exchange chromatography (Figure 2.1).

These chromatographic methods are normally followed by ammonium sulphate fractionation and dialysis, in order to reduce the volume of solution containing the enzyme.

During the extraction and purification phases is very important to assess enzyme purity, to determine if the methods that are being used are fulfilling their objective. But this has to be done without consuming the enzyme that is being purified, or else it will be none left to work with. Technically there are several options to surpass this problem, e.g. comparing UV-visible spectrum for the enzyme in pure state, with a spectrum from the solution that is being purified. The closer the profile of both spectrums, purer the solution is. Other, less precise, option is to measure the absorption of the solution that is being purified at 280 nm, which corresponds to the tryptophan maximum absorption wavelength [10]. Tryptophan is a common amino acid in enzymes primary structure. The higher the absorption at 280 nm, higher the proportion of this amino acid in the solution, therefore less non amino acids molecules are present in the mixture that is being purified.

Another method of enzyme purification is electrophoresis. Electrophoretic separation is based on the movement of charged molecules under the influence of an applied potential difference. The rate of movement of a molecule is governed by the charge it carries and also by its size and shape. Although the technique can be used on a large preparative scale, it is usually performed on a small analytical scale [7]. For a researcher constrained by few resources or low budget, electrophoresis can be mainly used to determine if the purified enzyme is the correct one, i.e. its molecular weight can be compared to an expected theoretical value.

Besides these methods a lot of other can be used for extraction and purification, depending on the specificities of the enzyme. They were chosen because of their technical easiness and low budget.

2.3 Fruits and Vegetables Concerning Enzymes: The Effect of Thermal and Non Thermal Treatments

Pre-treatments are common in most processing operations to improve product quality or process efficiency [11]. The blanching process is usually one of the pre-treatments used before freezing, canning or dehydration to inactivate undesirable enzymes, destroy microorganisms and prevent biochemical reactions that might contribute for the development of off-flavours and discoloration in the frozen product [12]. Blanching is a heat treatment and consists heating the food rapidly to a desired temperature for a specific time, and then cooling it rapidly or going to another process operation [13]. The blanching process typically uses temperatures around 75-95 °C at different times of exposure, depending on the type of product. This process may also remove tissues gases, shrink the product, clean and stabilise colour [14]. However, cellular tissues are also affected by high temperature, and effects similar to those caused by freezing may be observed, particularly on texture. The quality loss through blanching can be minimised by using high temperatures and short time exposure instead of longer times at relatively lower temperatures. Several factors such as the type and size of the vegetable and blanching temperature influence the blanching processing time [15].

Considerable losses in quality are achieved if the vegetables are not blanched or if the blanching process is not totally efficient. This under blanching may cause more damage to food than the absence of blanching, since heat promotes the tissues disruptions and enzymes release, causing accelerated damage by mixing the enzymes and substrates. One heat-resistant enzyme which is found in most vegetables and frequently used as indicator enzyme to determine the effectiveness of blanching is peroxidase. Peroxidase is one of the most stable enzymes, so the absence of residual peroxidase activity would indicate that other less heat-resistant enzymes are also destroyed. In order to achieve an efficient enzyme inactivation, the product should be heated rapidly to a pre-set temperature, held for a pre-set time and then cooled rapidly [15].

2.3.1 Peroxidase

Peroxidase (POD, E.C. 1.11.1.7) is an enzyme commonly found in fruits and vegetables and it is a heme-containing enzyme, which can catalyse a large number of reactions in which a hydrogen peroxide is reduced while an electron donor is oxidized, and it is considered to have an empirical relationship to off-flavours and off-colours in raw and unblanched frozen vegetables [16-17].

PODs have been essentially classified in three classes, supported by comparison of aminoacid sequence and crystal structure data: class I, intracellular prokaryotic PODs; class II, extracellular fungal PODs; and class III, secretory plant PODs [18]. In plant extracts, POD is present in soluble, ionically bound and covalently bound forms; the first one is distributed within both intra- and extracellular environment, while the bound ones are considered to be associated with plant cell walls and possibly with certain organelles, for example mitochondria [19-20].

PODs are present in plant tissues as a combination of various isoenzymes with different heat resistances and their differences vary considerably with the vegetable source and origin

[21-24]. In the presence of peroxide, POD produces phytotoxic free radicals which react with a wide range of organic compounds (ascorbic acid, carotenoids and fatty acids), leading to losses in the colour, flavour and nutritional value of raw and processed foods [25-28].

Plant peroxidases can catalyse the oxidative coupling of phenolic compounds using H_2O_2 as the oxidizing agent. The reaction (peroxidatic) is a three-step cyclic reaction in which the enzyme is first oxidized by H_2O_2 and then reduced in two one-electron transfer steps by reducing substrates, typically a small-molecule phenol derivative. The oxidized phenolic radicals can polymerize, with the final product depending on the chemical character of the radical, the environment and the peroxidase used [21, 29-30].

Classically, the peroxidase level activity has been used to monitor quality changes in frozen vegetables, since increases in peroxidase are thought to indicate changes in flavour, colour, and texture in vegetables. Thus, measurement of peroxidase is often performed, as previously referred, as a reference for determining the effectiveness of the blanching process [31].

2.3.2 Polyphenoloxidase

Polyphenoloxidase (EC 1.14.18.1) is a copper-containing enzyme widely distributed in the plant kingdom, and is responsible for catalyzing the discoloration of fruits and vegetables, when plant tissues undergo physical damage such as bruising, cutting, ripening, or senescence, by the conversion of phenolic compounds to quinones and their products' polymerization, causing undesirable quality changes during handling, processing and storage [32-38].

PPO catalyzes two kinds of oxidative reactions in combination with molecular oxygen: the hydroxylation of monophenols to *o*-diphenol and the oxidation of *o*-diphenol to *o*-quinone [36].

2.3.3 Pectinmethylesterase

Pectinmethylesterase (EC 3.1.1.11) is an endogenous enzyme found in many fresh fruits and vegetables and catalyzes the deesterification of the methyl group of pectin (methyl ester of polygalacturonic acid) and converts it into low-methoxy pectin or pectic acid [39-41]. The resulting galacturonic units in pectin molecules are potentially substrates for polygalacturonase (PG). PG is also an endogenous enzyme that catalyzes hydrolytic cleavage of α-1,4 glycoside bonds between galacturonic acid residues, resulting in tissue softening of fruits and vegetables, viscosity decrease of fruit juices and cloud separation in citrus juices [42].

Several authors reported the effect of heat and alternative treatments on peroxidase, polyphenoloxidase and pectinmethylesterase inactivation in several fruits and vegetables. The following table (2.1) reports the experimental conditions in which several enzymes can be partial or totally inactivated.

Table 2.1. Processing conditions for POD, PPO and PME partial or total inactivation.

Vegetable/Fruit	Treatment	Enzyme/treatment effect (%)	References
Apple	Heat treatment 75 °C for 10 min	POD 85 reduction	[43]
Apple juice	ultraviolet irradiation for 30 min + pulsed electric field (PEF) treatment at 40 kV/cm for 100 pulses	POD 52.8 reduction	[44]
Apple juice	50 °C and PEF at 27 kV/cm for 58.7 μs	PPO 33 reduction	[45]
Apple slice	5 min heating under radiation intensity of 5000 W/m^2	PPO 99-100 reduction	[46]
Artichokes	Heat treatment 80 °C for 10 min	POD 90 reduction	[47]
Asparagus (stems)	Heat treatment 95 °C for 10 min	POD 87 reduction	[48]
Banana purée	Heat treatment 96.4 °C for 3 s	POD 96 reduction	[49]
Banana purée	Heat treatment 96.4 °C for 3 s	PPO 98 reduction	[49]
Broccoli (florets)	Heat treatment 95 °C for 1 min	POD 90 reduction	[48]
Butternut squash	Heat treatment 85 °C for 0.27 min	POD 90 reduction	[50]
Carrot	Heat treatment 75 °C for 10 min	POD 100 reduction	[51]
Carrot	Microwave treatment 10% level for 12 min	POD 100 reduction	[51]
Carrot	350 MPa and 20 °C for 30 min	POD 84 reduction	[52]
Coconut water	Microwave treatment 90.14 °C for 18 s	POD 90 reduction	[53]
Green bean	Heat treatment 90 °C for 3 min	POD >90 reduction	[54]
Green bean	250 MPa and 20 °C for 60 min + 50 °C for 15 min	POD 75 reduction	[52]
Kiwi	600 MPa and 50 °C for 30 min	POD 70 reduction	[55]
Mango purée	Heat treatment 86 °C for 30.8 s	POD 100 reduction	[56]
Mint leaves	Heat treatment 95 °C for 10s	POD 100 reduction	[57]
Mushroom	1 min microwave at 85 °C + Heat treatment 20 s 92 °C	PPO 99 reduction	[58]

Table 2.1. (Continued).

Vegetable/Fruit	Treatment	Enzyme/treatment effect (%)	References
Orange juice	PEF at 35 kV/cm for 59 μs	PME 88 reduction	[59]
Orange juice	Heat treatment 94.6 °C for 30 s	PME 98 reduction	[59]
Orange juice	PEF treatment of 35 kV/cm applied for a total treatment time of 1500 μs	POD 100 reduction	[60]
Pea purée	Heat treatment 100 °C for 300 s	POD 100 reduction	[61]
Pea	Heat treatment 96 °C for 12 min	POD 90 reduction	[21]
Pea	900 MPa for 10 min	POD 88 reduction	[62]
Peaches	Heat treatment 60 °C for 9 min	PPO 80 reduction	[63]
Pineapple purée	Heat treatment 90 °C for 5 min	PPO 98.8 reduction	[64]
Pinto bean	Heat treatment 70 °C for 5 min	POD 97 reduction	[65]
Potato	Heat treatment 83.2 °C for 5 min	POD 90 reduction	[66]
Pumpkin	Heat treatment 95 °C for 3.9 min	POD 90 reduction	[67]
Strawberry	600 MPa and 60 °C for 10 min	PPO 28.2 reduction	[68]
Strawberry	800 MPa for 10 min	PPO 100 reduction	[69]
Strawberry purée	285 MPa and 20 °C for 15 min	PPO 60 reduction	[70]
Strawberry	600 MPa and 60 °C for 10 min	POD 58 reduction	[68]
Sweet potato	Heat treatment 80 °C for 15 min	PPO 80 reduction	[71]
Tomato juice	Heat treatment 72 °C for 25.3 min	PME 90 reduction	[72]
Tomato juice	Thermosonication 72 °C and 20 kHz for 0.4 min	PME 90 reduction	[72]
Watercress	Thermosonication 90 °C and 20 kHz for 5 s	POD 90 reduction	[73]

2.4 Effect of Processing Treatments on Enzymes Stability

The enzymes have a region (called the substrate binding site, the active site or the catalytic site) that is complementary in size, shape and chemical nature to the substrate molecule. Today, it is recognized that the active site, rather than a rigid geometrical cavity, it is a very specific and precise spatial arrangement of amino acid residues R-groups that can interact with complementary groups on the substrate [74].

Presumable mechanism of enzyme inactivation is related to changes in the conformation its structure. The reduction of activity is related to the conformation changes in the tertiary structure, as in the active site three-dimensional structure affecting the enzyme-substrate interaction. Moreover, depending on the processing treatment conditions, protein denaturation may occur leading to the total enzyme inactivation [75, 76].

Conclusion

Several processing treatments are available for inactivating fruits and plants concerning enzymes. Depending on the processing treatment and the product main objective, it is possible to reduce the activity of concerning enzyme partial or totally. There are various methods reporting the activity determination of enzymes. Nevertheless, besides all the methodologies steps, the extraction and purification must be taken into account since they are essential for obtaining accurate results, and thus contributing for the selection of the most suitable processing treatment.

References

[1] McKee, T., & McKee, J.R. (2003). *Biochemistry: The molecular basis of life* (3rd ed.). New York, McGraw-Hill, 771 pp.

[2] Price, N.C., & Stevens, L. (2000). *Fundamentals of enzymology: The cell and molecular biology of catalytic proteins* (3rd ed. Reprinted). Oxford, Oxford University Press, 478 pp.

[3] Reed, G. (1975). Enzymes in food processing (2nd ed.) Academic Press, London.

[4] Nagodawithana, T., & Reed, G. (1993). *Enzymes in food processing* (3rd ed.).London, Academic Press, Inc., 480 pp.

[5] Wong, D. (1995). *Food enzymes: Structure and mechanism*. New York, Chapman & Hall, 390 pp.

[6] Stauffer, C.E. (1989). *Enzyme assays for food scientists*. New York, Avi Book, 317 pp.

[7] Scopes, R.K. (1994). *Protein purification: principles and practice* (3rd ed.). New York, Springer-Verlag, 380 pp.

[8] Price, N.C. (1996). Proteins Labfax. New York, Academic Press, 318 pp.

[9] Birnie, G.D., & Rickwood, D. (1978). Centrifugal separation in molecular and cell biology. London, Butterworths-Heinemann, 327 pp.

[10] Zeidan, H.M., & Dashek, W.V. (1996). *Experimental approaches in biochemistry and molecular biology*. Dubuque, Wm. C. Brown Publishers, 219 pp.

[11] Jha, S. N. & Prasad, S. (1996) Determination of processing conditions on gorgon nut (*Euryale ferox*). *Journal of Agricultural Engineering Research*, *63*, 103-112.

[12] Mountney, G.J., & Gould, W.A. (1988). Low-temperature food preservation. In *Practical Food Microbiology and Technology*, 3rd ed., Chapt. 7, (pp. 112-115). New York, Van Nostrand Reinhold Company.

[13] Grandison, A.S. (2006). Postharvest handling and preparation of foods for processing. In J.G. Brennan (Ed.), Food processing handbook. WILEY-VCH Verlag.

[14] Barret, D.M., & Theerakulkait, C. (1995). Quality indicators in blanched, frozen, stored vegetables. *Food Technology*, *49*, 62-65.

[15] Fellows, P.J. (2000). Processing by application of heat. In Food processing technology principles and practice 2nd Edition. pp. 233-240, CRC Press

[16] López, P., Sala, F.J., Fuente, J.L., Condón, S., Raso, J., & Burgos, J. (1994). Inactivation of peroxidase, lipoxygenase, and polyphenol oxidase by manothermosonication. *Journal of Agricultural and Food Chemistry*, *42*, 252-256.

[17] Vianello A., Zancani, M., Nagy, G., & Macri, F. (1997). Guaiacol peroxidase associated to soybean root plasma membranes oxidized ascorbate, *Journal of Plant Physiology*, *150*, 573-577.

[18] Welinder, K.G. (1992). Superfamily of plant, fungal and bacterial peroxidases. *Curr. Opin. Struct. Biol. 2*, 388-393.

[19] Ievinsh, G. (1992). Characterization of the peroxidase system in winter rye seedlings: Compartmentation and dependence on leaf development and hydrogen donors used. *Journal of Plant Physiology*, *140*, 257-263.

[20] Sergio, L., Pieralice, M., Di Venere, D., & Cardinali, A. (2007). Thermostability of peroxidases from artichoke, *Food Technology and Biotechnology*, *45*, 367-373.

[21] Günes, B., & Bayindirli, A. (1993). Peroxidase and lipoxygenase inactivation during blanching of green beans, green peas and carrots. *LWT-Food Science and Technology*, *26*, 406-410.

[22] Tijskens, L.M.M., Rodis, P.S., Hertog, M.L.A.T.M., Waldron, K.W., Ingham, L., Proxenia, N., & van Dijk, C. (1997). Activity of peroxidase during blanching of peaches, carrots and potatoes. *Journal of Food Engineering*, *34*, 355-370.

[23] Busto, M.D., Owusu Apenten, R.K., Robinson, D.S., Wu, Z., Casey, R., & Hughes, R.K. (1999). Kinetics of thermal inactivation of pea seed lipoxygenases and the effect of additives on their thermostability. *Food Chemistry*, *65*, 323-329.

[24] Garrote, R.L., Silva, E.R., Bertone, R.A., & Roa, R.D. (2004). Predicting the end point of a blanching process. *LWT-Food Science and Technology*, *37*, 309-315.

[25] Bruemmer, J.H., Roe, B., & Bowen, E.R. (1976). Peroxidase reactions and orange juice quality. *Journal of Food Science*, *41*, 186-189.

[26] Kampis, A., Bartuczkovacs, O., Hoschke, A., & Aosvigyazo, V. (1984). Changes in peroxidase-activity of broccoli during processing and frozen storage. *LWT-Food Science and Technology*, *17*, 293-295.

[27] Nebesky, E.A., Esselen, W.B., Kaplan, A.M., & Felleres, C.R. (1950). Thermal destruction and stability of peroxidase in acid foods. *Food Research*, *15*, 114-124.

[28] Robinson, D.S. (1987). Scavenging enzyme and catalases. In D. S. Robinson (Ed.), Biochemistry and nutritional value (pp. 459-465). Harlow U.K: Longman Scientific and Technical.

[29] Whitaker, J.R. (1994). Catalase and peroxidase. In *Principles of enzimology for the food sciences* (pp. 591-604). New York: Marcel Dekker.
[30] Hemeda, H.M., & Klein, B.P. (1991). Inactivation and regeneration of peroxidase activity in vegetable extracts treated with antioxidants. *Journal of Food Science*, *56*, 68-71.
[31] Blond, G., & Le Meste, M. (2004). Principles of frozen storage. In Y.E., Hui, P., Cornillon, I.G., Legaretta, M.H., Lim, K.D., Murrell, & W., Nip (Eds.), *Handbook of Frozen Foods* (pp. 25-53, 30, 40, 48). Marcel Dekker, New York.
[32] Arogba, S.S., Ajiboye, O.L., Ugboko, L.A., Essienette, S.Y., & Afolabi, P.O. (1998). Properties of polyphenol oxidase in mango (*Mangifera indica*) kernel. *Journal of the Science of Food and Agriculture*, *77*, 459-462.
[33] Nicoli, M.C., Elizalde, B.E., Pitotti, A., & Lerici, C.R. (1991). Effect of sugars and Maillards reaction products on polyphenol oxidase and peroxidase activity in food. *Journal of Food Biochemistry*, *15*, 169-184.
[34] Park, Y.K., Sato, H.H., Almeida, T.D., & Moretti, R.H. (1980). Polyphenol oxidase of mango (*Mangifera indica* var. Haden). *Journal of Food Science*, *45*, 1619-1621.
[35] Prabha, T.N., & Patwardhan, M.V. (1982). Purification and properties of polyphenoloxidase of mango peel (*Mangifera indica*). *Journal of Bioscience*, *4*, 69-78.
[36] Tomás-Barberán, F.A., & Espín, J.C. (2001). Phenolic compounds and related enzymes as determinants of quality in fruits and vegetables. *Journal of the Science of Food and Agriculture*, *81*, 853-876.
[37] Mayer, A.M., & Harel, E. (1979). Polyphenol oxidases in plants. *Phytochemistry*, *18*, 193-215.
[38] Golbeck K.H., & Cammarata, K.V. (1981). Spinach thylakoid polyphenol oxidase: isoltion, activation and properties of the native choloroplast enzyme. *Plant Physiology*, *67*, 977-984.
[39] Evans, R., & McHale, D. (1968). Multiple forms of pectinesterase in limes and oranges. *Phytochemistry, 17,* 1073-1075.
[40] Puri, A., Solomos, T., & Kramer, A. (1982). Partial purification and characterisation of potato pectinesterase. *Food Chemistry, 8,* 203-213.
[41] Castaldo, D., Quagliciolo, L., Servillo, L., Balestrien, C., & Govane, A. (1989). Isolation and characterisation of pectin methylesterase from apple fruit. *Journal of Food Science*, *54*, 653-673.
[42] Pilknik, W., & Voragen, A.G.J. (1991). The significance of endogenous and exogenous pectic enzymes in fruit and vegetable processing. In P. F. Fox (Ed.), *Food Enzymology*. England, Elsevier Applied Science.
[43] Valderrama, P., Marangoni, F., & Clemente, E. (2001). Efeito do tratamento térmico sobre a atividade de peroxidase (POD) e polifenoloxidase (PPO) em maçã (*Mallus comunis*). *Ciência e Tecnologia de Alimentos*, *21*, 321-325.
[44] Noci, F., Riener, J., Walkling-Ribeiro, M., Cronin, D.A., Morgan, D.J., & Lyng, J.G. (2008). Ultraviolet irradiation and pulsed electric fields (PEF) in a hurdle strategy for the preservation of fresh apple juice. *Journal of Food Engineering*, *85*, 141-146.
[45] Liang, Z., Cheng, Z., & Mittal, G.S. (2006). Inactivation of spoilage microorganisms in apple cider using a continuous flow pulsed electric field system. *LWT-Food Science and Technology*, *39*, 51-357.

[46] Zhu, Y., & Pan, Z. (2009). Processing and quality characteristics of apple slices under simultaneous infrared dry-blanching and dehydration with continuous heating. *Journal of Food Engineering*, *90*, 441-452.

[47] Sergio, L., Pieralice, M., Di Venere, D., & Cardinali, A. (2007). Thermostability of soluble and bound peroxidases from artichoke and a mathematical model of its inactivation kinetics. *Food Technology and Biotechnology*, *45*, 367-373.

[48] Morales-Blancas, E.F., Chandia, V.E., & Cisneros-Zevallos L. (2002). Thermal inactivation kinetics of peroxidase and lipoxygenase from broccoli, green asparagus and carrots. *Journal of Food Science*, *67*, 146-154.

[49] Ditchfield, C. & Tadini, C.C. (2006). Polyphenol oxidase and peroxidase thermal inactivation kinetics used as indicators for the pasteurization of acidified banana purée (*Musa cavendishii*, Lamb.). *Brazilian Journal of Food Technology*, *9*, 77-82.

[50] Agüero, M.V., Ansorena, M.R., Roura, S.I., & del Valle, C.E. (2008). Thermal inactivation of peroxidase during blanching of butternut squash. *LWT-Food Science and Technology*, *41*, 401-407.

[51] Soysal, Ç., & Söylemez, Z. (2005). Kinetics and inactivation of carrot peroxidase by heat treatment. *Journal of Food Engineering*, *68*, 349-356.

[52] Akyol, Ç., Alpas, H., & Bayındırlı, A. (2006). Inactivation of peroxidase and lipoxygenase in carrots, green beans, and green peas by combination of high hydrostatic pressure and mild heat treatment. *European Food Research and Technology*, *224*, 171-176.

[53] Matsui, K.N., Gut, J.A.W., Oliveira, P.V., & Tadini, C.C. (2008). Inactivation kinetics of polyphenol oxidase and peroxidase in green coconut water by microwave processing. *Journal of Food Engineering*, *88*, 169-176.

[54] Bahçeci, K.S., Serpen, A., Gökmen, V., & Acar J. (2005). Study of lipoxygenase and peroxidase as indicator enzymes in green beans: change of enzyme activity, ascorbic acid and chlorophylls during frozen storage. *Journal of Food Engineering*, *66*, 187-192.

[55] Fang L., Jiang, B., & Zhang, T. (2008). Effect of combined high pressure and thermal treatment on kiwifruit peroxidase. *Food Chemistry*, *109*, 802-807.

[56] Sugai, A.Y., & Tadini, C.C. (2006). Thermal inactivation of mango (*Mangifera indica* L., variety Palmer) purée peroxidase. CIGR Section VI International Symposium on Future of Food Engineering. Warsaw, Poland, 26-28 April.

[57] Shalini, G.R., Shivhare, U.S., & Basu, S. (2008). Thermal inactivation kinetics of peroxidase in mint leaves. *Journal of Food Engineering*, *85*, 147-153.

[58] Devece, C., Rodríguez-López, J.N., Fenoll, L.G., Tudela, J., Catalá, J. M., Reyes, E., & García-Cánovas, F. (1999). Enzyme inactivation analysis for industrial blanching applications: comparison of microwave, conventional, and combination heat treatments on mushroom polyphenoloxidase activity. *Journal of Agricultural and Food Chemistry*, *47*, 4506-4511.

[59] Yeom, H.W., Streaker, C.B., Zhang, Q.H., & Min, D.B. (2000). Effects of pulsed electric fields on the quality of orange juice and comparison with heat pasteurization. *Journal of Agricultural and Food Chemistry*, *48*, 4597-4605.

[60] Elez-Martínez, P., Aguiló-Aguayo, I., & Martín-Belloso, O. (2006). Inactivation of orange juice peroxidase by high-intensity pulsed electric fields as influenced by process parameters. *Journal of the Science of Food and Agriculture*, *86*, 71-81.

[61] Icier, F., Yildiz, H., & Baysal, T. (2006). Peroxidase inactivation and colour changes during ohmic blanching of pea purée. *Journal of Food Engineering*, *74*, 424-429

[62] Quaglia, G.B., Gravina, R., Paperi, R., & Paoletti, F. (1996). Effect of high pressure treatments on peroxidase activity, ascorbic acid content and texture in green peas. *LWT-Food Science and Technology*, *29*, 552-555.

[63] Toralles, R. P., Vendruscolo, J. L., Vendruscolo, C. T., Del Pino, F. A. B., & Antunes, P. L. (2005). Properties of polyphenoloxidase and peroxidase from Granada Clingstone peaches. *Brazilian Journal of Food Technology*, *8*, 233-242.

[64] Chutintrasri, B. & Noomhorm, A. (2006). Thermal inactivation of polyphenoloxidase in pineapple purée. *LWT-Food Science and Technology*, *39*, 492-495.

[65] Yemenicioğlu, A., Özkan, M., Velioğlu, S., & Cemeroğlu, B. (1998). Thermal inactivation kinetics of peroxidase and lipoxygenase from fresh pinto beans (*Phaseolus vulgaris*). *Z Lebensm Unters Forsch A*, *206*, 294-296.

[66] Anthon, G.E., & Barrett, D.M. (2002). Kinetic parameters for the thermal inactivation of quality-related enzymes in carrots and potatoes. *Journal of Agricultural and Food Chemistry*, *50*, 4119-4125.

[67] Gonçalves E.M., Pinheiro, J., Abreu, M., Brandão, T.R.S., & Silva, C.L.M. Modelling the kinetics of peroxidase inactivation, colour and texture changes of pumpkin (*Cucurbita maxima* L.) during blanching. *Journal of Food Engineering*, *81*, 693-701.

[68] Terefe, N.S., Matthies, K., Simons, L., & Versteeg, C. (2009). Combined high pressure-mild temperature processing for optimal retention of physical and nutritional quality of strawberries (*Fragaria* × *ananassa*). *Innovative Food Science and Emerging Technologies*, *10*, 297-307.

[69] Garcia-Palazon, A., Suthanthangjai, W., Kajda, P., & Zabetakis, I. (2004). The effects of high hydrostatic pressure on β-glucosidase, peroxidase and polyphenoloxidase in red raspberry (*Rubus idaeus*) and strawberry (*Fragaria* × *ananassa*). *Food Chemistry*, *88*, 7-10.

[70] Cano, M.P., Hernandez, A., & De Ancos, B. (1997). High pressure and temperature effects on enzyme inactivation in strawberry and orange products. *Journal of Food Science*, *62*, 85-88.

[71] Lourenço, E.J., Neves, V.A., & Da Silva, M.A. (1992). Polyphenol oxidase from sweet potato: purification and properties. *Journal of Agricultural and Food Chemistry*, *40*, 2369-2373.

[72] Raviyan, P., Zhang Z., & Feng, H. (2005). Ultrasonication for tomato pectinmethylesterase inactivation: effect of cavitation intensity and temperature on inactivation. *Journal of Food Engineering*, *70*, 189-196.

[73] Cruz, R.M.S., Vieira, M.C., & Silva, C.L.M. (2006). Effect of heat and thermosonication treatments on peroxidase inactivation kinetics in watercress (*Nasturtium officinale*). *Journal of Food Engineering*, *72*, 8-15.

[74] Segel, I.H. (1993). Enzymes as biological catalysts. In *Enzyme Kinetics*. (pp. 7-14). John Wiley and Sons, USA.

[75] Lemos, M.A., Oliveira, J.C., & Saraiva, J. A. (2000). Influence of pH on the thermal inactivation kinetics of horseradish peroxidase in aqueous solution. *LWT-Food Science and Technology*, *33*, 362-368.

[76] Vámos Vigyázó, L. (1981). Polyphenol oxidase and peroxidase in fruits and vegetables. *Critical Reviews in Food Science and Nutrition*, *15*, 49-127.

In: Practical Food and Research
Editor: Rui M. S. Cruz, pp. 67-88

ISBN: 978-1-61728-506-6

Chapter III

MICROORGANISMS AND SAFETY

Célia Quintas
Departamento de Engenharia Alimentar, Instituto Superior de Engenharia, Universidade do Algarve, Campus da Penha, 8005-139 Faro, Portugal, cquintas@ualg.pt

ABSTRACT

The consumption of fresh food of plant origin (fruits and vegetables) has been increasing over the recent past mainly due to changes in the dietary patterns and the year-round availability through global food supply chains. These trends stimulated the growing demand for "quick" and convenient fresh food products and the rapid growth of the fresh fruits and vegetables' industry. At the same time, the number of outbreaks of human infections attributed to the consumption of fresh fruits and vegetables and unpasteurized fruit juices has increased during the last years.

Fruits and vegetables consumed as fresh contain a microbial population originating from the natural microbiota present in the field of production and the microorganisms introduced during harvesting, processing, packaging, storage and distribution. In most situations these microorganisms do not have a visible effect and the food is consumed without any adverse consequence. However, some of those organisms may grow and cause spoilage and foodborne diseases while others can originate food products through fermentation processes. Microorganisms of concern in food of plant origin are bacteria, fungi, parasitic protozoa and virus. The most relevant human pathogenic bacteria causing diseases through the consumption of fruits and vegetables belong to the groups *Aeromonas*, *Bacillus*, *Campylobacter*, *Clostridium*, *Escherichia* (pathogenic strains), *Listeria*, *Salmonella*, *Shigella*, *Staphylococcus*, *Yersinia* and *Vibrio*. The parasitic protozoa associated worldwide to fresh produce and foodborne illness are *Cryptosporidium*, *Giardia*, *Cyclospora* and *Toxoplasma*. The groups of virus currently recognised as the most important human foodborne pathogens with regard to the number of outbreaks and people affected are norovirus and hepatitis A virus.

In the present chapter the main microorganisms causing foodborne illness through the ingestion of fruits and vegetables will be studied as well as the major sources of contamination in each phase of the production chain. The principal factors affecting the levels of contamination will also be discussed.

3.1 INTRODUCTION

Fruits and vegetables represent a special challenge to food safety since they can be consumed raw with little or no treatment to reduce/eliminate microbial hazards. In the last few years, fresh produce emerged as a new vehicle for the transmission of food-borne diseases associated with etiological agents that in the past were attributed to animal reservoirs [1-2]. In fact, the number of outbreaks reported has been increasing in the last decades all over the word [1-8].

The factors that mostly contributed to the emergence of fresh produce as a cause of outbreaks are listed in Table 3.1.

Table 3.1. Some factors causing the increasing numbers of outbreaks related to fresh produce [1, 6, 9].

Factors
Changes in the industry:
Intensification and centralization of production
Large distribution networks
Introduction of minimally processed fruits and vegetables
Increased importation of fresh fruits and vegetables
Changes in consumer habits
Increased consumption of meals outside the house
Increased popularity of salad bars
Increased consumption of fresh fruits and vegetables
Increased consumption of fresh fruit juices
Lifestyle changes
Increased size of at-risk population (elderly, immunocompromised)
Enhanced epidemiological surveillance
New methods to identify and track pathogens
Emerging pathogens with low infection doses
Microbial evolution affecting virulence or pathogenicity

A vegetable is the edible part of a plant including leaves, stalks, roots, tubers, bulbs and flowers. The fruiting bodies of the basidiomycetes fungi, mushrooms, are, in general, included in the vegetable group. Nutritionally, vegetables are rich in water, fibre, vitamins and minerals and contain starch and some lipids which can favour microbial growth. The presence and growth of microorganisms is also facilitated by the pH of vegetable tissues ranging from 5 to 7. Items sometimes thought of as vegetables such as tomatoes, bell peppers, cucumbers, are in fact fruits [10].

Fruits, botanically, are the structures formed from the ovary of a flower after the fertilization and consist of a fruit wall, the pericarp, enclosing the seeds. In true fruits, the middle layer of the pericarp becomes succulent (peaches, berries, grapes, oranges, and banana). In false fruits the receptacle of the flower develops and becomes fleshy (apples, strawberries). In terms of their nutritional value, fruits are relevant sources of vitamins, minerals and sugars. They are also rich in organic acids which justify their low pH ranging from 1.9 to 5.6. Melons and some tropical fruits are an exception with higher pH values (6.2-6.5). However, in some cases organic acids are contained inside compartmented areas of the fruit, as in oranges, leaving the surrounding tissues with pH values closer to neutrality. This may facilitate the colonization of fruits by microorganisms [11].

3.2 Microbiota in Fruits and Vegetables

When an intact part of a plant is commercialized and consumed, the microbial contamination reflects the environment through which the product has passed including pre-harvest conditions, harvest practices, post-harvest processing and preparation for consumption [5, 12].

The microbial communities found in fruits and vegetables are diverse and include many different genera of bacteria, filamentous fungi, yeasts, algae, protozoa, nematodes and virus. The dominant groups are bacteria, yeasts and filamentous fungi which are part of the epiphytic microbiota and some of those microorganisms are present at the time of consumption. Epiphytic microbiota includes microorganisms that colonize the surfaces of plants without causing disease.

The majority of bacteria found on the surface of plants is usually Gram-negative and belongs to the genera, *Pectobacterium*, *Xanthomonas*, *Pseudomonas* and *Enterobacteriacea*, namely *Erwinia*, however most of them have not been recognized as pathogenic to humans. Some species of these genera are pectinolytic and can cause softening of the vegetable tissues. The presence of lactic acid bacteria and of spore formers has also been described. The most frequently found fungi belong to the genera: *Penicillium*, *Sclerotinia*, *Botrytis*, *Rhizopus*, *Fusarium* and *Alternaria* and may be involved in spoilage. Yeasts have also been isolated from fresh vegetables namely representatives of the following genera: *Rhodotorula*, *Candida*, *Kloeckera* [10].

On raw fruits, filamentous fungi and yeasts often form the majority of microbiota, mainly due to the pH values of fruit tissues. Species of the genera *Penicillium*, *Aspergillus*, *Mucor*, *Alternaria*, *Cladosporium* and *Botrytis* are the most frequently isolated from fruits and some fungi species are associated to softening and weakening of plant structures. The more common genera of yeasts isolated from fruits are *Saccharomyces*, *Hanseniaspora*, *Pichia*. *Kloeckera*, *Candida* and *Rhodotorula*. Both groups of Fungi may be involved in spoilage processes of fruits or fruit juices. The bacteria's genera better represented in fruits are: *Pseudomonas*, *Xanthomonas*, *Enterobacter* and *Corynebacterium* [11].

The survival and growth of contaminating microorganims in fresh produce is affected by intrinsic, extrinsic and processing factors. Intrinsic parameters such as the nature of the epithelium and cuticule, tissue pH and buffering capacity, nutrient composition, water activity (a_W), redox potential and the presence of antimicrobial compounds or antimicrobial structures, determine which microorganisms are able to survive and grow in each type of produce. On the other hand, the environment [geographical location, precipitation, temperature, wind, the presence of vectors (insects, nematodes)] and agricultural practices in which plants grow, impose extrinsic factors that interfere with the number and diversity of the microbial population present in the produce. Moreover, the processing factors may change the intrinsic and extrinsic factors, increasing or decreasing the contamination or proliferation of microbes [10, 11].

Although the majority of microorganisms found in food of plant origin are not human pathogens, a few are of potential concern for the fresh produce industry, and may become a serious problem of food safety.

3.2.1 Human-Pathogens in Fruits and Vegetables

Human pathogens can contaminate food of plant origin throughout the various stages of production: in the fields, during harvesting, processing, distribution, marketing and preparation for consumption. In all these steps the inappropriate human handling is an important cause of contamination [5, 13-16].

Some of the microbial human pathogens belonging to bacteria, protozoa and virus, that have been isolated from fruits and vegetables, are listed in Table 3.2. The most frequent bacterial etiologic agents associated to foodborne outbreaks due to fresh produce are *Salmonella enterica*, *Escherichia coli* (pathogenic strains) *Shigella* spp., *Campylobacter* spp., *Listeria monocytogenes*, *Staphylococcus aureus*, *Yersinia* spp., and *Bacillus cereus* [6]. Tables 3.3.1 and 3.3.2 present the main characteristics of human pathogenic bacteria isolated from fresh produce.

Salmonellosis is among the most frequent cases of outbreaks in the United States [17] and several of those have been associated with cantaloupe and lettuce.

The second most important etiological agent of outbreaks from fresh produce is *E. coli* (pathogenic strains). *E. coli* 0157:H7 was the causal agent of the largest outbreak, with more than 6000 cases, related to contaminated radish sprouts in Japan in 1996 [18]. In 2006 an outbreak caused by *E. coli* 0157:H7 in spinach, in the United States and Canada was reported [7].

Members of the genus *Citrobacter*, *Enterobacter* and *Klebsiella* have also been isolated from fruits and vegetables. In general they occur naturally on food of plant origin and are responsible by spoilage but have occasionally been associated to illness. *Enterobacter sakazakii*, an emergent pathogen that now is included in the genus *Cronobacter* spp. [19] was isolated from lettuce [20].

In terms of food safety, the protozoa with significance, in the preparation and consumption of fresh produce are *Cryptosporidium*, *Giardia* and *Cyclospora* [14] and their major features are mentioned in Table 3.4. These groups are especially relevant in ready-to-eat foods not receiving heat treatment. Raw raspberries were associated to some large *Cyclospora cayetanensis* outbreaks in North America [14] and unpasteurized fruit juices were related with *Cryptosporidium parvum* outbreaks [21].

In respect to viruses, the most frequent foodborne illness caused by viruses associated to fruits and vegetables are norovirus and hepatitis A virus which are highly infectious and may lead to widespread outbreaks [13, 22]. Noroviruses are referred as a major cause of nonbacterial gastroenteritis. These viruses can survive on refrigerated ready-to-eat lettuce for at least 10 days [23].

Table 3.2. Microbial human pathogens isolated from fruits and vegetables [5, 7, 13, 14, 24-26].

Categories	Human Pathogen	Produce
Bacteria	*Aeromonas*	Alfalfa sprouts, asparagus, broccoli, cauliflower, celery, lettuce, pepper, spinach
	Bacillus cereus	Alfalfa sprouts, cress sprouts, cucumber, mustard sprouts, soybean sprouts
	Campylobacter jejuni	Green onion, lettuce, mushroom, potato, parsley, pepper, spinach
	Clostridium botulinum	Cabbage, mushroom, pepper, garlic, vegetables salad
	Escherichia coli 0157:H7	Alfalfa sprouts, apple juice, cabbage, celery, cilantro, coriander, cress sprouts, lettuce, spinach
	Listeria monocytogenes	Bean sprouts, cabbage, chicory, cucumber, eggplant, lettuce, mushroom, potato, radish, vegetables salad, tomato
	Salmonella	Bean sprouts, artichoke, beet leaves, celery, cabbage, cantaloupe, cauliflower, chilli, cilantro, eggplant, endive, fennel, green onion, lettuce, mung bean sprouts, mustard cress, orange juice, parsley, pepper, spinach, strawberry, tomato, watermelon
	Shigella	Celery, cantaloupe, lettuce, parsley, scallion
	Staphylococcus	Alfalfa sprouts, carrot, lettuce, onion sprouts, parsley, radish
	Vibrio cholerae	Cabbage, coconut milk, lettuce
	Y. enterocolitica	Cabbage, carrot, lettuce
Viruses	Hepatitis A Norovirus	Celery, green onion, lettuce, strawberry, tomato
Protozoan parasites	*Cryptosporidium* *Cyclospora* *Giardia* *Toxoplasma*	Basil, blackberry, lettuce, onion, raw vegetables, raspberry, tomato, spinach, watercress

Table 3.3.1. Main characteristics of human pathogens isolated from fresh produce [5, 9, 27-32].

Organisms	Characteristics
Aeromonas	**Reservoir**: Aquatic environment: freshwater lakes, streams and wastewater systems; isolated from food and water Facultative anaerobe Non spore former **Dose causing illness**: 10^9-10^{10} cells **Temperature**: Optimum 28 °C; Range for growth <5-45 °C; Grow at refrigerating/chill temperatures on raw vegetables **pH**: Sensitive to low pH (<5.5) **Modified atmosphere**: Growth was observed

Table 3.3.1. (Continued).

Organisms	Characteristics
Bacillus cereus	**Reservoir**: Widespread in nature and frequently isolated from soil and plants. Easily spread to foods of plant origin Facultative anaerobe Spore former Toxin production **Dose causing illness**: 10^5-10^7 cells (Diarrhoeal syndrome) and 10^5-10^8 cells/g (emetic syndrome) **Temperature**: Optimum 28-35 °C; Range for growth: 8-55 °C; Some strains can growth at refrigerating temperatures which raise concern about the safety of refrigerated foods with extended shelf lives Pasteurization kills vegetative bacteria but not spores **pH**: Minimum 5.0-6.0 (depending on the acidulant) **Resistances**: Spores resist to desiccation and heating **Note**: Spores of *B. cereus* are very hydrophobic, adhere to several types of surfaces and form biofilms. Difficult to clean and disinfect **Outbreaks**: seed sprouts
Campylobacter jejuni	**Reservoir**: Gastrointestinal tract of many animals (wild or domestic); frequently isolated from water Microaerophiles sensitive to oxygen Non-spore former **Dose causing illness**: <1000 cells (depends on virulence of strain and the susceptibility of the individual) **Temperature**: Optimum 42-45 °C; Don't grow at temperatures<30 °C; Survival at refrigeration and freezing temperatures **pH**: Sensitive to low pH (killed at pH 2.3) **Sensitivities**: Drying and high-oxygen conditions **Modified atmosphere:** Grow best in an atmosphere containing 5-10% carbon dioxide and 3-5% oxygen **Outbreaks**: lettuce, sweet potato, cucumber, melon
Clostridium botulinum	**Reservoir**: Soil (soil saprophyte) and mud; can grow in the gastrointestinal tract of birds and mammals; spores found in raw fruits and vegetables Obligate anaerobic Spore former Produce botulinum toxins: Toxin lethal dose 10^{-8} g **Temperature**: Minimum temperature for growth 10-12 °C (proteolytic strains), 3.0 °C (non-proteolytic strains) **pH**: Minimum for growth: 4.6; Maximum: 8.5-8.9 **Foods of concern**: a) Foods receiving a moderate heat process (75 °C-95 °C), cooled rapidly and stored at refrigeration temperatures; b) Food packed under vacuum or an anaerobic atmosphere, restricting growth of aerobic but not anaerobic microorganisms **Note**: Avoid vacuum packing or wrapping in aluminium foil; Avoid temperature abuses: inadequate refrigeration after pasteurization (vegetable juices); Provide adequate acidification and heat treatment **Outbreaks**: potato salad, garlic in oil, mushrooms, cabbage, carrot juice

Escherichia coli (pathogenic strains)	**Reservoir**: Gastrointestinal tract of humans and other warm-blooded animals, transmitted to humans via contaminated food Facultative anaerobe Non spore former **Dose causing illness**: 2-2000 cells - *E. coli* 0157:H7 (outbreaks) **Temperature**: Optimum 37 °C; Range for growth 7-50 °C **pH**: growth 4.4-9; survives at pH:3.5-5.5; Survives at low pH at refrigeration temperatures **Modified atmosphere**: Has no effect on survival or growth **Outbreaks** (*E. coli* 0157:H7): lettuce, spinach, apple juice, cucumber salad, seed sprouts, carrot
Listeria monocytogenes	**Reservoir**: Ubiquitous: fresh and salt water; soil; sewage sludge; decaying vegetation and silage; asymptomatic human and animal faeces Facultative anaerobe Non spore former **Dose causing illness**: Not determined but estimates vary from 10^2-10^9 cells depending on immunological status of the host **Temperature**: Optimum: 30-35 °C; Range for growth: 0-42 °C (Psychrotroph); Inactivated by heating at 50 °C **pH**: Tolerant to low pH and high salt concentration Produces biofilms on food processing surfaces **Foods of concern**: Minimally processed foods and ready–to-eat **Packing under modified atmosphere**: Has no effect on growth rates **Outbreaks**: lettuce, tomato, cabbage

Table 3.3.2. Main characteristics of human pathogens isolated from fresh produce [5, 9, 33-37].

Organisms	Characteristics
Salmonella	**Reservoir**: Gastrointestinal tract of vertebrates (reptiles, mammals and birds) Non spore former Facultative anaerobe **Dose causing illness**: 10^6 cells, varying with the virulence of the strain and the susceptibility of people (infective doses of 10-100 and 1.5-9.1 cells have been reported) **Temperature**: Optimum 37 °C; Range for growth 5 °C-47 °C; Heat sensitive destroyed by pasteurization temperatures **pH**: Optimum 6.5-7.5; Minimum 4.05-5.4, depending on the acidulant **Outbreaks**: lettuce, tomato, melon, watermelon, seed sprouts, orange juice
Shigella	**Reservoir**: Gastrointestinal tract of humans and monkeys; Contaminate raw fruits and vegetables by contaminated water; insects, hands of persons Non spore formers Facultative anaerobes **Dose causing illness**: 10-100 cells-200 cells **Temperature**: Range for growth 10-45 °C; *S. sonnei*: survive under refrigeration temperatures in vegetables; *Shigella* spp. can survive at a temperature of -20 °C **pH**: Optimum 6-8, Minimum 4.5 **Vacuum or modified atmosphere**: Growth and survival were observed **Outbreaks**: lettuce, parsley

Table 3.3.2. (Continued).

Organisms	Characteristics
Staphylococcus aureus	**Reservoir**: Skin, skin glands and mucous membranes (nose) of warm blooded animals; Can be isolated from faeces and environmental sites: soil, marine and fresh water, plant surfaces, dust and air Produces enterotoxins **Dose causing illness**: 10^5 cells/g produces level of toxin that will elicit symptoms of food poisoning (<1µg) Non spore former Facultative anaerobe **Temperature**: Optimum 35-37 °C; Range 7-48 °C **pH**: Optimum 6.0-7.0; Range 4.0-9.8 **NaCl**: Optimum 0.5-4.0%; Range 0-20%
Vibrio cholerae	**Reservoir:** Marine and estuarine environments; Sources of infections: human faeces from infected individuals; It is isolated from food previously in contact with contaminated water (washed fruits and vegetables) Facultative anaerobes Non spore former **Dose causing illness:** 10^3-10^4 cells (with food) 10^{10} cells (without food) **Temperature**: Optimum 37 °C; Range 5-43 °C **pH**: Does not survive at low pH **NaCl**: Optimum 1-3%; Range 0.5-8% **Outbreaks**: salad crops, vegetables
Yersinia spp.	**Reservoir**: Gastrointestinal tract of many animals, including humans and domestic animals; Widespread occurrence in terrestrial and freshwater ecosystems such as soil, vegetation, lakes, rivers and wells where it can persist for long periods of time at low temperatures Non spore former Facultative anaerobic **Dose causing illness**: Not known but probably higher than 10^4 CFU **Temperature**: *Y. enterocolitica* growth range:-1 °C - +40 °C Psychrotroph: Growth at refrigerating temperatures (<4 °C); Survives in refrigerating and frozen foods **pH**: Optimum 7-8; Minimum 4.1 or 5.1 **Vacuum or modified atmosphere:** Growth and survival were observed

Table 3.4. Main characteristics of parasitic protozoa isolated from fresh produce [14, 21].

Organisms	Characteristics
Cryptosporidium	**Reservoir**: Human and Animal faeces **Species infecting humans**: *C. parvum* and *C. meleagridis* **Life cycle**: After ingestion, each oocysts excysts 4 sporozoites in the small intestine and life cycle is completed in the gut cells of a single host. **Dose causing illness:** 16-100 oocysts **Resistances**: Chlorine and drug treatment **Outbreaks:** Unpasteurized apple juice **Can be removed by filtration**

Table 3.4. (Continued).

Giardia	**Reservoir**: Pet, livestock and wild animals faeces **Species infecting humans**: *Giardia duodenalis* (*Giardia lamblia*) **Life cycle**: Two stage life cycle: Reproductive trophozoite and a resistant cyst; Ingested cysts release in the duodenum two trophozoite which multiply asexually and disease occurs **Dose causing illness:** 10-100 cysts **Resistances:** Chemical resistance [Chlorine (> than bacteria)] **Outbreaks:** Fruit salad, sliced vegetables **Can be removed by filtration**
Cyclospora	**Reservoir:** It is not clear **Species infecting humans:** *C. cayetanensis* **Life cycle**: Not known **Dose causing illness:** Not known (probably low) **Resistances:** Chemical disinfectants (chlorine) **Outbreaks:** Raspberries, salads, lettuce, basil

The most important single produce items implicated in epidemics of foodborne illness are lettuce, melon, tomato, seed sprouts and fruit juice [2]. With respect to potozoans, raw raspberries were associated to some large *Cyclospora cayetanensis* outbreaks in North America [14].

3.3 Sources and Mechanisms of Contamination in Fruits and Vegetables

Food of plant origin may be contaminated with human pathogens at any stage from the field to the consumption. In fact, conditions and practices taken during pre-harvest, harvest, post–harvest, affect the microbial presence in fruits and vegetables.

3.3.1 Pre-harvest

The soil is a reservoir for a variety of pathogenic and non-pathogenic microorganisms.

Bacillus cereus, *Listeria monocytogenes* and *Clostridium botulinum* occur naturally in soil and decaying vegetable matter and can be present on the surface of fresh produce. Other pathogens from human/animal source can be found in the soil due to irrigation with contaminated water and fertilization with incompletely treated organic fertilizers (manure) or due to faeces and droppings of animal origin in the cultivating area. The occurrence of contamination with human pathogens due to the utilization of contaminated organic fertilizing and watering has been described by several authors. Ibenyassine et al. [38] observed high incidences of microbial contamination, including species of *E. coli*, *Enterobacter cloacae* and *Klebsiella pneumoniae*, in several vegetable species irrigated with untreated wastewater. Organically grown lettuce revealed the presence of *Listeria monocytogenes* and *E. coli* [39]. Melloul et al. [40] found the same strains of *Salmonella* in both the edible parts of vegetables and the raw wastewater used in the irrigation of plants. The same study revealed that vegetables growing on the surface of the ground, such as lettuce or parsley, were more contaminated than those that developed above the surface, such as tomato. In another study, it was observed that *E. coli* 0157:H7 persists in soil for 154-196 days after addition of contaminated manure or irrigation water and was detected on onions and carrots for 74 and 168 days, respectively [41]. Another possible source of contamination of plants with pathogens is the runoff from livestock pastures [42].

The contamination of produce seems to be enhanced via some vectors such as insects, nematodes and protozoans. Sela et al. [43] showed that the transmission of *E. coli* can occur by *Drosophila melanogaster* in apples. The transfer of *Salmonella* spp. from the soil to fruits and vegetables via the nematode *Diploscapter* sp. was observed by Gibbs et al. [44]. Additionally, protozoa, present on leafy vegetables (lettuce), seem to play a special role in the protection and survival of foodborne pathogens (*E. coli* and *S. enterica*) [45].

The main factors responsible by pathogenic contamination of fresh produce on the field are represented in Figure 3.1.

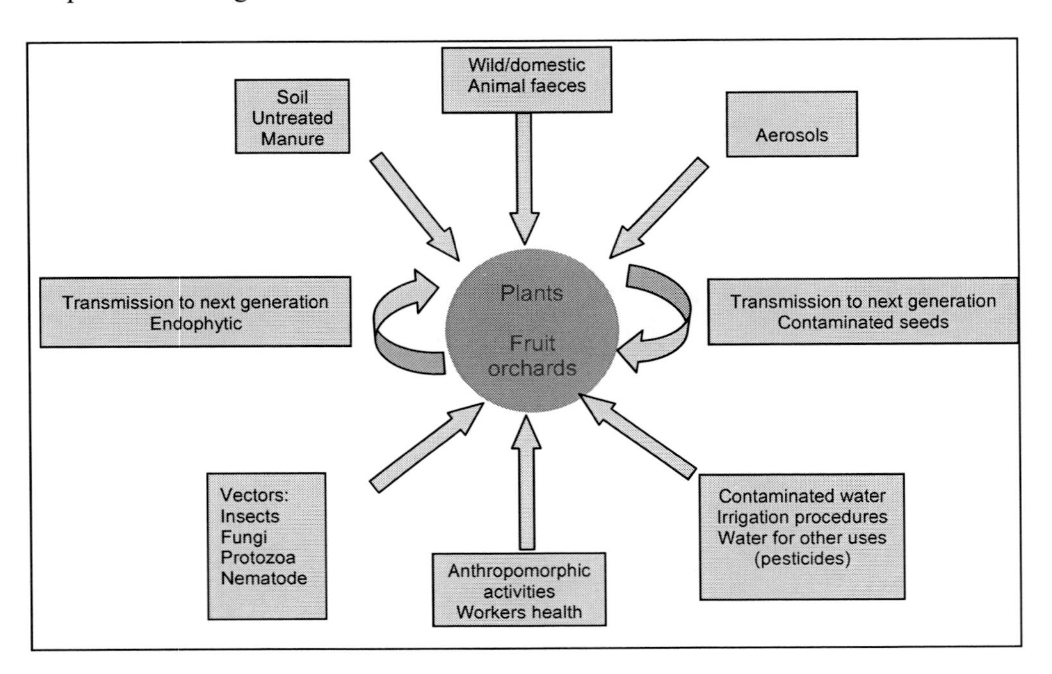

Figure 3.1. Pre-harvest main sources of human pathogens on fruits and vegetables.

Brandl and Amundson [46] demonstrated that *E. coli* and *Salmonella enterica* are able to multiply on the surface of young plants. On the other hand, the endophytic colonization and seed contamination with human pathogens facilitates the transmission of those microbes to next generations and the prevalence of the contaminations [47] (Figure 3.1).

The adoption of good agricultural practices and good sanitary conditions to reduce potential contamination by foodborne disease causing agents, in the field stages, are mandatory to produce safer food.

3.3.2 Harvest

Fodborne disease causing agents can contaminate foods of plant origin during the harvesting steps (collecting, sorting, packing and transportation) through cross contamination from faecal material, harvesting equipment, transport containers, transport vehicles. The presence of wild and domestic animals, aerosols, and the use of contaminated ice or water will increase the levels of contamination of fruits and vegetables [5, 15]. Additionally, poor human handling can cause injuries in produce leading to the release of nutrients and promoting microbial growth/survival and points of entry for microorganisms (human pathogens or spoilage). Moreover, workers health and personal hygiene and good sanitation practices have to be taken into account to reduce the potential contamination of pathogenic microorganisms on fresh produce. Containers and all the equipment used in harvest and transport may become an additional source of contamination if not correctly cleaned.

Finally, the level of contamination on produce prior to processing is also influenced by environmental conditions such as temperature, humidity and transportation time.

3.3.3 Post-harvest

Fruits and vegetables may receive some type of handling or processing before they are commercialized and consumed. Those post-harvest steps or treatments are storage, washing/ rinsing packaging and transportation. Additional operations such as cutting, slicing, chopping, shredding, peeling, blending, extracting can be introduced in the case of minimally processed vegetables, pre-cut fruits or production of juices. During these steps conditions may arise which increase the likelihood of contamination of the produce through cross contamination from other materials (utensils, surfaces, equipment, transport containers, other foods) or from the food handlers. Post-harvest sources of contamination include faeces, wild and domestic animals, insects, aerosols/environmental dust, rinse/wash water and ice, among others. In order to prevent additional contaminations and cross contamination during the post-harvest steps, a strict hygienic level has to be maintained in all food-contact materials which have to be subjected to regular cleaning as well as food handlers' hygiene [8, 15].

Produce normally undergo washing and rinsing treatments to remove dust, cell debris, reduce microbial levels, increasing the shelf life and quality of products. After washing/ rinsing foods of plant origin, the rinse water should be let to evaporate. However, it has been observed that the level of microorganisms can increase after the wash step [15]. Any residual water on the surface of processed or unprocessed fruits and vegetables will allow for multiplication of microbes. It is essential to adopt strategies to maintain appropriate quality of wash and rinse water cleaned, using proper sanitizers, and, on the other hand, the produce's surface should be left to dry. Another recommendation consists in the fact that washes and rinse water should be at least 10 ºC warmer than the product, to prevent the eventual internalization of microorganisms [10].

Due to increasing demand of convenience foods, such as ready to eat products, fresh produce is often cut, sliced, chopped, and mixed. These operations eliminate the protection offered by peels and skins influencing the microbial populations on the cut/damaged produce's surfaces. Microbes grow faster on cut produce due to greater availability of nutrients and water. For example, the mechanical damage caused to cantaloupe melon during slicing allows microorganisms to enter the tissues and grow to high numbers when stored at non-refrigeration temperatures as was observed with *Salmonella* by Ukuku and Sapers [48].

Packaging, storage and transportation are other relevant processing steps which influence the development of microorganisms on fruits and vegetables. Microorganisms can originate and increase during the packing shed phase and affect the shelf life of the product. On cantaloupe melons the microbial load increased significantly during the packing process. The surface topography of this fruit may favour microbial attachment and the pulp pH (6.1-7.1) is suitable for microbial growth [15]. Gagliardi et al. [49] concluded that a significant amount of contamination on cantaloupe occurs at the packing shed and during washing, rather than in the field or during harvest. If a limited number of products is contaminated, contamination may be spread over an entire lot during washes (water dips) commonly used in packing sheds.

Modified Atmosphere Packaging (MAP) is used to control the rate of senescence of fruits and vegetables and the growth rate of certain microorganisms, thus increasing the shelf life of those foods. Since raw produce cells keep respiring, precise atmosphere control is difficult to achieve. Additionally, anaerobic conditions in MAP should be avoided to prevent *Clostridium botulinum* growth [30].

Microorganisms can be present and grow on food processing surfaces and form biofilms which may contaminate fresh produce through mechanisms of cross contamination if adequate programmes of cleaning and disinfection are not applied. Environments where food is processed prior to packaging, (during exposition time to sale or the preparation before consumption), are at a particular risk for the development of biofilms due to the availability of humidity and nutrients. Prazak et al. [50] found that packing sheds provided a suitable environment for the survival and proliferation of *Listeria* spp. especially conveyor belts where cross contamination can occur between processing surfaces and produce. Particular attention should be given to adequate programs of cleaning and disinfection of facilities and equipment to minimize the opportunities for cross contamination through food contact surfaces. However, during the cleaning and disinfecting operations, the formation of aerosols has to be avoided since the small drops may recontaminate cleaned surfaces or even food [10, 11].

Control of temperatures, especially through refrigeration, along post-harvest procedures is essential to minimise microbial growth and assure quality and safety. However, the presence of psychrotrophs has to be controlled.

Figure 3.2 summarises the main sources and conditions that facilitate the presence and proliferation of microorganisms in fresh produce during the post-harvest procedures.

If the food is to be frozen, a heating step, the blanching, must be applied. The blanching times vary with the size and variety of vegetables. This processing step is especially important if the vegetable is to be frozen for more than a few months. The heat from blanching slows or destroys the enzymes action which can cause the vegetables to lose flavour, colour, texture and nutritive value. Additionally, it also helps to destroy microorganisms and to clean dirt on the surface of vegetables. However the microbiota of the frozen product is a reflection of the handling the product received after blanching (air-borne microorganisms, cross contamination on post blanch surfaces, equipment and handling) [10].

In order to control the contamination of fresh produce with foodborne pathogens during post-harvest operations, the recommendations summarized in Table 3.5 should be taken into account.

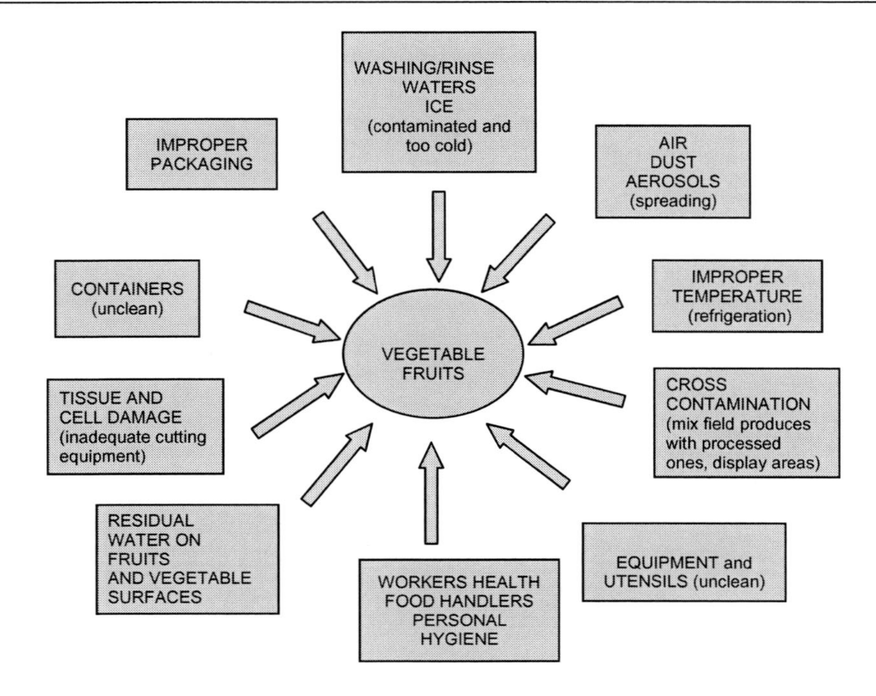

Figure 3.2. Post-harvest main sources of human pathogens on fruits and vegetables.

Table 3.5. Control measures to reduce foodborne pathogens during the production (post-harvest) of fresh produce.

Use water supplies of good quality
Washing
Rinsing
Disinfection
Avoid temperature abuses
Storage
Transportation
Distribution
Marketing
At home
Avoid moisture in the processing environment and on the surface of produce
Avoid condensation in the packages
Provide adequate drying after washing and rinsing
Provide adequate sharp cutting equipment
Minimize tissue damage
Avoid cross contamination
Separate field produce from processed produce
Separate food of animal origin from food of plant origin
Provide adequate cleaning and disinfecting programmes to food contact surfaces and equipments
biocides
Microbes may acquire virulence

Table 3.5. (Continued).

Provide adequate cleaning and disinfecting programmes
Adequate frequency
Facilities
Equipment
Biofilms
Avoid aerosol formation
Good hygienic practices
Prevent faecal-oral spread
Stringent personnel hygiene
Education to food handlers
Be aware of microbial evolution
Microbes can always change
Microbes can adapt to stress and biocides
Microbes may acquire virulence

3.4 Production of Biofilms

The interactions among microorganisms and plants are probably dependent on their ability to form microbial biofilms [51]. Biofilms are defined as assemblages of micro-organisms' adherent to each other and/or to a surface and embedded in a matrix of exopolymers [52]. Exopolyssacarides secreted by microorganisms can form a layer when associated with the plant cells creating a matrix structure that behaves as an ion-exchange resin. This matrix may harbour bacteria, yeasts and molds in aggregates [53].

The biofilms contribute to increase resistance to certain environmental stresses as well as antimicrobial tolerance, protection from protozoan predation and consortial metabolism. The high population density in a biofilm provides the opportunity to perform certain processes that single cells can not accomplish efficiently (Quorum sensing). Formation of biofilms is a way of maintaining a critical mass of cells in a specific location for periods sufficient to initiate beneficial or antagonistic interactions with host plants. Biofilms form a protective environment for pathogens and spoilage organisms [51].

Some human pathogens come into contact with plants through contaminated water or in processing facilities. In other cases, the plant is an alternative host. Biofilms present on fresh produce, seeds and sprouts for human consumption are a potential health concern, as they may harbour pathogens. Morris et al. [54] estimated that about 10-40% of the bacteria on endive and parsley leaves form biofilms. Rayner et al. [55] described the presence of biofilms on the surface of tomatoes, carrots, lettuce and mushrooms. *Salmonella* strains formed biofilms in cantaloupe melons after inoculation [56]. A variety of commercial sprouts showed biofilms with thickness ranging from 5 to 12 μm [57]. Surface attachment and biofilm formation enable the bacteria to persist and withstand washing and other antimicrobial treatments. *Salmonella typhimurium* embedded in a biofilm matrix resisted sodium hypochlorite at concentrations above 500 mg/l, while planktonic cells were sensitive to less than 50 mg/l [58]. Additionally, biofilm formation also plays a role in the persistence of *Salmonella* after chlorination treatment of parsley [59]. *Listeria monocytogenes* in a multispecies biofilm containing *Pseudomonas fragi* and *Staphylococcus xylosus* is unaffected by treatment with 500 ppm free chlorine [60].

Other aspects of concern are the containers used to harvest, transport and display raw fruit and vegetables, especially when they are not effectively cleaned and sanitized. Those situations can lead to the development of biofilms. Contamination of fresh produce with pathogens may result from contact with surfaces harbouring biofilms. If pathogens attach to the surfaces and form biofilms during transport or processing, their survival and growth may be enhanced. Growth of pathogens incorporated into biofilms would increase the probability of cross-contamination of produce.

During growth and maturation of fruits and vegetables as well as during harvesting, transport, processing and storage after processing, opportunities arise for the development of biofilms. The occurrence of these biofilms is one of the aspects that influence the effectiveness of strategies to control food borne pathogens in fresh produce.

3.5 Internalization of Microorganisms

The internalization of human pathogens into fresh produce has been observed in various studies, but how the bacteria enter the plant and where they are localized within the edible parts of the plants is not yet understood. The access of microorganisms to the interior of fresh produce is particularly important since in these conditions they are protected from washing and sanitizing procedures. Microbes can be internalized through damage induced to plant tissue, through stomata or by establishing an endophytic relationship with the plant. Microorganisms can be also present in the inner tissue of fresh produce as a result of the uptake of water through certain irrigation or washing procedures. If these waters are contaminated with human pathogens these may be introduced into the tissue produce [47].

Pathogens can be internalized during slicing, through surface breaks or porous sites and in processing procedures when surfaces are bruised or broken or have greater porosity. Results obtained with spinach illustrate this situation. Bacteria were not detected on the surface of unbroken spinach leaves but they were present on those leaves where the cuticle was broken [61].

On the other hand, some human pathogens seem to be able to establish an endophytic relationship with plants being inherited from generation to generation. *E. coli* was observed to be internalized through the root system of maize grown in contaminated hydroponic media [62].

In tomato, *Salmonella* survived both in and on tomato throughout plant growth, from inoculation on the stem, to flowering and fruit ripening [63]. In apples, *E. coli* 0157:H7 was observed attached to seed integuments, with infiltration occurring through the blossom's calyx and travelling up the floral tube to the internal parts of apple [64]. Endophytic seed colonization may be employed by bacteria to make the transition across plant generations. An efficient colonization of the interior of a plant depends on the capacity to form a biofilm on the seedling and in the rizosphere.

These observations support the hypothesis that enteric pathogens colonizing produce can spread to the seeds and be transmitted to the next generation. Within this context, plants can be seen as alternative hosts for human pathogens and as vehicles in the colonization of animals once ingested [47].

Apart from the mechanism through which internalization occurs, several studies confirm the presence of enteric pathogens within vegetables' tissues. *E. coli* was observed to be internalized through the root system of maize grown in contaminated hydroponic media [62]. In carrots *E. coli* was found at cell junctions and intracellular spaces within carrot tissue but did not penetrate the carrot cells [65]. *E. coli* and *Salmonella* [66] were found within lettuce. In oranges, *E. coli* 0157:H7 and *Salmonella* were able to colonize and grow in the interior of oranges at 24 °C [67]. The internalization of human pathogens into fruit and vegetable tissues is dependent on temperature, time and pressure.

All the situations of internalization provide pathogens with higher levels of nutrients and protection from desiccation, washing and disinfecting and, at the time of consumption, microbial cells may survive and grow to high population. Moreover, once internalized, cells may form aggregates or even biofilms making their elimination almost impossible to achieve. Internalization is an additional factor that causes an increase of the risk of consuming raw fruits and vegetables.

3.6 Decontamination

Reduction of the microbial load and the inhibition of undesired growth to control human pathogens and spoilage microorganisms on fruits and vegetables may be achieved by utilization of disinfection and methods of decontamination such as: ultraviolet C, modified atmospheres, ozone treatments, acidic or alkaline electrolyzed water, ultrasounds and application of bacteriocins.

The disinfectants most frequently used (sodium hypochlorite, chlorine dioxide, hydrogen peroxide, peracetic acid) are not capable of reducing microbial population by more than 90 or 99% [68, 69]. The lack of efficacy of sanitizers used to decontaminate the surface of raw fruits and vegetables has been attributed to the inability of active components to reach the site of microbial cells during treatments [68]. The capacity to produce biofilms and/or the capacity to become internalized may explain the reason by which the washing and disinfections are not effective. Within this context Beuchat [68] suggested that prevention of contamination at all points of the food chain is preferred over the application of disinfectants.

Combinations of several disinfection agents with weak organic acids (sodium lactate, citric acid, acetic acid) have been reported as good solutions to reduce microbial levels on produce. Some other ways to decontaminate fresh produce may be the application of plant extracts or essential oils [70].

Conclusion

The association between disease cases and foods of plant origin contaminated with human pathogens has been rising in the last years. Outbreaks involving a range of bacteria, viruses and protozoans are increasing due to changing in farming practices, in food processing, in social lifestyle and the internationalization of food trading. Microbial evolution (adaptation to stress and biocides, acquisition of virulence) and the emergence of newly found pathogens also contribute to this fact. Fruits and vegetables can be contaminated with human pathogens from the environment, from animal and human faeces, or through handling, processing and storage, distribution and preparation for consumption. When growing in the field they are exposed to major sources of contamination such as, non treated organic fertilizer (manure) and/or contaminated water. During harvesting and post harvesting different situations may conduce to the contamination of fresh produce with human pathogens. A correlation between poor personal hygiene and foodborne illness has been established, pointing to the importance of inappropriate worker handling as a major way to introduce disease causing microbes on fresh produce. Worker's health and hygiene and good sanitation practices along the production chain are critical in minimizing the potential for contamination of fresh produce.

The survival and/or growth of human pathogens on fresh produce in order to cause illness are determined by the nature of the microorganisms, the food product and environmental conditions in the field and subsequent steps of production and preparation for consumption. If temperature and humidity levels are adequate, human pathogens will proliferate on fruits and vegetables. Additionally survival and growth of foodborne pathogens are enhanced when protective tissue barriers are destroyed by physical damage, inappropriate handling, insects, and post harvest procedures (cutting, slicing, skinning, shredding and others). Some foods cause especial concern such as fresh-cuts and fruit juices, where the natural tissue barriers are destroyed during processing and are not subjected to any treatments to kill or remove the microbial population.

Temperature abuses may favour microbial growth. However, refrigerating temperatures are commonly used as a way of controlling microbes either by retarding or inhibiting growth. Special attention should be given to psychrotrophs, such as non proteolitic *Clostridium botulinum*, *Listeria monocytogenes* and *Yersinia enterocolitica* which are able to proliferate at refrigerating temperatures.

The modified atmosphere packaging (MAP) used to reduce the respiration rate and thus retarding senescence process on produce, can interfere in the microbial growth on fresh produce. Microorganisms respond to MAP differently, depending on the composition of gaseous mixture. The effects of MAP on the microbial ecology of fresh produce are not easily predictable since it changes the growth rate of pathogens (preventing or enhancing).

The presence of human pathogens on/in food is unacceptable. Foods of plant origin should be grown under good agricultural practices and handled under good or even stringent sanitary conditions to avoid introduction of human pathogen microorganisms onto the raw material and into the food-manufacturing environment. Producers, processors, distributors, retailers and consumers must respect the good manufacturing practices in order to prevent contamination and microbial growth. The HACCP may help to assure an adequate management of contaminants present during the manufacturing process. Finally, the prevention of microbial contamination is a better strategy to produce safe food than the implementation of corrective actions after contamination occurs.

Acknowledgment

Thanks are given to Professor Maria Nelma Gaspar for suggestions and for the review of the manuscript.

References

[1] Tauxe, R., Kruse, H., Hedberg, C., Potter, M., Madden, J., & Wachsmuth, K. (1997). Microbial hazards and emerging issues associated with produce: a preliminary report to the National Advisory Committee on Microbiologic Criteria for Foods. *Journal of Food Protection*, *60*, 1400-1408.

[2] Sivapalasingam, S., Friedman, C. R., Cohen, L., & Tauxe, R.V. (2004). Fresh produce: a growing cause of outbreaks of foodborne illness in the United States, 1973 through 1997. *Journal of Food Protection*, *67*, 2342-2353.

[3] Sewell, A.M., & Farber, J.M. (2001). Foodborne outbreaks in Canada linked to produce. *Journal of Food Protection*, *64*, 1863-1877.

[4] European Commission Scientific Committee on Foods (ECSCF) (2002). Risk profile on the microbiological contamination of fruits and vegetables eaten raw. Adapted April 2002 http://europa.eu.int/comm/food/fs/sc/scf/out125_en.pdf).

[5] Beuchat, L.R. (2002). Ecological factors influencing survival and growth of human pathogens on raw fruits and vegetables. *Microbes and Infection*, *4*, 413-423.

[6] Brandl, M.T. (2006). Fitness of human enteric pathogens on plants and implications for food safety. *Annual Review of Phytopathology*, *44*, 367-392.

[7] Calvin, L. (2007). Outbreak linked to spinach forces reassessment of food safety practices. *Amber Waves*, *5*, 24-31.

[8] Little, C. L., & Gillespie, I. A. (2008). Prepared salads and public health. *Journal of Applied Microbiology*, *105*, 1729-1743.

[9] Adams, M.R., & Moss, M.O. (2008). Food Microbiology (3. Ed.). Cambridge: RSC Publishing.

[10] International Commission on Microbiological Specifications for Foods (ICMSF) (2005). Vegetables and Vegetables Products. In Microorganisms in Foods 6: Microbial Ecology of Food Commodities (2 ed., pp. 277-325). SpringerLink (http://www.springerlink.com/content/uvu73v032515p845/).

[11] International Commission on Microbiological Specifications for Foods (ICMSF) (2005). Fruits and Fruits Products. In Microorganisms in Foods 6: Microbial Ecology of Food Commodities (2nd ed., pp. 326-359). SpringerLink.

[12] De Roever, C. (1998). Microbiological safety evaluations and recommendations on fresh produce. *Food Control*, *9*, 321-347.

[13] Koopmans, M., & Duizer, E. (2004). Foodborne viruses: an emerging problem. *International Journal of Food Microbiology*, *90*, 23-41.

[14] Dawson, D. (2005). Foodborne protozoan parasites. *International Journal of Food Microbiology*, *103*, 207-227.

[15] Johnston, L.M., Jaykus, L-A., Moll, D., Martinez, M.C., Anciso, J., Mora, B., & Moe, C.L. (2005). A field study of the microbiological quality of fresh produce. *Journal of Food Protection*, *68*, 1840-1847.

[16] Abadias, M., Usall, J., Anguera, M., Solsona, C., & Viñas, I. (2008). Microbiological quality of fresh, minimally-processed fruit and vegetables, and sprouts from retail establishments. *International Journal of Food Microbiology*, *123*, 121-129.

[17] Mead, P.S., Slutsker, L., Dietz, V., McGaig, L.F., Bresee, J.S., Shapiro, C., Griffin, P.M., & Tauxe, R.V. (1999). Food-related illness and death in the United States. *Emerging Infectious Diseases*, *5*, 607-625.

[18] Watanabe, Y., Ozasa, K., Mermin, J.H., Griffin, P.M., Masuda K., Imashuku, S., & Sawada, T. (1999). Factory outbreak of *Escherichia coli* O157:H7 infection in Japan. *Emerging Infectious Diseases*, *5*, 424-428.

[19] Iversen, C., Lehner, A., Mullane, N., Bidlas, E., Cleenwerck, I., Marugg, J., Fanning, S., Stephan, R., & Joosten, H. (2007). The taxonomy of *Enterobacter sakazakii*: proposal of a new genus *Cronobacter* gen. nov. and descriptions of *Cronobacter sakazakii* comb. nov., *Cronobacter sakazakii* subsp. *sakazakii*, comb. nov. *Cronobacter sakazakii* subsp. *malonaticus* subsp. nov. *Cronobacter turicensis* sp. nov., *Cronobacter muytjensii* sp. nov., *Cronobacter dublinensis* sp. nov. and *Cronobacter* genomospecies 1. *BMC Evolutionary Biology*, *7*, 64.

[20] Soriano, J.M., Rico, H., Moltó, J.C., & Mañes, J. (2001). Incidence of microbial flora in lettuce, meat and Spanish potato omelette from restaurants. *Food Microbiology*, *18*, 159-163.

[21] Ortega, Y. (2005). Foodborne and Waterborne Protozoan Parasites. In P.M., Fratamico, A.K., Bhunia, & J.L., Smith (Eds.), *Foodborne Pathogens: Microbiology and Molecular Biology* (1st ed., pp. 145-162). Norfolk: Caister Academic Press.

[22] Carter, J.M. (2005). Enterically infecting viruses: pathogenicity, transmission and significance for food and waterborne infection. *Journal of Applied Microbiology*, *98*, 1354-1380.

[23] Lamhoujeb, S., Fliss, I., Ngazoa, S.E., & Jean, J. (2008). Evaluation of the persistence of infectious human noroviruses on food surfaces by using real-time nucleic acid sequence-based amplification. *Applied and Environmental Microbiology*, *74*, 3349-3355.

[24] Buck, J.W., Walcott, R.R., & Beuchat, L.R. (2003). Recent trends in microbiological safety of fruits and vegetables. [On-line serial]. Plant Health Progress.

[25] Naimi, T.S., Wicklund, J.H., Olsen, S.J., Krause, G., Wells, J. G., Bartkus, J.M., Boxrud, D.J., Sullivan, M., Kassenborg, H., Besser, J.M., Mintz, E.D., Osterholm, M.T., & Hedberg, C.W. (2003). Concurrent outbreaks of *Shigella sonnei* and enterotoxigenic *Escherichia coli* infections associated with parsley: implications for surveillance and control of foodborne illness. *Journal of Food Protection*, *66*, 535-541.

[26] Cook, N., Nichols, R.A.B., Wilkinson, N., Paton, C.A., Barker, K., & Smith, H.V. (2007). Development of a method for detection of *Giardia duodenalis* cysts on lettuce and for simultaneous analysis of salad products for the presence of *Giardia* cysts and *Cryptosporidium* oocysts. *Applied and Environmental Microbiology*, *73,* 7388-7391.

[27] Kirov, S.M. (2001). *Aeromonas* and *Plesiomonas* species. In M.P., Doyle, L.R., Beuchat, & T.J., Montville (Eds.), *Food Microbiology: Fundamentals and Frontiers* (2nd ed., pp. 301-328). Washington, D.C: ASM Press.

[28] Granum, P.E. (2005). *Bacillus cereus*. In P.M., Fratamico, A.K., Bhunia, & J.L., Smith (Eds.), *Foodborne Pathogens: Microbiology and Molecular Biology* (1st ed., pp. 409-420). Norfolk: Caister Academic Press.

[29] Nachamkin, I., & Guerry, P. (2005). *Campylobacter* infections. In P.M., Fratamico, A.K., Bhunia, & J.L., Smith (Eds.), *Foodborne Pathogens: Microbiology and Molecular Biology* (1st ed., pp. 285-294). Norfolk: Caister Academic Press.

[30] Novak, J.S., Peck, M.W., Juneja, V.K., & Johnson, E.A. (2005). *Clostridium botulinum* and *Clostridium perfringens*. In P.M., Fratamico, A.K., Bhunia, & J.L., Smith (Eds.), *Foodborne Pathogens: Microbiology and Molecular Biology* (1st ed., pp. 383-408). Norfolk: Caister Academic Press.

[31] Smith, J.L., & Fratamico, P.M. (2005). Diarrhea-inducing *Escherichia coli*. In P.M., Fratamico, A.K., Bhunia, & J.L., Smith (Eds.), *Foodborne Pathogens: Microbiology and Molecular Biology* (1st ed., pp. 357-382). Norfolk: Caister Academic Press.

[32] Paoli, G.C., Bhunia, A.K., & Bayles, D.O. (2005). *Listeria monocytogenes*. In P.M., Fratamico, A.K., Bhunia, & J.L., Smith (Eds.), *Foodborne Pathogens: Microbiology and Molecular Biology* (1st ed., pp. 295-326). Norfolk: Caister Academic Press.

[33] Andrews, H.L., & Bäumier, A.J. (2005). *Salmonella* Species. In P.M., Fratamico, A.K., Bhunia, & J.L., Smith (Eds.), *Foodborne Pathogens: Microbiology and Molecular Biology* (1st ed., pp. 327-340). Norfolk: Caister Academic Press.

[34] Lampel, K.A. (2005). *Shigella* Species. In P.M., Fratamico, A.K., Bhunia, & J.L., Smith (Eds.), *Foodborne Pathogens: Microbiology and Molecular Biology* (1st ed., pp. 340-356). Norfolk: Caister Academic Press.

[35] Stewart, G.C. (2005). *Staphylococcus aureus*. In P.M. Fratamico, A.K., Bhunia, & J.L., Smith (Eds.), *Foodborne Pathogens: Microbiology and Molecular Biology* (1 ed., pp. 273-284). Norfolk: Caister Academic Press.

[36] Nishibuchi, M., & DePaola, A. (2005). *Vibrio* species. In P.M., Fratamico, A.K., Bhunia, & J.L., Smith (Eds.), *Foodborne Pathogens: Microbiology and Molecular Biology* (1 ed., pp. 251-272). Norfolk: Caister Academic Press.

[37] Nesbakken, T. (2005). *Yersinia enterocolitica*. In P.M., Fratamico, A.K., Bhunia, & J.L., Smith (Eds.), *Foodborne Pathogens: Microbiology and Molecular Biology* (1 ed., pp. 227-250). Norfolk: Caister Academic Press.

[38] Ibenyassine, K., Mhand, R.A., Karamoko, Y., Anajjar, B., Chouibani, M.M., Ennaji, M. (2007). Bacterial pathogens recovered from vegetables irrigated by wastewater in Morocco. *Journal of Environmental Health*, *69*, 47-51.

[39] Loncarevic, S., Johannessen, G.S., & Rørvik, L.M., (2005). Bacteriological quality of organically grown leaf lettuce in Norway. *Letters in Applied Microbiology*, *41*, 186-189.

[40] Melloul, A.A., Hassani, L., & Rafouk, L. (2001). *Salmonella* contamination of vegetables irrigated with untreated wastewater. *World Journal of Microbiology and Biotechnology*, *17*, 207-209.

[41] Islam, M., Doyle, M.P., Phatak, S.C., Millner, P., Xiuping, J. (2005). Survival of *Escherichia coli* O157:H7 in soil and on carrots and onions grown in fields treated with contaminated manure composts or irrigation water. *Food Microbiology*, *22*, 63-70.

[42] Muirhead, R.W., Collins, R.P., & Bremer, P.J. (2006). Interaction of *Escherichia coli* and soil particles in runoff. *Applied and Environmental Microbiology*, *72*, 3406-3411.

[43] Sela, S., Nestel, D., Pinto, R., Nemny-Lavy, E., & Bar-Joseph, M. (2005). Mediterranean fruit fly as a potential vector of bacterial pathogens. *Applied and Environmental Microbiology*, *71*, 4052-4056.

[44] Gibbs, D.S., Anderson, G.L., Beuchat, L.R., Carta, L.K., & Williams, P.L. (2005). Potential role of *Diploscapter* sp. strain LKC25, a bacteriovorous nematode from soil, as a vector of food-borne pathogenic bacteria to preharvest fruits and vegetables. *Applied and Environmental Microbiology*, *71*, 2433-2437.

[45] Gourabathini, P., Brandl, M.T., Redding, K.S., Gunderson, J.H., Berk, S.G. (2008). Interactions between food-borne pathogens and protozoa isolated from lettuce and spinach. *Applied and Environmental Microbiology*, *74*, 2518-2525.

[46] Brandl, M.T., & Amundson, R. (2008). Leaf age a risk factor in contamination of lettuce with *Escherichia coli* O157:H7 and *Salmonella enterica*. *Applied and Environmental Microbiology*, *74*, 2298-2306.

[47] Tyler, H.L., & Triplett, E.W. (2008). Plants as a habitat for beneficial and /or human pathogenic bacteria. *Annual Review of Phytopathology*, *46*, 53-73.

[48] Ukuku, D.O., & Sapers G.M. (2001). Effect of sanitizer treatments on *Salmonella* Stanley attached to the surface of cantaloupe and cell transfer to fresh-cut tissues during cutting practices. *Journal of Food Protection*, *64*, 1286-1291.

[49] Gagliardi, J.V., Millner, P.D., Lester, G., & Ingram, D. (2003). On farm and postharvest processing sources of bacterial contamination to melon rinds. *Journal of Food Protection*, *66*, 82-87.

[50] Prazak, A.M., Murano, E.A., Mercado, I., & Acuff, G.R. (2002). Prevalence of *Listeria monocytogenes* during production and postharvest processing of cabbage. *Journal of Food Protection*, *65*, 1728-1734.

[51] Danhorn, T., & Fuqua, C. (2007). Biofilm formation by plant-associated bacteria. *Annual Review of Microbiology*, *61*, 401-422.

[52] Costerton, J.W., Stewart, P.S., & Greenberg, E.P. (1999). Bacterial biofilms: a common cause of persistent infections. *Science*, *284*, 1318-1322.

[53] Fett, W.F. (2000). Naturally occurring biofilms on alfalfa and other types of sprouts. *Journal of Food Protection*, *63*, 625-632.

[54] Morris, C.E., Monier, J.M., & Jacques, M.A. (1998). A technique to quantify the population size and composition of the biofilm component in communities of bacteria in phyllosphere. *Applied and Environmental Microbiology*, *64*, 4789-4795.

[55] Rayner, J., Veeh, R., & Flood, J. (2004). Prevalence of microbial biofilms on selected fresh produce and household surfaces. *International Journal of Food Microbiology*, *95*, 29-39.

[56] Annous, B.A., Solomon, E.B., Cooke, P.H., & Burke, A. (2005). Biofilm formation by *Salmonella* spp. on cantaloupe melons. *Journal of Food Safety*, *25*, 276-287.

[57] Fett, W.F., & Cooke, P.H. (2005). A survey of native microbial aggregates on alfalfa, clover and mung bean sprout cotyledons for thickness as determined by confocal scanning laser microscopy. *Food Microbiology*, *22*, 253-259.

[58] Scher, K., Romling, U., & Yaron, S. (2005). Effect of heat, acidification, and chlorination on *Salmonella enterica* serovar Typhimurium cells in a biofilm formed at the air-liquid interface. *Applied and Environmental Microbiology*, *71*, 1163-1168.

[59] Lapidot, A., Romling, U., & Yaron, S. (2006). Biofilm formation and the survival of *Salmonella* Typhimurium on parsley. *International Journal of Food Microbiology*, *109*, 229-233.

[60] Norwood, D.E., & Gilmour, A. (2000). The growth and resistance to sodium hypochlorite of *Listeria monocytogenes* in a steady-state multispecies biofilm. *Journal of Applied Microbiology*, *88*, 512-520.

[61] Babic, I., Roy, S., Watada, A.E., & Wergin, W.P. (1996). Changes in microbial population on fresh cut spinach. *International Journal of Food Microbiology*, *31*, 107-119.

[62] Bernstein, N., Sela, S., Pinto, R., & Ioffe, M. (2007). Evidence for internalization of *Escherichia coli* into the aerial parts of maize via the root system. *Journal of Food Protection*, *70*, 471-475.

[63] Guo, X., Chen, J., Brackett, R.E., & Beuchat, L.R. (2001). Survival of Salmonellae on and in tomato plants from the time of inoculation at flowering and early stages of fruit development through fruit ripening. *Applied and Environmental Microbiology*, *67*, 4760-4764.

[64] Burnett, S.L., Chen, J., & Beuchat, L.R. (2000). Attachment of *Escherichia coli* O157:H7 to the surfaces and internal structures of apples as detected by confocal scanning laser microscopy. *Applied and Environmental Microbiology*, *66*, 4679-4687.

[65] Auty, M., Duffy, G., O'Beirne, D., McGovern, A., Gleeson, E., & Jordan, K. (2005). In situ localization of *Escherichia coli* O157:H7 in food by confocal scanning laser microscopy. *Journal of Food Protection*, *68*, 482-486.

[66] Franz, E., Visser, A.A., van Diepeningen, A.D., Klerks, M.M., Termorshuizen, A.J., & van Bruggen, A.H.C. (2007). Quantification of contamination of lettuce by GFP-expressing *Escherichia coli* O157:H7 and *Salmonella enterica* serovar Typhimurium. *Food Microbiology*, *24*, 106-112.

[67] Eblen, B.S., Walderhaug, M.O., Edelson-Mammel, S., Chirtel, S.J., De Jesus, A., Merker, R.I., Buchauan, R.L., & Miller, A.J. (2004). Potential for internalization, growth, and survival of *Salmonella* and *Escherichia coli* O157:H7 in oranges. *Journal of Food Protection*, *67*, 1578-1584.

[68] Beuchat, L.R. (1998). Surface decontamination of fruits and vegetables eaten raw: a review. Food Safety Unit, World Health Organization, WHO/FSF/98.2, pp. 1-42.

[69] Brackett, R.E. (1999). Incidence, contributing factors, and control of bacterial pathogens in produce. *Postharvest Biology and Technology*, *15*, 305-311.

[70] Allende, A., Tomás-Barberán, F.A., & Gil, M.I. (2006). Minimal processing for healthy traditional foods. *Trends in Food Science and Technology*, *17*, 513-519.

In: Practical Food and Research
Editor: Rui M. S. Cruz, pp. 89-113

ISBN: 978-1-61728-506-6

Chapter IV

TEXTURE AND MICROSTRUCTURE

Netsanet S. Terefe and Cornelis Versteeg
CSIRO Food Science and Nutrition, 671 Sneydes Road, Werribee, VIC 3030, Australia,
Netsanet.Shiferawterefe@csiro.au and Kees.versteeg@csiro.au

ABSTRACT

Texture of fruits and vegetable products is fundamentally related to their structure, of which the cell wall polymers form the basic structural units. The structural integrity (and texture) of plant foods can be mainly attributed to the primary cell wall, the middle lamella and the turgor generated within cells by osmosis. The basic structure of the primary cell wall consists of a cellulose-hemicellulose network with pectic polymers interwoven with this network. The network of these three polysaccharides forms the basis for the structural integrity of the cell. As other quality attributes, food processing and preservation processes affect texture. Postharvest processing and storage may affect both the structural integrity of the cell and the biochemical composition of the cell wall. During most food processing operations, the structural integrity of the cell is lost leading to loss of turgor pressure and crispiness. On the other hand, the three different component polysaccharides of plant cell wall respond differently to postharvest storage and processing giving rise to two components of the firmness of plant materials. The firmness generated by the cellulose-hemicellulose domain of plant cell walls is not significantly affected by processing or storage while the pectin component is affected both by enzyme catalyzed reactions and β-eliminative degradation during thermal processing. This chapter deals with the effects food processing operations on the texture and microstructure of fruit and vegetable products. Thus, the structure and microstructure of plant materials as related to texture, the role of enzymes on cell wall biochemistry and the effects of the most important conventional and emerging processing techniques on the structural integrity of plant materials and cell wall biochemistry are discussed.

4.1 INTRODUCTION

Texture is an important quality attribute of foods. Sensorial texture is hard to define precisely, since it arises from a complex manifestation of perceptions by the senses of touch, vision, hearing and kinaesthesia (the sensation of presence, movement, and position resulting from the stimulation of nerve endings) [1]. Consumers use descriptive terminologies such as 'crispy', 'crunchy', 'pulpy', and 'juicy' to describe the texture of food materials, which are highly subjective. Bourne [2] defined texture as the group of physical characteristics that arise from the structural elements of food that are sensed by the feelings of touch, are related to deformation, disintegration and flow of food under force, and measured objectively by functions of mass, time and distance [2]. Texture of fruits and vegetable products is fundamentally related to their structure, of which the cell wall polymers form the basic structural units. The structural integrity (and texture) of plant foods can be mainly attributed to the primary cell wall, the middle lamella and the turgor generated within cells by osmosis [3]. Depending on the produce, special compounds within the cell like starch, overall structure and shape of individual cells and the structure and shape of tissues such as the presence of strong vascular tissue may also contribute to texture [4].

The basic structure of the primary cell wall consists of a cellulose-hemicellulose network with pectin interwoven with this network. The network of these three polysaccharides forms the basis for the structural integrity of the cell. Pectin, as the main constituent of the middle lamella, cements cell walls together and gives firmness and elasticity to tissues [5]. Postharvest processing and storage may affect both the structural integrity of the cell and the biochemical composition of the cell wall. During most food processing operations, the structural integrity of the cell is lost leading to loss of turgor pressure and 'fresh' texture. The three different component polysaccharides of plant cell wall respond differently to postharvest storage and processing giving rise to two components of the firmness of plant materials. The firmness generated by the cellulose-hemicellulose domain of plant cell walls is hardly affected by processing or storage while the pectin component is affected both by enzyme catalyzed reactions and β-eliminative breakdown during thermal processing [4]. This chapter deals with the effects food processing operations on the texture and microstructure of fruit and vegetable products. Thus, the structure and microstructure of plant materials as related to texture, the role of enzymes on cell wall biochemistry and the effects of the most important conventional and emerging processing techniques on the structural integrity of plant materials and cell wall biochemistry are discussed. Furthermore, the methods that are commonly used in the evaluation of the textural quality of plant materials are briefly reviewed.

4.2 The Structural and Biochemical Basis of Texture of Plant Materials

Texture is fundamentally related to structure [6]. The texture of plant materials is derived from a structural hierarchy with the physicochemical properties exhibited at each level dependent on the properties of the preceding level, their relative proportion and the physical forces involved in their interaction and the spatial arrangements of the different elements. The basic structural unit of plant materials is the cell. Cells are organised into tissues. Different types of tissue may combine to form the edible part of a plant i.e. fruits, leaves, nuts, grains, tubers or seeds. The type of cells that constitute the tissue, the way the cells are organised and the proportion of intercellular air space, the turgor generated within the cells by osmosis and the composition of the polysaccharides that constitute the cell wall all contribute to the texture of fruits and vegetable products [1, 3].

4.2.1 The Plant Cell

Plant cells are classified into several different groups depending on their structure, function and shape. Parenchyma cells, which are thin walled and primarily devoted to the manufacture and accumulation of food materials, the epidermal cells which protect the parenchyma cells, the guard cells of the stomata, the xylem and phloem cells of the vascular tissue and sclerenchyma and collenchyma cells which are thick walled and provide rigidity and strength to the stems. Most of the edible part of fruits and vegetables consist of parenchyma cells. However there are some exceptions as in the case of asparagus spear which consists of vascular tissue and sclerenchyma cells typical of an immature stem. Parenchyma cells are generally isodiametric polyhedrons with diameters ranging from 50 to 500 μm except in products such as citrus where they are larger and less regularly shaped [1, 3]. Figure 4.1 shows the schematic representation of a parenchyma cell.

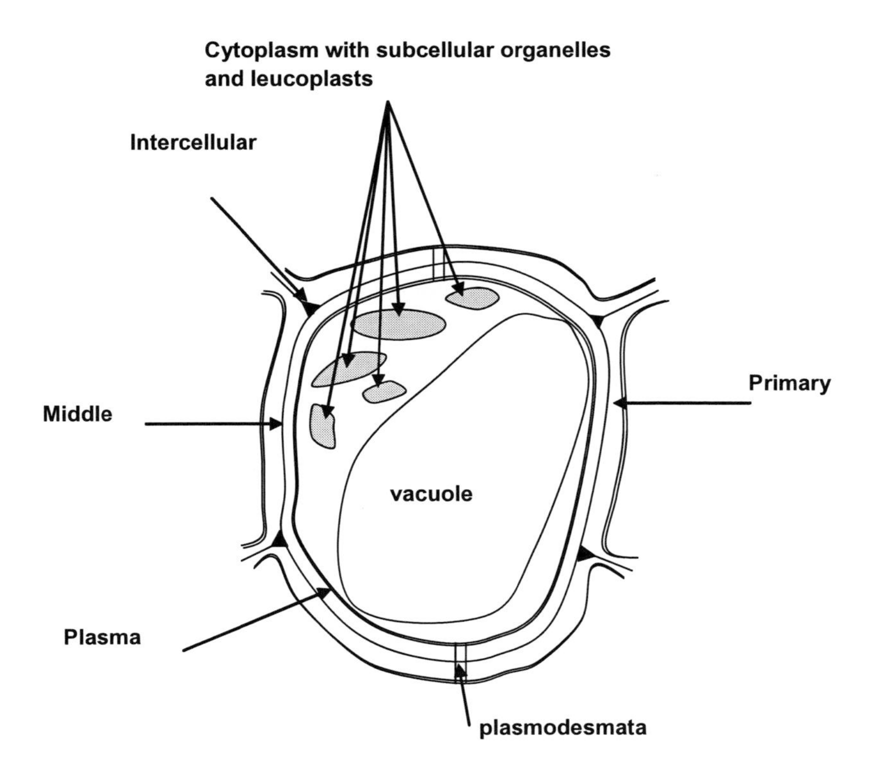

Figure 4.1. Schematic representation of a parenchyma cell.

Each cell is surrounded by a cell wall, which is permeable to water. The semi permeable plasma membrane (plasmalemma) lies within the cell wall and serves as a primary barrier to the movement of water and solutes into and out of the cell. Inside the cell are a large central vacuole, which contains much of the water in the cell together with sugars, acids and salts and the cytoplasm. The cytoplasm is an organic fluid, in which are found various subcellular organelles which control cell metabolism, storage structures called leucoplasts, which contain food materials such as lipids, starch or protein granules and a matrix of organised polymeric proteins which is commonly known as the cytoskeleton. The vacuole and the different subcellular organelles are surrounded by semi-permeable lipoprotein membranes. During the latter stages of plant growth, secondary cell walls may form inside the primary cell walls giving additional strength and turgidity to the cell. However, their formation is limited to scelerenchyma cells. Lignin may also impregnate the cell walls and the intracellular substance of mature cells. Nevertheless, lignification occurs only in specific types of cells such as the stone and fibre cells of fruits and vegetables and the xylem of woody plants. Adjacent plant cells are cemented together by the middle lamella, which is an amorphous layer containing pectic materials. In immature fruits, the middle lamella is comprised of water-insoluble protopectin, which has high molecular weight and imparts strength to the tissue. During ripening, the long chain protopectin molecules are enzymatically converted into short chain pectin molecules, soluble in water resulting in cell separation and softening [1].

The main factors that determine the texture of parenchymatous tissue are the mechanical strength of the cell wall, the cell to cell adhesion (which is largely determined by the pectin matrix in the middle lamella) and turgor pressure. The turgor pressure is the hydrostatic pressure generated against the cell wall, which arises due to osmotically driven water influx into the cell through the semi-permeable cell membrane. In non-growing cells, the turgor generated against the wall is balanced by the inward counterforce exerted by the cell wall on the cell contents. As water enters the cell, the stress in the wall increases and the elastic energy stored in the strained polymer bonds compresses the protoplasm preventing change in volume resulting in turgor pressure of typically 3 to 4 atmospheres. In growing tissue, when the turgor pressure exceeds a threshold value, the wall expands irreversibly resulting in an increase in cell volume and cell growth [7]. In the absence of turgor pressure, the load bearing capacity of the thin cell walls of parenchyma cells is relatively low [8]. Tissues that contain turgid cells are perceived as fresh and crispy. They are also characterised by greater stiffness and lower toughness compared to flaccid tissues that contain cells of low turgidity [1]. Thus, in plant tissues primarily composed of parenchyma cells, the absorption and retention of water as a result of osmosis can increase the turgor pressure leading to an increase in crispiness or 'freshness'. During mastication of edible plant tissues, tissue failure may occur due to cell separation, cell breakage or a combination of the two. The mode of tissue failure determines the perceived texture of the tissue, which is in turn determined by the relative strength of the cell wall and the cell to cell adhesion. If cell to cell adhesion is stronger than the cell wall, tissue failure occurs in the cell walls resulting in the rupture of the cell and the release of tissue contents. This occurs in uncooked vegetables and fresh 'juicy' fruits of optimal maturity. In such tissues, mastication increases the release of flavour and their high turgor pressure makes them to be perceived as brittle and crispy. On the other hand, if the cell wall is stronger than cell to cell adhesion, failure occurs through cell separation [1, 9]. This occurs in cooked or senescent tissues. The texture of such tissues is perceived as soft and pulpy.

Factors such as intercellular air space and the types of cells that constitute a given tissue may also affect the textural characteristics of fruits and vegetables. Plant tissues contain a significant amount of intercellular air space, which is dependent on the type and shape of the constituent cells and their spatial arrangement. For instance, the intercellular space in apples ranges from 20 to 25% while it is 15% in peach and 1% in potato. The difference in the ratio of intercellular space of different fruits and vegetables gives rise to differences in texture. For instance, fruits with a large proportion of intercellular space such as apples have spongy texture. The texture of fruit and vegetable products may also be affected by the development of other types of cells. Celery and mango, which are comprised of a higher proportion of phloem and xylem cells, have distinct textural characteristics. The formation of sclereids in pears during the maturation of the fruit causes the slightly gritty texture of pear. The development of secondary cell walls in plant tissue as in the case of asparagus is associated with 'woodiness'. Lignification of vascular elements of fruits and vegetables that are allowed to stay on the plant past their maturity, make them 'tough' and more difficult to chew [10].

4.2.2 Cell Wall Composition and Architecture

The chemical composition and the physical structure of the cell wall play a major role in the texture of edible fruits and vegetables [11]. The plant cell wall is a highly complex structure composed of chemically and physically interacting polysaccharides and structural proteins. It is involved in almost all aspects of the developmental activities of the plant including the regulation of cell growth and division in a growing tissue, intercellular signalling and transport, protection against biotic and abiotic stresses and storage of food reserves in addition to its obvious function of providing support and shape to the cell [1]. The structure and composition of the cell wall varies depending on the type of cell and its stage of maturity. There is also a spatial variation in composition within individual cell walls since the cell wall consists of different structural components such as the plasmodesmata and the middle lamella with different functionalities. The cell wall is composed of three main polysaccharides associated with varying amount of proteins and phenolic compounds and form the underlying building block in the structural hierarchy of plant based products. The three main cell wall polysaccharides are cellulose, hemicelluloses (xyloglucan, xylan, mixed glucans), and pectins (homogalacturonan, rhamnogalacturonan I and II).

Cellulose, β(1→4)-D-polyglucan, forms about 30% of the dry weight of the cell walls of dicotlydones. Cellulose microfibrils of about 5 to 15nm diameter and 2000 to 6000 units long form the skeletal scaffolding of the cell wall matrix. Hemicelluloses are ridged and highly branched rod-shaped polymers of neutral sugars such as xylan, xyloglucan and β(1→3) and β(1→4) mixed glucans, which are 200nm in length. They link with cellulose, pectin and lignin by hydrogen bonding. Hemicelluloses comprise about 30% of the dry weight of the cell wall of dictoyledons. Pectins are present as the polygalacturonase (PG)-labile α-galacturonic acid residues (homoglacturonans) and PG-resistant rhamnogalacturonans of varying degree of polymerisation. They form about 35% of the dry weight the cell wall of dictoyledons. Most of the pectin is found in the middle lamella. The cell wall also contains significant proportion of proteins including enzymes and structural proteins [1, 3] The structural proteins are mainly glycoproteins, which comprise about 5 to 10% of the dry weight of the cell wall of dictoyledons. The most abundant structural glycoproteins are those that are rich in hydroxyproline including extensins and arabinogalactan proteins. They form cross-links with other cell wall polymers and are thought to contribute to the structural integrity of the cell wall. These proteins contain tyrosine, which provides the possibility for intermolecular and intramolecular cross-linking through the formation of isodityrosine by peroxidation [1]. The extensins are considered to be involved in locking the cell wall into fixed shape after growth is completed [3].

The phenolic compounds in the cell wall can be broadly classified into two; lignin and phenolic esters. Lignin, if present, provides compressive strength and support to tissues [1]. However, at optimal maturity, lignification does not occur in most edible fruit and vegetable tissues. Simple phenolic esters such as cinnamic acids are found in the cell wall of many vegetables attached to wall polysaccharides, which can form covalent crosslinks through peroxidation providing additional strength to the tissue. There are also other phenolic compounds in the cell wall with unknown functions [1, 11].

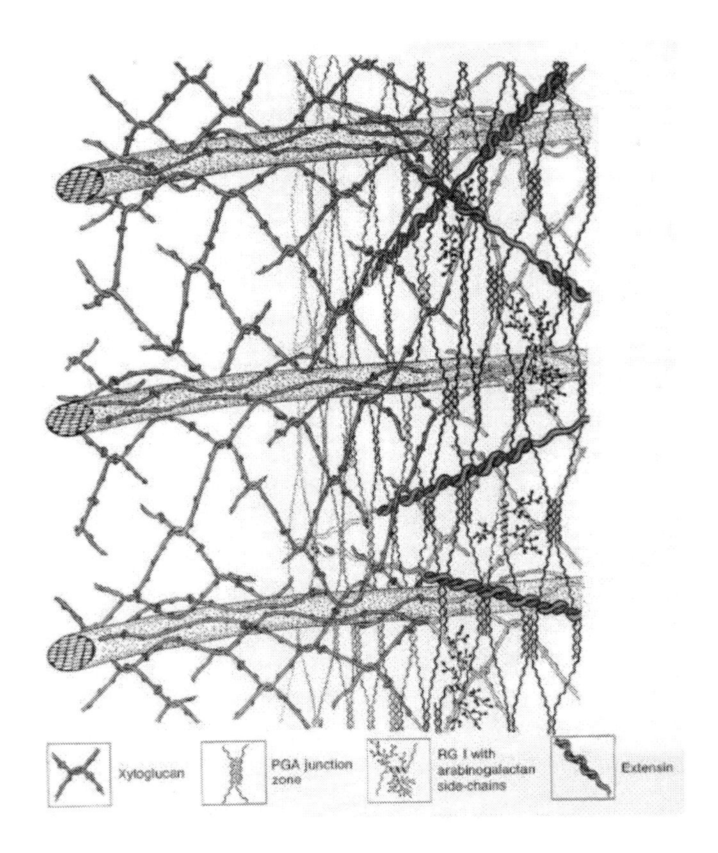

Figure 4.2. A structural model for a single layer of a growing cell wall of flowering plants reproduced from Carpita and Gibeaut [12].

Major insights into the functionality of the different polysaccharides have been gained in recent times through sequential extraction of polysaccharides. Advances in microscopic techniques have also led to better understandings of the cell wall microstructure. Based on these new insights, a number of models have been proposed to describe the cell wall architecture [3, 11, 12]. One such model is the generalised cell wall model for a growing cell proposed by Carpita and Gibeaut [12]. In this model, the cell wall is depicted as a composite consisting of three independent but interacting structural domains. The first domain consists of cellulose-hemicellulose framework which is embedded in a second domain of cross-linked pectic polysaccharides. The pectin matrix may interact with the cellulose-hemicellulose framework.

However, it may also form cross-linking without interacting with the cellulose-hemicellulose framework. Pectin cross-linking usually occurs by the formation of Ca^{2+} bridges formed between adjacent de-esterified polygalacturonic acid units. It may also occur through oxidative coupling of phenolic moieties such as ferulic acid. The third domain consists of the structural proteins of which the extensins are the major components. The extensins can also form intermolecular bridges with other proteins without necessarily binding to the polysaccharides. The cross-linking of extensins is believed to be one of the mechanisms through which the primary cell wall is locked into shape once growth is completed [3, 12]. The schematic representation of this cell wall model is given in Figure 4.2. This model agrees with insights gained from investigations on the mechanism of cell growth and expansion. However, considering the complexity of the cell wall structure, it is highly simplified. The role of cell wall metabolism and the location of the various enzymes involved in the synthesis, cross-linking, hydrolysis and turnover of the different polysaccharides are not explained. The role of the cytoskeleton and the various cell wall structural proteins is not fully described. In addition, liginification in mature tissue is not considered in the model making it incomplete [3]. Nevertheless such models are very useful in visualising the relationship between cell wall microstructure and texture of plant based food products.

4.2.3 Enzymes Involved in the Biochemical Evolution of the Cell Wall

The main enzymes which are believed to be responsible for the evolution of textural properties of plant materials during processing and storage are pectin methylesterase, polygalacturonase and peroxidase. The synergistic activity of pectin methylesterase and polygalacturonase results in pectin modification and subsequent change in texture. The consistency and cloud stability of fruit and vegetable juices is also affected by the activity of pectin methylesterase and polygalacturonase. Peroxidase is also considered to be important in the modification of the texture of plant materials. Peroxidase catalyses the oxidation of cinnamic acids and tyrosin containing cell wall proteins, which are involved in the oxidative cross linking of cell wall polymers [4].

4.2.3.1 Pectin methylesterase

Pectin methylesterase (PME) (EC 3.1.1.11) catalyzes the de-esterification of pectin to acidic pectin with a lower degree of esterification and methanol. PME is found in all higher plants and is also produced by phytopathogenic fungi and bacteria. The physiological role of PME is well established. PME is involved in fruit ripening as well as cell wall extension during cell growth [13]. PME exists in several isoforms differing greatly in molecular weight, isoelectric point, biochemical activity and stability. For instance up to ten different PME isoenzymes have been detected in grape fruit [14]. The activity of pectin methylesterase (PME) affects pectin and the firmness of plant tissue in three different ways. Plant PME demethylates pectin blockwise, which increases the probability that two adjacent polygalacturonic polymer chains form cross-linkings in the presence of divalent cations such as calcium leading to an apparent increase in firmness. The second effect is that demethylated pectin resists β-eliminative degradation of pectin. These two effects are the basis for the application of low temperature (50-80 °C) blanching with or with out calcium infiltration pre-treatments to reduce tissue softening during subsequent thermal processing, since the treatment activates PME resulting in increased pectin demethylation and pectin cross-linking [4, 5, 15, 16]. On the other hand demethylated pectin is the preferred substrate for the action of polygalacturonase (PG). PG attacks the glycosidic linkages between adjacent demethylated galacturonic acid units of pectin resulting in pectin degradation and consequent decrease in firmness. Thus low temperature blanching may not be suitable for fruits and vegetables with significant PG activity since the slightly elevated temperature may also activate PG in addition to PME potentially resulting in enhanced textural degradation. A suitable alternative in this regard is high pressure processing. PME's from several sources including tomato [17], carrot [18, 19], banana [20], plums [21] and strawberry [22] have been found to be highly resistant to high pressure inactivation while PG (at least in tomato and Chinese cabbage) have been found to be pressure labile [23, 24]. In addition the PME catalysed de-esterfication of pectin is accelerated under pressure [25-27]. A detailed discussion of high pressure processing and its effect on texture related enzymes is given in section 4.6.

4.2.3.2 Polygalacturonase

Polygalacturonase (PG) (EC 3.2.1.15) catalyses the cleavage of the α-(1-4) glycosidic bonds between two galacturonic acid residues in pectin resulting in pectin depolymerisation. As mentioned in section 2.3.1, the preferred substrate for the action of PG is demethylated pectin produced by the action of pectin methylesterase. The synergistic action of PME and PG results in the modification of pectin leading to the degradation of textural quality of many plant materials during processing. The same phenomenon is responsible for the softening of plant tissues during ripening and senescence [13]. PG is a cell wall bound enzyme, which is present in many fruits and vegetables. It is also produced by plant fungal and bacterial pathogens, which plays a major role in plant pathogenesis. Polygalacturonase exists in many isoforms with variable substrate preference, specific activity, optimal pH, and stability. For example, polygalacturonase in tomato exists as a mixture of two isoenzymes; PG1 and PG2. PG2 is heat labile which is totally inactivated after 5 min treatment at 65 °C. PG1 is a dimer formed from a structural association of PG2 with a heat stable glycoprotein β subunit, which confers heat stability on it [28].

4.2.3.3 Peroxidase

Peroxidase (EC 1.11.1.7) is found in almost all living organisms. Its principal physiological function is to control the level of peroxides generated in oxygenation reactions to avoid excessive formation of radicals which are harmful to all living organisms [4]. Peroxidase catalyses single-electron oxidation of a wide variety of compounds in the presence of hydrogen peroxide [29]. Plant peroxidase consists of a complex spectrum of isoenzymes existing both in soluble and bound forms with different substrate specificity and functionality [4]. Peroxidases are involved in colour and flavour degradation of horticultural products. Peroxidase catalyses the oxidation of phenolic compounds in the presence of hydrogen peroxide leading to the formation of brown degradation products.

In general due to the degradation of pectin and the resulting cell separation, significant tissue softening occurs during thermal processing of parenchyma-rich plant tissues. However there are some exceptions. No significant softening occurs during cooking of some vegetables including Chinese water chestnut and to a lesser extent sugar beet and beet root. This is mainly attributed to the activity of cell wall-bound peroxidase, which catalyses the oxidation of phenolic compounds such as ferulic acid in the cell wall in the presence of peroxides. This results in the formation of ferulic acid cross-links between cell wall polysaccharides enhancing cell to cell adhesion [1, 11, 30]. The role of peroxidase has been confirmed in beet root where incubation of beet root tissues in the presence of hydrogen peroxide resulted in much less rate of textural degradation during subsequent thermal processing compared to tissues that were incubated without hydrogen peroxide [30]. Thus peroxidase mediated phenolic cross-linking can potentially be used for modulating the texture of plant based products. For this to happen, phenolic moieties such as ferulic acid, which are capable of forming crosslinks, must be present at appropriate location in the cell wall for instance at the perimeters of cell faces, attached to pectic polysaccharides. Furthermore, peroxidase and hydrogen peroxide or enzymes involved in the production of hydrogen peroxide must be present in the tissue [11]. Peroxidases are ubiquitously found in plant cell walls. Nevertheless, the activity of peroxidase is limited by the availability of hydrogen peroxide. Hydrogen peroxide is generated as a by-product of normal plant metabolism and as a response to physical and biological stress [31]. It is also synthesised through several biochemical path ways including those involving plasmalemma-localised NADPH oxidase and cell wall peroxidases [32]. Except in the case of *Beta vulgaris* (sugar beet and beetroot), the level of ferulic acid dimers attached to pectin in the cell walls of dicotyledons is much less than what is found in monocotyledons. Genetic engineering of phenolic metabolism may be employed to increase the synthesis and incorporation of ferulic acid moieties in the cell wall of dicotyledons [11].

Tyrosine rich structural proteins in the cell wall are also capable of forming peroxidase mediated cross-linking in the cell wall. Oxidative cross-linking of cell wall proteins and the consequent toughening of the cell wall is one of the defence mechanisms of plants to invasion by pathogens. Oxidative burst, which is one of the early events that follow elicitation, leads to the accumulation of hydrogen peroxide in the tissue. This activates the peroxidase catalysed oxidation of tyrosine rich structural proteins in the cell wall resulting in increased cross-linking and a stiffer tissue [31]. Clearly, further research is needed to fully utilise the potential for enhancing the textural quality of plant based products through modulation of peroxidase mediated crosslinking of phenolic compounds in the cell wall. This must also envisage a strategy for controlling the reaction since peroxidase catalysed reaction also causes undesirable quality changes such as browning.

4.3 Evaluation of Texture

Both sensory and instrumental measuring techniques are commonly used in the evaluation of food texture. Sensory evaluation can be either 'hedonic' or 'analytical'. In the hedonic testing, largely untrained consumers state their opinion on the level of acceptability of a sample. In analytic assessment, trained panels provide more objective assessment of the quality attributes of the product. In general, instrumental measuring techniques are preferred to sensory evaluations in many commercial and research applications since instruments reduce variations resulting from bias among individuals, are more precise and provide a common language among researchers, industry and consumers. Nevertheless, instrumental measurements need to be meaningful and related to sensory textural attributes as consumers are the ultimate judges of quality. Another important issue in relation to the evaluation of texture and other quality parameters of fruits and vegetable products is the typically very large sample to sample variability. Therefore, suitable statistical approaches need to be used to determine the number of pieces required per sample as well as the number of samples required per lot so as to ensure representative sampling [33].

Instrumental texture measurement techniques are mainly based on evaluation of mechanical properties. Food products including fruits and vegetables undergo deformation in response to applied force. The amount of force required to produce a given amount of deformation can be used for quantitative evaluation of the texture of food materials [10]. The nature of the deformation depends on several factors including the rate at which the force is applied, the previous loading history and the structure, composition and moisture content of the material. This forms the basis for most instrumental texture evaluation techniques. Instrumental texture evaluation techniques can be broadly classified into two; destructive and non-destructive. The destructive measuring techniques include compression, shear, extrusion, and tensile tests. Fruits and vegetable tissues show viscoelastic behaviour under mechanical loading. Thus, both the applied force and the rate at which the force is applied determine the observed deformation. This is more so for soft fruits such as tomatoes, cherries, berry fruits and citrus where creep (measurement of deformation for a fixed time under constant load condition) or relaxation (measurement of the decrease in force as a function of time at constant deformation) measurements are considered more suitable compared to puncture test. However, there are no commonly accepted standards for creep and relaxation measurements [33]. The non-destructive techniques include resonance frequency (acoustic, dynamic oscillation, vibration), magnetic resonance imaging (MRI), impact tests and elastic modulus tests that use small scale compressive tests and involve only elastic deformation. The non-destructive techniques are mainly developed for sorting and grading of fruits and vegetables. For a more detailed discussion on the subject one may refer to the reviews by Harker et al. [8] and Abbott [33]. Further discussion in this section will be limited to some of the destructive techniques, which are commonly used for the evaluation of the firmness of fruit and vegetable products and those with the potential for wider application.

4.3.1 Compression Tests

Compression tests include the most common methods of texture evaluations of fruits and vegetables such as compression tests by the Instron universal testing machine and puncture tests using penetrometers such as the Magness – Taylor fruit firmness tester. In these methods, whole fruit and vegetables or samples that are cut into precise geometry such as cylinders or cubes are subjected to deformation by compression and shear at constant and relatively low loading rates [33]. Figure 4.3 shows examples of typical force deformation curves of two different types of fruit samples subjected to compression. The initial linear portion of the curves up to the inflection point represents elastic deformation. After that a distinct bioyield point where a clear change in slope indicates the beginning of cell rupture followed by the rupture point at which substantial tissue failure occurs. In most force deformation curves of biological materials, the bioyield point may not be distinguishable from that of the rupture point. After the rupture point, the force may decrease, increase again or level off as the deformation increases [1, 33].

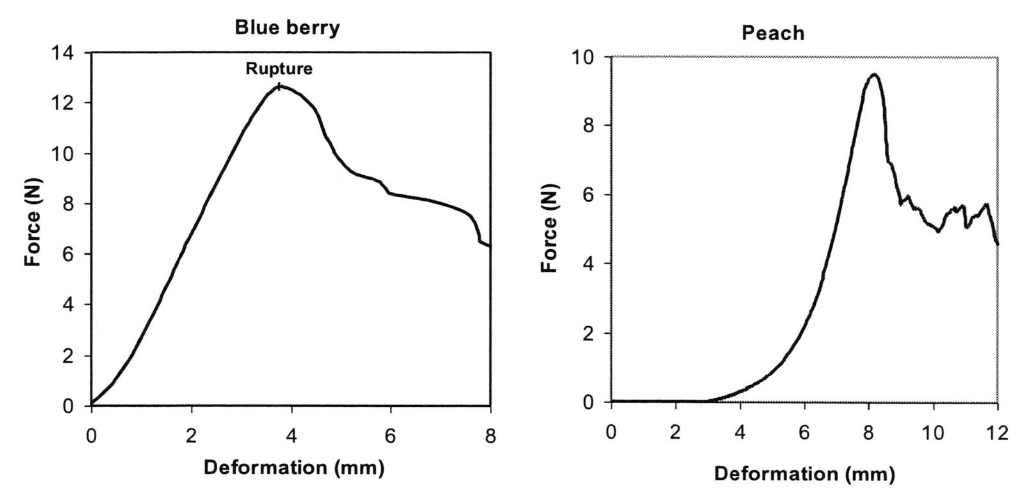

Figure 4.3. Examples of force deformation curves during compression of blueberries and peach cubes (10 mm) using an Instron universal testing machine at strain rate of 0.5 mm/s.

Firmness of horticultural products is usually defined as the maximum force attained in force deformation curves regardless of where it occurs, although the bioyield or the rupture point are used sometimes [33]. Another parameter, which is sometimes used to describe the texture of fruits and vegetables, is toughness. Toughness is defined as the area under the force-deformation curve. Parameters borrowed from material engineering such as the modulus of elasticity (E), which is defined as the ratio of stress (the force per unit area applied perpendicular to the plane) to strain (the change in length per unit length in the direction of the applied normal stress) are also used to describe the firmness of fruits and vegetables. For many engineering materials, the stress-strain curve is linear up to the proportional limit. Therefore, a single E value describes the linear region and is commonly used as an index of firmness. In contrast, only the initial part of the stress strain curve is linear in biological materials (see Figure 4.3). In addition, sample dimensions and therefore stress and strain are seldom known precisely in food tests. Thus only apparent modulus of elasticity value, calculated at a specific range, is defined for biological materials. The modulus of elasticity can be measured by non-destructive tests. However, the best relationship to sensory hardness and crispiness are obtained with forces at or beyond deformation. Therefore non-destructive mechanical tests generally do not predict sensorial texture as accurately as destructive testing devices such as penetrometers [33].

4.3.2 Shear and extrusion tests

The most commonly used device that measures the shear and extrusion properties of fruits is the Kramer shear cell. The Kramer shear cell is a box with horizontal slots both at the top and at the bottom. A fixed amount of sample is loaded into the box and a multi-blade shearing device is driven through the sample and eventually the slots in the box. Deformation of the samples involves compression, shear and extrusion. The firmness of the sample is expressed as the maximum force per gram of samples or the area under the force-deformation curve [8]. In most food materials, the maximum force per gram of sample decreases with an increase in sample weight. Thus sample weight must be kept constant during extrusion testing [34].

4.3.3 Tensile Tests

In this method, a block of fruit tissue, with notches cut at each side through the middle to provide a weakened zone, is fixed to metal stripes and placed between Instron claws. The claws move apart stretching the tissue until it breaks. After tissue failure, the fracture surface of the samples is examined by low temperature scanning electron microscopy so as to determine the mechanism of failure i.e. whether failure occurred by cell rupture or cell to cell debonding [8]. Harker et al. [9] compared measurement of texture of a number of fruits and vegetables by penetrometers, Kramer shear test and tensile test with sensory analysis. Hardness as measured by tensile strength, showed the closest correlation with sensory hardness evaluation.

4.3.4 Acoustic Emission Test

This test is based on the fact that sound generated during chewing or mechanical deformation of food materials is correlated with crispiness. In this method, sound emission is measured during instrumental or manual deformation of a sample or during chewing by holding a microphone near the ear. The resulting data is analyzed using a frequency analyzer to determine the number peaks, mean peak height and duration of the sound from the amplitude time plots, which are then used to define the crispiness and the crunchiness of the product [1, 8, 35].

4.4 Effects of Processing on the Texture of Fruit and Vegetables

During processing, fruits and vegetables undergo structural and biochemical changes, which significantly alter their texture. Processes such as freezing and heating kill cells destroying the differential permeability of their membranes. This allows the transport of solutes in and out of the cell resulting in turgor loss. Unit operations such as heating also affect cell wall polysaccharides resulting in changes in the structural integrity of the cell wall and the middle lamella. Pectin is the main cell wall polysaccharide susceptible to change both by enzyme catalysed reactions and chemical reaction (β-elimination) at high temperature. In vegetables rich in starch, heating may also cause gelatinization of starch which affects texture.

4.4.1 Minimal Processing

Minimal processing of fruits and vegetables involve washing, peeling, slicing, dicing or shredding prior to packaging often under modified atmospheric condition and storage with the objective of providing convenient and ready to use food products while maintaining nutritional and sensory quality [36]. In contrast to other food preservation processes, the unit operations involved in minimal processing enhance physiological and biochemical degradation processes rendering the products highly perishable. This arises from the fact that minimally processed products are essentially living respiring tissues which have been subjected to mechanical injury and wounding. The response to wounding triggers a complex set of metabolic processes which are designed to repair the wound. These include localised increase in the rate of respiration at the site of the wound, stress ethylene production, accumulation of secondary metabolites and cellular disruption, which together cause detrimental changes on quality attributes such as colour, flavour and texture [37].

Textural changes in minimally processed fruits and vegetables are mainly caused by biochemical reactions that are associated with wounding response. One of the reactions that occur following wounding is the enzymatic degradation of lipid components in the cell membrane and the membranes of subcellular organelles, which are lipoprotein structures. Wound-induced changes in the membrane also lead to the release of wound ethylene [37]. The effects of wound ethylene are similar to the effect of ethylene during ripening. Ethylene activates cell wall degrading enzymes such as pectin methylesterase, polygalacturonase and β-galactosidase [38], whose synergistic activity results in the degradation of cell wall polysaccharides leading to cell to cell separation and tissue softening. Microbial growth on minimally processed products may also lead to softening due to the activity of cell wall degrading enzymes of microbial origin. Dehydration caused by the removal of the peel also causes softening in minimally processed products. At the cellular level, dehydration leads to loss of turgor pressure resulting in flaccid cells that negatively affect the overall texture of the product [39].

The degree of susceptibility of fresh cut products to such deterioration processes depends on the type of fruit or vegetable, the variety, growing condition, postharvest and post process handling and packaging as well as the cutting equipment used in the processing. Depending on the product, different approaches may be used to inhibit or reduce the rate of deteriorative processes that affect texture and other quality attributes of minimally processed products. In addition to refrigerated storage, modified atmosphere packaging and natural and chemical additives are commonly used to extend the shelf life of fresh cut products. Reducing water migration through the application of edible coating, dipping in divalent cation solution such as calcium chloride to increase the strength of the cell wall through pectin cross-linking and the use of ethylene absorbents in packaging can all be used to reduce or slow down textural degradation [38, 39].

4.4.2 Thermal Processing

During thermal processing of fruits and vegetables, structural changes occur both at the cellular and cell wall polysaccharides level that affect texture. Thermal processing cause an initial loss of firmness due to the disruption of the cell membrane, which leads to free movement of solutes and water into and out of the cell resulting in turgor loss. This initial softening process is enhanced by the dissolution of the cell wall and the middle lamella, which increase the ease of cell to cell separation. Expulsion of trapped air in the intercellular space due to expansion and displacement by sap leaking from damaged cells may also affect the texture of those products with significant intercellular space [40, 41].

Among the cell wall polysaccharides, mainly the pectic polymers change during thermal processing [15]. β-elimination is mainly responsible for pectin degradation during thermal processing. Pectin degradation through β-elimination results in the thinning of middle lamella, cell separation and softening of the tissue [3-5]. β-elimination is a hydroxyl ion catalysed breakage of glycosidic linkages in pectin under nonacid conditions. The kinetics of β-elimination depends on temperature, pH and the degree of esterification of pectin. The reaction is enhanced at alkaline pH conditions. Ions such as Ca^{2+}, Mg^{2+}, K^{+} ions also enhance the reaction. The reaction is also dependent on the degree of methylation of pectin with pectins with higher degree of methylation more susceptible to β-elimination.

Turgor loss and the consequent change in texture due to thermal processing are unavoidable. However, different approaches may be used to modulate the textural changes that occur due to pectin degradation, which causes a significant part of the softening. Since pectins with lower degrees of methylation resist β-elimination, low temperature blanching pre-treatment can be used to reduce tissue softening during subsequent processing, as it activates pectin methylesterase (PME), which de-methylates pectin [5, 15, 16]. This approach may have limited application since low temperature blanching may also activate heat resistant enzymes such as polygalacturonase (PG) in tissues with significant PG activity [42, 43], and result in enzymatic pectin degradation. As mentioned in section 2.3.1, the application of high pressure processing for PME activation may be a suitable alternative in such instances. Another way of counteracting tissue softening due to pectin depolymerisation is through cross-linking of demethylated pectin by divalent cations such as Ca^{2+} and hence the basis for the use of calcium infiltration pre-treatments coupled with pectin demethylation [3]. The pH of the cooking media may also be controlled so as to reduce the rate of tissue softening since the rate of β-elimination is dependent on pH. In addition, controlling the intracellular concentration of divalent cations such as Ca^{2+} preharvest through fertilizer application can also effectively decrease tissue softening during postharvest thermal processing [44].

4.4.3 Freezing

In general, freezing preserves the nutritional and most organoleptic quality attributes of food materials. However, texture is significantly affected by freezing. The textural quality of frozen fruits and vegetables may be affected by pre-treatment processes such as blanching, the freezing process and the frozen storage conditions. Blanching causes both mechanical and biochemical changes affecting the microstructure of plant tissue while the effects of freezing and frozen storage are mainly mechanical. Freezing induced tissue disruption and decompartmentalisation of plant tissue leads to increased contact between quality degrading enzymes and their substrates resulting in detrimental quality changes. To prevent this, the enzymes are usually inactivated prior to freezing especially in vegetables, through blanching. However, blanching has consequences other than enzyme inactivation. As discussed in section 4.2, heat treatment processes including blanching cause turgor loss and degradation of pectin leading to texture alteration, which increases with increasing severity of heat treatment.

The main factor responsible for the deterioration of texture during freezing is mechanical stress. Mechanical stress can arise from ice crystallization, changes in volume as ice is formed (pure water expands during ice formation), dislocation of water that accompanies slow freezing (freeze dehydration), recrystallization or by the temperature gradients within the product. The volume change during freezing is due to the expansion of water during freezing. Pure water expands by about 9% on freezing at 0 °C. The low temperature together with the ice crystals can damage the cell membrane resulting in textural change due to the loss of osmotic integrity and consequently turgor pressure [45].

Ice crystallization can cause substantial microstructural change in tissue foods including fruits and vegetables. The extent of this change depends on the location of the ice crystals, which in turn depends on the freezing rate [46]. Ice crystallization generally starts at the extracellular space due to the lower solute concentration. But if the rate of heat removal is high enough to produce high supercooling inside the cells, the result is the formation of tiny ice crystals both in the intracellular and extracellular space. In fact the existence of intracellular ice is an indicator of high freezing rate [47, 48]. In plant tissue, ice forms in the extracellular space during slow freezing resulting in an ice-rich matrix of low temperature surrounding the cells. Thus its osmomolality continues to rise with solute concentration as the temperature is reduced and more ice is formed. This facilitates transfer of water through the cell membrane in an attempt to reach osmotic equilibrium and minimize undercooling. This phenomenon, called freeze induced dehydration, is undesirable in frozen food products since the translocated water in the extracellular space results in large ice crystals, a shrunken appearance of cells (extensive dehydration), drip loss and tissue shrinkage during thawing. Under these conditions, where osmotic transfer of water is high, cell walls may tear and buckle while membranes may rupture and/or fail collectively resulting in loss of firmness. Rapid freezing helps to avoid this problem since it limits the time for exosmosis, thus causing the cells to be extensively supercooled leading to intracellular ice formation which avoids exosmosis since the unfrozen fraction inside the cell will have the same solute concentration as the one outside the cell. Therefore, generally rapid freezing results in a better quality retention with the formation of small ice crystals resulting in a more homogeneous structure [49, 50].

The degree of the damage from mechanical stress depends on the nature and state of the material, the final temperature reached and the freezing rate [47, 51]. The damage is more severe in plant tissues with their rigid structure and poorly aligned cells than in for example muscles, which have a pliable consistency and parallel arrangement of cells [51]. It also depends on whether the tissue has been blanched or not. In unblanched tissue, the cell walls are intact, and osmotic exchange of water is possible. Thus freezing causes more damage to blanched tissue than unblanched [45]. In fruits such as berries, that are not normally blanched to avoid excessive quality damage, the effect of freezing on texture is greater than that of vegetables since their texture is mainly ensured by turgor [46]. The rate of freezing can also affect the stress severity during freezing. Although rapid freezing usually results in better retention of firmness, rapid freezing to a very low final temperature almost always results in severe cracking of foods that contain large amount of water [46, 47]. This is attributed to the non-uniform contraction following solidification and the Langham-Mason effect (rupture of the outer frozen shell when the interior freezes and expands). In contrast to pure water, the overall increase in volume of aqueous cellular systems is less than 9%. As the constituents other than water contract during freezing and since freezing occurs first on the surface layer, local areas of expansion and shrinkage are created. This may cause local stress and structural damage. The formation of ice within the cells, during rapid freezing, may also result in damage to cell structures and organelles [46, 51].

Change in microstructure and degradation of firmness also occurs during frozen storage, which is mainly attributed to ice recystallization. Ice crystals are relatively unstable and undergo changes in number, size and shape during frozen storage [50]. In general, two processes can occur; 'Oswald ripening' in which larger ice crystals grow at the expense of smaller ones at constant temperature and recrystallization as a result of temperature fluctuation [49]. Recrystallization occurs because systems tend toward a state of equilibrium wherein free energy is minimized and the chemical potential is equalized among all phases. Although the amount of ice remains relatively constant, over time this phenomenon can be extremely damaging to the texture of frozen products. This deterioration in quality becomes more prevalent with higher frozen storage temperatures, greater temperature fluctuations, and increased storage times. Recrystallization can reverse the advantage of fast freezing. Significant biochemical changes that affect texture are not expected to take place during frozen storage since the PME catalyzed de-esterification of pectin is very slow at subfreezing temperature conditions [52].

Different approaches may be used to minimize textural quality loss of frozen fruit and vegetable products. These include optimizing the freezing process, ensuring constant temperature during frozen storage and transport to retails stores to avoid recrystallization and the application of pre-treatment processes such as low temperature blanching with or without calcium infiltration. In general rapid freezing results in a better textural quality retention. Satisfactory results have been obtained when ultrarapid freezing rates were used, for products yielding only poor quality, when they are frozen by the conventional freezing method. Fast freezing resulted in better quality retention of green beans, carrot slices, potato and cauliflower [46]. However, it should be pointed out that a higher freezing rate is not always the better one, especially if it is to a very low temperature. As mentioned earlier, many products crack when they are submitted to a very high freezing rate as in cryogenic fluids. Only small objects can be frozen without the formation of cracks. Pre-cooling prevents freeze-cracking as it reduces the difference in temperature between the product and the freezing medium. It also reduces the time delay between the freezing of the surface and the center of the product [48].

Another alternative is the use of high-pressure assisted freezing. In the case of high-pressure assisted freezing, a higher degree of supercooling is obtained due to the freezing point depression as a function of pressure for pressures up to 220 MPa. For each degree Kelvin of supercooling, the rate of ice nucleation increases ten times [53]. As the pressure is uniformly distributed (Pascal's law) throughout the material, the nucleation and growth of ice crystals is uniform with no stress inducing ice front moving through the material, which may result in cracking. Thus high pressure facilitates rapid freezing and results in improved product quality. High pressure freezing was found to be effective in improving both the texture and histological characteristics of frozen carrots [54;55]. To date there has not been any commercial application of high pressure assisted freezing perhaps due to the potentially higher cost.

4.4.4 High Pressure Processing

High pressure processing is one of the emerging non-thermal food processing and food preservation technologies, which is being increasingly applied commercially for the processing of fruit and vegetable products. The fact that high pressure inactivates vegetative microbial cells and reduces the activity of many quality degrading enzymes without significant effects on the nutritional and organoleptic quality of food products has stimulated considerable interest in the technology both from academia and industry over the last 20 years. Over the years, a number of high pressure processed food products including jams, fruit preparations, fruit juices, vegetable products, sea food and processed meat products have been introduced into the market by food companies in Japan, USA, Europe and Australia [56-58]. A four fold increase in the number of commercial high pressure installations has occurred since 2000 with about 120 industrial scale high pressure equipments in operation in 2007 [59]. Consumers are increasingly looking for and are willing to pay more for safe 'fresh' like food products processed with minimal heat and chemical preservatives, which is possible with high pressure processing. Therefore the growth in the commercial application of high pressure processing is set to continue. Currently, the cost of high pressure processing ranges from 8 to 20 cents (US $) per kg including depreciation and operating costs, which is higher but not an order of magnitude higher than thermal processing. With increasing demand and innovations in equipment design, further decrease is expected in both capital and operating costs [60].

High pressure processing affects the microstructure and texture of fruits and vegetables in two ways; mechanical and biochemical. The compression and decompression steps during processing lead to deformation of cells, cell wall and membrane damage and tissue disruption resulting in turgor loss and loss of firmness [61-63]. As a consequence, high pressure treated tissues look like soaked and slightly cooked. The extent of cell disruption is dependent both on the level of the applied pressure and the type of cell. Prestamo and Arroyo [63] studied the effects of high pressure processing on the microstructure of cauliflower and spinach leaves using scanning electron microscopy. After 30 min treatment at 400 MPa and 5 °C, cell disruption, migration of soluble components, presence of fluid in the intercellular space displacing gasses that filled the space prior to treatment and turgor loss were observed both in cauliflower and spinach leaves.

However, the extent of cell disruption was more extensive in the case of spinach. In the parenchyma tissue of spinach leaves, the cell collapsed and new cavity formation was observed whereas the cells were cemented and maintained a ridged structure in the vascular tissue of the spinach leaves. Cauliflower tissues maintained a firm structure close to the untreated tissue even though the treatment caused significant cell disruption [63]. Similarly significant microstructural changes were observed in high pressure processed carrot cortex tissue with cell wall buckling and folding, reduction of cell to cell contact and cell deformation. The effect was dependent on the applied pressure. At 100 to 200 MPa, the cell walls were only slightly disrupted and disorganised while at pressures ranging from 300 to 550 MPa, definite increase in cell wall thickness was observed due to folding [61]. Light microscopy images of raw and high pressure treated carrot tissue are presented in Figure 4.4. Basak and Ramaswamy [62] studied the effect of high pressure processing at 100 to 400 MPa on the firmness of apple, pear, orange, pineapple, green pepper, red pepper, celery, and carrot. In all cases, they observed initial loss of firmness dependent on the magnitude of the applied pressure, which they called instantaneous pressure softening followed by gradual decrease or recovery of firmness during the pressure holding phase. The level of the pressure induced softening was dependent on the type the product and the pressure applied. Among the fruits, pear was the most sensitive to pressurization at 100 MPa, followed by apple, pineapple and orange. At 200 MPa, apple was found to be more sensitive than pear. In the case of the vegetables, red pepper was found to be the most sensitive followed by celery, carrot and green pepper. At higher pressure, celery was found to be the most sensitive followed by carrot and red pepper [62].

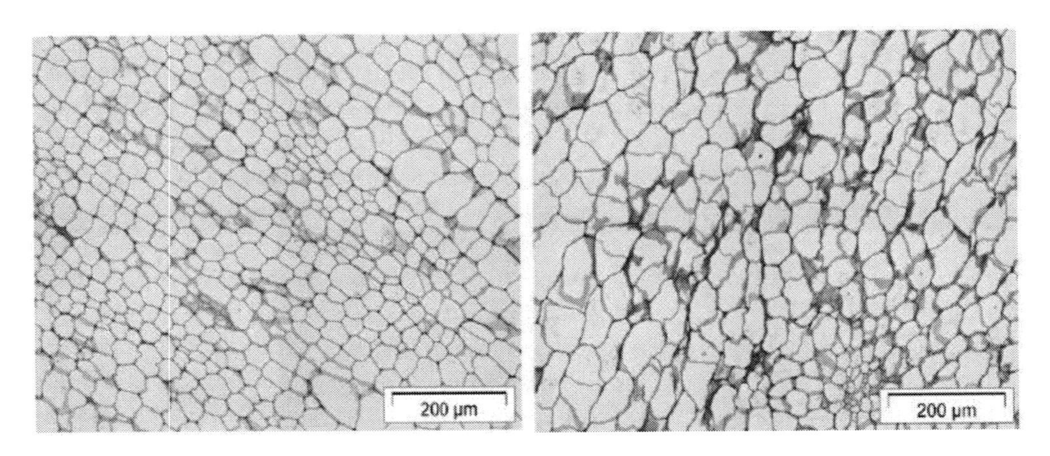

Figure 4.4. Light microscopy images of untreated (left) and high pressure treated (300 MPa/2 min) (right) carrot cortex tissue reproduced by permission from Araya et al. [61].

Degradation of pectin by β-elimination does not occur during high pressure processing at ambient condition. Kato et al. [64] compared the effect of thermal and high pressure processing on the pectin matrix of carrot. Heating alone above pH 5 resulted in the degradation of pectin possibly by β-elimination. The degradation increased with pH up to pH=8 and decreased afterwards. After 30 min cooking, an increase in galacturonic acid content indicating the breakdown of pectin by β-elimination and a decrease in high methoxy pectin were observed which resulted in softening. In contrast, no β-eliminative degradation of the pectin chain was observed during pressurization. However, the degree of methylation of pectin decreased during pressurization, apparently due to the activation of PME [64]. In general, high pressure enhances the PME catalysed de-esterification of pectin in plant tissue, which counteracts the negative effects of cell disruption and turgor loss due to pressurization on the texture of plant materials.

Nevertheless, the effect high pressure processing on overall hardness and firmness of plant materials depends on whether the mechanical or biochemical effects of the process are dominating. In the study of Kato et al. [64], the firmness of raw and pressurized carrots processed at 100 to 700 MPa as measured by the maximum force during compression were similar. However, the rupture strain was higher for samples treated at pressures higher than 200 MPa. Although the overall firmness of the product was maintained after high pressure processing due to enhanced pectin demethylation, crispiness was lost at pressures higher than 200 MPa [64]. Texture recovery following the initial softening due to pressurization has been observed in a number of fruits and vegetables especially after treatment at lower pressure conditions [61, 62]. Basak and Ramaswamy [62] observed slow recovery of firmness that was lost due to pressurization (due to instantaneous pressure softening) during the pressure holding phase for a number of fruits and vegetables after treatments at 100 to 400 MPa and ambient temperature. This observation was attributed to the PME catalysed de-methylation of pectin during the holding phase. The effect was found to be dependent on the type of product as well as the applied pressure. For instance, no recovery of firmness was observed in apples, carrot and green pepper at pressures higher than 100 MPa. Rather a continuous decrease in firmness was observed during the pressure holding phase. In contrast, recovery of firmness during the holding phase was observed for pressures up to 400 MPa in red pepper and celery [62]. Compared to thermal sterilization, substantially higher retention of firmness was observed in high pressure processed green beans both for continuous and pulsed application. In the continuous treatment, the samples were treated for 60 s at 600 MPa and 45 °C. In the pulsed treatment, samples were pre-heated at 75 °C for 2 min followed by 80 s treatment at 1000 MPa and 105 °C, depressurisation and holding for 30 s at 0.1 MPa and a second high pressure treatment at 1000 MPa. The firmness retention in the thermally sterilized green beans was 3% compared to 60% in the high pressure processed products at both conditions [65]. As discussed in section 4.2.3.1, PME is very resistant to high pressure inactivation, while PG is sensitive to pressure inactivation. In addition, the PME catalysed de-esterification of pectin is accelerated by high pressure [66]. Increased enzyme-substrate contact due to membrane disruption and decompartmentalization during high pressure processing also contributes to the acceleration of the PME catalysed de-esterification in situ [67]. This is the basis for the observed high retention of firmness and firmness recovery in high pressure treated plant products. High pressure processed products nearly always show better firmness retention compared to thermally processed products. These observations also led to a number of investigations on the possibility of using high pressure processing with/without infusion of exogenous PME and/or $CaCl_2$ as a pre-treatment prior to freezing or thermal processing.

Duvutter et al. [68] observed that vacuum-assisted (10 hPa) infusion of strawberry halves with PME (100 U/ml) and calcium chloride (0.5% w/w) prior to thermal (60 to 80 °C) and high pressure (400-550 MPa) treatment significantly improved the texture retention of processed strawberries. The greatest increase in firmness was correlated with the highest decrease in the degree of methylation of pectin, showing that the increase in firmness is mainly due to the action of the infused PME. The high pressure treated samples showed higher retention of texture regardless of treatment pressure and time while treatment temperature had significant effect on the thermally treated samples. Vacuum infused samples treated at 60 °C for up to 20 min were as firm as the untreated samples while much higher firmness was observed in the high pressure treated samples. Regardless of the pressure and treatment time, vacuum infused samples treated by high pressure were about twice firmer than the untreated samples. Without pre-treatment, the strawberries lost 80% of their firmness after high pressure processing. Both tomato and fungal PME were used in the experiments, however, fungal PME had the greatest effect [68], perhaps due to its random mode of de-esterification of pectin. Similarly vacuum infusion of fungal PME (100 U/ml) and calcium (0.5%) combined with rapid cryogenic or high pressure shift freezing significantly improved the texture of frozen strawberries. Vacuum infused strawberry halves showed 58% retention of their firmness after cryogenic freezing and thawing compared to 12% in untreated samples. Vacuum infused samples frozen by high pressure shift freezing were firmer than the fresh samples [69].

Sila et al. [70, 71] studied the effect of high pressure pre-treatment on the textural degradation kinetics of carrot during thermal processing at 90 to 110 °C. High pressure pre-treatment of carrots for 15 min at pressures ranging from 200 to 400 MPa caused up to a four fold reduction in the rate of textural degradation during subsequent thermal processing. The greatest effect was observed at 60 °C and 400 MPa. High pressure pre-treatment resulted in a larger decrease in the degree of methylation of pectin as well as texture degradation kinetics compared to low temperature blanching at the same temperature (60 °C, 40 min). This was attributed to tissue disruption by high pressure allowing contact between the enzyme and the substrate and increased PME activity under pressure. The reduced rate of textural degradation during the subsequent thermal processing is mainly due to the reduced rate of β-elimination as it its rate decreases with decrease in the degree of methylation of pectin. The decrease in the rate of texture degradation was correlated with the decrease in the degree of methylation of pectin in carrot tissue [70, 71]. Likewise Buggenhout et al. [72] observed improved texture of frozen carrots by combining pre-treatments at high pressure (300 MPa, 60 °C, 15 min) or mild temperature (30 min, 60 °C) with high pressure shift freezing. This was accompanied by decrease in the degree of methylation of pectin. In the untreated samples, high pressure-shift freezing did not have a significant advantage compared to conventional freezing in terms of texture. Interestingly, pre-treatment also did not have a significant effect on the texture of conventionally frozen carrots [72].

CONCLUSION

The texture of plant based products is highly dependent on their structure at different levels including the type of cells that make up the edible tissue, the intercellular air space, the turgor pressure which is dependent on the integrity of the cell membrane, and the structural integrity of the cell wall and the middle lamella which is highly dependent on the composition of the cell wall polysaccharides in general and pectin in particular. Processing operations affect the structure and hence the textural attributes of fruits and vegetables. The extent of the effect at different structural levels is dependent on the type and intensity of the process as well as the type of tissue. In general, processing operations such as heating, freezing and high pressure processing physically destroy the differential permeability of the cell resulting in turgor loss and loss of crispiness in processed products. Thermal processing also causes β-eliminative degradation of pectin in the cell wall contributing to cell separation and tissue softening. High pressure on the other hand inhibits the β-eliminative degradation of pectin since it enhances the PME catalysed de-esterification of pectin, which can be utilised to improve the firmness retention of fruit and vegetable products. In general pre-treatment processes such as low temperature blanching and high pressure processing with/without the infusion of calcium and/or exogenous PME can be used to counteract tissue softening during processes such as freezing and thermal processing.

REFERENCES

[1] Waldron, K.W., Parker, M.L., & Smith, A.C. (2003). Plant cell walls and food quality. *Comprehensive Reviews in Food Science and Food Safety*, *2*, 128-146.

[2] Bourne, M.C. (2002). *Food Texture and Viscosity: Concept and Measurement* (2nd edition, pp. 1-32). London: Academic Press.

[3] Jackman, R.L., & Stanley, D.W. (1995). Perspectives in the textural evaluation of plant foods. *Trends in Food Science and Technology*, *6*, 187-194.

[4] Van Dijk, C., & Tijskens, L.M.M. (2000). Mathematical modelling of enzymatic reactions as related to texture after storage and mild preheat treatments. In S.M., Alzamora, M.S., Tapia, & A., Lopez-Malo (Eds.), *Minimally Processed Fruits and Vegetables: Fundamental Aspects and Applications* (pp. 127-152). Gaithersberg, Maryland: Aspen Publishers, Inc.

[5] Fuchigiami, M. (1987). Relationship between pectic compositions and softening of the texture of Japanese radish roots during cooking. *Journal of Food Science*, *52*, 1317-1320.

[6] Ilker, R., & Szczesniak, A. (1990). Structural and Chemical bases for texture of plant foodstuffs. *Journal of Texture Studies*, *21*, 1-36.

[7] Alzamora, S.M., Castro, M.A., Vidales, S.L, Nieto, A.B., & Salvatori, D. (2000). The role of tissue microstructure in the textural characteristics of minimally processed fruits. In S.M., Alzamora, M.S., Tapia, & A., Lopez-Malo (Eds.), *Minimally Processed Fruits and Vegetables: Fundamental Aspects and Applications* (pp. 153-169). Gaithersberg, Maryland: Aspen Publishers, Inc.

[8] Harker, F.R., Redgwell, R.J., Hallet, I.C., & Murray, S.H. (1997). Texture of fresh fruit. *Horticultural Reviews*, *20*, 121-224.

[9] Harker, F.R., Stec, M.G.H., Hallett, I.C., & Bennett, C.L. (1997). Texture of parenchymatous plant tissue: a comparison between tensile and other instrumental and sensory measurements of tissue strength and juiciness. *Postharvest Biology and Technology*, *11*, 63-72.

[10] Mohsenin, N.N. (1984). Physical Properties of Food and Agricultural Materials: a teaching manual (1st edition). New York: Gordon and Breach Publishers.

[11] Waldron, K.W., Smith, A.C., Parr, A.J., Ng, A., & Parker, M.L. (1997). New approaches to understanding and controlling cell separation in relation to fruit and vegetable texture. *Trends in Food Science and Technology*, *8*, 213-221.

[12] Carpita, N.C., & Gibeaut, D.M. (1993). Structural models of primary cell walls in flowering plants: consistency of molecular structure with physical properties of the walls during growth. *The Plant Journal* 3, 1-30.

[13] Giovane, A., Sevillo, L., Balestrieri, C., Rajola, A., D'Avino, R., Tamburrini, M., Ciardiello, M.A., & Camardella, L. (2004). Pectin methylesterase inhibitor. *Biochimica et Biophysica Acta*, *1696*, 245-252.

[14] Versteeg, C. (1979). Pectinesterases from the orange fruit. Agricultural Research Report 892. Wageningen, Netherlands: Centre for Agricultural Publishing and Documentation.

[15] Stolle-Smits, T., Beekhuizen, J.G., Recourt, K., Voragen, A.J., & Van Dijk, C. (2000). Preheating effects on the textural strengths of green beans. 1. Cell wall chemistry. *Journal of Agricultural and Food Chemistry*, *48*, 5269-5277.

[16] Ng, A., & Walderon, K.W. (1997). Effects of cooking and pre-cooking on cell wall chemistry in relation to firmness of carrot tissues. *Journal of the Science of Food and Agriculture*, *73*, 503-512.

[17] Fachin, D., Van Loey, A., Ly-Nguyen, B., Verlent, I., Indrawati & Hendrickx, M. (2002). Comparative study of the inactivation kinetics of pectinmethylesterase in tomato juice and purified form. *Biotechnology Progress*, *18*, 739-744.

[18] Balogh, T., Smout, C., Ly-Nguyen, B.N., Van Loey, A., & Hendrickx, M. (2004). Thermal and high-pressure inactivation kinetics of carrot pectinmethylesterase: From model system to real foods. *Innovative Food Science and Emerging Technologies*, *5*, 429-436.

[19] Ly-Nguyen, B., Van Loey, A., Smout, C., Ozcan, S.E., Fachin, D., Verlent, I., Truong, S.V., Duvetter, T., & Hendrickx, M. (2003). Mild-heat and high-pressure inactivation of carrot pectin methylesterase. *Journal of Food Science*, *68*, 1377-1383.

[20] Ly-Nguyen, B., Van Loey, A., Smout, C., Verlent, I., Duvetter, T., & Hendrickx, M. (2003). Effect of mild-heat and high-pressure processing on banana pectin methylesterase: a kinetic study. *Journal of Agricultural and Food Chemistry*, *51*, 7974-7979.

[21] Nunes, C.S., Castro, S., Saraiva, J., Coimbra, M.A., Hendrickx, M., & Van Loey, A. (2006). Thermal and high pressure stability of purified pectin methylesterase from plums (*Prunes Domestica*). *Journal of Food Biochemistry*, *30*, 138-154.

[22] Ly-Nguyen, B., Van Loey, A., Smout, C., Ozcan, S., Fachin, D., Verlent, I., Truong, S.V., Duvetter, T., & Hendrickx, M. (2002). Strawberry pectin methylesterase (PME): Purification, characterization, thermal and high pressure inactivation. *Biotechnology Progress*, *18*, 1447-1450.

[23] Choi, E.-M., Hong, S.-I., & Park, W.-S. (1999). Effect of high-pressure treatment on polygalacturonase from Chinese cabbage. *Food Science and Biotechnology*, *8*, 197-203.

[24] Fachin, D., Van Loey, A., Ludikhuyze, L., & Hendrickx, M. (2002). Thermal and high-pressure inactivation of tomato polygalacturonase: a kinetic study. *Journal of Food Science*, *67*, 1610-1615.

[25] Castro, S., Van Loey, A., Saraiva, J., Smout, C., & Hendrickx, M. (2005). Process stability of Capsicum annuum pectin methylesterase in model systems, pepper purée and intact pepper tissue. *European Food Research and Technology*, *221*, 452-458.

[26] Sila, D., Smout, C., Satara, Y., Vu, T., Van Loey, A., & Hendrickx, M. (2007) Combined thermal and high pressure effect on carrot pectinemthylesterase stability and catalytic activity. *Journal of Food Engineering*, *78*, 755-764.

[27] Verlent, I., Hendrickx, M., Verbeyst, L., & Loey, A. (2007). Effect of temperature and pressure on the combined action of purified tomato pectinmethylesterase and polygalacturonase in presence of pectin. *Enzyme and Microbial Technology*, *40*, 1141-1146.

[28] Fachin, D., Smout, C., Verlent, I., Ly-Nguyen, B., Van Loey, A., & Hendrickx, M. (2004). Inactivation kinetics of purified tomato polygalacturonase by thermal and high-pressure processing. *Journal of Agricultural and Food Chemistry*, *52*, 2697-2703.

[29] Tomas-Barberan, F.A., & Espin, J.C. (2001). Phenolic compounds and related enzymes as determinants of quality in fruits and vegetables. *Journal of the Science of Food and Agriculture*, *81*, 853-876.

[30] Ng, A., Harvey, A.J., Parker, M.L., Smith, A.C., & Waldron, K.K. (1998). Effect of oxidative coupling on the thermal stability of texture and cell wall chemistry of beet root (*Beta vulgaris*). *Journal of Agricultural and Food Chemistry*, *46*, 3365-3370.

[31] Brisson, L.F., Tenhaken, R., & Lamb, C. (1994). Function of oxidative cross-linking of cell wall structural proteins in plant disease resistance. *The Plant Cell*, *6*, 1703-1712.

[32] Bolwell, G.P., Buti, V.S., Davies, D.R., & Zimmerlin, A. (1995). The origin of the oxidative burst in plants. *Free Radical Research*, *23*, 517-532.

[33] Abbott, J.A. (1999). Quality measurement of fruits and vegetables. *Postharvest Biology and Technology*, *15*, 207-225.

[34] Barrett, D.M., Garcia, E., & Wayne, J.E. (1998). Textural Modification of Processing Tomatoes. *Critical Reviews in Food Science and Nutrition*, *38*, 173-258.

[35] Edmister, J.A., & Vickers, Z.M. (1985). Instrumental acoustical measures of crispiness in foods. *Journal of Texture Studies*, *16*, 153-167.

[36] Rico, D., Martín-Diana, A.B., Barat, J.M., & Barry-Ryan, C. (2007). Extending and measuring the quality of fresh-cut fruit and vegetables: a review. *Trends in Food Science and Technology*, *18*, 373-386.

[37] Rolle, R.S., & Chism, G.W. (1987). Physiological consequences of minimally processed fruits and vegetables. *Journal of Food Quality*, *10*, 157-177.

[38] Perera, C.O., & Baldwin, E. (2001). Biochemistry of fruits and its implication on processing. In D., Arthey, & P.R., Ashurst (Eds.), *Fruit Processing: Nutrition, Products and, Quality Management* (pp. 19-36). Gaithersburg, Maryland: Aspen Publishers, Inc.

[39] Olivas, G.I., & Barbosa-Canovas, G.V. (2005). Edible Coatings for Fresh-Cut Fruits. *Critical Reviews in Food Science and Nutrition*, *45*, 657-670.

[40] Del valle, J.M., Aranguiz, V., & Leon, H. (1998). Effects of blanching and calcium infiltration on PPO activity, texture, microstructure and kinetics of osmotic dehydration of apple tissue. *Food Research International*, *31*, 557-569.

[41] Dobias, J., Voldrich, M., & Curda, D. (2006). Heating of canned fruits and vegetables: Deaeration and texture changes. *Journal of Food Engineering*, *77*, 421-425.

[42] Anthon, G.-E., & Barrett, D.-M. (2002). Kinetic parameters for the thermal inactivation of quality-related enzymes in carrots and potatoes. *Journal of Agricultural and Food Chemistry*, *50*, 4119-4125.

[43] Lopez, P., Sanchez, A.C., & Vercet, A.B.J. (1997). Thermal resistance of tomato polygalacturonase and pectinmethylesterase at physiological pH. *Zeitschrift für Lebensmittel-Untersuchung und -Forschung A*, *204*, 146-150.

[44] Sams, C.E. (1999). Preharvest factors affecting postharvest texture. *Postharvest Biology and Technology*, *15*, 249-254.

[45] Reid, D.S. (1990). Optimizing the quality of frozen foods. *Food Technology*, *44*, 78-82.

[46] Partmann, W. (1975). The effects of freezing and thawing on food quality. In: R.B. Duckworth (Ed.), Water Relations of Foods (pp. 505-537). London: Academic press.

[47] Singh, R.P., & Wang, C.Y. (1977). Quality of frozen foods-a review. *Journal of Food Processing and Preservation*, *1*, 97-127.

[48] Zartizky, N.E. (2000). Factors affecting the stability of frozen foods. In C.J. Kennedy (Ed.), Managing Frozen Foods (pp. 111-135). Boca Raton, California: CRC Press.

[49] George, R.M. (1993). Freezing processes used in the food industry. *Trends in Food Science and Technology*, *4*, 134-138.
[50] Sahagian, M.E., & Goff, H.D. (1995). Fundamental aspects of the freezing process. In L.E. Jeremiah (Ed.), Freezing Effects on Food Quality (pp. 1-43). New York: Marcel Dekker, Inc.
[51] Fennema, O.R. (1973). Nature of the freezing process. In O.R., Fennema, W.D., Powrie, & E.H., Marth (Eds.), *Low Temperature Preservation of Foods and Living Matter* (pp. 151-227). New York: Marcel Dekker, Inc.
[52] Terefe, N.S., & Hendrickx, M. (2002). Kinetics of the pectin methylesterase catalyzed de-esterification of pectin in frozen food model systems. *Biotechnology Progress*, *18*, 221-228.
[53] Kalichevsky, M.T., Knorr, D., & Lillford, P.J. (1995). Potential food application of high-pressure effects on ice water transitions. *Trends in Food Science and Technology*, *6*, 253-259.
[54] Fuchigami, M., Kato, N., & Teramoto, A. (1997). High pressure freezing effects on textural quality of carrots. *Journal of Food Science*, *62*, 804-808.
[55] Fuchigami, M., Miyazaki, K., Kato, N., & Teramoto, A. (1997). Histological changes in high-pressure-frozen carrots. *Journal of Food Science*, *62*, 809-812.
[56] Hugas, M., Garcia, M., & Monfort, J.M. (2002). New mild technologies in meat processing: high pressure treatment on the sarcoplasmic reticulum of red and white muscles. *Meat Science*, *62*, 359-371.
[57] Mermelstein, N.H. (1997). High pressure processing reaches the US market. *Food Technology*, *51*, 95-96.
[58] Thakur, B.R., & Nelson, P.E. (1998). High pressure processing and preservation of foods. *Food Reviews International*, *14*, 427-447.
[59] Saiz, A.H., Mingo, S.T., Balda, F.P., & Samson, C.T. (2008). Advances in design for successful commercial high pressure food processing. *Food Australia*, *60*, 154-156.
[60] Balasubramaniam, V.M., Farkas, D., & Turek, E.J. (2008). Preserving foods through high-pressure processing. *Food Technology*, *62*, 32-38.
[61] Araya, X.I.T., Hendrickx, M., Verlinden, B.E., Van Buggenhout, S., Smale, N.J., Stewart, C., & Mawson, A.J. (2007). Understanding texture changes of high pressure processed fresh carrots: A microstructural and biochemical approach. *Journal of Food Engineering*, *80*, 873-884.
[62] Basak, S., & Ramaswamy, H.S. (1998). Effect of high pressure processing on the texture of selected fruits and vegetables. *Journal of Texture Studies*, *29*, 587-601.
[63] Prestamo, G., & Arroyo, G. (1998). High hydrostatic pressure effects on vegetable structure. *Journal of Food Science*, *63*, 878-881.
[64] Kato, N., Teramoto, A., & Fuchigami, M. (1997). Pectic substance degradation and texture of carrots as affected by pressurization. *Journal of Food Science*, *62*, 359-363.
[65] Krebbers, B., Matser, A.M., Koets, M., & Van den Berg, R.W. (2002). Quality and storage-stability of high-pressure preserved green beans. *Journal of Food Engineering*, *54*, 27-33.
[66] Sila, D.N., Duvetter, T., De Roeck, A., Verlent, I., Smout, C., Moates, G.K., Hills, B.P., Waldron, K.K., Hendrickx, M., & Van Loey, A. (2008). Texture changes of processed fruits and vegetables: potential use of high-pressure processing. *Trends in Food Science and Technology*, *19*, 309-319.

[67] Hendrickx, M., Ludikhuyze, L., Vanden Broeck, I., & Weemaes, C. (1998). Effect of high pressure on enzymes related to food quality. *Trends in Food Science and Technology*, *9*, 197-203.

[68] Duvetter, T., Fraeye, I., Van Hoang, T., Van Buggenhout, S., Verlent, I., Smout, C., Van Loey, A., & Hendrickx, M. (2005). Effects of pectinmethylesterase infusion methods and processing techniques on strawberry firmness. *Journal of Food Science*, *70*, S383-S388.

[69] Buggenhout, S.V., Messagie, I., Maes, V., Duvetter, T., Van Loey, A., & Hendrickx, M. (2006). Minimizing texture loss of frozen strawberries: effect of infusion with pectinmethylesterase and calcium combined with different freezing conditions and effect of subsequent storage/thawing conditions. *European Food Research and Technology*, *223*, 395-404.

[70] Sila, D., Smout, C., Vu, T., & Hendrickx, M. (2004). Effects of high-pressure pretreatment and calcium soaking on the texture degradation kinetics of carrots during thermal processing. *Journal of Food Science*, *69*, E205-E211.

[71] Sila, D., Smout, C., Elliot, F., Van Loey, A., & Hendrickx, M. (2006). Non-enzymatic depolymerisation of carrot pectin: toward a better understanding of carrot texture during thermal processing. *Journal of Food Science*, *71*, E1-E9.

[72] Buggenhout, S.V., Messagie, I., Van Loey, A., & Hendrickx, M. (2005). Influence of low-temperature blanching combined with high-pressure shift freezing on the texture of frozen carrots. *Journal of Food Science*, *70*, S304-S308.

In: Practical Food and Research
Editor: Rui M. S. Cruz, pp. 115-152

ISBN: 978-1-61728-506-6

Chapter V

VITAMINS

Pablo A. Ulloa[1,2], Javiera F. Rubilar[1,3], Rui M. S. Cruz[3,4], Igor Khmelinskii[1,3], Amadeu F. Brigas[1,3], Ana C. Figueira[2,4] and Margarida C. Vieira[3,4]

[1]Departamento de Química e Farmácia, Faculdade de Ciências e Tecnologia, Universidade do Algarve, Campus de Gambelas, 8005-139 Faro, Portugal
[2]CIEO - Centro de Investigação sobre Espaço e Organizações, Universidade do Algarve, Campus de Gambelas, 8005-139 Faro, Portugal
[3]CIQA - Centro de Investigação em Química do Algarve, Universidade do Algarve, Campus de Gambelas, 8005-139 Faro, Portugal
[4]Departamento de Engenharia Alimentar, Instituto Superior de Engenharia, Universidade do Algarve, Campus da Penha, 8005-139 Faro, Portugal

ABSTRACT

Nowadays, due to a continuous consumer demand there is an increase in new food processing technologies, some of those are known as non-thermal technologies. One of the main goals of a food processor is to obtain safe food products with high nutritional and sensorial properties. This chapter will present the most important vitamins in fruits and vegetables and address the different effects of processing on their final content. Common processes such as blanching, pasteurization and sterilization; alternative processes such as high pressure processing (HPP), ultrasound and high intensity pulsed electric fields (HIPEF) will be discussed.

5.1 INTRODUCTION

In the processing of fruits and vegetables, changes in nutritional quality are strongly related to compounds such as vitamins. A major challenge is how to use the recent advances in processing technologies and how to adjust raw materials, ingredients and processes to improve the nutritional quality of processed foods [1].

Vitamins are organic compounds that are essential in a diet, that are highly responsible for the good functioning of the human body in general and participating very actively in processes such as growth or reproduction. Vitamins have specific functions such as: regulating metabolic processes, controlling cellular functions and preventing diseases (e.g. scurvy). Thirteen vitamins are well recognized and divided into two categories, the fat-soluble group constituted by four vitamins and the water-soluble group represented by nine vitamins.

The fruits and vegetables vitamin content is affected by several factors such as: species, variety, maturity and growing conditions. Vitamin values variation, even after harvest, is due to many factors including processing procedures, storage conditions (time, humidity and temperature), sample preparation and variation in analytical methods. The losses of water-soluble nutrients are related to processing operations such as selection, washing, peeling and cutting, blanching, pasteurization and/or other treatments. Vitamin losses can range from 0 to 100% depending on the type of vitamin, processing conditions (e.g., temperature, oxygen, light, moisture, pH and length of exposure) and type of food [2-5].

Many people have no opportunity to eat fresh vegetables and fruits every day and frequently use processed foods that have undergone some treatment for preservation, mainly for convenience, time-saving and practical reasons. In the last decade, non-thermal technologies have received attention by the food industry, driven by an increase in consumer demand for nutritious, fresh-like food products with a high sensory quality and an acceptable shelf-life [6].

5.2 Vitamins and Their Properties

5.2.1 Vitamin A and carotenoids

The term vitamin A was first used in 1920. The designation "provitamin A" is accepted to differentiate carotenoid precursors of vitamin A from carotenoids without vitamin A activity [7]. Food sources of carotenoids include carrot and green leafy vegetables, such as spinach and amaranth, generally containing large concentrations of these substances. Fruits like papaya and orange also have appreciable quantities of carotenoids [8].

5.2.1.1 Properties and chemical structure

Provitamins or carotenoids' precursors (Figure 5.1) is a group including more than 500 individual compounds with β-carotene being the most important, followed by others such as β-apo-8′-carotenal, cryptoxanthin, α-carotene, etc. [9].

The biological activity of vitamin A is quantified by conversion of the vitamin A active components to retinol equivalents (RE). One RE is defined as 1 μg of retinol. For calculation of RE values in foods, 100% efficiency of absorption of retinol is assumed. Incomplete absorption and conversion of β-carotene is taken into account by the relationship of 1 RE = 6 μg of β-carotene. The conversion factor of other provitamin A carotenoids is 1 RE = 12 μg. International units (IU) are defined by the relationship of 1 IU = 0.3 μg of all-trans-retinol or 0.63 μg of β-carotene. Therefore, 1 RE = 3.33 IU based on retinol. RE and IU designations are both used to define vitamin A activity; however, usage of RE is preferred. Initial studies that estimated the vitamin A activity of various carotenoids and defined IUs did not account for absorption and bioavailability differences compared to all-trans retinol [7].

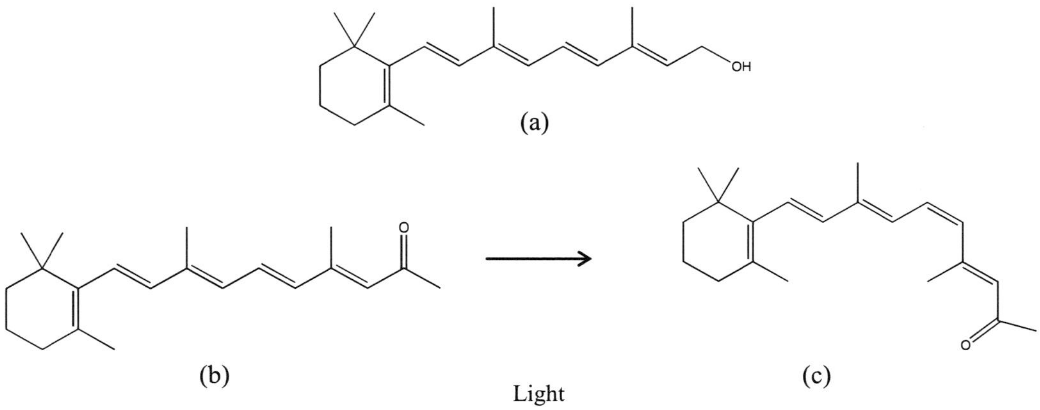

(a)

(b)

Light

(c)

Figure 5.1. Structure of all-*trans* retinol (a), all-*trans*-retinal (b) and 11-*cis*-retinal (c).

Vitamin A has its maximum biological activity in all-*trans* configuration. It is commercialized as an acetate, *trans*-retinyl and palmitate because these forms are more stable, active and soluble in oil. As to its chemical stability, vitamin A and its precursors, being unsaturated isoprenoid hydrocarbons with double bonds, are sensitive to oxidation, same as fats and oils, particularly at high temperatures and in the presence of enzymes and transition metals (Fe and Cu), subject to electromagnetic radiation and in systems with low water activity [9].

5.2.2 Vitamin B_1 (thiamin)

Vitamin B_1 or thiamin was the first vitamin, from the water-soluble group, to be structurally characterized. It exists naturally as free thiamin and phosphorylated as thiamin monophosphate (TMP), thiamin diphosphate or thiamin pyrophosphate (TPP), and thiamin triphosphate (TTP) [7]. It is found in a large variety of vegetables (soy bean, carrot, cabbage, asparagus green, etc) and in lower amounts in fruits (banana, apple, orange, etc.) [8].

5.2.2.1 Properties and chemical structure

The structure of thiamin in the free base form is given in Figure 5.2. The vitamin has a pyrimidine ring (4′amino-2′- methylpyrimidinyl-5′-ylmethyl) linked by a methylene bridge to the 3 nitrogen atom in a substituted thiazole (5-(2-hydroxyethyl)-4-methyl-thiazole) [7]. Thiamin hydrochloride forms a strongly acidic solution in water; thiamin is stable to heat in strongly alkaline solutions. In neutral, alkaline and slightly acidic solutions, thiamin decomposes to its constituent pyrimidine and thiazole moieties. Thiamin is sensitive to both oxidation and reduction. Upon oxidation, it is converted to thiochrome and on mild reduction a dihydro compound is formed [10]. The most distinguishing difference between the hydrochloride salt and the mononitrate is their water-solubilities. The hydrochloride is soluble in water (1 g/ml) and the mononitrate is only slightly water-soluble (0.027 mg/ml). This solubility difference leads to differentiation in industrial uses for the two thiamin forms. The thiamin hydrochloride is nearly insoluble in methanol, ethanol and glycerol. It is also insoluble in ether, acetone, benzene, hexane and chloroform [7].

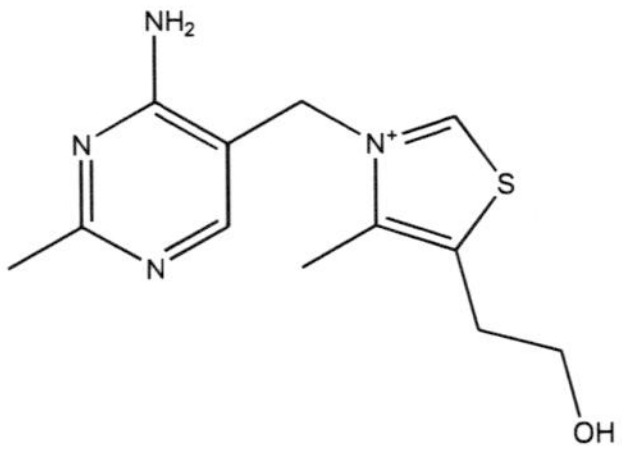

Figure 5.2. Thiamin (free base) structure.

5.2.3 Vitamin B_2 (riboflavin)

Vitamin B_2 is a water-soluble (only 10-13 mg/100 ml at 25 °C), yellow and fluorescent compound with a low solubility in most other solvents. This vitamin is present in food or food-derived products in three forms, namely, free riboflavin (RF), flavin mononucleotide (FMN) and flavin adenine dinucleotide (FAD). These molecules are coenzymes for redox reactions in numerous metabolic pathways and in energy production. Therefore, inadequate amounts of riboflavin may lead to an unbalanced intermediate metabolism [11]. The major source of riboflavin intake are foods derived from animals, legumes and green leafy vegetables (broccoli, tomato, collard green and turnip green), but not from fruits.

5.2.3.1 Properties and chemical structure

This vitamin is formed by a heterocyclic ring of isoalloxazine combined with a molecule of sugar-alcohol ribitol; ribose derivative (Figure 5.3). The primary form of this vitamin is an integral component of the coenzymes FMN and FAD.

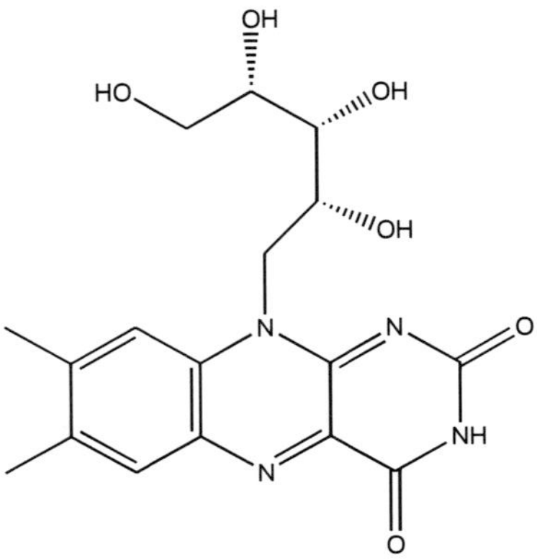

Figure 5.3. Riboflavin structure.

5.2.4 Vitamin B_3 (niacin)

Vitamin B_3 is also called niacin and this term is accepted as a broad descriptor of vitamins having the biological activity associated with nicotinamide, nicotinic acid and a variety of pyridine nucleotide structures. It is one of the most stable of all the vitamins, being resistant to heat, air, and oxidants; it is, however, hydrolysed in strong acids and alkaline solutions. This vitamin is found mainly in vegetables (beans, lentils, peas, etc) [8].

5.2.4.1 Properties and chemical structure

Niacin refers to nicotinic acid ($C_6H_5O_2N$) and nicotinamide ($C_6H_6ON_2$) which have equal biological activities. Structures are given in Figure 5.4. Nicotinic acid is, chemically, pyridine-3-carboxylic acid and nicotinamide is pyridine 3-carboxylic acid amide. The acid and amide forms are readily interconvertible and nicotinic acid is converted to the amide in formation of nicotinamide adenine dinucleotide (NAD) and nicotinamide adenine dinucleotide phosphate (NADP) [7].

Figure 5.4. Structures of nicotinamide (a) and nicotinic acid (b).

5.2.5 Vitamin B_5 (pantothenic acid)

Vitamin B_5 or pantothenic acid (PA), is a water-soluble vitamin of the B-group, which occurs in all types of animal and plant tissues, but is rarely found in the free state. Most of the vitamin is present in the bound form, essentially as coenzyme A (CoA) or acyl carrier protein (ACP), both biologically active complexes. It is found in a number of vegetables (broccoli, mushroom, cauliflower) and fruits (avocado, orange). The conjugates of PA are physiologically active in steroid, fatty acid, and phosphatide metabolism [12-14].

5.2.5.1 Properties and chemical structure

Pantothenic acid (Figure 5.5) is a yellow, viscous, hygroscopic and optically active oil, with only its dextro-rotary forms having vitamin activity. It is stable in neutral solutions but rapidly decomposes in acid or alkaline solutions.

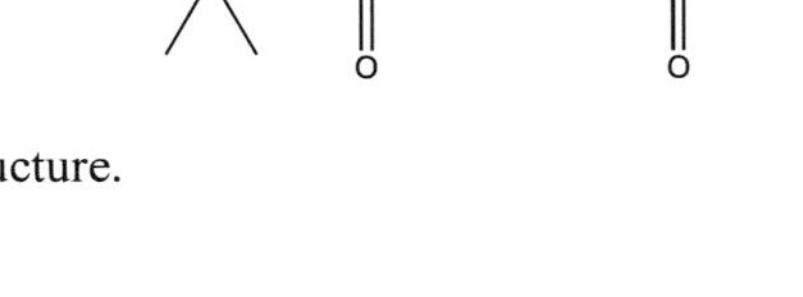

Figure 5.5. Pantothenic acid structure.

5.2.6 Vitamin B_6

Vitamin B_6 is the recommended term for the generic descriptor of all the 3-hydroxy-2-methylpyridine derivatives. The trivial names and abbreviations commonly used for the three principal forms of vitamin B_6, their phosphoric esters, and analogues are as follows: pyridoxine, PN; pyridoxal, PL; pyridoxamine, PM (Figure 5.6). As will be discussed later, other forms of vitamin B_6 exist, particularly bound forms [7]. It is found in a large variety of vegetables (carrots, cabbages, asparagus, green peas, potatoes, tomatoes, spinach, broccoli, cauliflower, corn, etc) and such fruits as apples, apricots, avocados, bananas, oranges, grapefruits, etc [8].

5.2.6.1 Properties and chemical structure

Vitamin B_6 refers to all 2-methyl-3-hydroxy-5-hydroxymethyl pyridine compounds that possess the biological activity of pyridoxine (PN) (2-methyl-3-hydroxy-4,5-bis(hydroxymethyl)-pyridine). PN, substituted at the 4-position with a hydroxymethyl group, has been commonly referred to as pyridoxol; however, the preferred name is pyridoxine. Other vitamin B_6 forms present in nature include pyridoxal (PL) and pyridoxamine (PM) with an aldehyde and aminoethyl substituents at the 4-position of the pyridine ring, respectively. The PN·hydrochloride is readily soluble in water (22 g/100 ml), alcohol, and propylene glycol. It is sparingly soluble in acetone and practically insoluble in ethyl ether and chloroform. PN is more stable than PL and PM, providing a highly usable form for food fortification and formulation of pharmaceuticals. PN·HCl is used as an US Pharmacopeia reference standard [8].

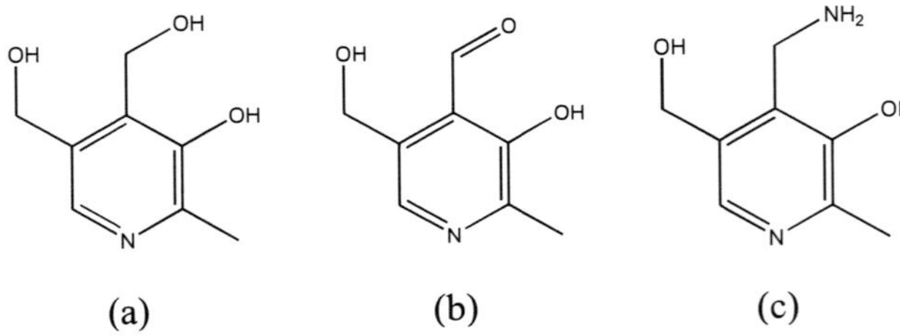

(a) (b) (c)

Figure 5.6. Structures of B_6 Vitamins: (a) Pyridoxine (PN), (b) Pyridoxal (PL), (c) Pyridoxamine (PM).

5.2.7 Vitamin B_9 (folates)

Folates are members of the water-soluble vitamin B complex having a great nutritional importance related to pteroyl monoglutamic acid (folic acid, Figure 5.7). They normally exist in polyglutamate form composed by five to seven molecules of glutamic acid and show biological activity as folic acid. Naturally occurring folate is present in a wide range of foods, especially leafy green vegetables such as cabbages, onions, legumes and citrus fruits (orange) [15].

5.2.7.1 Properties and chemical structure

Tetrahydrofolate (THF) and 5-methyltetrahydrofolate (5-MTHF) are the main forms present in plant foods, whereas 5-MTHF is the major form in humans [16, 17].

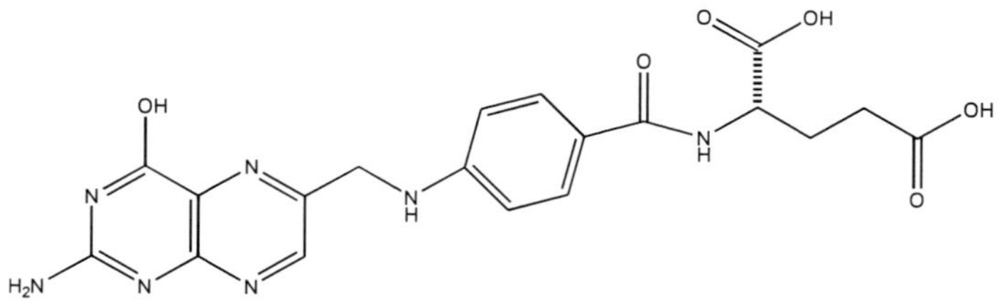

Figure 5.7. Folate structure.

5.2.8 Vitamin C

Vitamin C is a water-soluble and the most important vitamin for human nutrition that is supplied by fruits (especially citrus and some tropical) and vegetables. L-ascorbic acid (AA) is widely distributed in plant cells and plays many crucial roles in growth and metabolism. As a potent antioxidant, AA has several roles such as: eliminating different reactive oxygen species, keeping the membrane bound antioxidant α-tocopherol in the reduced state, keeping metal ions in the reduced state and thus, acting as a cofactor in several enzymes and an important role in stress resistance. [18].

5.2.8.1 Properties and chemical structure

AA is the main biologically active form of vitamin C. AA is reversibly oxidized to form L-dehydroascorbic acid (DHAA), which also exhibits biological activity (Figure 5.8). Further oxidation generates 2,3-diketogulonic acid (DKGA), which has no biological function [18].

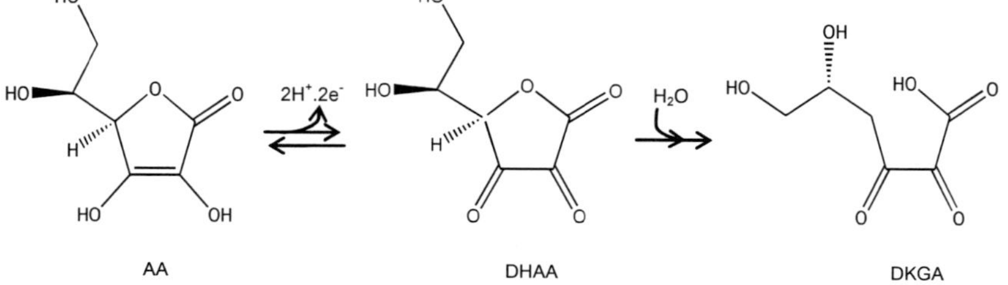

Figure 5.8. Oxidation of L-ascorbic acid.

5.2.9 Vitamin E

Vitamin E is lipid soluble and is represented by a family of structurally related compounds, eight of which are known to occur in nature. These compounds are α-, β-, γ-, and δ-tocopherol, and α-, β-, γ-, and δ-tocotrienol; the most biologically active form is α-tocopherol, with other vitamers having lower activities: γ-tocopherol has 10% of the activity of α-tocopherol, and δ-tocopherol has 1% of the activity of α-tocopherol [19, 20].

Most plant-derived foods, especially fruits and vegetables contain low to moderate levels of vitamin E activity, providing a significant and consistent source of this vitamin [8].

5.2.9.1 Properties and chemical structure

The physiological role of vitamin E centres on its ability to react with and quench free radicals in cell membranes and other lipid environments, thereby preventing polyunsaturated fatty acids from being damaged by lipid oxidation. Vitamin E also influences cellular responses to oxidative stress. The structures of α-, β-, γ-, and δ-tocopherol and the corresponding tocotrienols are described in Figure 5.9.

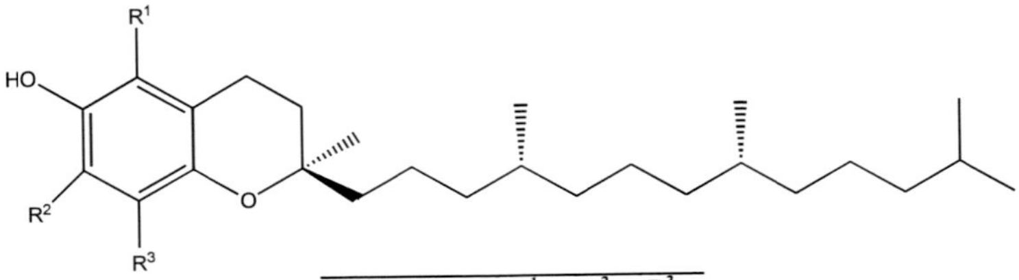

Compound	R^1	R^2	R^3
α - tocopherol	CH_3	CH_3	CH_3
β - tocopherol	CH_3	H	CH_3
γ - tocopherol	CH_3	CH_3	H
δ - tocopherol	CH_3	H	H

Figure 5.9. Tocopherol structure.

5.2.10 Vitamin B_7 (biotin)

Vitamin B_7, also called biotin, is a water-soluble vitamin and an essential co-factor for five biotin-dependent carboxylases: acetyl-CoA carboxylase-α, acetyl-CoA carboxylase-β, propionyl-CoA carboxylase, pyruvate carboxylase, and β-methylcrotonyl-CoA carboxylase [21]. The vitamin D-biotin is widely distributed in plants but in small concentrations. It can occur both in the free state (oranges and apple juice, legumes etc.) and in a form bound to protein (yeast). It is commercially available as a white crystalline powder [7].

5.2.10.1 Properties and chemical structure

Biotin is cis-hexahydro-2-oxo-1H-thieno (3,4-d) imidazole-4-pentanoic acid. Different stereoisomers exist, of these D-biotin is the biologically active form and no other isomers are found in nature. The biotin structure is shown in Figure 5.10. The bicyclic ring structure contains an ureido ring which is fused to a tetrahydrothiophene ring with a valeric acid side chain. Biocytin is formed by proteolysis of the biotin-enzyme complex. Biotin is released from biocytin by biotinidase to preserve biotin in the cellular pool. In nature, biotin occurs free or bound through e-amino linkages with lysine as biocytin or to carrier proteins and peptides. It exists in the free state in plants and is soluble in alkali solutions, but sparingly soluble in water (20 mg/100 ml at 25 °C) and 95% ethanol [7].

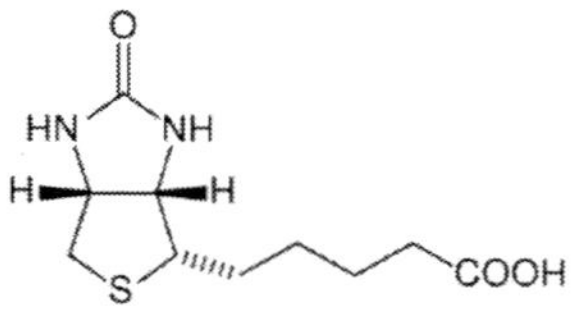

Figure 5.10. Biotin structure.

5.2.11 Vitamin K

The term vitamin K is used as a generic descriptor of 2-methyl-1,4-naphthoquinone [8]. Vitamin K is a fat-soluble vitamin that occurs in nature as a series of molecular forms [22]. Phylloquinone, or vitamin K_1 (Figure 5.11), is predominately found in photosynthetic plants and is the major source of dietary vitamin K [23], including green beans, broccoli, cabbages, cauliflower, carrots, white potatoes, spinach, and tomatoes and in fruits such as grapes, bananas, apples and oranges [8].

5.2.11.1 Properties and chemical structure

Chemically, vitamin K are derivatives of 2-methyl-1,4-naphthoquinone [7]. The compound 2-methyl-3-phytyl-1,4-naphthoquinone (K_1) is preferably called phylloquinone (phytonadione). Menaquinone is chemically 2-methyl-3-farnesyl geranylgeranyl-1,4-naphthoquinone and has 7 isoprenoid units, or 35 carbons in the side chain; it was once called vitamin K_2, but now it is called menaquinone-7 (MK-7). The parent compound of the vitamin K series, 2-methyl-1,4-naphthoquinone, has often been called vitamin K_3 but is more commonly and correctly designated as menadione [8]. The water soluble menadione forms show better stabilities to environmental and food processing conditions compared to free menadione [7].

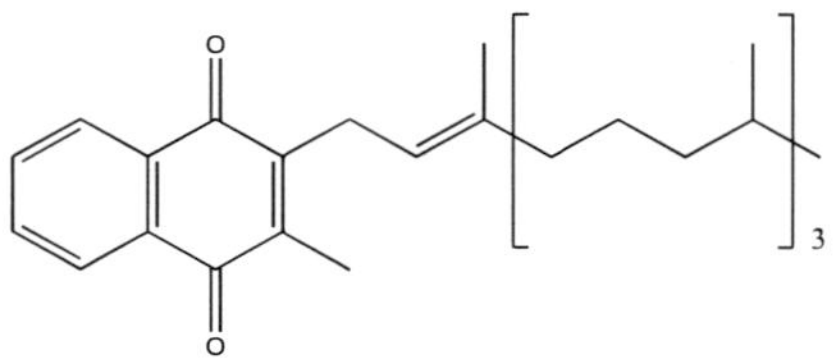

Figure 5.11. Vitamin K_1 structure.

5.3 Effects of Processing Technologies in Fruits and Vegetables

5.3.1 Conventional processing

The nutritional quality of processed food is mainly related to the severity of the applied heat treatments (blanching, pasteurization and sterilization). These thermal processes are designed to inactivate the enzymes responsible for generating off-flavours and odours and to achieve the stabilization of texture, nutritional quality and the destruction of microorganisms. Moreover, blanching as a heat treatment, may contribute for the loss of turgor in cells, due to destruction of membrane integrity and partial degradation of the cell wall [6]. The vitamin losses vary widely according to processing treatment and type of food. The degradation depends on specific parameters during the process [5].

5.3.1.1 Vitamin A and Carotenoids

It is well known that carotenoids are extremely susceptible to degradation. Their highly unsaturated structure is stable under an inert atmosphere; however, it rapidly loses its activity when heated in the presence of oxygen, especially at higher temperature and light intensities. Their instability is very important when the objective is to minimize these losses. Their biological activity can be lost when isomerized, as the *cis* forms are not as active as *trans*. Heating in the absence of oxygen (canned products), induces such changes. Isomerization can also take place when exposed to strong acids or if irradiated with light of different wavelengths, especially near ultraviolet [7]. Table 5.1 shows the effect of different treatments on vitamin A and carotenoids in fruits and vegetables. Some of the reported studies show higher values of γ-carotene and β-carotene after each treatment. These values are related to the higher extractability from the cells, after each processing treatment, and thus presenting an enhanced bioavailability of carotenes.

5.3.1.2 Vitamin B1 (thiamin)

This water-soluble vitamin may leach out during the blanching process into the blanch water, and this is a common cause of vitamin B_1 loss [35]. In food matrices, some components, such as proteins, are known to protect thiamine, though the protective mechanism is not clear [36]. Table 5.2 shows the effects of different processing treatments on vitamin B_1 in vegetables.

5.3.1.3 Vitamin B_2 (riboflavin)

Although, water soluble vitamins are generally sensitive to heat treatments, riboflavin is stable to heat, oxygen and acid medium. Riboflavin is regarded as being one of the more stable vitamins but unstable to alkali and light. In food processing, the losses occur mainly due to leaching, especially during blanching processes [41]. In the processing of lentils (with soaking), riboflavin increase is related to the microorganisms action (*Lactobacillus*). Riboflavin is synthesized by the microorganisms after the soaking process of the seeds [42]. Table 5.3 shows the effects of processing on vitamin B_2 in vegetables.

Table 5.1. Effect of different processing treatments on vitamins A and carotenes in fruits and vegetables.

Fruits and Vegetables	Vitamin A and carotenoids	Processing	Treatment effect (%)	References
Amaranth	β-Carotene	Sterilization	33-84 retention	[24]
		Pressure-cooking	29-79 retention	
Amaranth (spineless)	β-Carotene	Steaming (100 °C/8 min)	15±0.9 increase	[25]
Carrot juice	β-Carotene	Canning (2.67-8.0 min)	77 retention	[26]
Carrot	β-Carotene	Open-pan boiling	33-84 retention	[24]
		Pressure-cooking	29-79 retention	
		Steam cooking (115-120 °C/15 min)	84 retention	[27]
		Water-cooking without pressure/with pressure (99 °C/21 min, 100 °C/17 min)	80 retention /89 retention	[27]
		Canning (2.67-8.0 min)	44 retention	[28]
		Water-cooking without pressure (99 °C/21 min)	89 retention	[27]
	α-Carotene	Water-cooking without pressure (99 °C/21 min)	78 retention	[27]
		Steam cooking (115-120 °C/15 min)	67 retention	[27]
		Water-cooking without pressure/with pressure (99 °C/21 min, 100 °C/17 min)	61 retention /78 retention	[27]
		Moist/dry-cooking (99-200 °C/21-15 min)	56 retention	[27]
	Vitamin A	Water-cooking without pressure/with pressure (99 °C/21 min; 100 °C/17 min)	86 retention /75 retention	[27]
		Moist/dry-cooking (99-200 °C/21-15 min)	66 retention	[27]

Table 5.1. (Continued).

Fruits and Vegetables	Vitamin A and carotenoids	Processing	Treatment effect (%)	References
		Steam-cooking (115-120 ºC/15 min)	80 retention	[27]
Cassava cultivars	Provitamin A activity	Coking (115-120 ºC/15 min)	45-80 retention	[29]
Chinese cabbage	β-Carotene	Blanching (100 ºC/5 min)	93 retention	[25]
Chive leaf	β-Carotene	Blanching (94-96 ºC/90 s)	80 retention	[30]
Green bean	β-Carotene	Sterilization (3.0-5.5 min)	57-79 retention	[31]
		Microwave steaming (3.25-5 min)	74-87 retention	[31]
Ivy gourd	β-Carotene	Blanching (100 ºC/5 min)	16 increase	[25]
Jalapeño pepper	Provitamin A activity	Thermal processing (100 ºC/30 min)	75 retention	[32]
Pumpkin	β-Carotene	Open-pan boiling	33-84 retention	[24]
		Pressure-cooking	29-79 retention	[24]
Spinach	β-Carotene	Sterilization (3.0-5.5 min)	57-79 retention	[31]
		Microwave steaming (3.25-5 min)	74-87 retention	[31]
Swamp cabbage	β-Carotene	Blanching (77-100 ºC/5 min)	89 retention	[25]
		Stir-frying (175-180 ºC/2 min)	82 retention	[25]
Tomato purée	Lycopene	High pasteurization (90 ºC/1 min)	95 retention	[34]
	γ-Carotene	High pasteurization (90 ºC/1 min)	34 increase	[34]
	β-Carotene	High pasteurization (90 ºC/1 min)	0.26 increase	[34]
Water spinach	β-Carotene	Sterilization (3.0-5.5 min)	57-79 retention	[31]
		Microwave steaming (3.25-5 min)	74-87 retention	[31]

Table 5.2. Effect of different processing on Vitamin B_1 in vegetables.

Vegetables	Vitamins B_1	Processing	Treatment effect (%)	References
Bean	Thiamin	Cooking (75 min/soaking 15 h)	73 retention	[37]
Cauliflower		pressure-cooking (10 min)	45 retention	[38]
		Cooking in oil (4 min)	42 retention	[38]
Faba bean		Cooking (45 min)	25 retention	[39]
		Sterilization (121 °C/15 min)	33 retention	[39]
Fresh cabbage		pressure-cooking (5 min)	51 retention	[38]
		Cooking in oil (4 min)	46 retention	[38]
Pea		Cooking (45 min/soaking 16 h)	81 retention	[37]
		pressure-cooking (50 min)	42 retention	[37]
Potato large fries		Blanching 1- Blanching 2 (T<100 °C and very short t)	80-93 retention	[40]
		Par-frying (T<100 °C and very short t)	87 retention	[40]
		Finish deep-frying (T<100 °C and very short t)	87 retention	[40]
Soybean		Cooking (165 min/soaking 16 h)	53 retention	[37]
		pressure-cooking 60 min)	32 retention	[37]

Table 5.3. Effect of different processing on Vitamin B_2 in vegetables.

Vegetables	Processing	Treatment effect (%)	References
Cabbage	Pressure-cooking (5 min)	34 retention	[37]
	Pressure-cooking (10 min)	57 retention	[37]
Chickpea	Citric acid soaking	86 retention	[42]
	Water soaking	92 retention	[42]
Faba bean	Citric acid soaking	97 retention	[42]
	Water soaking	100 retention	[42]
Lentil	Citric acid soaking	95 increase	[42]
	Water soaking	98 increase	[42]
	Water soaking + cooking	13 increase	[42]
Pea	Pressure-cooking 40 min	96 retention	[37]
Spinach	Different cooking 50 to 120 °C/0-60 min	70 retention	[41]

5.3.1.4 Vitamin B_3 (niacin)

The losses of niacin are caused mainly by leaching into the cooking water. Retention of niacin is in the range of 45-90% within the various treatments of legumes. In comparison with the classical procedure, short pressure-cooking may increase niacin retention in legumes, except in some beans (soybean) [5]. Table 5.4 shows the effect of different processing treatments on vitamin B_3 in vegetables.

5.3.1.5 Vitamin B_5 (pantothenic acid)

PA is the most stable vitamin during thermal processing in the pH range of 5-7. Large losses can occur through leaching into cooking water during preparation of vegetables. It is resistant to light and air presence. Some of the factors that significantly influence the retention of the pantothenic acid in legumes are the pre-soaking and processing times [12, 13]. Table 5.5 presents the effect of different processing treatments in the amount of vitamin B_5 in fruits and vegetables.

Table 5.4. Effect of different processing treatments on vitamin B_3 in vegetables.

Vegetables	Vitamins B_3	Processing	Treatment effect (%)	References
Bean	Niacin	Cooking (75 min)	79 retention	[37]
		Pressure-cooking (50 min)	90 retention	[37]
Chickpea		Water soaking + cooking	28 retention	[42]
Faba bean		Cooking (45 min)	6 retention	[39]
		Autoclaving (121 °C/30 min)	7 retention	[39]
		Water soaking + cooking	68 retention	[42]
Lentil		Water soaking + cooking	39 retention	[42]
Pea		Cooking (45 min/soaking 16 h)	67 retention	[37]
		Pressure-cooking (40 min)	87 retention	[37]
Soybean		Cooking (165 min/soaking)	86 retention	[37]
		Pressure-cooking (60 min)	49 retention	[37]

Table 5.5. Effect of different processing on pantothenic acid in fruits and vegetables.

Fruits and vegetables	Vitamin	Processing	Treatment effect (%)	References
Broccoli	B_5	Steam-blanching	87 retention	[43]
Legumes	B_5	Cooking 20 min	76 retention	[12, 13]
		Cooking 20 min/soaking 1 h	33 retention	[12, 13]
		Cooking 20 min/soaking 16 h	44 retention	[12, 13]
		Cooking 90 min/soaking 16 h	58 retention	[12, 13]
		Cooking 150 min/soaking 16 h	55 retention	[12, 13]
Orange juice fortified	B_5	Heat 95 °C/45 s	77-82 retention	[44]
Seabuckthorn juice	B_5	High-temperature short time 90 °C/45 s	93-94 retention	[45]
Seabuckthorn juice concentration		High-temperature short time 90 °C/45 s	77 retention	[45]
Spinach	B_5	Steam-blanching	36 retention	[43]

5.3.1.6 Vitamin B_6

The relatively good retention of pyridoxal and the increase in pyridoxamine after cooking in some vegetables is related to the partial conversion of pyridoxine into pyridoxamine and pyridoxal. Vitamin B_6 is resistant to heat, acid, and alkaline media, but sensitive to light in neutral and alkaline solutions. Pyridoxal and pyridoxamine are more resistant to heat, oxygen, and light than pyridoxine. Holmes et al. [46] mentioned that blanching reduced the vitamin B_6 content of raw green beans by approximately 11%. The dried blanched green beans were approximately 33% lower in vitamin B_6 than the dried unblanched ones. Cooking or canning of vegetables results in losses of 20-40% [46]. Thermal degradation of vitamin B_6 increases as pH rises. The higher losses of vitamin B_6 in vegetables may result from leaching. Pyridoxal and pyridoxamine are more heat labile than pyridoxine, the primary vitamin found in plants [7]. The higher values of vitamin B_6 in some fruits are probably related to higher extractability of the compounds after each treatment and the results reported in a dry basis [46]. Table 5.6 presents the effect of different processing treatments on vitamin B_6 in fruits and vegetables.

5.3.1.7 Vitamin B_9 (folates)

The solubility and reactivity of folates make them susceptible to potentially large losses during food processing and storage. The chemical stability of folates in foods can be adversely influenced by temperature, exposure to oxygen, pH of the medium, folate derivatives, pressure, metal ions, light intensity, antioxidants and heating exposure. Because of their solubility, a large effect of processing may also occur by leaching of folates into surrounding water used for washing, blanching and cooking [48]. However, in some treatments [17], higher values of folates are obtained due to the breakdown of cellular organelles and the polyglutamate conversion to monoglutamate derivatives of the folate. Table 5.7 presents the effect of different processing treatments on vitamin B_9 in vegetables.

Table 5.6. Effect of different processing treatments on vitamin B_6 in fruits and vegetables.

Fruits and Vegetables	Vitamins B_6	Processing	Treatment effect (%)	References
Bean	Pyridoxine	Cooking (75 min)	100 retention	[37]
	Pyridoxine	Pressure-cooking (50 min)	90 retention	[37]
Boysenberry leather	Pyridoxine	Dried (4.5 h)	757 increase	[46]
	Pyridoxidal	Dried (4.5 h)	460 increase	[46]
	Pyridoxamine	Dried (4.5 h)	900 increase	[46]
Broccoli	Pyridoxine	Sterilization (96-99 °C, 150 min)	39 retention	[47]
	Pyridoxine	Steaming	76 retention	[47]
Raspberry leather	Pyridoxine	Dried (4.5 h)	594 increase	[46]
	Pyridoxidal	Dried (4.5 h)	509 increase	[46]
	Pyridoxamine	Dried (4.5 h)	15900 increase	[46]

Table 5.7. Effect of different processing treatments on vitamin B_9 in vegetables.

Vegetables	Vitamin	Processing	Treatment effect (%)	References
Beet	B_9	Canned 105 °C/30 min	76 retention	[49]
	B_9	Canned 115 °C/30 min	69 retention	[49]
	B_9	Canned 125 °C/30 min	61 retention	[49]
Broccoli (florets)	B_9	Boiled (3.5 min)	43 retention	[50]
Broccoli	B_9	Blanching (water)	40 retention	[51]
	B_9	Blanching (steaming)	91 retention	[51]
	5-MTHF	Blanching 90 °C/2 min	90 retention	[52]
	5-MTHF	Blanching 90 °C/4 min	87 retention	[52]
	5-MTHF	Blanching 90 °C/8 min	80 retention	[52]
Brussels sprout	5-MTHF	Blanching 90 °C/2 min	100 retention	[52]
	5-MTHF	Blanching 90 °C/4 min	92 retention	[52]
	5-MTHF	Blanching 90 °C/8 min	89 retention	[52]
Cabbage	B_9	Blanching 96 °C/3 min	40-60 retention	[51]
Carrot	5-MTHF	Blanching 90 °C/4 min	53 retention	[52]
	5-MTHF	Blanching 90 °C/4 min	55 retention	[52]
Cauliflower	B_9	Blanching (8 min)	90 retention	[16]
	B_9	Steaming (7 min)	92 retention	[16]
	B_9	Blanching 96 °C/3 min	40-60 retention	[51]
	5-MTHF	Blanching 90 °C/2 min	89 retention	[52]
	5-MTHF	Blanching 90 °C/4 min	82 retention	[52]
	5-MTHF	Blanching 90 °C/12 min	78 retention	[52]
Green bean	B_9	Canned 100 °C/20 min	97 retention	[49]
	B_9	Canned 115 °C/20 min	95 retention	[49]
	B_9	Blanching (6 min)	79 retention	[16]
	B_9	Steaming (6 min)	90 retention	[16]

Table 5.7. (Continued).

Vegetables	Vitamin	Processing	Treatment effect (%)	References
Leek	B_9	Blanching (5 min)	72 retention	[16]
	B_9	Steaming (5 min)	74 retention	[16]
Pea	B_9	Blanching 95 °C/2 min	65-88 retention	[51]
Potato	B_9	Boiled (60 min)	82 retention	[50]
Savoy cabbage	5-MTHF	Blanching 90 °C/2 min	60 retention	[52]
	5-MTHF	Blanching 90 °C/4 min	57 retention	[52]
	5-MTHF	Blanching 90 °C/8 min	52 retention	[52]
Spinach	B_9	Boiled (10 min)	49 retention	[50]
	B_9	Blanching 96 °C/3 min	30 retention	[51]
	5-MTHF	Blanching 90 °C/2 min	92 retention	[52]
	5-MTHF	Blanching 90 °C/4 min	64 retention	[52]
	5-MTHF	Blanching 90 °C/8 min	40 retention	[52]
Tomato purée	B_9	Pasteurization 98 °C/40 s	25 increase	[17]
	B_9	Pasteurization 108 °C/40 s	6.2 increase	[17]
	B_9	Pasteurization 128 °C/40 s	75 retention	[17]

5.3.1.8 Vitamin C

Vitamin C is the least stable of all vitamins and presents a high rate of destruction during processing and storage. The loss of vitamin is enhanced by the presence of metals (Cu, Fe and Zn) and enzymes. Prolonged heating with oxygen and light exposure are all factors detrimental to vitamin C content in foods. With oxygen present, the contribution of the anaerobic degradation of the total loss of vitamin C is small or negligible, when compared to aerobic degradation that occurs at high rates [53].

Because of its lability, AA is routinely used as an index to measure processing effects on nutrient retention [32]. Nevertheless, Beirão da Costa et al. [57] reported higher values of AA, in minimal processed kiwi, due to the stress imposed by the treatment which led to AA synthesis. Table 5.8 shows the effect of different processing treatments on vitamin C in fruits and vegetables.

5.3.1.9 Vitamin E

Vitamin E is less affected than water-soluble nutrients by processing treatments such as washing, blanching, or cooking, however, it is very susceptible to oxidation.

Table 5.8. Several effects of different processing treatments on vitamin C in fruits and vegetables.

Fruits and vegetables	Vitamin	Processing	Treatment effect (%)	References
Beet	AA	Canned 115 °C/45 min	84 retention	[49]
	AA	Canned 115 °C/30 min	89 retention	[49]
	AA	Canned 125 °C/30 min	81 retention	[49]
	DHAA	Canned 115 °C/45 min	92 retention	[49]
	DHAA	Canned 115 °C/30 min	92 retention	[49]
	DHAA	Canned 125 °C/30 min	95 retention	[49]
Broccoli (florets)	C	Steaming	100 retention	[54]
	C	Conventional 5 min	73 retention	[54]
	AA	Blanching 92-96 °C/60 s	49 retention	[2]
	AA	Blanching 92-96 °C/120 s	50 retention	[2]
	AA	Blanching 92-96 °C/150 s	49 retention	[2]
	AA	Canned 121 °C/30 min	16 retention	[2]
	AA	Canned 121 °C/30 min	27 retention	[55]
	AA	Canned 121 °C/30 min	38 retention	[56]
Broccoli	AA	Blanching 90±2 °C/110 s	75 retention	[32]
Carrot	AA	Blanching 90±2 °C/60 s	87 retention	[32]
Chive	AA	Blanching	71 retention	[30]
Green bean	AA	Canned 115 °C/10 min	88 retention	[49]
	AA	Canned 115 °C/20 min	94 retention	[49]
	AA	Canned 115 °C/40 min	96 retention	[49]
	AA	Canned 115 °C/20 min	94 retention	[49]
	AA	Canned 121 °C/20 min	97 retention	[49]
	AA	Blanching 97±2 °C/120 s	83 retention	[32]
Hot pepper	AA	Blanching 65 °C/120 s	99 retention	[3]
	AA	Blanching 75 °C/120 s	92 retention	[3]
	AA	Blanching 85 °C/120 s	86 retention	[3]
	AA	Blanching 65 °C/240 s	99 retention	[3]
	AA	Blanching 75 °C/240 s	87 retention	[3]
	AA	Blanching 85 °C/240 s	84 retention	[3]
Kiwi	AA	Immersion 45 °C/25 min	12 increase	[57]
	AA	Immersion 45 °C/75 min	123 increase	[57]
Orange juice	C	Low pasteurization 70 °C/30 s	1 retention	[58]
	C	High pasteurization 90 °C/1 min	91 retention	[58]
Paprika	AA	Blanching 100 °C/3 min	72 retention	[59]
	AA	Blanching 100 °C/3 min (hot air)	73 retention	[59]

Table 5.8. (Continued).

Fruits and vegetables	Vitamin	Processing	Treatment effect (%)	References
	AA	Blanching 100 °C/3 min (N_2 gas)	63 retention	[59]
Potato	AA	Blanching 100 °C/3 min (hot air)	44 retention	[59]
	AA	Blanching 100 °C/3 min (N_2 gas)	46 retention	[59]
Red bell pepper	AA	Blanching 70 °C/1 min	8.9 increase	[4]
	AA	Blanching 80 °C/1 min	77 retention	[4]
	AA	Blanching 98 °C/1 min	62 retention	[4]
Strawberry nectar	AA	Pasteurization 85 °C/5 min	35 retention	[60]
	AA	Pasteurization 85 °C/5 min	40 retention	[60]
Strawberry purée	AA	Thermal 70 °C/20 min	77 retention	[61]
Strawberry	AA	Pasteurization 76 °C/20 s	91 retention	[62]
	AA	Sterilization 120 °C/20 min/1.5 atm	77 retention	[62]
Sweet green pepper	AA	Blanching 70 °C/2.5 min	95 retention	[4]
	AA	Blanching 80 °C/2.5 min	73 retention	[4]
	AA	Blanching 98 °C/2.5 min	51 retention	[4]
Sweet pepper	AA	Blanching 65 °C/120 s	92 retention	[3]
	AA	Blanching 75 °C/120 s	91 retention	[3]
	AA	Blanching 85 °C/120 s	89 retention	[3]
Tomato	AA	Baked 220 °C/15 min	98 retention	[63]
Tomato soup	AA	Cooking 25 min	93 retention	[63]
	AA	Cooking 35 min	84 retention	[63]
Tomato purée	C	Low pasteurization 70 °C/30 s	73 retention	[34]
	C	High pasteurization 90 °C/1 min	74 retention	[34]

Several authors have studied the effect of different preservation treatments (blanching, pasteurization, canning, etc) on vitamin E. These studies showed that many of these treatments generate some losses and sometimes increases on the vitamin content, through large losses of water-soluble compounds (hydrophilic) or increasing the extractability and thereby improving the bioavailability of vitamin from the fruit and vegetable matrix [56, 64, 65]. Table 5.9 shows the content of vitamin E for different fruits and vegetables and types of processing.

Table 5.9. Effect of different processing on vitamins E in fruits and vegetables.

Fruits and vegetables	Vitamin	Processing	Treatment effect (%)	References
Asparagus	α-tocopherol	Cooked	33 increase	[56]
	α-tocopherol	Canned (frozen)	27 retention	
Broccoli	E	Boiled	37 increase	[65]
	E	Steamed	1 retention	[65]
	α-tocopherol	Boiled 100 °C/16 min	381 increase	[64]
	α-tocopherol	Steaming 100 °C/10 min	393 increase	[64]
	α-tocopherol	Pressure steaming 120 °C/2 min	431 increase	[64]
Cherry	α-tocopherol	Canned	283 increase	[65]
Onion	E	Boiled	6 retention	[65]
Peach	α-tocopherol	Canned	103 increase	[48]
Pear	α-tocopherol	Dried	29 retention	[65]
Red sweet pepper	α-tocopherol	Boiled 100 °C/6 min	91 retention	[64]
	α-tocopherol	Steaming 100 °C/10 min	86 retention	[64]
	α-tocopherol	Pressure steaming 120 °C/1 min	94 retention	[64]
Spinach	α-tocopherol	Cooked	2.5 increase	[56]
	α-tocopherol	Canned (frozen)	95 retention	[56]
Sweet potato	E	Baked	56 retention	[65]
	α-tocopherol	Cooked	173 increase	[56]
	α-tocopherol	Canned (frozen)	284 increase	[56]
Tomato	E	Boiled	12 increase	[65]
	E	Canned	27 increase	[65]
	E	Juice canned	22 increase	[65]
	E	Purée canned	205 increase	[65]
Tomato slice	α-tocopherol	Baked 180 °C/15 min	1116 increase	[66]
	α-tocopherol	Baked 200 °C/15 min	314 increase	[66]
	α-tocopherol	Baked 220 °C/15 min	286 increase	[66]
Tomato	α-tocopherol	Cooked	9 retention	[56]
	α-tocopherol	Canned (frozen)	13 retention	[56]

5.3.1.10 Vitamin B_7 (biotin)

Biotin is stable when heated in the presence of light and in neutral or even strong acid solutions, but it is labile in alkaline solutions. Biotin retention is less affected by leaching and heat destruction during the preparation of dried legumes than other water soluble vitamins [12]. Biotin is generally regarded as having a good stability, being fairly stable in air, heat and daylight. It can, however, be gradually decomposed by ultraviolet radiation. Biotin in aqueous solutions is relatively stable if the solutions are weakly acidic or weakly alkaline. In strong acid or alkaline solutions the biological activity can be destroyed by heating [7]. Table 5.10 shows the effect of different processing treatments on vitamin B_7 in fruits and vegetables.

Table 5.10. Effect of different processing on vitamin B_7 in fruits and vegetables.

Fruits and Vegetables	Vitamins B_7	Processing	Treatment effect (%)	References
Apple juice	Biotin	Canning, concentration	160 increase	[21]
Legumes	Biotin	Sterilization (20 min)	95 retention	[12, 13]
	Biotin	Boiling and soaking (20, 90, and 150 min)	88, 95, 88 retention	[12, 13]
Orange juice	Biotin	Canning, concentration	743 increase	[21]
Orange juice (fortified)	Biotin	Heat (95 ºC/45 s)	77-82 retention	[44]

5.3.1.11 Vitamin K

The various forms of vitamin K are relatively stable to heat and are retained after most processing treatments. This vitamin is destroyed by sunlight and decomposed by alkalis. Vitamin K_1 (phyloquinone) is only gradually decomposed by atmospheric oxygen [67]. According to Ruan and Chen [68], more vitamin K was retained when microwave heating was used (cooking in a minimum of water). In general, only a few studies deal with the stability of this vitamin. This situation could be result from the good stability of the vitamin during processing treatments combined with sufficient human intake of this nutrient world-wide [7]. Table 5.11 shows the effect of cooking and sterilization treatments on vitamins K in vegetables. Although some increases of vitamin K are shown, differences are mainly related to sampling variation rather than treatment effect [69].

Table 5.11. Effect of different processing on vitamins K in vegetables.

Vegetables	Vitamin K	Processing	Treatment effect (%)	References
Broccoli	Phylloquinone	Cooking	38 increase	[69]
Broccoli raab	Phylloquinone	Cooking	95 retention	[69]
Carrot	Phylloquinone	Cooking	65 increase	[69]
Yellow onion	Phylloquinone	Cooking	10750 increase	[69]
Green pepper	Phylloquinone	Cooking	201 increase	[69]
Red pepper	Phylloquinone	Cooking	237 increase	[69]
Red potato	Phylloquinone	Cooking	91 retention	[69]
Russet potato	Phylloquinone	Cooking	11 increase	[69]
White potato	Phylloquinone	Cooking	56 increase	[69]
Spinach	Phylloquinone	Boiling	47 increase	[69]
Sweet potato	Phylloquinone	Cooking	22 increase	[69]

5.3.2 Emerging technologies

Conventional thermal sterilization processes are the most commonly used methods of food preservation and involve heat transfer from a processing medium to the slowest heating zone of a product and subsequent cooling. Thus, although being effective mechanisms for microbial inactivation, thermal processes can allow changes in product quality and cause off-flavour generation, textural softening and destruction of colours and vitamins, the extent of which is dependent on the product being treated and the temperature gradients between food and process boundaries [70]. In the present days, consumers judge food quality based on its sensory and nutritional characteristics (e.g. flavour, aroma, shape and colour, calorie content, vitamins, etc). With increasing consumer demand for "fresh-like" foods, these now determine an individual's preference for specific high-quality food products; there has been a growing interest in the development of alternative food processing methods such as pulsed electric fields, irradiation, and high-pressure processing [71, 72]. Table 5.12 shows the effect of some emerging technologies on different vitamins present in fruits and vegetables.

5.4 Vitamins Determination in Fruits and Vegetables

There are several studies reporting different methodologies for the determination of vitamins present in different fruits and vegetables. Older procedures for quantification such as colorimetric methodologies, spectrophotometric, spectrofluorometric, polimetric, enzyme protein binding and microbiological assays can be used, but the ones which may deliver higher reliability and precision are the methods using High Performance Liquid Chromatography (HPLC). For the quantification of vitamins it is necessary to submit the samples to various processes before analyzing, which facilitates extraction of the vitamin from the matrix [7]. Table 5.13 shows some of the HPLC methods, currently in use, describing several important features (column, mobile phase and flow rate) for vitamins determination.

Table 5.12. Effect of emerging technologies on vitamins in fruits and vegetables.

Fruits and Vegetables	Vitamins	Processing	Treatment effect (%)	References
Apricot	E	Dried Infrared 50-60 Hz/80 °C	114 increase	[73]
	E	Dried Microwave 2.450 GHz	146 increase	[73]
Bean	Niacin	Microwaving (60 min/soaking 17 h)	56 retention	[37]
	Pyridoxine	Microwaving (60 min/soaking 17 h)	69 retention	[37]
	Thiamin	Microwaving (60 min/soaking 17 h)	49 retention	[37]
Broccoli (florets)	AA	HPP 3 min	70 retention	[54]
		Microwaving 1000 W/5 min	51 retention	[54]
	DHAA	HPP 3 min	17 increase	[54]
		Microwaving 1000 W/5 min	82 retention	[54]
	Folates	HPP 22 °C/200 MPa/5 min	19-57 retention	[74]
	Folates	HPP 25-45 °C/100-600 MPa	22-52 retention	[74]
Cauliflower	Folates	HPP 200 MPa/5 min	57 retention	[16]
	Phylloquinone	Microwaving in small amount of water (14 min)	57 retention	[38]
	Phylloquinone	Microwaving in oil (3 min)	47 retention	[38]
Carrot	Phylloquinone	Microwaved	82 retention	[69]
Fresh cabbage	Phylloquinone	Microwaving in small amount of water (8 min)	71 retention	[38]
	Phylloquinone	Microwaving in oil (3 min)	62 retention	[38]
Green bean	Folates	HPP 200 MPa/5 min	53 retention	[16]
Kiwi	C	γ-irradiation	86 retention	[75]
	C	γ-irradiation	50 retention	[76]
Leek	Folates	HPP 200 MPa/5 min	19 retention	[16]
Lettuce	C	γ-irradiation	60-100 retention	[77]
Multivitamin model system	Pyroxidal	UHPP 200 MPa/30 min	99.8 retention	[62]
		UHPP 400 MPa/30 min	99.4 retention	[62]
		UHPP 600 MPa/30 min	99.6 retention	[62]

Orange juice	C	HPP 100 MPa/5 min/60 °C	90 retention	[33]
		HPP 300 MPa/2.5 min/30 °C	98 retention	[33]
		HPP 400 MPa/1 min/40 °C	92 retention	[33]
		HPP 400 MPa/1 min/40 °C	96 retention	[33]
		Pulsed electric fields (PEF) 35 kV/cm/750 μs	93 retention	[33]
		HIPEF 15, 25 to 35 kV/cm for 100, 400 and 1000 μs pulses 4 μs with 200 Hz in mono-bipolar mode	88-98 retention	[78]
	Carotenoids	HPP 400 MPa/40 °C/1 min	54 increase	[58]
		PEF 35 kV cm^{-1} / 750 μs	91 retention	[58]
	Lutein	HPP 400 MPa/40 °C/1 min	75 increase	[58]
		PEF 35 kV cm^{-1} /750 μs	94 retention	[58]
	Vitamin A	HPP 400 MPa /40 °C/1 min	39 increase	[58]
		PEF 35 kV cm^{-1} /750 μs	91 retention	[58]
	Zeaxanthin	HPP 400 MPa/40 °C/1 min	45 increase	[58]

Table 5.12. (Continued).

Fruits and Vegetables	Vitamins	Processing	Treatment effect (%)	References
Orange juice	α- cryptoxanthin	PEF 35 kV cm^{-1} /750 µs	86 retention	[58]
		HPP 400 MPa/40 °C/1 min	46 increase	[33]
	α-Carotene	PEF 35 kV cm^{-1} /750 µs	88 retention	[58]
		HPP 100 MPa/60 °C/5 min	5 increase	[33]
		HPP 400 MPa/40 °C/1 min	34 increase	[58]
	α-Carotene	PEF 35 kV cm^{-1} /750 µs	89 retention	[58]
	β- cryptoxanthin	HPP 400 MPa/40°C/1 min	43 increase	[58]
	β-Carotene	PEF 35 kV cm^{-1} /750 µs	98 retention	[58]
		HPP 100 MPa/60 °C/5min	8 increase	[33]
		HPP 400 MPa /40 °C/1 min	30 increase	[58]
		PEF 35 kV cm^{-1} /750 µs	83 retention	[58]
Orange, lemon and carrot mixed juice	Carotenoids	HPP 14 kbar /up to 4 °C /1 to 2 min	4 increase	[79]
Persimmon Flesh (Cv. Rojo brillante)	Vitamin A	HPP 50 MPa/15 min/25 °C	99 retention	[80]
		HPP 400 MPa/ 15 min/25 °C	81 retention	[80]
	Carotenoids	HPP 50 MPa/15 min/25 °C	5 increase	[80]
		HPP 400 MPa/15 min/25 °C	89 retention	[80]

Persimmon Flesh (Cv. Sharon)	Vitamin A	HPP 50 MPa/15 min/25 °C	16 increase	[80]
		HPP 400 MPa/15 min/25 °C	16 increase	[80]
	Carotenoids	HPP 50 MPa/15 min/25 °C	20 increase	[80]
		HPP 400 MPa/15 min/25 °C	16 increase	[80]
Red bell pepper	AA	HPP 100 MPa/10 min	16 increase	[4]
		HPP 100 MPa/20 min	16 increase	[4]
		HPP 200 MPa/10 min	26 increase	[4]
		HPP 200 MPa/20 min	14 increase	[4]
Spinach	Phylloquinone	Microwaved	24 increase	[69]
Strawberry purée	AA	HPP 400 MPa/15 min/20 °C	91 retention	[61]
		HPP 500 MPa/15 min/20 °C	91 retention	[61]
		HPP 600 MPa/15 min/20 °C	95 retention	[61]
Strawberry	AA	HPP 400 MPa/30 min	89 retention	[62]
Strawberry juice	C	HIPEF 35 kV/cm for 1000 μs pulses between 1-7 μs width 50 to 250 Hz frecuency in ono-bipolar mode	98-99 retention	[81]
Sweet green pepper	AA	HPP 100 MPa/10 min	84 retention	[4]
		HPP 100 MPa/20 min	88 retention	[4]
		HPP 200 MPa/10 min	94 retention	[4]
		HPP 200 MPa/20 min	79 retention	[4]

Table 5.12. (Continued).

Fruits and Vegetables	Vitamins	Processing	Treatment effect (%)	References
Tomato	Carotenoids	HPP (14 kbar/up to 4 °C/1 to 2 min)	97 retention	[79]
Tomato purée	C	HPP 400 MPa/25 °C/15 min	70 retention	[34]
	Carotenoids	HPP 400 MPa/25 °C/15 min	48 increase	[34]
	Lutein	HPP 400 MPa/25 °C/15 min	71 increase	[34]
	Vitamin A	HPP 400 MPa/25 °C/15 min	39 increase	[34]

Table 5.13. Methodologies for vitamins determination in fruits and vegetables.

Analytes	Methods	Conditions	References
α- and β-carotene	HPLC	**Column:** Lichrospher (5 μm RP-18 25 cm x 4 mm), **Mobile phase:** Methanol:Acetronitrile:ethyl acetate (80:10:10)	[27]
Vitamin B_1 (thiamin)	HPLC	**Column:** Ultrasphere reversed phase column (4.6 mm x 25 cm, 5 μm) **Mobile Phase:** Methanol/0.01 M citric acid buffer (35/65), pH 7.0, 1 ml/min	[82]
Vitamin B_2 (riboflavin)	HPLC	**Column:** μ-Bondapck C_{18}, **Mobile Phase:** 45% methanol in H_2O 0.005 M PIC B7	[83]
		Column: LiChrosorb C_{18} **Mobile phase:** Isocratic 0.005 M 1-Hex sulfonic acid:MeOH (60:40), 1 ml/min	[7]
Vitamin B_3 (niacin)	HPLC	**Column:** μBondapak C_{18} (30 cm x 3.9 mm, 10 μm) **Mobile Phase:** Solution A (methanol/0.01M sodium acetate buffer (pH 4.66, 1/9, v/v) and 5 mM tetrabutyl ammonium bromide] and solution B (methanol/0.01M sodium acetate buffer (pH 4.66, 9/1, v/v) and 5 mM tetrabutyl ammonium bromide, Detection: 254 nm	[84]
		Column: Spherisorb ODS2 (30 cm x 3.9 mm, 10 μm) **Mobile Phase:** Isocratic 0.005M TBAB in MeOH : 0.01M NaOAC (1:9), pH 4.72, 1.5 ml/min	[85]
Vitamin B_6 (PN, PL, PM)	HPLC	**Column:** RP-18, 5 μm, 12.5 cm x 4mm **Mobile Phase:** Gradient A: MeOH, B: 0.03 M KH_2PO_4, pH 2.7 with 4 mM, octane sulfonic acid, 0-2 min; 90% B, 10% A, 2-12 min to 60% B; 12-17 min; 60% B; 40% A; 17-19 min; To 90% B;1.5 ml/min	[7]
PL		**Column:** C_{18}, 5 μm, 250 x 4.6 mm. **Mobile Phase:** Isocratic solvent system consisting of 95% of 50 mM dipotassium phosphate and 5% acetonitrile, the flow rate being 1 ml/min	[62]
Vitamin B_9 (folates)	HPLC	Extraction and deconjugation of folate, purification of extract, chromatographic separation of folate isomers, detection and quantification against folate standards	[86]

Table 5.13. (Continued).

Analytes	Methods	Conditions	References
Vitamin C	HPLC	**Column:** Spherisorb ODS2 (125 x 4 mm, 5 μm) **Mobile phase:** Isocratic 0.08 M KH_2PO_4 : MeOH (80:20) pH 7.8, 1.0 ml/min	[7]
		Column: Supelcosil LC-18 (250 x 4.6 mm, 5 μm) **Mobile phase:** Isocratic, acidified deionized water, (pH 2.7), 0.5 ml/min	[49]
		Column: Eurospher 100 (250 x 4 mm, 5 μm) **Mobile phase:** Water (pH 2.2), 1.0 ml/min	[63]
		Column: Rainin Dynamax-60Å amine (250 x 4.6 mm protected by Rainin amine guard module 8 μm 1.5 cm) **Mobile phase:** Isocratic Acetonitrile: 0.05 M KH_2PO_4 (75:25; pH 5.95), 1.5 ml/min	[32]
		Column: NH_2- Spherisorb S5 (250 x 4.6 mm, 5 μm) **Mobile phase:** Acetonitrile:5 mM potassium dihydrogen phosphate (40:60, pH 3.5), 1.0 ml/min	[78]
		Column: Hypersil ODS **Mobile phase:** Water : metaphosphoric acid (pH 2.2), 0.5 ml/min	[87]
Carotenoids	HPLC	**Column:** Reversed-phase C_{18} Hypersil ODS (250 x 4.6 mm, 5 μm) **Mobile phase:** Methanol/water (75:25, solution A); acetonitrile/dichloromethane/methanol (70:5:25 solution B)	[33]
Vitamin E	HPLC	**Column:** Lichrosorb Si60 (250 x 4 mm, 5 μm) **Mobile phase:** Isocratic 0.9% isopropanol in n-hexane 1.0 ml/min	[65]
		Column: Lichrosorb (250 x 4.6 mm, 10 μm) **Mobile phase:** n-hexane-ethanol, (99.5:0.5), 1.5 ml/min	[88]
		Column: Supelcosil LC-18 (250 x 4.6 mm, 5 μm) **Mobile phase:** Methanol : H_2O (98:2), 1.5 ml/min	[19]

Table 5.13. (Continued).

Analytes	Methods	Conditions	References
Vitamin B_7 (biotin)	HPLC	**Column:** RP 18 endcapped (5 mm x 250 mm, octadecylsilyl, 5 μm) and guard column RP 18 (4 mm x 4 mm, octadecylsilyl, 5 μm) **Mobile Phase:** Isocratically, 0.1 M phosphate buffer (pH 6) methanol (81:19 v/v), 0.4-1 ml/min **Column:** RP 18 endcapped (5 mm x 250 mm, octadecylsilyl, 5 μm) and guard column RP 18 (4 mm x 4 mm, octadecylsilyl, 5 μm) **Mobile Phase:** Isocratically, 0.1 M phosphate buffer (pH 6) methanol (81:19 v/v), at 0.4-1 ml/min, Detection: 490-520 nm	[90]
		Column: Reverse-phase C_{18} (150 x 2.1 mm, 0.3 μm) **Mobile Phase:** 0.1% formic acid (solvent A) and methanol (solvent B). The linear gradient 0-5 min: 20% B; 5-15 min 20-80% B; 15-15.1 min: 80-20% B; 15.1-25 min: 20% B. Flow 0.2 ml/min, injection volume 20 μl	[44]
Vitamin K	HPLC	**Column:** Hypersil ODS (15 cm x 4.6 mm, 3 μm,) **Mobile Phase:** Isocratic, MeOH : CH_2Cl_2 (90:10), add 5 ml of a solution of 2 M ZnCl, 1 M HAC and 1 M NaOAC per 100 ml, 1 ml/min	[91]
		Column: C_{18} analytical column (150 x 3mm), packed with 5 mm **Mobile Phase:** Solvent A: methanol with 10 mM $ZnCl_2$, 5.0mM CH_3COOH, and 5.0 mM CH_3COONa. Solvent B: Methylene chloride	[69]

CONCLUSION

All vitamins exhibit a degree of instability and the rate at which they are degraded is affected by a number of factors. Naturally-occurring vitamins in foods are susceptible to many of these factors during processing and shelf-life. The most common factor during processing is the application of heat (blanching, sterilization and cooking). That is why the application of new technologies (HPP, HIPEF and others), whose principle of processing is based not only on the application of heat, but a combination of factors, is becoming a strong research priority, since they have shown promising results, both in safety and nutritional quality.

REFERENCES

[1] Sila, D.N., Duvetter, T., De Roeck, A., Verlent, I., Smout, C., Moates, G., Hills, B., Waldron, K.K., Hendrickx, M., & Van Loey, A. (2008). Texture changes of processed fruits and vegetables: potential use high-pressure processing. *Trends in Food Science and Technology*, *19*, 309-319.

[2] Murcia, M.A., López-Ayerra, B., Martínez-Tomé, M., Vera, A.M., & García-Carmona, F. (2000). Evolution of ascorbic acid and peroxidase during industrial processing of broccoli. *Journal of the Science of Food and Agriculture*, *80*, 1881-1886.

[3] Orak, H.H., & Demirci, M. (2005). Effect of different blanching methods and period of frozen storage on enzyme activities and some quality criterias of hot and sweet red peppers (*Capsicum annuum* L). *Pakistan Journal of Biological Sciences*, *8*(4), 641-648.

[4] Castro, S.M., Saraiva, J.A., Lopes-da-Silva, J.A., Delgadillo, I., Loey, A.V., Smout, C., & Hendrickx, M. (2008). Effect of thermal blanching and of high pressure treatments on sweet green and red bell pepper fruits (*Capsicum annuum* L.). *Food Chemistry*, *107*, 1436-1449.

[5] Lešková, E., Kubíková, J., Kováčiková, E., Kosická, M., Porubská, J., & Holčíková, K. (2006). Vitamin losses: Retention during heat treatment and continual changes expressed by mathematical models review. *Journal of Food Composition and Analysis*, *19*, 252-276.

[6] Olivera, D.F., Viña, S.Z., Marani, C.M., Ferreyra, R.M., Mugridge, A., & Chaves, A.R. (2008). Mascheroni RH. Effect of blanching on the quality of Brussels sprouts (*Brassica oleracea* L. *gemmifera* DC) after frozen storage. *Journal of Food Engineering*, *84*, 148-155.

[7] Eitenmiller, R.R., & Laden, W.O. (1999). Vitamin A and β-carotene, ascorbic acid, Thiamin, Vitamin B_6, Folate. In R., Eitenmiller, & W.O., Laden (Eds.), *Vitamin Analysis for the Health and Food Science*. CRC Press, Boca Raton, FL (pp. 15-19, 226-228, 275, 375, 411-465).

[8] Rucker, R.B., Suttie, J.W., Cornick, M.C., Cornick, D.B., & Machlin, L.J. (2001). Handbook of vitamins. Marcel Dekker. New York. NY.

[9] Badui, S. (2006). "Química de los alimentos: vitaminas y nutrimentos inorgánicos". Grupo Herdez, Ed. Pearson Educación de México. S.A. pp. 363-399.

[10] Bhat, S.V., Nagasampagi, B.A., & Sivakumar, M. (2005). Chemistry of natural products. Chapter 12: Vitamins. *Narosa Publishing House*. Pag. 756-794.

[11] Golbach, J.L., Chalova, V.I., Woodward, C.L., & Ricke, S.C. (2007). Adaptation of *Lactobacillus rhamnosus* riboflavin assay to microtiter plates. *Journal of Food Composition and Analysis*, *20*, 568-574.

[12] Hoppner, K., & Lampi, B. (1993). Folate retention in dried legumes after different methods of meal preparation. *Food Research International*, *26*, 45-48.

[13] Hoppner, K., & Lampi, B. (1993). Pantothenic acid and biotin retention in cooked legumes. *Journal of Food Science*, *58*(5), 1084-1089.

[14] Smith, C.M., & Song, W.O. (1996). Review. Comparative nutrition of pantothenic acid. *Nutritional Biochemistry*, *7*, 312-321.

[15] Stea, T.H., Johansson, M., Jägerstad, M., & Frølich, W. (2006). Retention of folates in cooked, stored and reheated peas, broccoli and potatoes for use in modern large-scale service systems. *Food Chemistry*, *101*, 1095-1107.

[16] Melse-Boonstra, A., Verhoef, P., Konings, E.J.M., Dusseldorp, M., Matser, A., Hollman, P.C.H., Meyboom, S., Kok, F.J., & West, C.E. (2002). Influence of processing on total monoglutamate and polyglutamate folate contents of leeks, cauliflower, and green beans. *Journal of Agricultural of Food Chemistry*, *50*, 3473-3478.

[17] Iniesta, M.D., Pérez-Conesa, D., García-Alonso, J., Ros, G., & Periago, M.J. (2009). Folate content in tomato (*Lycopersicon esculentum*). Influence of cultivar, ripeness, year of harvest, and pasteurization and storage temperatures. *Journal of Agricultural and Food Chemistry*, *57*, 4739-4745.

[18] Hernández, Y., Lobo, M.G., & González, M. (2006). Determination of vitamina C in tropicals fruits: A comparative evaluation of methods. *Food Chemistry*, *96*, 654-664.

[19] Wyatt, J.C., Carballido, S.P., & Méndez RO. (1998). α- and γ-Tocopherol contento f selected foods in the Me-ican diet: effect of cooking losses. *Journal of Agricultural and Food Chemistry*, *46*, 4657-4661.

[20] Bramley, P.M., Elmadfa, I., Kafatos, A., Kelly, F.J., Manios, Y., Ro-borough, H.E., Schuch, W., Sheehy, P.J.A., & Wagner, K-H. (2000). Vitamin E Review. *Journal of the Science of Food and Agriculture*, *80*, 913-938.

[21] Staggs, C.G., Sealey, W.M., McCabe, B.J., Teague, A.M., & Mock, D.M. (2004). Determination of the biotin content of select foods using accurate and sensitive HPLC/avidin binding. *Journal of Food Composition and Analysis*, *17*, 767-776.

[22] Shearer, M.J., Bach, A., & Kohlmeier, M. (1996). Chemistry, nutritional sources, tissue distribution and metabolism of vitamin K with special reference to bone health. *Journal of Nutrition*, *126*, 1181S-1186S.

[23] Booth, S.L., Webb, D.R., & Peters, J.C. (1999). Assessment of phylloquinone and dihydro-vitamin K dietary intakes among a nationally representative sample of U.S. consumers using 14-day food diaries. *Journal of American Dietetic Association*, *99*, 1072-1076.

[24] Gayathri, G.N., Platel, K., Prakash, J., & Srinivasan, K. (2004). Influence of antioxidant spices on the retention of β-carotene in vegetables during domestic cooking processes. *Food Chemistry*, *84*(1), 35-48.

[25] Sungpuag, P., Tangchitpianvit, S., Chittchang, U., & Wasantwisut, U. (1999). Retinol and β-carotene content of indigenous raw and home prepared foods in Northeast Thailand. *Food Chemistry*, *64*, 163-167.

[26] Kim, H.Y., & Gerber, L.E. (1988). Influence of processing on quality of carrot juice. *Korean Journal of Food Science and Technology*, *20*, 683-690.

[27] Sant'ana, H.M.P., Stringheta, P.C., Brandão, S.C.C., & Cordeiro de Azeredo, R.M. (1998). Carotenoid retention and vitamin A value in carrot (*Daucus carota* L.) prepared by food service. *Food Chemistry*, *61*, 145-151.

[28] Olunlesi, A.T., & Lee, C.Y. (1979). Effect of thermal processing on the stereoizomerization of major carotenoids and vitamin A value of carrots. *Food Chemistry*, *4*, 311-318.

[29] Penteado, M.V.C., & Almeida, L.B. (1988). Occurrence of carotenoids in roots of five cassava (*Manihot esculenta crantz*) cultivars of Sao Paulo State. *Revista de Farmacia e Bioquimica da Universidade de Sao Paulo*, *24*(1), 39-49.

[30] Kmiecik, W., & Lisiewska, Z., (1999). Effect of pretreatment and conditions and period of storage on some quality indices of frozen chive (*Allium schoenoprasum* L.). *Food Chemistry*, *67*, 61-66.

[31] Masrizal, M.A., Giraud, D.W., & Driskell, J.A. (1997). Retention of vitamin C, iron and b-carotene in vegetables prepared using different cooking methods. *Journal of Food Quality*, *20*, 403-418.

[32] Howard, L.A., Wong, A.D., Perry, A.K., & Klein, B.P. (1999). β-Carotene and ascorbic acid retention in fresh and processed vegetables. *Journal of Food Science*, *64*(5), 929-936.

[33] Sánchez-Moreno, C., Plaza, L., De Ancos, B., & Cano, M.P. (2003). Vitamin C, provitamin A carotenoids, and other carotenoids in high-pressurized orange juice during refrigerated storage. *Journal Agricultural and Food Chemistry*, 51, 647-653.

[34] Sánchez-Moreno, C., Plaza, L., De Ancos, B., & Cano, M.P. (2006). Impact of high-pressure and traditional thermal processing of tomato purée on carotenoids, vitamin C and antioxidant activity. *Journal of the Science of Food and Agricultural*, *86*, 171-179.

[35] Tannenbaum, S. R., Archer, M. C., & Young, V. R. (1985). Vitamins and minerals. In O. R. Fennema (Ed.), *Food Chemistry* (2nd ed., pp. 499, 523-524). New York: Marcell Dekker Inc.

[36] Dwiveldi, B.K., & Arnold, R.G. (1973). Chemistry of thiamine degradation in food products and model systems: a review. *Journal Agricultural Food Chemistry*, *21*, 54-60.

[37] Mašková, E., Rysová, J., Fiedlerová, V., Holasová, M., & Vavreinová, S. (1996). Stability of selected vitamins and minerals during culinary, treatment of legumes. *Potravinářské Vědy*, *14*(5), 321-328.

[38] Vargas, F.E., Villanueva, O.M.T., & Marquina, D.A. (1995). Influences of cooking on hydrosoluble vitamins in cabbage: cabbage and cauliflower I. Thiamin and riboflavin. *Alimentaria*, *261*, 111-117.

[39] Khalil, A.H., & Mansour, E.H. (1995). The effect of cooking, autoclaving and germination on the nutritional quality of faba beans. *Food Chemistry*, *54*, 177-182.

[40] Fillion, L., & Henry, C.J.K. (1998). Nutrient losses and gains during frying: A review. *International Journal of Food Sciences and Nutrition*, *49*, 157-168.

[41] Nisha, P., Singhal, R.S., & Pandit, A.B. (2005). A study on degradation kinetics of riboflavin in spinach (*Spinacea aleracea* L). *Journal of Food Engineering*, *67*, 407-412.

[42] Prodanov, M., Sierra, I., & Vidal-Valverde, C. (2004). Influence of soaking and cooking on the thiamin, riboflavin and niacin content of legumes. *Food Chemistry*, *84*, 271-277.

[43] Cheng, T.S., & Eitenmiller, R.R. (2007). Effects of processing and storage on the pantothenic acid content of spinach and broccoli. *Journal of Food Processing and Preservation*, *12*(2), 115-123.

[44] Rivas, A., Rodrigo, D., Company, B., Sampedro, F., & Rodrigo, M. (2007). Effects of pulsed electric fields on water-soluble vitamins and ACE inhibitory peptides added to amixed orange juice and milk beverage. *Food Chemistry*, *104*, 1550-1559.

[45] Gutzeit, D., Klaubert, B., Rychlik, M., Winterhalter, P., & Jerz, G. (2007). Effects of processing and of storage on the stability of pantothenic acid in sea buckthorn products (*Hippophaë rhamnoides* L. ssp. *rhamnoides*) assessed by stable isotope dilution assay. *Journal of Agricultural and Food Chemistry*, *55*, 3978-3984.

[46] Holmes, Z.A., Miller, L., Edwards, M., & Benson, E. (1979). Vitamin retention during home drying of vegetables and fruits. *Home Economics Research Journal*, *7*(4), 258-264.

[47] Bognár, A. (1993). Studies on the influence of cooking on the vitamin B_6 content of food. Bioavailability '93—nutritional, chemical and food processing implications of nutrient availability. Part II, pp. 346-351. ISSN: 0933-5463.

[48] Piironen, V., Syvãoja, E.L., Varo, P., Salminen, K., & Koivistoinen, P. (1986). Tocopherols and tocotrienols in Finnish foods: vegetables, fruits and berries. *Journal of Agricultural and Food Chemistry*, *34*, 742-746.

[49] Jiratanan, T., & Liu, R.H. (2004). Antioxidant activity of processed table beets (*Beta vulgaris var, conditiva*) and green beans (*Phaseolus vulgaris L*). *Journal of Agricultural and Food Chemistry*, *52*, 2659-2670.

[50] McKillop, D.J., Pentieva, K., Daly, D., McPartlin, M., Hughes, J., Strain, J.J., Scott, J.M., & McNulty, H. (2002). The effect of different cooking methods on folate retention in various foods that are amongst the major contributors to folate intake in the UK diet. *British Journal of Nutrition*, *88*, 681-688.

[51] Puupponen-Pimiä, R., Häkkinen, S.T., Aarni, M., Suortti, T., Lampi A.M., Eurola, M., Piironen, V., Nuutila, A.M., & Oksman-Caldentey, K.M. (2003). Blanching and long-term freezing affect various biactives compounds of vegetables in different ways. *Journal of the Science of Food and Agriculture*, *83*, 1389-1402.

[52] Holasová, M., Fiedlerová, V., & Vavreinová, S. (2008). Determination of folates in vegetables and their retention during boiling. *Czech J. Food Sci.*, 26, 31-37.

[53] Cruz, R.M.S., Vieira, M.C., & Silva, C.L.M. (2008). Effect of heat and thermosonication treatments on watercress (*Nasturtium officinale*) vitamin C degradation kinetics. *Innovative Food Science and Emerging Technologies*, *9*, 483-488.

[54] Vallejo, F., Tomás-Barberán, F.A., & García-Viguera, C. (2002). Glucosinolates and vitamin C content in edible parts of broccoli florets alter domestic cooking. *European Food Research and Technology*, *215*, 310-316.

[55] Weits, J.M., Lessche, M.A., Meyer, J.B., Steinbuch, J.C., & Gersons, L. (1970). Nutritive value and organoleptic properties of three vegetables fresh and preserved in six different ways. *International Journal Vitamin and Nutrition Research*, *40*, 648-658.

[56] Rickman, J.C., Bruhn, C.M., & Barrett, D.M. (2007). Review Nutritional comparison of fresh, frozen, and canned fruits and vegetables II. Vitamin A and carotenoids, vitamin E, minerals and fiber. *Journal of the Science of Food and Agriculture*, *87*, 1185-1196.

[57] Beirão-da-Costa, S., Steiner, A., Correia, L., Leitão, E., Empis, J., & Moldão-Martins, M. (2008). Influence of moderate heat pre-treatments on physical and chemical chararcteristics of kiwifruits slices. *European Food Research and Technology*, *226*, 641-651.

[58] Sánchez-Moreno, C., Plaza, L., Elez-Martinez, P., De Ancos, B., Martín-Belloso, O., & Cano, M.P. (2005). Impact of high pressure and pulsed electric fields on bioactive compounds and antioxidant activity of orange juice in comparison with traditional thermal processing. *Journal of Agricultural and Food Chemistry*, *53*, 4403-4409.

[59] Ramesh, M.N., Wolf, W., Tevini, D., & Jung, G. (1999). Studies on inert gas processing of vegetables. *Journal of Food Engineering, 40*, 199-205.

[60] Klopotek, Y., Otto, K., & Bohm, V. (2005). Processing strawberries to different products alters contents of vitamin C, total phenolics, total anthocyanins, and antioxidant capacity. *Journal of Agricultural and Food Chemistry, 53*, 5640-5646.

[61] Patras, A., Brunton, N.P., Da Pieve, S., & Butler, F. (2009). Impact of high pressure processing on total antioxidant activity, phenolic, ascorbic acid, anthocyanin content and colour of strawberries and blackberries purées. *Innovative Food Science and Emerging Technologies, 10*, 308-313.

[62] Sancho, F., Lambert, Y., Demazeau, G., Largeteau, A., Bouvier, J.M., & Narbonne, J.F. (1999). Effect of ultra-high hydrostatic pressure on hydrosoluble vitamins. *Journal of Food Engineering, 39*, 246-253.

[63] Gahler, S., Otto, K., & Böhm, V. (2003). Alterations of vitamin C, total phenolics, and antioxidant capacity as affected by processing tomatoes to different products. *Journal of Agricultural and Food Chemistry, 51*, 7962-7968.

[64] Bernhardt, S., & Schlich, E. (2006). Impact of different cooking methods on food quality: Retention of lipophilic vitamins in fresh and frozen vegetables. *Journal of Food Engineering, 77*, 327-333.

[65] Chun, J., Lee, J., Ye, L., Exler, J., & Eitenmiller, R.R. (2006). Tocopherol and tocotrienol contents of raw and processed fruits and vegetables in the United States diet. *Journal of Food Composition and Analysis, 19*, 196-204.

[66] Seybold, C., Fröhlich, K., Bitsch, R., Otto, K., & Böhm, V. (2004). Changes in content of carotenoids and vitamin E during tomato processing. *Journal of Agricultural and Food Chemistry, 52*, 7005-7010.

[67] Ottaway, P.B. (2002). The stability of vitamins during food processing: vitamin K. In C.J.K., Henry, & C., Chapman (Eds.), *The Nutrition Handbook for Food Processors*. CRC Press, Boca Raton, FL, pp. 247-264.

[68] Ruan, X.Y.R., Chen, P., Donna, C., & Taub, I. (2002). University of Minnesota. Ohmic heating: the effect of ohmic heating on nutrient loss: thermal destruction. In C.J.K., Henry, & C., Chapman (Eds.), *The Nutrition Handbook for Food Processors*. CRC Press, Boca Raton, FL, p. 413.

[69] Damon, M., Zhang, N.Z., Haytowitz, D.B., & Booth, S.L. (2005). Phylloquinone (Vitamin K_1) content of vegetables. *Journal of Food Composition and Analysis, 18*, 751-758.

[70] Norton, T., & Sun, D.W. (2008). Recent advances in the use of high pressure as an effective processing technique in the food industry. *Food and Bioprocess Technology, 1*, 2-34.

[71] Caner, C., Hernandez, R.J., Pascall, M., Balasubramaniam, V.M., & Harte, B.R. (2004). The effect of high-pressure food processing on the sorption behaviour of selected packaging materials. *Packaging Technology and Science, 17*, 139-153.

[72] Galotto, M.J., Ulloa, P.A., Hérnandez, D., Fernández-Martín, F., Gavara, R., & Guarda, A. (2008). Mechanical and thermal behaviour of flexible food packaging polymeric films materials under high pressure/temperature treatments. *Packaging Technology and Science, 21*(5), 297-308.

[73] Karatas, K., & Kamişli, F. (2007). Variations of vitamins (A, C and E) and MDA in apricots dried in IR and microwave. *Journal of Food Engineering, 78*, 662-668.

[74] Verlinde, P., Oey, I., Hendrickx, M., & Loey, A.V. (2008). High-pressure treatments induce folate polyglutamate profile changes in intact broccoli (*Brassica oleraceae* L. *cv. Italica*) tissue. *Food Chemistry*, *111*, 220-229.

[75] Kyoung-Hee, K., & Hong-Sun, Y. (2009). Effect of gamma irradiation on quality of kiwifruit (*Actinidia deliciosa* var. *deliciosa* cv. Hayward). *Radiation Physics and Chemistry*, *78*, 414-421.

[76] Harder, M.N., De Toledo, T.C., Ferreira, A.C., & Arthur, V. (2009). Determination of changes induced by gamma radiation in nectar of kiwi fruit (*Actinidia deliciosa*). *Radiation Physics and Chemistry*, *78*, 579-582.

[77] Zhang, L., Lu, Z., Lu, F., & Bie, X. (2006). Effect of γ irradiation on quality-maintaining of fresh-cut lettuce. *Food Control*, *17*, 225-228.

[78] Elez-Martínez, P., & Martín-Belloso, O. (2007). Effects of high intensity pulsed electric field processing conditions on vitamina C and antioxidante capcity of orange juice and *gazpacho*, a cold vegetable soup. *Food Chemistry*, *102*, 201-209.

[79] Butz, P., Fernández García, A., Lindauer, R., Dieterich, S., Bognár, A., & Tauscher, B. (2003). Influence of ultra high pressure processing on fruit and vegetable products. *Journal of Food Engineering*, *56*, 233-236.

[80] De Ancos, B., Gonzales, E., & Cano, P. (2000). Effect of high-pressure treatment on the carotenoid composition and the radical scavenging activity of persimmon fruit purees. *Journal Agricultural Food Chemistry*, *48*, 3542-3548.

[81] Odriozola-Serranom, I., Soliva-Fortuny, R., & Martín-Belloso, O. (2009). Impact of high-intensity pulsed Electric fields varibles on vitamin C, anthocyanins and antioxidant capacity of strawberry juice. *LWT Food Science and Technology*, *42*, 93-100.

[82] Duodu, K.G., Minnaar, A., & Taylor J.R.N. (1999). Effect of cooking and irradiation on the labile vitamins and antinutrient content of a traditional African sorghum porridge and spinach relish. *Food Chemistry*, *66*, 21-27.

[83] Kumar, S., & Aalbersberg, B. (2006). Nutrient retention in foods after earth-oven cooking compared to other forms of domestic cooking 2. Vitamins. *Journal of Food Composition and Analysis*, *19*, 311-320.

[84] Prodanov, M., Sierra, I., & Vidal-Valverde, C. (1997). Effect of germination on the thiamine, riboflavin and niacin contents in legumes. *Z Lebensm Unters Forsch.*, *205*, 48-52.

[85] Vidal-Valverde, C., & Redondo, P. (1993). Effect of microwave heating on the thiamin content of cows' milk. *Journal of Dairy Research*, *60*, 259-262.

[86] Arcot, J., & Shrestha, A. (2005). Fotale: methods of analysis. *Trends in Food Science and Technology*, *16*, 253-266.

[87] Polydera, A.C., Stoforos, N.G., & Taoukis, P.S. (2003). Comparative shelf life study and vitamin C loss kinetics in pasteurized and high pressure processed reconstituted orange juice. *Journal of Food Engineering*, *60*, 21-29.

[88] Daood, H.G., Vinkler, M., Márkus, F., Hebshi, E.A., & Biacs, P.A. (1996). Antioxidant vitamin A content of Spice red pepper (paprika) as affected by technological and varietal factors. *Food Chemistry*, *55*(4), 365-372.

[89] Sundl, I., Murkovic, M., Bandoniene, D., & Winklhofer-Roob, M. (2007). Vitamin E content of foods: comparison of results obtained from food composition tables and HPLC analysis. *Clinical Nutrition*, *26*,145-153.

[90] Lahély, S., Ndaw, S., Arella, F., & Hasselmann, C. (1999). Determination of biotin in foods by high-performance liquid chromatography with post-column derivatization and fluorimetric detection. *Food Chemistry, 65*, 253-258.
[91] Booth, S.L., Davidson, K.W., & Sadowski, J.A. (1994). Evaluation of an HPLC method for determination of phylloquinone (vitamin K_1) in various food matrices. *Journal of Agricultural and Food Chemistry*, *42*, 295-300.

PART II

MEAT AND FISH

In: Practical Food and Research
Editor: Rui M. S. Cruz, pp. 155-180

ISBN: 978-1-61728-506-6

Chapter VI

COLOUR AND PIGMENTS

Beyza Ersoy
Department of Fishing and Fish Processing Technology, Faculty of Fisheries, Mustafa Kemal University, Hatay, Turkey, beyzaersoy@gmail.com

ABSTRACT

The colour of muscle foods is critically appraised by consumers and often is their basis for product selection or rejection. Colour of uncooked meat and meat products is usually described as pink or red, but colours range from nearly white to dark red. Discoloration of these products often involves tan, brown, gray, green or yellow colours. Important pigments in meat and fish, chemical and physical properties of these pigments, meat colour stability during processing and storage, and colour measurement are reviewed.

6.1 INTRODUCTION

The principal aim of this chapter is to provide a brief overview of the major colour changes that occur during processing and storage of meat and seafood products as well as some of the evaluation techniques commonly used.

The meat industry is concerned with turning an animal carcass into many different end-products. The end-products are derived from all parts of the animal (muscle, bone, fat, cartilage, skin, fluids and glands), and are produced through a range of physical, chemical and biological processes. Meat purchasing decisions are influenced by colour more than any other quality factor because consumers use discoloration as an indicator of freshness and wholesomeness. Meat colour is due to the heme pigments (myoglobin and hemoglobin) and their derivatives, their chemical states and the light-scattering properties of the meat [1, 2]. Myoglobin has been known to be a major contributor to the colour of muscle, depending upon its derivatives and concentration [3, 4]. Myoglobin concentration generally depends on species, breed, sex and age of animal, training and nature of nutrition, muscular activity, oxygen availability, blood circulation and muscle type, as well as the way the meat is treated [4-7]. Hemoglobin is lost rather easily during handling and storage, while myoglobin is retained by the muscle intracellular structure [6]. Therefore, colour changes in meat are mainly due to the reaction of myoglobin with other muscle components, especially myofibrillar proteins. During improper handling, especially at abusive temperatures, or processing in the presence of high salt concentrations, the interaction between myoglobin and other muscle proteins can occur and affect colour [8].

People prefer to fish flesh having their own characteristic colours, and sometimes find an unconventional colour unattractive or even repulsive. The flesh of white fish such as cod and haddock is expected to be white, and even when it is only slightly darkened or coloured it may be rejected. Yellow discoloration can be caused by a carotenoid pigment called zeaxanthin that may be present in some parts of the diet, such as algae. Pink discoloration is usually due to the carotenoid pigment astaxanthin which is naturally present in shellfish such as prawns and shrimp [9, 10].

6.2 Important Pigments in Meat and Fish

6.2.1 Pigments in meat

6.2.1.1 Myoglobin and hemoglobin

The red colour in meat is due mainly to a protein pigment called myoglobin and, to a lesser extent, hemoglobin. Both hemoglobin and myoglobin are complex proteins consisting of a protein moiety, globin, and a nonpeptide component, heme. Heme is composed of an iron atom and tetrapyrrole, or porphyrin. In myoglobin, the heme is attached to globin, while hemoglobin may be viewed simply as four myoglobin molecules linked together.

Red meats such as mutton and beef contain more myoglobin than do lamb, veal, and pork. There is less myoglobin in turkey and chicken breasts than in the legs and thighs. Meat that has only a small amount of myoglobin is called "light" or "white" meat; "dark" or "red" meat is significantly higher in this pigment protein. When the iron is in the ferrous state, myoglobin can bind oxygen. In living muscle, this is very important, because this enables myoglobin to bind or take oxygen from hemoglobin of blood. However, when the animal dies, the supply of oxygen in cell tissue is depleted, and the colour of meat (muscle) turns to the purplish red colour of myoglobin. When fresh meat is cut and myoglobin is again exposed to oxygen in the air, the iron in the ferrous state again binds oxygen and meat colour is a desirable cherry red colour due to the formation of oxymyoglobin. After a period of time, or due to environmental conditions, the iron in the interior of this pigment is oxidized. It is no longer able to bind oxygen, and a brown-coloured pigment (metmyoglobin) is formed [11-13].

6.2.2 Pigments in fish

The body colours of fish are predominantly dependent on the presence of special cells in the skin, called chromatophores. These contain pigments or lightscattering or light-reflecting organelles. In biology, any substance that can impart colour to the tissues or cells of animals or plants can be called a pigment. There are four main groups of pigments that can be used to provide colour in these cells: melanins, carotenoids, pteridines, and purines. Melanins are responsible for the dark coloration seen in fishes. Carotenoids, which are lipid soluble, dominate in giving the yellow to red colours. Pteridines are water-soluble compounds and result in bright colouration like the carotenoids. Pteridines play a small role in colouration when compared to carotenoids. In the purine compounds, guanine predominates, and large amounts of guanine can be found in the silvery belly skin of most species of fish. These basic compounds can be combined with other components, like proteins, to produce the blue, violet, and green colour ranges seen in fishes [14].

6.2.2.1 Carotenoids

The carotenoids are the dominant pigment in fish. These pigments contribute to the yellow, orange and red colour of fish and shellfish. Carotenoids in seafoods are oxygenated forms of β carotene referred to as xantophylls [15]. Most of the muscle pigments of fish and shellfish are xantophylls. Commonly occuring xantophylls in fishes include tunaxanthin, zeaxanthin, lutein and astaxanthin [16]. Where carotenoids occur in fish, they are not just a playful diversion of nature as was often thought in the oast, but a benefit to the fish [17, 18]. In addition to pigmentation, carotenoids are known to play an important potential role in human health by acting as biological antioxidants, protecting cells and tissues from the damaging effects of free radicals and singlet oxygen [19]. Other health benefits of carotenoids that may be related to their antioxidative potential include enhancement of immune system function [20], protection from sunburn [21], and inhibition of the development of certain types of cancers [22]. In this context, shrimps may be a good alternative food for humans when carotenoids are necessary for health [23]. The main pigment material of shrimps is astaxanthin, one of the main carotenoid pigments. It provides the tissue with red–orange pigmentation [23-26]. Gopakumar and Nair found a general average of 13.3 mg/kg total carotenoid content in four penaeid species (*Metapenaeus affinis*, *M. dopsoni*, *Penaeus indicus*, *Parapenaeopsis stylifera*) and 4.2 mg/kg in *M. monoceros* from brackish water [27]. Clarke stated that the total carotenoid content of *Pandalus montagui* varies from 21 to 72 mg/kg [28]. The carotenoid contents of shrimps vary depending on their native habitat or manufactured diets. Mean carotenoid contents of *P. semisulcatus* and *M. monoceros* were 14.1-0.45 and 16.9-0.26 mg/kg, respectively. These values are quite high compared to other seafoods or even to terrestrial animal meats. For example, the highest level of carotenoid deposition in rainbow trout is reported to be 10.2–13.7 mg/kg [29]. Carotenoids are the major pigmenting compounds and can not typcially be synthesized by fish; therefore it must be suplemented by synthetic or natural carotenoid sources to the fish feed. However addition of carotenoids to fish feeds increase the cost of fish feed about 20-25%. Therefore the use of native carotenoid sources for fish pigmentation. In this review, the use of natural carotenoid sources are summarized for fish feed in aquaculture [30].

The main pigment of wild salmon is astaxanthin. However, smaller amounts of canthaxanthin, β-carotene, lutein, tunaxanthin and zeaxanthin are found [29, 31]. The concentration of astaxanthin in salmon and trout has been found to vary between ~3 and 37 mg/kg [29, 32]. Cultured salmon and trout are typically reared on diets containing either astaxanthin or canthaxanthin. Scientific trials show that commercial astaxanthin and canthaxanthin are absorbed, transported and deposited in the same way as the carotenoids consumed by wild fish [33-38]. Astaxanthin is, however, more effectively absorbed and deposited than canthaxanthin [29, 33-35, 39, 40]. Salmonid products are exposed to different conditions during processing and storage which may, in turn, result in changes in the carotenoid content and which may lead to colour modifications. While investigations into the absorption, deposition and utilisation of various carotenoids are numerous, little is known about the effect of storage and processing on their stability. Astaxanthin and canthaxanthin were affected by frozen storage and smoking of salmon fillets [41, 42].

6.3 Chemical and Physical Properties of Myoglobin and Carotenoids

6.3.1 Properties of myoglobin

Myoglobin is a metalloprotein composed of globin and an iron-containing heme prosthetic group (Figure 6.1). Globin can exist in the native or denatured state; the iron atom can exist in variousoxidation states, and the porphyrin ring can be intact, oxidized, polymerized or opened.

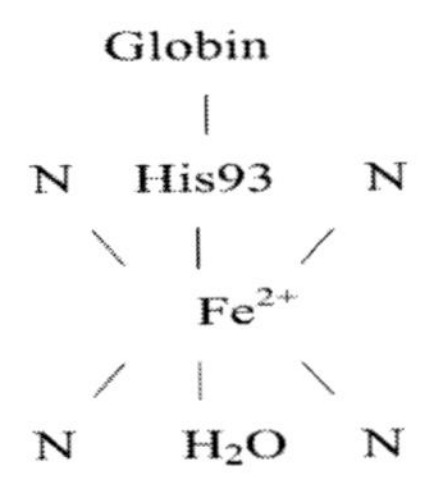

Figure 6.1. Iron coordination points in heme. N=nitrogen atom of the imidazole ring of the heme [43].

Myoglobin behavior is related to its biological function–storing oxygen until needed by living tissues and the chemistry required to optimizing that function. Oxygen storage is permitted by heme's ability to undergo oxidation–reduction and electron transfer reactions. Recent work on myoglobin chemistry has examined the regeneration of postmortem reducing equivalents, the relationship between pigment and lipid oxidation, and factors involved in myoglobin stability. A key ingredient in meat colour life is metmyoglobin reduction, a process that requires NADH. This requirement has, to some extent, questioned the importance of postmortem metmyoglobin reduction because NADH is available in such small amounts and mechanisms involved in its regeneration are not straightforward. Other work has concluded that there is little dependence of bovine muscle colour stability on metmyoglobin reducing ability [44]. However, when evaluating the relation between colour stability and muscle reducing capacity, location of metmyoglobin reducing activity within extracts (sarcoplasmic versus particulate fractions) was critical and therefore, may influence results [45]. Enzymatic reduction of metmyoglobin has been studied in various fish muscle such as dolphin, bluefin, tuna and mackerel, and blue white dolphin [46-48].

Table 6.1. Characteristics of various states of myoglobin [4, 43, 49-57].

Pigment	Ligand	Type of bonding	Conditions	State of globin	State of iron	State of heme	Colour
Deoxymyoglobin	H_2O	Ionic	Very low oxygen tension peak <5 mm Hg	Native	Fe^{++}	Intact	Purple-red
Oxymyoglobin	$:O_2$	Covalent	High O_2 tension (peak 70–80 mm Hg)	Native	Fe^{++}	Intact	Bright red
Metmyoglobin	-	Ionic	Low O_2 tension (peak 10 mm Hg). Loss of electron	Native	Fe^{+++}	Intact	Brown
Ferrylmyoglobin	O	Covalent	Breakdown of perferrylmyoglobin (MbFe(IV)=O)	Native	Fe^{+4}	Intact	Red
Globin myohemochromogen	-	Covalent	Heating MbO_2. Myohemichromogen in a reducing environment irradiation establishes a reducing environment	Denatured	Fe^{+++}	Intact	Dull red-crimson
Globin myohemochromogen	H_2O	Ionic	Heating of Mb, MbO_2, MetMb, hemochromogen	Denatured	Fe^{++}	Intact	Brown-grey
Carbon monoxymyoglobin	CO:	Covalent	Preferentially bound (to O_2); not consumed in vacuum	Native	Fe^{++}	Intact	Bright red
Sulfmyoglobin	HS	Covalent	Reaction of H_2S and O_2 with myoglobin	Native	Fe^{++}	Intact; 1 double bond saturated	Green
Metsulfmyoglobin	HS	Ionic	Oxidation of sulfmyoglobin. Loss of electron	Native	Fe^{+++}	Intact; 1 double bond saturated	Red
Choleglobin	-OOH	Covalent or ionic	H_2O_2 reaction with Mb or MbO_2	Native	Fe^{++} or Fe^{+++}	Intact; 1 double bond saturated	Green
Verdohaem	H_2O	Ionic	Heat, denaturation effects on Mb, MbO_2, MetMb, hemochromogen, hemichromogen	Absent	Fe^{+++}	Porphyrin ring opened	Green

6.3.2 Properties of carotenoids

The carotenoids are hydrophobic molecules with little or no solubility in water. They are thus expected to be restricted to hydrophobic areas in the cell. Polar functional groups obviously alter tile polarity of carotenoids and affect their interactions with other molecules. We have known a great deal about the physical properties of carotenoids with respect to their ability to absorb light and transfer radiant energy. Some of these physical and chemical properties are described below [58].

6.3.2.1 Light absorbtion

The best known property of carotenoids is their ability to absorb light. Carotenoids are therefore intensely yellow, orange, or red. The relationship between chromophore and light-absorption properties, widely used in the identification of carotenoids, is developed more fully elsewhere [59-60].

6.3.2.2 Energy transfer

In addition to absorbing light directly, carotenoids can also be excited by an energy transfer reaction to form the triplet state species. The best known energy transfer from triplet-state chlorophyll or other porphyrins to carotenoids occurs much more readily than the alternative energy transfer to oxygen to form the highly reactive and destructive singlet oxygen $^1O_2$ [61-62].

6.3.2.3 Chemical reactivity

Carotenoids can undergo many reactions with chemical reagents. The process that draws much attention in biological systems is the oxidation of carotenoids, during which time they may function as biological antioxidants. The ready ability of carotenoids to react with radical species has served as the basis for the determination of the activity of various lipoxygenases. The basis for the assay involves an oxidative interruption of the conjugated double bond system which is invariably accompanied by a loss of the visible absorption, or a "bleaching" of the carotenoid [58].

6.3.2.4 Solubility

Most of the carotenoid pigments present in nature are quite water-insoluble. Lipids are one class of hydrophobic compounds, but even proteins can have strong hydrophobic regions which can play an important role in protein-protein interactions. The specificity of the latter interaction can range from very high, as in the case of enzymes binding hydrophobic compounds tightly at catalytically active sites [58].

6.3.2.5 Protein-carotenoid interactions

Interactions between carotenoids and proteins are common in all kinds of living organisms and are of vast physiological importance. The carotenoids may be greatly stabilized by the protein. Commonly, carotenoids in vivo are much more stable than when they are isolated and in organic solution. Carotenoids associate with hydrophobic areas in the protein or with the lipid components of lipoproteins. The interactions with the protein can alter the physical or chemical properties of the carotenoid. Thus, the binding of astaxanthin by protein in carotenoprotein complexes like the blue α-crustacyanin from lobster carapace causes a redistribution of electron density in the molecule, resulting in a large bathochromic shift in the light absorption spectrum from 488 to 632 nm [29, 60].

6.3.3 Enzymatic and non-enzymatic browning in meat and fish

Enzymes are important components of the edible tissues from fish and other aquatic organisms and they influence seafood quality in many ways [43]. The enzymes that may influence the colour and appearance of seafood are summarized in Table 6.2.

Melanosis in crustacean species like krill, lobster, shrimp and crab is one of the most studied enzyme problems in seafood technology. The occurance of "blackspot" gives the product an unacceptable appearance and thus lowers market value. Polyphenol oxidase (PPO) in these species catalyses a two-step reaction: (i) hydroxylation of monophenols like tyrosine to o-diphenols (Dopa) and (ii) oxidation of o-diphenols to di-quinones.

A series of non-enzymic reactions are involved with the conversion of di-quinones to black melanin. Zymogens from crustacean species are activated by proteolytic enzymes such as trypsin-like enzymes in the tissue [62].

Table 6.2. Enzymes that may influence seafood appearance [48].

Enzyme(s)	Component(s)	Results
ATPases	H^+, Myoglobin	Low pH favor oxidation of myoglobin
Calpain, cathepsins	Cytoskeletal proteins	Increased flesh opacity
Glycolytic enzymes	Glycogen	Formation of reducing hexose sugars contributes to Maillard browning
Lipoxygenase-like enzyme; Meyeloperoxidase	Carotenoids	Bleaching of yellow-red epithelial and flesh pigments
Metmyoglobin reductase	Metmyoglobin	Rate of brown heme discoloration reduced
Nucleosido phosphorylase, Inosine nücleosidase	Ino	Formation of reducing pentose sugars contributes to Maillard browning
Polyphenoloxidase	Tyrosine	Melanin

Nonenzymatic browning reactions, such as the Maillard reaction, can occur during the processing and storage of foodstuffs or animal feeds. The nutritional value of proteins may be considerably decreased by the Maillard reaction [63-66] and there are reports suggesting toxicity of browning products [67-69]. Maillard reaction has been identified as the cause of discoloration in fishery products. Brown discoloration is considered undesirable when it occurs in products like frozen, dried, salted and canned seafoods. Nevertheless, in smoked or cooked fish, the browning reaction contributes for a desirable appearance [70].

6.4 Colour Stability During Meat and Fish Processing

6.4.1 The colour of meat and meat products

6.4.1.1 Dry-cured meat colour

The meat products, such as dry-cured ham and dry-fermented sausages are traditional foods that have been produced and consumed in different areas of the world.

The colour of meat and meat products is influenced by its moisture and fat content and also by the content of hemoprotein, particularly myoglobin. The colour of the dry-cured meat products depends on the concentration of myoglobin, the degree of conversion to the nitrosyl pigment (usually 10-40%) and the state of the protein (case of heat treatment). The myoglobin concentration in pork is between 4 and 9 mg/g. However, myoglobin concentration depends on the type of muscle [71], being higher in muscles with oxidative patterns than in muscles with glycolytic patterns like *M. semimembranosus* (about 4.2 mg/g). Most of the muscles in ham exhibit glycolytic or intermediate metabolism [72, 73].

A change occurs in the myoglobin when meats are "cured". Sodium nitrite, salt, and heat are used during the curing process. The nitrite combines with myoglobin to form the stable pink colour of cured meats, nitroso-myoglobin. This cured meat pigment is unstable in the presence of light, and photo-induced fading will occur [11, 13]. Some brown or green colour may appear as a consequence of oxidation by certain peroxide-producing bacteria. The product needs to be hygienically handled to avoid microbial contamination and cut surfaces (especially when sliced) protected from external light [73].

6.4.1.2 Cooked meat colour

The traditional hot water immersion cooking is still by far the most extensively used and practical thermal process to cook meat [74-77]. Hot water cooking directly impacts the physiochemical properties of meat in many ways including altering meat colour, providing protein hydrolysis and denaturation, improving texture, and causing loss of water-binding capacity [78]. Many studies have been performed to relate different types of thermal processing to meat quality. In frying experiments, a few research groups demonstrated the propagation of meat physiochemical properties including water loss, browning of the product and alteration of lipid profile [79]. The effect of heat treatment on browning of food material is a well-recognized development of cooked products as suggested by many authors [80].

Thipayarat reported that different isothermal cooking in a water immersion system strongly affected the CIE colour indices of cooked pork. High cooking temperature activated a few chemical reactions including Maillard reaction contributing to the development of brown colour and caused dehydration of the cooked pork sample. The increase of browning as indicated by the CIE colour indices was corresponding well with the decrease of moisture content. Low cooking temperatures resulted in less browning and more stable colour retention than higher temperature cooking [81].

6.4.1.3 Ionising radiation and meat colour

Ionising radiation has been proposed as a suitable method for the prolongation of storage life of meat and meat products [82]. The usefulness of the technology for the inactivation of pathogenic micro-organisms in beef and poultry has been demonstrated [83,84]. Commercialisation of irradiated food, including fresh meat, has been limited due mainly to consumer fears about the safety of the process. The formation of undesirable or unexpected colours from the consumer's viewpoint is critical as meat colour is of vital importance in the decision to buy a certain piece of meat. In general, light meat such as pork loin and poultry breast meat produced pink colour, while dark meat such as beef became brown or gray after irradiation [85-87]. The pink colour compounds in irradiated light meat was characterized as a carbon monoxide-heme pigment complex [88, 89], but the mechanisms and cause of colour changes in irradiated beef is not known yet. The colour changes induced by irradiation are different depending on animal species, muscle type, irradiation dose, and packaging type [90]. The earliest work in this area was performed on beef extracts and showed that ionising radiation caused the formation of a red colour in metmyoglobin extracts and the formation of metmyoglobin in oxymyoglobin extracts [82, 91, 92]. Later work on fresh meat irradiated in a nitrogen atmosphere also showed the formation of a red colour and it was postu- lated that this compound was oxymyoglobin formed from metmyoglobin via ferrylmyoglobin [92, 93]. Further studies confirmed the presence of a red colour in meat and meat extracts following irradiation [82, 94, 95].

6.4.1.4 High pressure and meat colour

Meat colour characteristics are modified by high pressure processing. As a consequence, bovine meat appears lighter and less red [90, 91]. This modification of lightness is due both to myoglobin denaturation [98] and myofibrillar proteins denaturation [99]. Redness decreases when meat has been pressurized at higher pressures (400-500 MPa), because of the increase of the metmyoglobin content [100] and the meat becomes brown. Cheah and Ledward showed that high pressure (800 MPa, 20 min) treated pork mince samples revealed faster oxidation than control samples, and that pressure treatment at greater than 300-400 MPa caused conversion of reduced myoglobin/oxymyoglobin to the denatured ferric form [101]. Cheah and Ledward also demonstrated that iron released from metal complexes during pressure treatment catalysed lipid oxidation in meat [102].

6.4.2 The colour of fish and fish products

6.4.2.1 Surimi products

Surimi products are being consumed in increasing levels every year. The commercial surimi production of word in 1987 ranged between 350,000 and 400,000 metric tons. Whiteness is a critical factor determining the quality of surimi gels, and myoglobin and hemoglobin can influence this important parameter [103]. In general, both heme proteins are removed during the washing process of fish mince, leading to increase whiteness.

Dark fleshed-fish species have a high content of dark muscle, which comprises a considerable amount of lipids and sarcoplasmic proteins [104, 105]. Myoglobin is the predominant pigment protein in the sarcoplasmic fraction of fish dark muscle [106] and contributes to the lowered whiteness of surimi gel [103]. Generally, it is difficult to wash all the myoglobin from dark fleshed-fish muscle because it resides within the muscle cells [107,108]. Normally, myoglobin in fresh fish can be removed during the washing process, leading to increased whiteness of the resulting surimi. However, heme proteins become less soluble as the fish undergo deterioration [7, 109]. Denaturation of the myoglobin and/or myofibrillar proteins, before or during processing, can also cause their cross-linking, resulting in the discoloration of the surimi [108]. The interaction between fish myoglobin and natural actomyosin can be enhanced at higher ionic strength and higher temperature and the binding can be augmented with increasing incubation times. After capture, fish are normally kept in ice prior to unloading [7, 110]. During this stage, discoloration of muscle and binding of pigments to muscle generally occur [108, 111]. Interaction between tuna myoglobin and myosins from tuna and sardine was investigated in a model system at 4 °C for up to 24 h. Both sardine and tuna myosins bound progressively with tuna myoglobin as the storage time increased ($P<0.05$). The soret absorption peak was noticeable in the myoglobin–myosin mixture. The oxidation of oxymyoglobin in the presence of myosin was generally greater than that found in the absence of myosin ($P<0.05$). Oxymyoglobin underwent oxidation to a greater extent in the presence of tuna myosin than sardine myosin ($P<0.05$). The interaction between fish myoglobin and myosin also caused changes in reactive sulfhydryl content and altered the tryptophan fluorescent intensity. The loss in Ca^{2+}-ATPase activity of myosin varied with fish species and was governed by the myoglobin added. Thus, the interaction between fish myoglobin and myosin most likely occurred as a function of time and was species-specific. Effect of ionic strength and temperature on interaction between fish myoglobin and myofibrillar proteins was also invesigated in a model system by Chaijan et al. [112].

6.4.2.2 Dried fish colour

Drying has been shown to be an efficient and cheap method of conservation. However it is also responsible for a severe deterioration in the quality of food [113-115] and specially fish, particularly with respect to their colour [116-119]. The colour of fish can change markedly when it is dried. Although the Maillard reaction is an obvious candidate to explain the yellow–brown colour after processing, the low amounts of reducing sugar, particularly at the beginning of processing, makes this hypothesis less convincing. However, a small degree of browning results from the degradation of L-methylhistidine, even in the absence of sugar. Furthermore, oxidation of lipids is an important factor that also leads to browning of the skin of dried fish by interacting with the proteins [120].

6.4.2.3 Thawed fish colour

Nowadays, fish is kept frozen for relatively long periods before being consumed. Frozen storage is an important preservation method for seafood. Its effectiveness stems from internal dehydration or immobilization of water and lowering of temperature. However, meat and fish may undergo quality losses such as protein denaturation, colour deterioration, weight decrease, oxidation of lipids and textural changes due to the freezing and thawing processes [121]. The extent of quality loss depends on careful prefreezing preparation, control of the freezing rate, storage conditions and thawing conditions [122]. During thawing, foods can be damaged by chemical, physical and microbiological changes. The freezing and thawing processes can have a profound effect on muscle physicochemical characteristics [123]. The thawing rate during conventional thawing processes is controlled by two main parameters outside the product: the surface heat transfer coefficient and the surrounding ambient temperature of the sample. This ambient temperature is supposed to remain below 15 °C during thawing to prevent the development of a microbial flora [124]. There are many commercial methods for thawing fish. Water thawing is suitable for a short-time batch or continuous process immediately prior to boning and gives a net weight gain which partially offsets the previous weight loss during cooling and freezing. Air thawing is suitable for a batch process with a cycle time between 8 h and 60 h, which determines the air temperature, required in the range 2-18 °C (35-65 °F). A high relative humidity is required to minimize weight loss [125]. The satisfactory techniques for thawing large portions of animal tissue include thawing in a refrigerator and microwave [126]. Thawing methods has a significant effect on the physical, chemical and microbiological quality of frozen eel. The redness (a*) value of water thawed samples did not changed comparing with the fresh fish values, according to Ersoy et al. Therefore, the water thawing method is most suitable for conservation of colour quality in frozen eel [127].

6.5 Colour Stability During Storage

During storage, processed meats deteriorate in the first instance because of discoloration, secondly because of oxidative rancidity of fat and thirdly on account of microbial changes [128].

In uncooked meat, the pigments responsible for its colour are the myoglobin, oxymyoglobin and metmyoglobin, which can be changed from one to the other, depending on the conditions at which the meat is stored. Storage time, temperature and handling are among the major influencing factors of meat colour and discoloration [129].

Packaging can be an effective method for meat shelf life extension that avoids the use of chemical preservatives [130, 131]. These methods of preservation include the use of modified atmosphere packaging (MAP) using gas mixtures containing variable O_2 and CO_2 concentrations in order to inhibit the different spoilage-related bacteria and are often associated with the use of low temperatures during storage [132]. Substantial fractions of CO_2 are used to retard the growth of organisms produced by aerobic spoilage, and a certain concentration of O_2 is employed for red meat MAP to preserve meat colour [133-135].

Modified atmosphere packed meat is a complex and dynamic system where several factors interact [136]. Models can be used to describe how the initial package atmosphere changes over time and how these changes affect product quality and shelf-life. The dynamic changes in headspace gas composition during storage can be modelled as a function of gas transmission rates of the packaging material, initial gas composition, product and package geometry, gas absorption in the meat etc. Combined with the knowledge from models on quality changes in the meat as a function of packaging and storage conditions such as storage temperature, gas composition and light exposure, predictions of product shelf life can be made. Pfeiffer et al. developed simulations of how product shelf life changes with different packaging and storage conditions for a wide range of food products (primarily dry products)

[137]. However, at present sufficient models for much quality deteriorative reactions are lacking and only a few attempts have been made to model chemical quality changes in meat products, in contrast to modelling of microbial shelf-life, where extensive work has been performed [138].

6.5.1 Frozen meat colour

The colour of frozen meat is affected by: freezing rate; storage temperature and fluctuation in temperature during storage; intensity of light during display; and method of packaging. Very slowly frozen meat is excessively dark, while meat frozen in liquid nitrogen is unnaturally pale. Such extremes are unlikely to be seen under commercial conditions, but faster freezing regimes will give a paler product than slower freezing regimes. The large variation in lightness is a result of differences in the rate of ice-crystal growth. Small crystals formed by fast freezing scatter more light than large crystals formed by slow freezing; so fast-frozen meat is opaque and pale, while slow-frozen meat is translucent and dark. To optimise frozen-beef colour, fresh beef should be exposed to air for 30 minutes before freezing to allow optimum bloom to develop prior to freezing. This can result in frozen beef that is similar in appearance to fresh beef. During storage of frozen meat, if it is exposed to air, slow thawing and refreezing of the exposed surface leads to dehydration of the meat. This leads to the development of freezer burn which appears as a grey-white area on the exposed surface where the fibbers of the meat are visible. Freezer burn areas stay dry and pale when the meat is defrosted, and is tough and dry to eat. The major colour problem during retail display of frozen meat is photooxidation. Frozen meat under direct illumination oxidises from the surface inwards (compared with fresh meat which oxidises from the subsurface outwards). Oxidation of frozen oxymyoglobin is temperature dependent-the rate increases from -5 to -12°C and then decreases to a minimum at -20°C. The rate of oxidation is affected also by muscle type-*Longissimus dorsi* (loin) will fade more slowly than *Gluteus medius* (rump). Frozen-beef colour remains attractive for at least 3 months in the dark, but only 3 days in the light [13].

The effects of refrigerated and frozen storage on the external coloration (bloom/ darkening) of meats was reported [140-142].

It was reported that the comparison of frozen and cooked beef extract with the corresponding portions analysed at the time of cooking did not change the L*, a*, b*, C (Chroma) or H (Hue angle) colour values during 3 weeks storage at -20 °C [143].

6.5.2 Iced and frozen fish colour

During the handling and storage of fish, a number of biochemical, chemical and microbiological changes occur, leading to discoloration [3, 145].

Discoloration of fish during frozen storage is caused by the formation of metmyoglobin [146]. Metmyoglobin formation can be influenced by factor, such as pH, temperature, ionic strength and oxygen consumption reaction [147]. Chen reported that iced or frozen storage decreased the myoglobin extracting efficiency in washed milkfish due to the insolubility of myoglobin by the oxidation of myoglobin to form metmyoglobin [109]. The storage of sardine and mackerel in ice causes the oxidation and denaturation of pigment [7].

Several studies have shown that the pigment stability of salmonids during frozen storage is uncertain. Christophersen et al. measured the astaxanthin content of rainbow trout fillets over a storage period of 180 days at -18 °C and found a slight decrease in pigment concentration during that time [148]. However, synthetic astaxanthin and canthaxanthin were found to be stable in rainbow trout fillets after 180 days at -20 °C [149], and for up to 180 days at -18 °C, -28 °C and -80 °C [150]. These latter studies suggest that both astaxanthin and canthaxanthin are stable during storage. Sheehan reported that canthaxanthin levels of raw and smoked Atlantic salmon decreased considerably during frozen storage. The process of smoking had a negative effect on the carotenoid concentration of fillets pigmented with astaxanthin, but this effect was not evident in fresh fillets and only in frozen fillets. In conclusion, astaxanthin and canthaxanthin were affected in slightly different ways by frozen storage and smoking of salmon fillets. Canthaxanthin-fed fish seem to be better for smoking although, when frozen, they lose colour more rapidly than astaxanthin-pigmented fish [41].

Generally, both heme proteins in fresh fish can be removed during the washing process, leading to increased whiteness of the flesh. However, heme proteins become less soluble as the fish undergoes deterioration. Chen reported that iced or frozen storage decreased the myoglobin extracting efficiency in washed milkfish due to the insolubility of myoglobin by the oxidation of myoglobin to form metmyoglobin. As a consequence, the surimi produced from unfresh fish is more likely to be discolored, especially dark-fleshed fish species. Due to the shortage of lean fish, which are commonly used for surimi production, more attention has been given to dark-fleshed fish, such as sardine and mackerel as the raw material for surimi. After capture, fish are normally kept in ice prior to unloading and during this stage discoloration of muscle can occur and binding of pigments to muscle can also take place [109].

6.6 Evaluation Techniques

6.6.1 Types of colour system

Currently, many colour systems are available for express colour data. For example, CIE Lab, Hunter Lab, Tristimulus, Munsell and Reflectance at specific wavelengths have been used to colour data [151].

CIELAB colour scale is approximately uniform colour space. In a uniform colour scale, the differences between the points plotted in the colour space correspond to visual differences between the colours plotted. The CIELAB colour scale is organized in a cube form (Figure 6.3). The L* axis runs from top to bottom. The maximum L* is 100, which represents a prefect reflecting diffuser. The minimum L* is zero and represents black. The a* and b* have no specific numerical limits. Positive a* is red. Negative a* is green. Positive b* is yellow. Negative b* is blue.

The HunterLab colour scale is visually uniform. In a uniform colour scale, the difference between the points plotted in the colour space corresponds to visual difference between the colours plotted. The maximum for L is 100, which would be a perfect reflecting diffuser. The minimum of L would be zero, which would be black. a and b have no specific numerical limits. Positive values of a are red and negative values are green, whereas positive values of b are yellow and negative ones are blue.

The lightness correlate in CIELAB is calculated using the cube root of the relative luminance, and using the square root in Hunter Lab (an older approximation). Except where data must be compared with existing Hunter *L*, *a*, *b* values, it is recommended that CIELAB be used for new applications [154]. The Hunter *L*, *a*, *b* scale may be used on any object whose colour may be measured. It is not used as frequntly today as in the past [153].

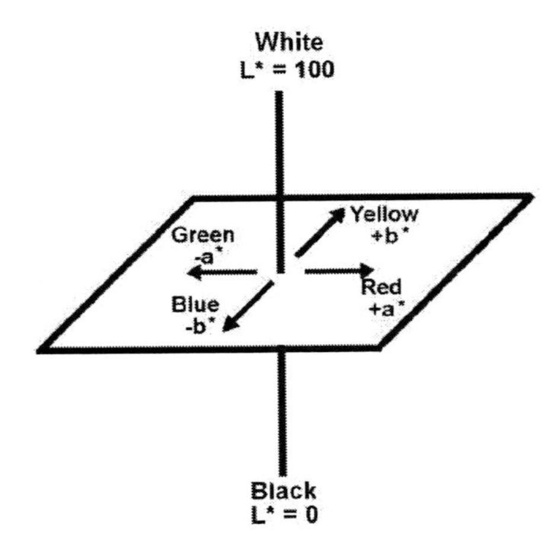

Figure 6.2. The CIELAB colour scale diagram [152].

Tristimulus colorimeters employ filters to simulate the response to the human eye. White light from a standard CIE source is shone on the sample. Light reflected at 45° is measured by a photo cell after it passes through an X, Y or Z filter. For example, at 650-700 nm in the red region of the visible spectrum, the human eye is not sensitive to blue or green light. The perceived red colour is due entirely to the quantity of red primary colour reaching the eye. For a colorimeter to simulate the human eye, the X filter must filter out all blue or green light reaching the detector. Tristimulus values (X, Y, Z), tristimulus coordinates (x, y, z), and mathematical expressions of these units are useful to describe accurately a particular meat colour, but they have not been used much for following colour change during display or for quality control measures [151, 155].

In Munsell colorimeters, there are five principal hues (red, yellow, green, blue and purple) equally spaced around the circle at the base of the colour solid [156]. The value notation in the vertical axis indicates the degree of lightness or darkness of a colour. Zero is absolute black and 10 is absolute white. Understanding the Munsell system is of value to meat colour measurement, because it was the basis for later colour measurement systems [151].

Selecting the most appropriate colour variable is project specific and dependent on the objectives of each experiment.

6.6.2 Instrumental methods

Currently, many options are available for instrumental colour analysis. For example, several types of instruments (colorimeters and spectrophotometers) are available. Three major considerations are: what instrumental methodology to use (pigment extraction or reflectance), how to express the data and how to use the data. For example, although CIE is commonly used by researchers to measure colour, Alcalde and Negueruela reported that tristimulus coordinates (XYZ) also were useful for measuring lamb carcass colour. Researchers often report several variables because they are easy to obtain and readily available with most instrumentation. However, depending on the experiments objective, increasing the number of variables measured may provide little additional information [157].

6.6.3 Computer vision

Computer vision based on analysis of digital camera images has distinct advantages over traditional colour evaluation. O'Sullivan et al. noted several benefits associated with digital camera derived j.peg images, including (1) compared with a colorimeter, only a single digital observation is needed for a representative assessment of colour, (2) digital images can converted to numerous colour measurement systems (Hunter, CIE, XYZ) [158, 159]. Lu et al. acquired digital images using a colour camera and sensory colour scores can be predicted with statistical and/or neural network models. Overall, computer vision is a promising method for predicting visual colour [160].

Instrumental colour measures derived from digital camera images can predict red and brown sensory terms. Because the camera measured the entire sample surface, it was more representative of sensory descriptors than the colorimeter, which was based on point-to-point measurements. Ringkob also has evaluated image analysis for measuring meat colour [161]. Ringkob found that image analysis was useful for assessing pork colour, particularly detecting differences between yellow and white fat [162].

A computer vision method was developed and used to assign colour score in salmon fillet according to SalmonFan™ card. The methodology was based on the transformation of RGB to $L^*a^*b^*$ colour space. In the algorithm, RGB values assigned directly to each pixel by the camera in the salmon fillet image, were transformed to $L^*a^*b^*$ values, and then matched with other $L^*a^*b^*$ values that represent a SalmonFan score (between 20 and 34). Colours were measured by a computer vision system (CVS) and a sensorial panel under the same illumination conditions in ten independent sets of experiments. The methodology is very versatile and can potentially be used by computer-based vision systems in order to qualify salmon fillets based on colour according to the SalmonFan card [163].

6.6.4 Visual colour

Visual colour are closely related to consumer evaluations and set the benchmark for instrumental measurement comparison. They are not easy to conduct with either trained or consumer panels, since human judgments may not be repeatable from day to day and are influenced by personal preference, lighting, visual deficiencies of the eye and appearance factors other than colour. Various scoring scales have been utilized for panel evaluations.

Many of these are descriptive and imply averaging the colour over the entire meat surface area. Others utilize a "worst-point" colour score for a single or cumulative discolored area of at least 2 cm in diameter, whereas still other systems are various combinations of these two methods. Some scales with hedonic terms have been misused with descriptive panels. Correctly structured scoring scales and appropriate pictorial standards substantially improve panel consistency and validity. Pictorial colour standards have been developed and are useful to support specific colour scales. Appropriate scales and standards for specific experiments are dependent upon project objectives, as is the decision of whether trained (descriptive) panels, consumer (acceptance) panels or both are required. The most suitable scales usually are best constructed through preliminary studies in which the product is treated in a similar manner, because scales reflecting the type of colour changes unique to specific experiments are essential. Scales developed in this way will encompass the spectrum of sample colours most likely to appear and can be used for panel training and as reference standards during visual appraisals conducted during the experiment [151].

Carpenter et al. noted a strong association between colour preference and purchasing intent with consumers discriminating against beef that is not red (i.e., beef that is purple or brown). Therefore, visual determinations are the gold standard for assessing treatment effects and estimating consumer perception. [164].

Several reports have recommended or suggested levels of flesh carotenoids over which visual assessment of pigment is unreliable. Foss et al. considered that visual assessment is unreliable in rainbow trout if the carotenoid concentration is greater than 6 mg/kg [39].

Numerous researchers have used L* a* b*-values to document treatment effects on colour. Ratios of a/b, hue angle and saturation index have been used for discoloration studies [165,151]. From these various other characteristics can be calculated like Hue angle, which is used for distinguishing colour families and chroma, which is the strength of a colour. a* represents the redness of the meat and is very dependent on the blooming time [166]. It is influenced by the pH of the meat because oxidation and reduction processes of myoglobin are pH-dependent. In comparison L* is independent of blooming time but still very dependent on pH. As about 65% of the variation in L* can be explained from variations in solubility of the sarcoplasmatic proteins [167]. Garcia-Esteben et al. reported that Hunter measurements for L*, a* and b* were best suited for measuring the colour of dry-cured ham. Hunter Lab and CIE L* a* b* provided more reproducible lightness data than L* u* v* and XYZ systems. Instrumental measures of L* and a* are straightforward and can easily be applied to muscle colour. On the other hand, the colours represented by b* (blue and yellow) are not typical or intuitively related to meat [168].

Conclusion

Colour stability and rates of discoloration varied among muscles. Maintenance of ideal meat colour during the process and storage can be provided by efficient refrigeration during chilling, holding, transportation, storage, preparation and display; and hygiene, packaging and meat selection.

New colour measurement methods such as digital imaging techniques are now available. Nevertheless, unanswered questions regarding meat colour remain. The researches should continue to develop novel ways of improving muscle colour and colour stability.

REFERENCES

[1] McDougall, D.B. (1983). Instrumental assessment of the appearance of foods: In A.A., Williams, & K.K., Atkin (Eds.), *Sensory Quality in Foods and Beverages: Definition, Measurement and Control* (pp. 121-139). UK, Chichester: Ellis Horvard.

[2] Lawrie, R.A. (2002). *Meat Science. The eating quality of meat* (5th ed.). New York: Pergamon Press.

[3] Faustman, C., Yin, M.C., & Nadeau, D.B. (1992). Colour stability, lipid stability, and nutrient composition of red and white veal. *Journal of Food Science, 57*, 302-304.

[4] Postnikova, G.B., Tselikova, S.V., Kolaeva, S.G., & Solomonov, N. G. (1999). Myoglobin content in skeletal muscles of hibernating ground squirrels rises in autumn and winter. *Comparative Biochemistry and Physiology: Part B, 124*, 35-37.

[5] Giddings, G.G. (1974). Reduction of ferrimyoglobin in meat. *Critical Reviews in Food Technology*, 143-173.

[6] Livingston, D J., & Brown, W.D. (1981). The chemistry of myoglobin and its reactions. *Food Technology, 25*(3), 244-252.

[7] Chaijan, M., Benjakul, S., Visessanguan, W., & Faustman, C. (2005). Changes of pigments and colour in sardine (*Sardinella gibbosa*) and mackerel (*Rastrelliger kanagurta*) muscle during iced storage. *Food Chemistry, 93*, 607-617.

[8] Hanan, T., & Shaklai, N. (1995). Peroxidative interaction of myoglobin and myosin. *European Journal of Biochemistry, 233*, 930-936.

[9] Love, R.N. (1978). *Dark colour in white fish flesh*. Ministry of Agriculture, Fisheries and Food. Torry Research Station, Torry Advisory Note, no 76.

[10] Archer, M. (2008). *Fillet discoloration*. The Sea Fish Industry, www.seafish.org.

[11] Cornforth, D., Calkins, C.R., & Faustman, C. (1991). Methods for identification and prevention of pink colour in cooked meat. *Reciprocal Meat Conference Proceedings, 44*, 53-58.

[12] Cornforth, D.P. (1994). Colour: Its basis and importance. In A.M., Pearson, & T.R., Dutson (Eds.), *Advances in Meat Research* (Vol. 9, pp. 34-78). London: Blackie Academic & Professional.

[13] Van Laack, R.J.L.M., Berry, B.W., & Solomon, M.B. (1996). Effect of precooking conditions on colour of cooked beef patties. *Journal of Food Protection, 59*(9), 976-983.

[14] Anderson, S. (2000). Salmon colour and the consumer. *International Institute of Fisheries Economics and Trade Proceedings*, 1-4.

[15] Simpson, B.K., & Haard, N.F. (1985). The use of proteolytic enzymes to extract carotenoproteins from shrimp wastes. *Journal of Applied Biochemistry, 7*, 212-222.

[16] Fox, D.L. (1979). Biochromy: Natural colouration of living things. Berkeley (pp. 248). CA: University of California Press.

[17] Deufel, J. (1975). Physiological effect of carotenoids on Salmonidae. *Hydrologie, 37*, 244-248.
[18] Love, R.M. (1992). Biochemical dynamics and the quality of fresh and frozen fish. In G. M. Hall (ed.), *Fish Processing Technology* (pp. 1-30). New York: Black Academic & Professional Published in North America by VCH Publishers.
[19] Di Mascio, P., Murphy, M.E., & Sies, H. (1991). Antioxidant defense systems: the role of carotenoids, tocopherols, and thiols. *American Journal of Clinical Nutrition*, *53*, 194-200.
[20] Bendich, A. (1989). Carotenoids and the immune response. *Journal of Nutrition*, *119*, 112-115.
[21] Mathews-Roth, M.M. (1990). Plasma concentration of carotenoids after large doses of beta-carotene. *American Journal of Clinical Nutrition*, *52*(3), 500-501.
[22] Nishino, H. (1998). Cancer prevention by carotenoids. *Mutation Research*, *402*, 159-163.
[23] Yanar, Y., Celik, M., & Yanar, M. (2004). Seasonal changes in total carotenoid contents of wild marine shrimps (*Penaeus semisulcatus* and *Metapenaeus monoceros*) inhabiting the eastern Mediterranean. *Food Chemistry, 88*(2), 267-269.
[24] Katayama, T., Hirata, K., & Chichester, C.O. (1972). The biosynthesis of astaxanthin. VI. The carotenoids in the prawn, *Penaeus japonicus* Bate (Part 2). *International Journal of Biochemistry*, *3*, 363-368.
[25] Tanaka, Y.H., Matsuguchi, T., Katayama, T., Simpson, K.L., & Chichester, C.O. (1976). The biosynthesis of astaxanthin-XVII. The metabolism of the carotenoids in Prawn, *Penaeus japonicas* Bate. *Bulletin of the Japanese Society of Scientific Fisheries*, *42*, 197-202.
[26] Okada, S., Nur-E-Borhan, S.A., & Yamaguchi, K.Y. (1994). Carotenoid composition in the exoskeleton of commercial black tiger prawn. *Fisheries Science*, *60*, 213-215.
[27] Gopakumar, K., & Nair, M.R. (1975). Lipid composition of five species of Indian prawn. *Journal of the Science of Food and Agriculture*, *26*, 319-325.
[28] Clarke, A. (1979). Lipid content and composition of the pink shrimp *Pandalus montagui* (Leach) (Crustacea: Decapoda). *Journal of Experimental Marine Biology and Ecology*, *38*, 1-17.
[29] Torissen, 0. J. (1989). Pigmentation of salmonids: Interactions of astaxanthin and canthaxanthin on pigment deposition in rainbow trout. *Aquaculture*, *79*, 363-374.
[30] Yeşilayer, N., Doğan, G., & Erdem, M. (2008). Balık Yemlerinde Doğal Karotenoid Kaynaklarının Kullanımı. *Journal of FisheriesSciences.com*, *2*(3), 241-151.
[31] Simpson, K.L., Katayama, T., & Chichester, C.O. (1981). Carotenoids in fish feeds. In J.C. Bauernfeind (Ed.), *Carotenoids as Colourants and Vitamin A Precursors* (pp. 463-538). Academic Pres: New York.
[32] Storebakken, T., & No, H.K. (1992). Pigmentation of rainbow trout. *Aquaculture*, *100*(1-3), 209-229.
[33] Schiedt, K., Leuenberger, F., Vecchi, M., & Glinz, E. (1985). Absorption, retention and metabolic transformations of carotenoids in rainbow trout, salmon and chicken. *Pure and Applied Chemistry*, *57*, 685-692.
[34] Storebakken, T., Foss, P., Schiedt, K., Austreng, E., Liaaen- Jensen, S., & Manz, U. (1987). Carotenoids in diets for salmonids. IV Pigmentation of Atlantic Salmon with astaxanthin, astaxanthin dipalmitate and canthaxanthin. *Aquaculture*, *65*, 279-292.

[35] Foss, P., Storebakken, T., Austreng, E., & Liaaen-Jensen, S. (1987). Carotenoids in diets for samonids I. Pigmentation of rainbow trout and sea trout with astaxanthin and astaxanthin dipalmitate in comparison with canthaxanthin. *Aquaculture*, *65*, 293-305.

[36] Mori, T., Makabe, K., Yamaguchi, K., Konosu, S., & Arai, S. (1989). Comparison between krill astaxanthin diester and synthesised free astaxanthin supplemented to diets in their absorption and deposition by juvenile coho salmon (*Oncorhynchus kisutch*). *Comparative Biochemistry and Physiology*, *93*, 255-258.

[37] Bjerkeng, B., Storebakken, T., & Liaaen-Jensen, S. (1990). Response to carotenoids by rainbow trout in the sea: resorption and metabolism of dietary astaxanthin and canthaxanthin. *Aquaculture*, *91*, 153-162.

[38] Liaaen-Jensen, S., & Storebakken, T. (1990). The pigmentation of salmonids. I. Carotenoids. In *Documentation Index 2265*. F. Hoffmann-La Roche Ltd, Basel.

[39] Foss, P., Storebakken, T., Schiedt, K., Liaaen-Jensen, S., Austreng, E., & Streiff, K. (1984). Carotenoids in diets for samonids I. Pigmentation of rainbow trout with the individual optical isomers of astaxanthin in comparison with canthaxanthin. *Aquaculture*, *41*, 213-226.

[40] Choubert, G., & Storebakken, T. (1989). Dose response to astaxanthin and canthaxanthin pigmentation of rainbow trout fed various dietary carotenoid concentrations. *Aquaculture*, *81*, 69-77.

[41] Sheehan, E. M., O'Connor, T. P., Sheehy, P. J. A., Buckley, D. J., & FitzGeraldd, R. (1998). Stability of astaxanthin and canthaxanthin in raw and smoked Atlantic salmon (*Salmo salar*) during frozen storage. *Food Chemistry*, *63*(3), 313-317.

[42] Yanar, M., Çelik, M., Yanar, Y., & Kumlu, M. (1998). Derin Dondurucuda Depolanan Gökkusagı Alabalıgı (*Oncorhynchus mykiss*) Filetolarında Karotenoyit Pigmentlerinin stabilitesi. *Turkish Journal of Biology*, *22*, 61-65.

[43] Brewer, M.S., Sosnicki, A., Field, B., Hankes, R., Ryan, K.J., Zhu, L. G., & McKeith, F.K. (2004). Enhancement effects on quality characteristics of pork derived from pigs of various commercial genetic backgrounds. *Journal of Food Science*, *69*(1), SNQ5-SNQ10.

[44] Bekhit, A.E.D., Geesink, G.H., Morton, J.D., & Bickerstaffe, R. (2000). Metmyoglobin reducing activity and colour stability of ovine longissimus muscle. *Meat Science*, *57*(4), 427-435.

[45] Bekhit, A.E.D., Geesink, G.H., Ilian, M.A., Morton, J.D., Sedcole, R., & Bickerstaffe, R. (2003). Particulate metmyoglobin reducing activity and its relationship with meat colour. *Journal of Agricultural and Food Chemistry*, *51*(20), 6026-6035.

[46] Shimizu, C., & Matsuura, F. (1971). Occurrence of new enzyme reduccing metmyoglobin in dolphin muscle. *Agricultural Biological Chemistry*, *35*, 485-490.

[47] Matsui, T., Shimizu, C., & Matsuura, F. (1975). Studies on metmyoglobin reducing systems in the muscle of blue white-dolphin. II. Purification and some physico-chemical properties of ferrimyoglobin reductase. *Bulletin of Japanese Society of Scientific Fisheries*, *41*(7), 771-777.

[48] Haard, N.F. (1992). Biochemistry and chemistry of colour and colour change in seafoods. In G.J., Flick, & R.E., Martin (Eds.), *Advances in Seafood Biochemistry* (pp. 305-360). Lancaster, Pa.: Technomic Publishing Co. Inc.

[49] DeMan, J.M. (1999). Colour. In M. D. Gaithersburg (ed.), *Principles of food chemistry* (3rd ed., pp. 239-242). Aspen Publishers, Inc.

[50] Fox, J.B., Jr., Strehler, T., Bernofsky, C., & Schweigert, B.S. (1958). Biochemistry of myoglobin, Production and identification of a green pigment formed during irradiation of meat extracts. *Journal Agriculture and Food Chemistry*, *6*, 692-696.

[51] Giddings, G.G., & Markakis, P. (1972). Characterization of the red pigments produced from ferrimyoglobin by ionizing radiation. *Journal of Food Science*, *37*, 361-364.

[52] Liu, Y., & Chen, Y.R. (2000). Two-dimensional correlation spectroscopy study of visible and near-infrared spectral intensity variations of chicken meats in cold storage. *Applied Spectroscopy*, *54*, 1458-1470.

[53] Liu, Y., Chen, Y.R., & Ozaki, Y. (2000). Two-dimensional visible/ near-infrared correlation spectroscopy study of thermal treatment of chicken meats. *Journal of Agricultural and Food Chemistry*, *48*, 901-908.

[54] Liu, Y., Chen, Y.R., & Ozaki, Y. (2000). Characterization of visible spectral intensity variations of wholesome and unwholesome chicken meats with two-dimensional correlation spectroscopy. *Applied Spectroscopy*, *54*, 587-594.

[55] Millar, S.J., Moss, B.W., & Stevenson, M.H. (1996). Some observations on the absorption spectra of various myoglobin derivatives found in meat. *Meat Science*, *42*, 277-288.

[56] Ostdal, H., Skibste, L.H., & Andersen, H.J. (1997). Formation of long-lived protein radicals in the reaction between H_2O_2-activated metmyoglobin and other proteins. *Free Radical Biology and Medicine*, *23*, 754-761.

[57] Whitburn, K.D., Shieh, J.J., Sellers, R.M., Hoffmann, M.Z., & Taub, I.A. (1982). Redox transformations in ferrimyglobin induced by radiation-generated free radicals in aqueous solution. *Journal of Biological Chemistry*, *257*, 1860-1869.

[58] Krinsky, N.I. (1994). The biological Properties of Carotenoids. *Pure and Applied Chemistry*, *66*, 1003-1010.

[59] Britton, G. (1995). UV/Visible spectroscopy. In G., Britton, S., Liaaen-Jensen, & H. Pfander (Eds.), *Carotenoids: Spectroscopy* (vol. 1B, pp. 13-62). Birkhäuser Verlag, Basel.

[60] Britton, G. (1995). Structure and properties of carotenoids in relation to function. *The FASEB Journal*, *9*, 1551-1558.

[61] Frank, H., & Cogdell, R.J. (1993). Photochemistry and function of carotenoids in photosynthesis. In A., Young, & G., Britton (Eds.), *Carotenoids in Photosynthesis* (pp.252-326). London: Chapmann and Hall.

[62] Savagaon, K.A., & Sreenivasan, A. (1978). Activation mechanism of pre-phenoloxidase in lobster and shrimp. *Fishery Technology*, *15*, 49-55.

[63] Henry, K.M., & Kon, S.K. (1950). Effect of reaction with glucose on the nutritive value of casein. *Biochimica et Biophysica Acta*, *5*, 455-456.

[64] Tanaka, M., Lee, T.-C., & Chichester, C.O. (1975). Effect of browning on chemical properties of egg albumin. *Agrie Biological Chemistry*, *39*, 863-866.

[65] Tanaka, M., Kimiagar, M., Lee, T.-C., & Chichester, C.O. (1977). Effect of Maillard browning reaction on nutritional quality of protein. *Advances in Experimental Medicine and Biology*, *86B*, 321-341.

[66] Hurrell, R.F., & Carpenter, K.J. (1974). Mechanisms of heat damage in proteins. 4. The reactive lysine content of heat-damaged material as measured in different ways. *Brazilian Journal of Nutrition*, *32*, 589-604.

[67] Erbersdobler, H.F., Brandt, A., Scharrer, E., & Von Wangenheim, B. (1981). Transport and metabolism studies with fructose amino acids. Prog. *Food Nutrition Science*, *5*, 257-263.

[68] Lee, T.-C., Pintauro, S.J., & Chichester, C.O. (1982). Nutritional and toxicologic effects of nonenzymatic Maillard browning. *Diabetes*, *31*, 37-46.

[69] Plakas, S.M., Lee, T.C., Wolke, R.E., & Meade, T.L. (1985). Effect of Maillard Browning Reaction on Protein Utilization and Plasma Amino Acid Response by Rainbow Trout (*Salmo gairdneri*). *The Journal of Nutrition*, *115*, 1589-1599.

[70] Ershov, A.M., Sabbotin, A.A., Rulev, N.N., & Baranov, V.V. (1980). Process for Browning Fish. *USSR patent*, *736*, 941.

[71] Aristoy, M.C., & Toldra, F. (1998). Pork meat quality affects peptide and amino acid profiles during the ageing process. *Meat Science*, *50*, 327-332.

[72] Laborde, D., Talmant, A., & Monin, G. (1985). Activités enzymatiques métaboliques et contractiles de 30 muscles du porc. Relations avec le pH ultime atteint après la mort. *Reproduction Nutrition Development*, *25*, 619-628.

[73] Toldrá, F. (2004). Manufacturing of Dry-Cured Ham: Dry-Cured Meat Products (pp. 27-60). USA: Blackwell Publishing.

[74] Bouton, P.E., Harris, P.V., & Shorthose, W.R. (1971). Effect of ultimate pH upon the water-holding capacity and tenderness of mutton. *Journal of Food Science*, *36*, 435-439.

[75] Hearne, L.E., Penfield, M.P., & Goertz, G.E. (1978). Heating effects on bovine semitendinosus: shear, muscle fibre measurements, and cooking losses. *Journal of Food Science*, *43*, 10-14.

[76] Shirsat, N., Lyng, J.G., Brunton, N.P., & McKenna, B. (2004). Ohmic processing: Electrical conductivities of pork cuts. *Meat Science*, *67*, 507-514.

[77] Zhang, Y.S., Yao, S., & Li, J. (2006). Vegetable-derived isothiocyanates: anti-proliferative activity and mechanism of action. *Proceedings of the Nutrition Society*, *65*, 68-75.

[78] Palka, K., & Daun, H. (1999). Changes in texture, cooking losses, and myofibrillar structure of bovine M. semitendinosus during heating. *Meat Science*, *51*, 237-243.

[79] Bastida, S., & Sánchez-Muniz, F.J. (2001). Thermal oxidation of olive oil, sunflower oil and a mix of both oils during forty discontinuous domestic fryings of different foods. *Food Science and Technology International*, *7*, 15-21.

[80] Swatland, H.J. (2002). On-line monitoring of meat quality. In J., Kerry, J., Kerry, & D., Ledward (Eds.), *Meat Processing: Improving Quality* (pp. 193-212). England, Cambridge: Woodhead Publishing.

[81] Thipayarat, A. (2006). Effect of Elevated Isothermal Cooking on Colour Degradation of Cooked Pork Ham. *IUFOST 13th World Congress of Food Sciences Technology, Food is Life Nantes, France, 17-21 September 2006*, 1203-1214.

[82] Millar, S.J., Moss, B.W., & Stevenson, M.H. (2000). The effect of ionizing radiation on the colour of beef, pork and lamb. *Meat Science*, *55*, 349-360.

[83] Diehl, J.F. (1995). *Radiological and toxicological safety of irradiated foods in safety of irradiated foods* (2nd ed., pp. 43-88). New York: Marcel Dekker.

[84] ICGFI. (1993). Irradiation of poultry meat and its products-a compilation of technical data for its authorization and control. *International Consultative Group on Food Irradiation, IAEA- TECDOC- 688 IAEA*, Vienna.

[85] Millar, S.J., Moss, B.W., MacDougall, D.B., & Stevenson, M.H. (1995). The effect of ionising radiation on the CIELAB colour cooordinates of chicken breast as measured by different instruments. *International Journal of Food Science & Technology*, *30*(5), 663-674.

[86] Nanke, K.E., Sebranek, J.G., & Olson, D.G. (1998). Colour characteristics of irradiated vacuum-packaged pork, beef and turkey. *Journal of Food Science*, *63*, 1001-1006.

[87] Kim, Y.H., Nam, K.C., & Ahn, D.U. (2002). Colour, oxidation reduction potential and gas production of irradiated meat from different animal species. *Journal of Food Science*, *61*, 1692-1695.

[88] Nam, K.C., & Ahn, D.U. (2002). Carbon monoxide-heme pigment is responsible for the pink colour in irradiated raw turkey breast meat. *Meat Science*, *60*, 25-33.

[89] Nam, K.C., & Ahn, D.U. (2002). Mechanisms of pink colour formation in irradiated precooked turkey breast meat. *Journal of Food Science*, *67*, 600-607.

[90] Ahn, D.U., Olson, D.G., Jo, C., Chen, X., Wu, C., & Lee, J. I. (1998). Effect of muscle type, packaging and irradiation on lipid oxidation volatile production and colour in raw pork patties. *Meat Science*, *49*(1), 27-39.

[91] Ginger, I.D., Lewis, U.J., & Schweigert, B.S. (1955). Changes associated with irradiating meat and meat extracts with gamma rays. *Journal of Agricultural and Food Chemistry*, *3*, 156-159.

[92] Brewer, S. (2004). Irradiation effects on meat colour – a review. *Meat Science, 68*, 1-17.

[93] Tappel, A. L. (1956). Regeneration and stability of oxymyoglobin in some gamma irradiated meats. *Food Research, 21*, 650-655.

[94] Bernofsky, C., Fox, J.B.Jr., & Schweigert, B. S. (1959). Biochemistry of myoglobin VI. The effect of low dosage gamma irradiation in beef myglobin. *Archives of Biochemistry and Biophysics*, *80*, 9-21.

[95] Satterlee, L. D., Brown, W. D., & Lycometros, C. (1972). Stability and characteristics of the pigmeat produced by gamma irradiation of metmyoglobin. *Journal of Food Science*, *37*, 213-217.

[96] Carlez, A., Veciana-Nouges, T., & Cheftel, J.C. (1995). Changes in colour and myoglobin of minced beef meat due to high pressure processing. *Lebensmittel-Wissenschaft und-Technologie*, *28*, 528-538.

[97] Shigehisa, T., Ohmori, T., Saito, A., Taji, S., & Hayashi, R. (1991). Effects of high hydrostatic pressure on characteristics of pork slurries and inactivation of microorganisms associated with meat and meat products. *International Journal of Food Microbiology*, *12*, 207-215.

[98] Carlez, A. Rosec., J.P., Richard, N., & Cheftel, J.C. (1994). Bacterial Growth During Chilled Storage of Pressure Treated Minced Meat. *Lebensmittel-Wissenschaft und-Technologie*, *27*, 48-54.

[99] Goutefongea, R., Rampon, V., Nicolas, N., & Dumont, J.P. (1995). Meat colour changes under high pressure treatment. *In* American Meat Science Association (ed.), *Proceedings of the 41st Annual International Congress of Meat Science and Technology* (p. 384). San Antonio, Texas, USA: American Meat Science Association.

[100] Angsupanich, K., & Ledward, D.A. (1998). High Pressure Treatment Effects on Cod (*Gadus Morhua*) Muscle. *Food Chemistry*, *63*(1), 39-50.

[101] Cheah, P.B., & Ledward, D.A. (1996). High Pressure Effects on Lipid Oxidation in Minced Pork. *Meat Science*, *43*(3), 123-134.
[102] Cheah, P. B., & Ledward, D. A. (1997). Catalytic mechanism of lipid oxidation following high pressure treatment in pork fat and meat. *Journal of Food Science*, *62*, 1135-1138.
[103] Chen, H. H. (2002). Discoloration and gel-forming ability of horse mackerel mince by air-flotation washing. *Journal of Food Science*, *67*, 2970-2975.
[104] Spinelli, J., & Dassow, J.A. (1982). Fish proteins: their modification and potential uses in the food industry. In R.E., Martin, G.J., Flick, C.E., Hebard, & D.R., Ward (Eds.), *Chemistry & Biochemistry of Marine Food Products* (pp. 13-25). Westport, CT: Avi.
[105] Sikorski, Z.E., Kolakowska, A., & Pan, B.S. (1990). The nutritive composition of the major groups of marine food organisms. In Z. E. Sikorski (Ed.), *Seafood: Resources, nutritional composition, and preservation* (pp. 30-54). Boca Raton, FL: CRC Press.
[106] Hashimoto, K., Watabe, S., Kono, M., & Shiro, K. (1979). Muscle protein composition of sardine and mackerel. *Bulletin of the Japanese Society of Scientific Fisheries*, *45*, 1435-1441.
[107] Haard, N.F., Simpson, B.K., & Pan, B.S. (1994). Sarcoplasmic proteins and other nitrogenous compounds. In Z.E., Sikorski, B.S., Pan, & F., Shahidi (Eds.), *Seafood Proteins* (pp. 13-39). New York: Chapman & Hall.
[108] Lanier, T.C. (2000). Surimi gelation chemistry. In J. W. Park (Ed.), Surimi and surimi seafood (pp. 237-265). New York: Marcel Dekker.
[109] Chen, H.H. (2003). Effect of cold storage on the stability of chub and horse mackerel myoglobins. *Journal of Food Science*, *68*, 1416-1419.
[110] . Emilia, E., & Santos-Yap, M. (1995). Fish and seafood. In L. E. Jeremiah (Ed.), *Freezing effects on food quality* (pp. 109-133). New York: Marcel Dekker.
[111] Sikorski, Z. E. (1994). The myofibrillar proteins in seafoods. In Z.E., Sikorski, B.S., Pan, & F., Shahidi (Eds.), *Seafood Proteins* (pp. 40-57). New York: Chapman & Hall.
[112] Chaijan, M., Benjakul, S., Visessanguan, W., & Faustman, C. (2007). Characterization of myoglobin from sardine (*Sardinella gibbosa*) dark muscle. *Food Chemistry*, *100*, 156-164.
[113] Chinnaswamy, R., & Hanna, M.A. (1988). Relationship between amylose content and extrusion expansion properties of corn starches. *Cereal Chemistry, 65*, 138-143.
[114] Lee, D.S., & Kim, H.K. (1989). Carotenoid destruction and nonenzymatic browning during red pepper drying as functions of average moisture content and temperature. *Korean Journal of Food Science and Technology*, *21*, 425-429.
[115] Garcia, P., Brenes, M., Romero, C., & Garrido, A. (1999). Colour and texture of acidified ripe olives in pouches. *Journal of Food Science*, *64*, 248-251.
[116] Sainclivier, M. (1985). L'industrie alimentaire halieutique. Tome 2. Des techniques ancestrales à leur reálisation contemporaines. *Sciences Agronomiques Rennes.* France, Rennes.
[117] Sainclivier, M. (1988). L'industrie alimentaire halieutique. Tome 3. La conservation par des moyens physiques. *Partie 1. Conserve de poissons. Sciences Agronomiques Rennes.* France, Rennes.
[118] Sainclivier, M. (1988). L'industrie alimentaire halieutique. Tome 3. La conservation par des moyens physiques. *Partie 2. L'ionisation-Le conditionnement. Sciences Agronomiques Rennes.* France, Rennes.

[119] Skonberg, D.I., Hardy, R.W., Barrows, F.T., & Dong, F.M. (1998). Colour and flavor analyses of fillets from farm-raised rainbow trout (*Oncorhynchus mykiss*) fed low-phosphorus feeds containing corn or wheat gluten. *Aquaculture (Amsterdam)*, *166*, 269-277.

[120] Pokorny, J. (1981). Browning from lipid–protein interactions. *Progress in Food Nutrition Science*, *5*, 421-428.

[121] Foegeding, E.A., Lanier, T.C., & Hultin, H.O. (1996). Characteristics of edible muscle tissues. In O. R. Fennema (Ed.), *Food Chemistry* (pp. 880-942). New York: Inc, Marcel Dekker.

[122] Giddings, G.G., & Hill, L.H. (1978). Relationship of freezing preservation parameters to texture-related structural damage to thermally processed crustacean muscle. *Journal of Food Processing and Preservation*, *2*, 249-264.

[123] Wagner, J. R., & Añon, M. C. (1985). Effect of freezing rate on the denaturation of myofibrillar proteins. *International Journal of Food Science and Technology*, *20*, 735-744.

[124] Chourot, J.M., Lemaire, R., Cornier, G., & Le Bail, A. (1996). Modelling of high-pressure thawing. In R., Hayashi, & C., Balny (Eds.), *High Pressure Bioscience and Biotechnology* (pp. 439-444).

[125] Vanichseni, S., Haughey, D.P., & Nottingham, P.M. (1972). Water-and air-thawing of frozen lamb shoulders. *International Journal of Food Science & Technology*, *7*(3), 259-270.

[126] Karel, M., & Lund, D.B. (2003). *Physical principles of food preservation* (2nd revised edition). New York: Marcel Dekker.

[127] Ersoy, B., Aksan, E., & Özeren, A. (2008). The effect of thawing methods on the quality of eels (*Anguilla anguilla*). *Food Chemistry*, *111*(2), 377-380.

[128] Pearson, A.M., & Tauber, F.W. (1984). Curing. In *Processed Meats* (2nd ed., pp. 47). Westport, Connecticut: AVI Publishing Company INC.

[129] Bekhit, A.E.D., & Faustman, C. (2005). Metmyoglobin reducing activity in fresh meat: A review. *Meat Science*, *71*, 407-439.

[130] Brody, A.L. (1996). Integrating aspect and modified atmosphere packaging to fulfill a vision of tomorrow. *Food Technology, 50*, 56-66.

[131] Nattress, F.M., & Jeremiah, L.E. (2000). Bacterial mediated off-flavours in retail-ready beef after storage in controlled atmospheres. *Food Research International*, *33*, 743-748.

[132] Faber, J.M. (1991). Microbiological aspect of modified atmosphere packaging technology. *Journal of Food Protection*, *54*, 58-70.

[133] Gill, C.O. (2003). Active packaging in practice: meat. In H. Ahvenainem (ed.), *Novel food packaging technology* (pp. 378-396). Boca Raton, Fla: Woodhead Publishing Limited and CRC Press LLC.

[134] Jeremiah, L. E. (2001). Packaging alternatives to deliver fresh meats using short- or long-term distribution. *Food Research International*, *34*, 749-772.

[135] Ercolini, D., La Storia, A., Villani, F., & Mauriello, G. (2006). Effect of a bacteriocin-activated polyethylene film on *Listeria monocytogenes* as evaluated by viable staining and epifluorescence microscopy. *Journal of Applied Microbiology*, *100*, 765-772.

[136] Zhao, Y., Wells, J.H., & Mcmillin, K.W. (1994). Applications of dynamic modified atmosphere packaging systems for fresh red meats: Review. *Journal of Muscle Foods, 5*, 299-328.

[137] Pfeiffer, C., D'Aujourd'Hui M., Walter, J., Nuessli, J., & Escher, F. (1999). Optimizing food packaging and shelf life. *Food Technology, 53*(6), 52-59.

[138] McDonald, K., & Sun, D. (1999). Predictive food microbiology for the meat industry: a review. *International Journal of Food Microbiology, 52*, 1-27.

[139] Meat Technology Update (2006). Colour defects in meat-Part 2: Greening, Pinking, Browning & Spots. *Food Science Australia Meat Technology Update, 6/06, December 2006.*

[140] Huffman, D.L. (1980). Processing effects on fresh and frozen meat colour. *Proceedings Reciprocal Meat Conference of the American Meat Science Association, 33*, 4-14.

[141] Kropf, D.H. (1980). Effects of retail display conditions on meat colour. *Proceedings Reciprocal Meat Conference of the American Meat Science Association, 33*, 15-32.

[142] Millar, S., Wilson, R., Moss, B.W., & Ledward, D.A. (1994). Oxymyoglobin formation in meat and poultry. *Meat Science, 36*, 397-406.

[143] Senter, S.D., Young, L.L., & Searcy, G.K. (1997). Colour values of cooked top-round beef juices as affected by endpoint temperatures, frozen storage of cooked samples and storage of expressed juices. *Journal of the Science of Food and Agriculture, 75*(2), 179-182.

[144] Pacheco-Aguilar, R., Lugo-Sánchez, M.E., & Robles-Burgueño, M.R. (2000). Postmortem biochemical and functional characteristic of Monterey sardine muscle stored at 0 °C. *Journal of Food Science, 65*, 40-47.

[145] O'Grady, M.N., Monahan, F.J., & Brunton, N.P. (2001). Oxymyoglobin oxidation and lipid oxidation in bovine muscle-mechanistic studies. *Journal of Food Science, 66*, 386-392.

[146] Haard, N.F. (1992). Control of chemical composition and food quality attributes of cultured fish. *Food Research International, 25*, 289-307.

[147] Renerre, M., & Labas, R. (1987). Biochemical factors influencing metmyoglobin formation in beef muscles. *Meat Science, 19*, 151-165.

[148] Christophersen, A.G., Bertelsen, G., Andersen, H.J., Knuthsen, P., & Skibsted, L.H. (1992). Storage life of frozen salmonids. Effect of light and packaging conditions on carotenoid oxidation and lipid oxidation. *Zeitschrzft fur Lebensmittel-Unterschung und-Forschung, 194*, 115-1 19.

[149] No, H.K., & Storebakken, T. (1991). Colour stability of rainbow trout fillets during frozen storage. *Journal of Food Science, 56*, 969-972.

[150] Scott, T.M., Rasco, B.A., & Hardy, R.W. (1994). Stability of krill meal, astaxanthin and canthaxanthin colour in cultured rainbow trout (*Oncorthynchus mykiss*) fillets during frozen storage and cooking. *Journal of Aquatic Food Product Technology, 3*, 53-64.

[151] AMSA. (1991). Guideline for meat colour evaluation. *Proceedings of the Reciprocal Meat Conference, 44*, 1-17.

[152] CIE L*a*b* Colour Scale (2008). *HunterLab Applications Note, 8 (7).*

[153] CIE L*a*b* Colour Scale (2008). *HunterLab Applications Note, 8 (9).*

[154] Hunter *L,a,b* Versus CIE 1976 L*a*b* (2008). *HunterLab Applications Note, 13*(2).

[155] Francis, F.J., & Clydesdale, F.M. (1975). *Food Colourimetry: Theory and Applications.* Westport, CT: AVI Publishing Co. Inc.

[156] Anon. (1988). *Munsell Colour*. Product information brochure of Macbeth division of Kollmorgen, Baltimore, Maryland.

[157] Alcalde, M.J., & Negueruela, A.L. (2001). The influence of final conditions on meat colour in light lamb carcasses. *Meat Science*, *57*(2), 117-123.

[158] O'Sullivan, M.G., Byrne, D.V., Martens, H., Gidskehaug, G.H., Andersen, H.J., & Martens, M. (2003). Evaluation of pork colour: Prediction of visual sensory quality of meat from instrumental and computer vision methods of colour analysis. *Meat Science*, *65*(2), 909-918.

[159] Mancini, R.A., & Hunt, M.C. (2005). Current research in meat colour. *Meat Science*, *71*, 100-121.

[160] Lu, J., Tan, J., Shatadal, P., & Gerrard, D.E. (2000). Evaluation of pork colour by using computer vision. *Meat Science*, *56*(1), 57-60.

[161] Ringkob, T.P. (2001). Image analysis to quantify colour deterioration on fresh retail beef. *In Proceedings 54th Reciprocal Meat Conference, 24-28 July 2001*. Indianapolis, Indiana.

[162] Ringkob, T.P. (2003). Comparing pork fat colour from barley and corn fed pork using image analysis. *In 56th Reciprocal Meat Conference, 28-31 July 2003*. Lansing, Michigan.

[163] Quevedo, R.A., Aguilera, J.M., & Pedreschi, F. (2008). Colour of salmon fillets by computer vision and sensory panel. *Food and Bioprocess Technology*, DOI 10.1007/s11947-008-0106-6.

[164] Carpenter, C.E., Cornforth, D.P., & Whittier, D. (2001). Consumer preferences for beef colour and packaging did not affect eating satisfaction. *Meat Science, 57*(4), 359-363.

[165] MacDougall, D.B. (1982). Changes in the colour and opacity of meat. *Food Chemistry*, *9*, 75-88.

[166] Hunt, R.W.G. (1991). *Measuring colour* (2nd ed.). New York: Ellis Horwood.

[167] Joo, Y.H., Chen, G., & Shieh, L.S. (1999). Hybrid state–space fuzzy modelbased controller with dual-rate sampling for digital control of chaotic systems. *IEEE Transactions on Fuzzy Systems*, *7*, 394-408.

[168] Garcia-Esteben, M., Ansorena, D., Gimeno, O., & Astiasaran, I. (2003). Optimization of instrumental colour analysis in dry-cured ham. *Meat Science*, *63*(3), 287-292.

In: Practical Food and Research
Editor: Rui M. S. Cruz, pp. 181-193

ISBN: 978-1-61728-506-6

Chapter VII

ENZYMES

Jaime M. C. Aníbal[1,2] and Rui M. S. Cruz[1,3]

[1]Departamento de Engenharia Alimentar, Instituto Superior de Engenharia, Universidade do Algarve, Campus da Penha, 8005-139 Faro, Portugal, janibal@ualg.pt and rcruz@ualg.pt.

[2]CIMA- Centro de Investigação Marinha e Ambiental, Universidade do Algarve, Campus de Gambelas, 8005-139 Faro, Portugal

[3]CIQA- Centro de Investigação em Química do Algarve, Universidade do Algarve, Campus de Gambelas, 8005-139 Faro, Portugal

ABSTRACT

The loss of freshness in meat and fish is often caused by a combination of physical, biochemical, and microbiological reactions. *Post mortem* tenderization is one of the most important attributes in meat and fish quality. In meat, the tenderization phenomenon is a desirable quality attribute, while in fish a soft texture is not attractive to the consumers. Several proteases have been isolated and the effects of proteolytic breakdown are normally related to the softening of the tissue, affecting textural properties. This chapter will present some enzymes, such as cathepsin and calpain, related to fish and meat softening. The determination of enzymatic activity and kinetic parameters by several methodologies will also be addressed.

7.1 INTRODUCTION

The consumers' acceptance of meat and fish products depends on several food quality attributes [1]. Freshness and firmness are two important aspects of raw fish which best contributes to define the quality of fish as food [2]. Thus, soft fillets are a problem for the fish industry [1, 3-5].

In contrast to fish, weakening of the mammalian muscle structure during *post mortem* storage is desirable since it improves tenderness [6]. This feature is the most important meat quality attribute for the consumer [7].

The loss of freshness is often caused by a combination of physical, biochemical, and microbiological reactions. Several proteases have been isolated from fish muscle and the effects of proteolytic breakdown are normally related to the softening of the tissue.

The enzymatic degradations introduce the *post mortem* softening of fish muscle, which allows a proliferation of bacterial flora. During the *post mortem* ageing of muscle under chilled conditions, degradation of muscle proteins contributes to the rapid softening of flesh. In the case of pelagic fish, the *rigor mortis* is completed between 3 and 18 h after the death of the animal [8].

In the case of bovine muscle the *rigor mortis* occurs 24 h after slaughter. Tenderness has been defined as that quality of cooked meat that is recognized by the characteristic of easy chewability without loss of desirable texture [9]. In bovine muscle, the acquisition of optimal tenderness requires at least 14 days [10]. Tenderization phenomenon has an enzymatic nature and the conditions such as the pH may vary the proteolytic action of endogenous enzymes on proteins and connective tissue. Meat toughness can be divided into actomyosin toughness, due to changes in myofibrillar proteins, and background toughness, related to the connective tissues [11]. Current studies have focused on the roles of connective tissues, collagen and elastin in meat and meat products [12].

Several studies indicate that calpain and calpastatin activities are related to the extent of tenderization in mammalian muscles, which creates the possibility to use those enzymes as final quality indicators [13-15].

7.2 Enzymes in Meat and Fish Processing

There are several chemically and physically methods that can lead to meat tenderization. The use of proteolytic enzymes is one of the most popular methods for meat tenderization [11].

Proteolytic enzymes such as papain (EC 3.4.22.2), and to a lesser extend bromelain [from pineapple juice (EC 3.4.22.33) or from the crushed stems of the plant (EC 3.4.22.32) and ficin (EC 3.4.22.3) are extracted from plant sources and have been widely used in meat tenderization [16]. However, these enzymes often degrade the texture of the meat, due to the wide substrate specificity, and develop undesired taste due to overtenderization [17]. Therefore, the ideal meat tenderizer would be a proteolytic enzyme with specificity for collagen and elastin in connective tissues, at the relatively low pH of meat that would work at a low temperature at which meat is stored or at a high temperature during cooking [7].

The critical point, when using proteolytic enzymes, is how the enzymes are introduced into the meat. There are several methods: 1) soaking in a solution containing proteolytic enzymes; 2) pumping enzyme solution into major blood vessels of meat cut and 3) rehydration of the freeze-dried meat in a solution containing a proteolytic enzyme [7]. Nevertheless, Lawrie [18] reported that the first two methods are unable to penetrate within the meat. The methods overtenderize the surface, producing a mushy texture, and, thus are not suitable.

The rehydration of the freeze-dried meat showed a much better distribution of enzymes than dipping or perfusion. Another approach is to apply a pre-slaughter injection of the enzymes into live animals, which was considered to be the most effective method of introducing the enzymes into meat [19]. In this case, an injection of an unsuitable enzyme may cause shock to the animal, resulting in the production of low quality meat. This method is not simple, because it deals with live animals.

Another aspect of meat tenderness is related to myofibrillar proteolysis, which can be attributed to endogenous protease activity. Currently, two characterized proteolytic systems are known to hydrolyze myofibrillar proteins during *post mortem* storage of meat and fish muscle: calpains and cathepsins [20].

7.2.1 Endogenous enzymes

Proteolytic enzymes are found in all tissue, although both the distribution of different enzymes and their activities show considerable variation. After death, the biological regulation of the enzymes is lost, and the enzymes hydrolyze muscle proteins and resolve the *rigor mortis* contraction [21].

Endogenous proteases are important in the deterioration process since they are able to hydrolyze different proteins in the muscle, by cleaving the peptide bonds of proteins and peptides [22-23].

The calpains (EC 3.4.22.17) are intracellular neutral calcium-dependent cysteine proteases and are found in meats, fish and crustaceans, and have been associated with muscle autolysis. The calpains are responsible for the early beginning of the proteolytic degradation of myofibrillar and cytoskeletal proteins, converting the muscle into meat [24]. On the other hand, fish softening through autolysis is a serious problem reducing its commercial value.

These enzymes are further subclassified into μ-calpain and m-calpain, which differ in sensitivity to calcium ions and are most active at neutral pH (6.9-7.5) and 30 °C, but are still quite active at pH 5. Calcium plays a key role in the activation mechanism of calpains, leading to dissociation and/or autoproteolysis, even in the presence of an alternative substrate. Calpastatin is known to be the endogenous specific inhibitor of the calpains [23, 25].

Cathepsins are acid proteases located in the lysosomes and responsible for protein breakdown at sites of injury, i.e. physical abuse or upon freezing and thawing of *post mortem* muscle, or any treatment that causes disruption of lysosomes, and thus involved in deterioration of muscle texture [23, 26-28]. The main types involved in muscle ageing are: cathepsins B (EC 3.4.22.1), L (EC 3.4.22.15), H (EC 3.4.22.16) and D (EC 3.4.22.5) [29].

The B and L cathepsins are the ones associated to the autolytic degradation of fish tissue. The other cathepsins have a relatively narrow pH range of activity. The temperature for optimum activity is about 40-50 °C and the pH for best performance is 3.0-4.0. The activity decreases with lower temperatures and some show high activity at pH 6.0-6.5 [23, 30-31].

Collagenases are other type of endogenous enzymes that are responsible for the break down of the connective tissue in the fish muscle and thus leading to undesirable textural changes [22, 27, 32-34].

7.3 Extraction and Isolation of Enzymes

The extraction and isolation of enzymes from animal tissues is, usually, much easier than from vegetable sources, because of their lack of a rigid cell wall, which makes homogenization relatively easy. The animal tissue is often cut up into small pieces and homogenized using a high-speed blender. Extraction is performed using isotonic or low ionic strength solutions, depending on the need of breaking cellular membranes or structures [35].

Extraction and enzymes isolation are conveniently discussed on Chapter II.

7.4 Enzymatic Activity and Kinetic Parameters Methodologies

Depending on the study objectives there are several methods for the determination of the enzymatic activity and the kinetic parameters. The following section will focus on several steps that can be run after a suitable enzyme extraction and isolation.

During a spectrophotometer assay to determine an enzyme specific activity or initial velocities is very important to follow the beginning of the reaction. When the substrate and enzyme react together, the instrument should be registering the results. This is not always easy to do, but a way to overcome it is to put one of the molecules (substrate or enzyme) inside the cuvette and then add the other directly with the pipette, using it to stir the solution by quickly filling and emptying it, and then starting the readings. If the enzyme and substrate are added outside the spectrophotometer and only after put it in a cuvette, when the reading starts already a few seconds have passed and the beginning of the reaction is missed.

When determining the relation between the enzyme and substrate at optimal conditions, or assaying to access the effect of temperature and pH over the enzyme, usually this determination units has three components, quantity of product formed (ex. micromoles) per unit of time (ex. minutes) and per quantity of enzyme present (milligrams), resulting in something like this: "$\mu mol.min^{-1}.mg^{-1}$". Using the following notation is not completely correct (although easier to write) because is not clear what is being arithmetical divided by what: "*µmol/min/mg*".

When the pH effect over an enzyme is assayed, the design of the experiment and the analysis of results are straight forward. Portions of the same enzyme are put at different buffer solution with several pH values, allowing the determination of the specific activity at each pH (Figure 7.1). But when doing temperature effect assays is necessary to really understand what is truly being measured. Two situations can be experimentally performed, but they don't give the same information. Firstly, several portions of the same enzyme can be placed at different temperatures and then their activities can be determined at room temperatures. In this case, what is being measured is the permanent effect of a given temperature over an enzyme that is stressed to a certain temperature. The enzyme return to room temperature sometimes neutralizes the effect of the exposure temperature. This kind of experiments is usually done when the spectrophotometer does not have the capacity to control the cuvette's reading chamber temperature. If the enzyme is placed directly in the spectrophotometer reading container at the assay given temperature, it will most probably cause condensation on the cuvette walls, and compromise the reading. The other experimental situation occurs when a spectrophotometer with reading chamber temperature control, mostly done by thermostatic bath, is available. The reaction between the enzyme and substrate is measured directly at a given temperature. Because all components of the experience are at the same temperature, there is no condensation on the cuvette walls. From this second type of experiments comes the result normally used to plot the theoretical curves of enzymatic activity versus temperature (Figure 7.1).

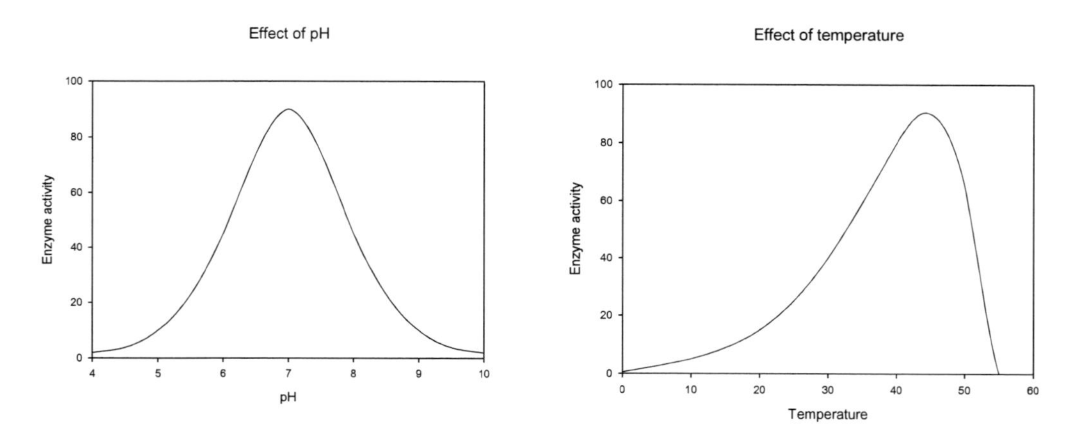

Figure 7.1. Effects of pH and temperature on enzyme activities.

The description of enzymatic characteristics involves kinetic analysis. In some cases this might be rather superficial, while in others many details need to be determined. The researcher should decide what level of information is required before embarking on large-scale experimentations. For instance, if the interest is just a surface response of peroxidase to bleaching in green-leaf vegetables, a detailed analysis of kinetic rate parameters would not be justifiable, saving a lot of time and money. On the other hand, a study might be ordered to measure all the parameters which would allow the prediction of the optimum temperature and exposure time on a blanching process to maximize the shelf life of lettuce sold in a package. Now, a more detailed experimental design is needed in order to obtain rate constants for the peroxidase reactions, with and without thermal treatments.

Once the level of sophistications of data is established, the experimental design requires knowledge and insight into the kinetics of enzyme action. More detailed knowledge on theoretical enzyme kinetics is available in many excellent texts [36-39]. Insight comes only from experience in working with enzymes, both in laboratory and in the plant.

An enzyme assay is performed through the measurement of product concentration formed (sometimes detected as a decrease in substrate concentration) which, expressed as a function of the time is the velocity rate at which the enzyme operated. This experimental principle is the basis of all kinetics determination, even when studying the effect of inhibitors. The effect of a specific inhibitor can be measured by the decrease of velocity in the enzymatic reaction, which is just a variant of previous experimental principal. That's why this chapter is focused on basics, once it is understood all the other issues are just different points of the same methodology.

In early investigations of enzyme reactions, made in the beginning of the XX centaury, it was found that the rate of product formation (v) was a hyperbolic function of the concentration of substrate ($[S]$). The key to this behavior was first proposed by V. Henry in 1902 and further elaborated by L. Michaelis and M. L. Menten in 1913. The basic concept is that the enzymatic reaction involves two steps. In the first step, there is a rapid and reversible combination of enzyme and substrate to form a complex. In the second step, there is a somewhat slower breakdown of the complex to give product and regenerate the enzyme. The key idea is that enzyme, substrate and complex are in rapid equilibrium characterized by an equilibrium dissociation constant, known as Michaelis-Menten constant (K_M). The equation is also definied by the maximum velocity (V_{max}) parameter, which correspond to the velocity at high concentrations of substrat (Figure 7.2). The Michaelis-Menten (M-M) equation is:

$$v = \frac{V_{max}\ [S]}{K_M +\ [S]}$$

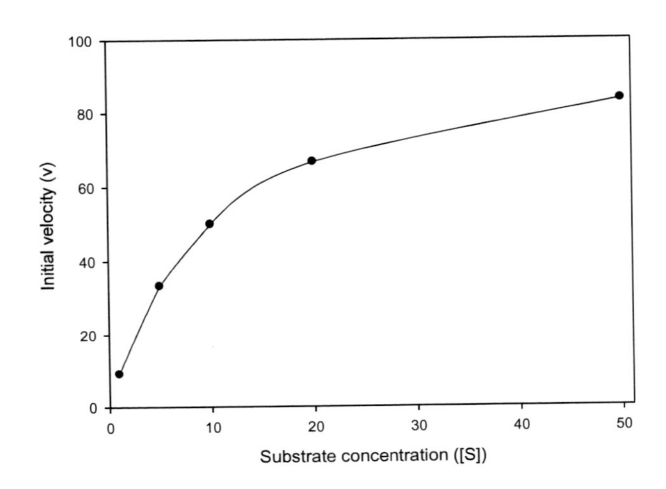

Figure 7.2. Michaelis-Menten plot according to the hyperbolic relation beween enzyme rate and substrate.

Not all enzymes follow this simple kinetic model, but this equation is the basis for almost all the other more complex methods. Because this book is also destined to researchers that are starting their practical work with food, this chapter will cover the more simple approaches to kinetics (one enzyme, one substrate systems), because this is probably what the researcher will do in the beginning. The advices below are of a general type, allowing their application to other kinetics models.

Once raw data is obtained from an enzyme kinetic experiment, the researcher will end up with a collection of numbers related to product or substrate concentrations determined at different times after the initiation of the reaction. The next phase is to mathematically manipulate the data in order to arrive at the kinetic parameters of interest. The methods have two aims: to minimize the effort required to perform them, and to maximize the reliability of the derived parameters. The main determinant of parameter reliability is the quality of the experimental data. Good mathematical analytical methods cannot transform poor data into reliable derived constants. GIGO (Garbage In, Gargabe Out) is still valid [40].

Modern personal computers have good calculation capacities, facilitating the use of very complex methods of kinetic parameters determination. If the basic assumptions behind any method are not really understood, the results, even those coming from reliable data, will induce wrong conclusions. A great variety of software (e.g. VisualEnzymics 2008; SigmaPlot®-Enzyme Kinetics Module) has been recently developed with user friendly interfaces. Nonetheless it is sometimes necessary to make calculations in the laboratory without an access to computers, moreover when using computers, the researcher should know which method should be used in a specific situation.

Most enzyme experiments are performed on the assumption that the observed reaction rate is correlated with the concentration of substrate present in the essay mixture at zero time, allowing the calculation of the reaction initial velocity to a certain substrate concentration. From this data it is possible to determine V_{max} and K_M.

The hyperbolic M-M equation does not allow accurate estimates of V_{max} and K_M from a direct plot of v versus *[S]*. However there are three linear transformations of M-M equation:

$$\frac{1}{v} = \frac{1}{V_{max}} + \left(\frac{K_M}{V_{max}}\right)\frac{1}{[S]} \quad \text{(Lineweaver-Burk)}$$

$$\frac{[S]}{v} = \frac{K_M}{V_{max}} + \left(\frac{1}{V_{max}}\right)[S] \quad \text{(Hanes)}$$

$$\frac{v}{[S]} = \frac{V_{max}}{K_M} - \left(\frac{1}{K_M}\right)v \quad \text{(Eadie-Hofstee)}$$

The algebraic manipulations to obtain these forms are straightforward. The approach of the Lineweaver-Burk method is a simple inversion of the M-M equation. Multiplying both sides of the Lineweaver-Burk equation by *[S]* gives the Hanes method. To obtain the Eadie-Hofstee equation, both sides of the M-M equation are multiplied by $(K_M+[S])/K_M[S]$, then the term are separated and reordered.

Which of the above transformations is the best one to use? The Lineweaver-Burk plot is the most often applied, yet it is by far the worst choice [40]. Experimental measurements of v are usually made at approximately equal increments of *[S]*. When the results are plotted with $1/[S]$ transformation on the x-axis, the data points tend to crowd together near the y-axis with a few point widely spaced towards the right side. This can be overcome by choosing *[S]* so that the reciprocals are more evenly spaced. This approach magnifies the second difficulty which is that rate measurements are less accurate at lower values of *[S]*. There is usually some inherent methodological error in enzyme assays. This can amount to as much as 25% of the rate at the lowest *[S]*, but is more acceptable 2% to 5% at higher *[S]* [40]. The points towards the right side of the Lineweaver-Burk plot have large vertical error component. In making a linear regression, these less accurate points have a much larger influence than those points nearer the origin which are more accurate.

The Hanes and the Eadie-Hofstee transformations are less sensible to these difficulties. The x-axis of the Hanes plot spaces the points according to the experimental spacing of *[S]*. If the error in determining v is inversely related to *[S]* then this plot is most reliable: the points containing higher experimental error are closer to the origin and have a reduced influence on the linear regression best-fit line. The Eadie-Hofstee plot is more reliable when inherent experimental difficulties result in rather large relative errors in v regardless of *[S]*. Since v is used directly on both axes the magnification of relative error introduced by taking the reciprocal is not apparent. Figure 7.3 shows the three transformation plots with the aim of helping visualizing the shape of the lines.

In a theoretical analysis of error propagation in the three methods, Atkins and Nimmo [41] concluded that the Hanes transformation was preferable to the Eadie-Hofstee in most situations, and was superior to the Lineweaver-Burk plot in all situations. Stauffer [40] "hoped that the use of the Lineweaver-Burk plot by enzyme researchers will soon cease".

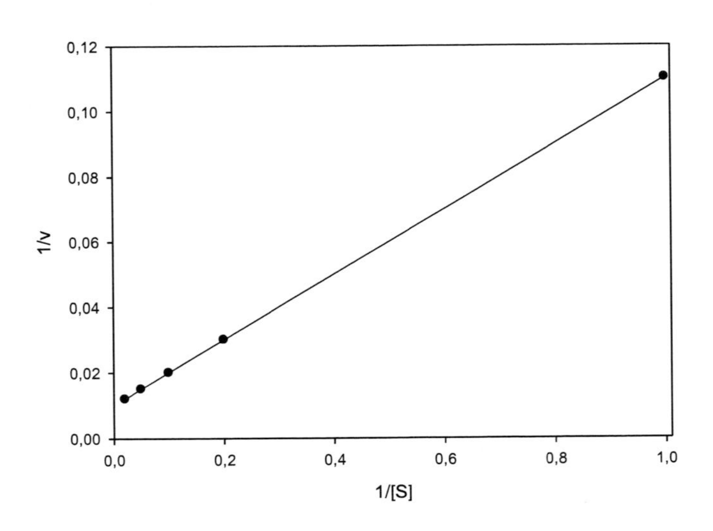

Hanes plot

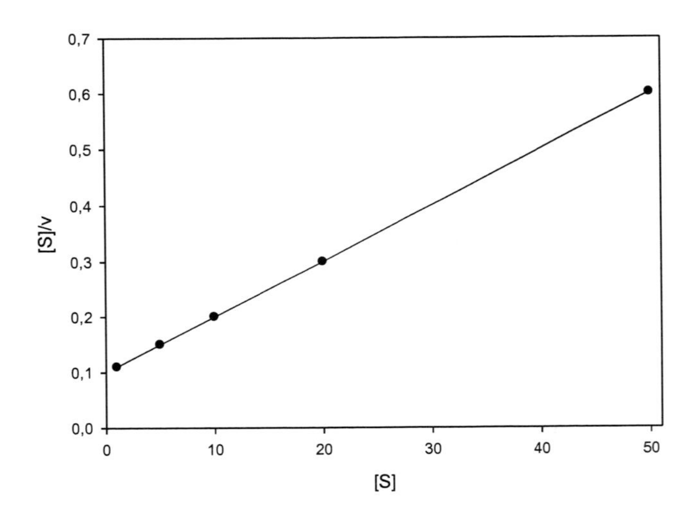

Figure 7.3. (Continued).

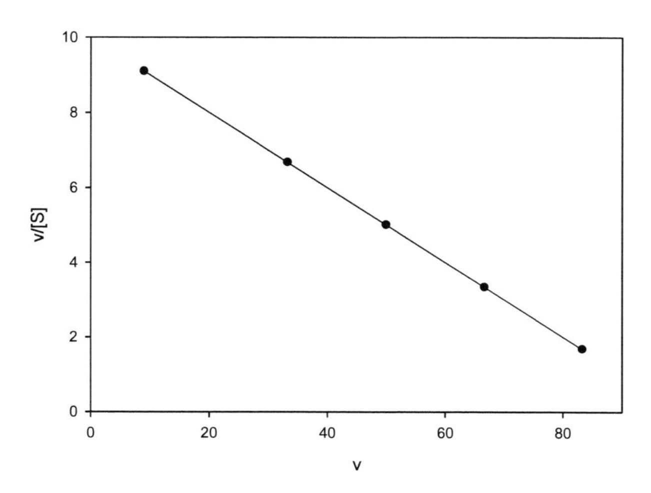

Figure 7.3. Transformation plot using the Lineaver-Burk, Hanes and Eadie-Hofstee methods.

The reciprocal plots described above transform the M-M equation into a linear equation $y=a+bx$. Least squares linear regression is a way of finding the values of a and b which define the straight line that best fits a set of data points such that the deviation of the actual y values from the line is minimized. While linear regression deals with the effect of normality distributed random errors in rate data, it is less successful in handling the occasional non-random error which gives rise to an "outlier".

The direct linear plot method, described by Cornish-Bowden and Eisenthal in 1974 and 1978 [42-43], circumvents this problem by using a different approach to minimizing deviations. Another way of writing an equation describing a straight line in Cartesian coordinates is:

$$\frac{y}{b}+\frac{x}{a}=1$$

Point a (on the x-axis) and point b (on the y-axis) will define a straight line. A second pair of points, a' and b', will define a second straight line, and so on. The point of intersection of the straight lines gives unique values for Vmax (y-axis) and KM (x-axis), as shown in Figure 7.4.

Additional data pairs define more straight lines; if there were no experimental error in either v or *[S]* all the lines would pass through the same intersection point. In real data the presence of error means that the intersection point of any given pair is different from that of any other line pair. The result is an "interception area" instead of a point. In this case, each point of that area is project in both axis, and the most likely value for V_{max} and K_M is taken as the median of those projected values. This procedure is statistically sound [44] and has the advantage of minimizing the effect of outliers on the final parameter values.

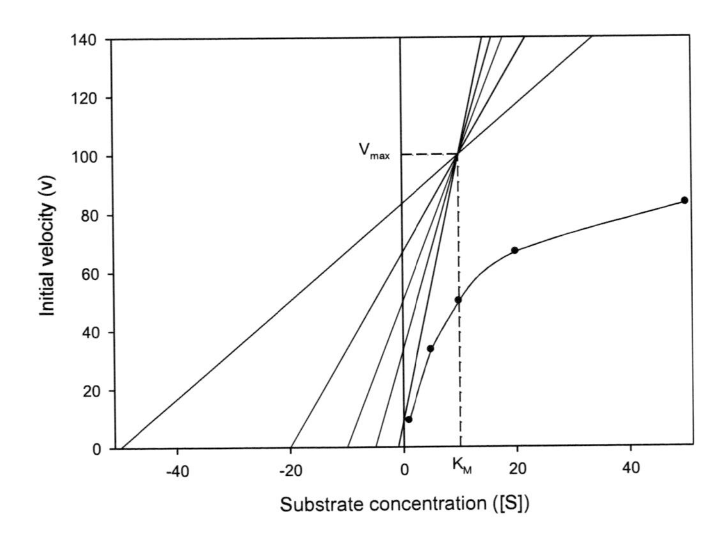

Figure 7.4. Representation of the Eisenthal and Cornish-Bowden direct linear plot method.

CONCLUSION

Understanding how different processing treatments and/or conditions may influence the proteins and proteolytic enzymes it is of extreme importance since they are related to changes in the quality of the fish and meat products.

Working with enzymes is more than a scientific work, is a form of art. Although we cannot consider enzymes to be alive, they are the functional basis of the living world, and sometimes they do unexpected things that are almost impossible to understand and justify. When that happens, please remember that working in chemistry is about patience and endurance.

"Even if we never have the time to do it right, we should always have the time to do it over". This comment is very important to understand. In the beginning of an enzymatic study is very difficult to predict what is going to happen (specially the bad parts). Often, we end up with months, if not years, of data that are very far away from what we intended at the start. But when we started we did not have the expert knowledge that we have after doing it. So most of the time, we end up with a pile of not so good data, but instead of starting all over again, we can repeat some assays with small changes, and then our data becomes, more easily, publishable.

REFERENCES

[1] Haard, N.F. (1992). Control of chemical composition and food quality attributes of cultured fish. *Food Research International*, *25*, 289-307.

[2] Chéret, R., Hernández-Andrés, A., Delbarre-Ladrat, C., Lamballerie, M., & Verrez-Bagnis, V. (2006). Proteins and proteolytic activity changes during refrigerated storage in sea bass (*Dicentrachus labrax* L.) muscle after high-pressure treatment. *European Food Research and Technology*, *222*, 527-535.

[3] Andersen, U.B., Thomassen, M.S., & Rørå, A.M.B. (1997). Texture properties of farmed rainbow trout (*Oncorhynchus mykiss*): Effects of diet, muscle fat content and time of storage on ice. *Journal of the Science of Food and Agriculture*, *74*, 347-353.

[4] Hallett, I.C., & Bremner, H.A. (1988). Fine structure of the myocommata- muscle fibre junction in hoki (Macruronus novaezelandiae). *Journal of the Science of Food and Agriculture*, *44*, 245-261.

[5] Sigholt, T., Erikson, U., Rustad, T., Johansen, S., Nordtvedt, T. S., & Seland, A. (1997). Handling stress and storage temperature affect meat quality of farmed-raised Atlantic salmon (*Salmo salar*). *Journal of Food Science*, *62*, 898-905.

[6] Chéret, R., Delbarre-Ladrat, C., Lamballerie-Anton M., & Verrez-Bagnis, V. (2007). Calpain and cathepsin activities in post mortem fish and meat muscles. *Food Chemistry*, *101*, 1474-1479.

[7] Gerelt, B., Ikeuchi, Y., & Suzuki A. (2000). Meat tenderization by proteolytic enzymes after osmotic dehydration. *Meat Science*, *56*, 311-318.

[8] Fauconneau, B. (2004). Diversification domestication et qualité des produits aquacoles. *INRA Productions Animales*, *17*, 227-236.

[9] Reed, G. (1975). *Enzymes in food processing.* 2nd edition. New York, Academic Press, 573 pp.

[10] Ouali, A. (1990). Meat tenderization: possible causes and mechanisms. A review. *Journal of Muscle Foods*, 129-165.

[11] Qihe, C., Guoqing, H., Yingchun, J., & Hui, N. (2006). Effects of elastase from a *Bacillus* strain on the tenderization of beef meat. *Food Chemistry*, *98*, 624-629.

[12] Takagi, H., Kondou, M., Tomoaki, H., Nakamori, S., Tsai, Y., & Yamasaki, M. (1992). Effects of an alkaline elastase from an alkalophilic *Bacillus* strain on the tenderization of beef meat. *Journal of Agricultural and Food Chemistry*, *40*, 2364-2368.

[13] Zamora, F., Debiton, E., Lepetit, A., Dransfield, E. & Ouali, A. (1996). Prediction variability of ageing and toughness in beef *M. Longissimus lumbarum et thoracis*. *Meat Science*, *13*, 321-333.

[14] Goll, D.E., Thompson, V.F., Taylor, R.G., & Ouali, A. (1998). The calpain system and skeletal muscle growth. *Canadian Journal of Animal Science*, *78*, 503-512.

[15] Toldra, F., & Flores, M. (2000). The use of muscle enzymes as predictors of pork meat quality. *Food Chemistry*, *69*, 387-395.

[16] Liu, L., & Tang, H. (2001). Study advancement of meat tenderization technology. *Meat Industry*, *11*, 40-42.

[17] Cronlund, A., & Woychik, J. (1987). Solubilization of collagen in restructured beef with collagenase and α-amylase. *Journal of Food Science*, *52*, 857-860.

[18] Lawrie, R. (1998). Lawrie's meat science (6th edition). London, Woodhead Publishing Ltd.

[19] Beuk, J., Hinsdale, S., Goeser, P.A., & Hogan, J.M. (1959). *Method of tendering meat.* US patent 2903362.

[20] Jiang, S.T. (2000). Effect of proteinases on the meat texture and seafood quality. *Food Science and Agriculture Chemistry*, *2*, 55-74.

[21] Foegeding, E.A., Lanier, T.C., & Hultin, H.O. (1996). Characteristics of edible muscle tissues. In O. R. Fennema (Ed.), *Food Chemistry. Third edition* (pp. 879-942). New York, USA: Marcel Dekker, Inc.

[22] Cepeda, R., Chou, E., Bracho, G., & Haard, N. F. (1990). An immunological method for measuring collagen degradation in the muscle of fish. In M.N., Voigt, & J.R., Botta (Eds.), *Advances in fisheries technology and biotechnology for increased profitability*: papers from the 34th Atlantic Fisheries Technological Conference and Biotechnological Workshop, August 27-September 1, 1989, St. Johns, NF, Canada (pp. 487-506). Lancaster, Pennsylvania, USA: Technomic Publishing Company.

[23] Kolodziejska, I., & Sikorski, Z. E. (1996). Neutral and alkaline muscle proteases of marine fish and invertebrates - A review. *Journal of Food Biochemistry*, *20*, 349-363.

[24] Delbarre-Ladrat, C., Verrez-Bagnis, V., Noel, J., & Fleurence, J. (2004). Relative contribution of calpain and cathepsins to protein degradation in muscle of sea bass (*Dicentrarchus labrax* L.). *Food Chem*istry, *88*, 389-395.

[25] Goll, D.E., Thompson, V.F., Li, H.Q., Wei, W., & Cong, J.Y. (2003). The calpain system. *Physiological Reviews*, *83*, 731-801.

[26] Ashie, I.N.A., Simpson, B.K., & Smith, J.P. (1996). Mechanisms for controlling enzymatic reactions in foods. *Critical Reviews in Food Science and Nutrition*, *36*, 1-30.

[27] Ashie, I.N.A., Smith, J.P., & Simpson, B.K. (1996). Spoilage and shelf-life extension of fresh fish and shellfish. *Critical Reviews in Food Science and Nutrition*, *36*, 87-121.

[28] Yamashita, M., & Konagaya, S. (1990). High activities of cathepsins B, D, H and L in the white muscle of chum salmon in spawning migration. *Comparative Biochemistry Physiology*, *95B*, 149-152.

[29] Goll, D.E., Otusuak, Y., Nagainis, P.A., Shannon, J.D., Sathe, A.K., & Mururuma, M. (1983). Role of muscle proteinases in maintenance of muscle integrity and mass. *Journal of Food Biochemistry*, *7*, 137-177.

[30] Aoki, T., Yamashita, T., & Ueno, R. (2000). Distribution of cathepsins in red and white muscles among fish species. *Fisheries Science*, *66*, 776-782.

[31] Jamdar, S.N., & Harikumar, P. (2002). Sensitivity of catheptic enzymes in radurized chicken meat. *Journal of Food Science and Technology*, *39*, 72-73.

[32] Ando, M., Yoshimoto, Y., Inabu, K., Nakagawa, T., & Makinodan, Y. (1995). Post-mortem change of three-dimensional structure of collagen fibrillar network in fish muscle pericellular connective tissues corresponding to post-mortem tenderization. *Fisheries Science*, *61*, 327-330.

[33] Bracho, G.E., & Haard, N.F. (1995). Identification of two matrix metalloproteinases in the skeletal muscle of Pacific rockfish (*Sebastes* sp.). *Journal of Food Biochemistry*, *19*, 299-319.

[34] Bremner, H.A., & Hallett, I.C. (1985). Muscle fiber–connective tissue junctions in the fish blue grenadier (*Macruronus novaezelandiae*). A scanning electron microscope study. *Journal of Food Science*, *50*, 975-980.

[35] Scopes, R.K. (1994). *Protein purification: principles and practice* (3rd ed.). New York, Springer-Verlag, 380 pp.

[36] Suckling, C.J. (1984). *Enzyme chemistry: Impact and applications*. London, Chapman and Hall, 255 pp.

[37] Page, M.I., & Williams, A. (1987). *Enzyme mechanisms*. London, The royal Society of Chemistry, 550 pp.

[38] Segel, I.R. (1993). *Enzyme kinetics: Behavior and analysis of rapid equilibrium and steady-state enzyme systems*. New York, John Wiley & Sons, Inc., 957 pp.
[39] Fukui, T., & Soda, K. (1994). *Molecular aspects of enzyme catalysis*. Tokyo, Kodansha, 244 pp.
[40] Stauffer, C.E. (1989). *Enzyme assays for food scientists*. New York, An Avi Book, 317 pp.
[41] Atkins, G.L., & Nimmo, A. (1980). Current trends in the estimation of Michaelis-Menten parameters. *Analytical Biochemistry*, *104*, 1-9.
[42] Cornish-Bowden, A., & Eisenthal, R. (1974). Statistical considerations in the estimation of enzyme kinetic parameters by the direct linear plot and other methods. *Biochemical Journal*, *139*, 721-730.
[43] Cornish-Bowden, A., & Eisenthal, R. (1978). Estimation of Michaelis constant and maximum velocity from the direct linear plot. *Biochimica et Biophysica Acta*, *523*, 268-272.
[44] Eisenthal, R., & Cornish-Bowden, A. (1974). The direct linear plot. A new graphical procedure for estimating enzyme parameters. *Biochemical Journal*, *139*, 715-720.

In: Practical Food and Research
Editor: Rui M. S. Cruz, pp. 195-217

ISBN: 978-1-61728-506-6

Chapter VIII

MICROORGANISMS AND SAFETY

Maria M. Gil and Ana L. Barbosa
Escola Superior de Turismo e Tecnologia do Mar - Instituto Politécnico de Leiria,
Santuário N.ª Sra. dos Remédios 2520-641 Peniche, Portugal,
mariamanuel@estm.ipleiria.pt

ABSTRACT

Although a large research effort has been given to attain safe fish and meat products, food-borne diseases are a constant concern both from a public health and/or economic perspectives. Consequently, the microbial spoilage of those foods and the presence of pathogens are of major importance to process industries and an appropriate control of growth and/or inactivation of microorganisms is crucial. Most of the microorganisms are originate from the microflora of the raw material, inadequate handling practices (catching or slaughtering) and improper hygienic procedures during processing, storage and distribution. The occurrence of pathogens in meat starts before slaughtering, with cattle shedding bacteria or parasites in their feces, and for this main reason, ongoing control should be directed to the primary sector. Therefore, treatments should be designed to provide an adequate margin of safety against microbiological risk of food poisoning and food spoilage throughout shelf life, thus leading to safer products with improved shelf life and quality. The proper and efficient development and application of these treatments, requires knowledge of the microbial kinetic behaviour. Studies concerning the effect of several stressing factors on bacterial spoilage and growth behaviour are fundamental to attain an appropriate control of foods'.

Proper cooking, freezing, and sanitary official controls, are vital and traditional strategies to control bacteria growth. However, proper standards, policies, promotion programs for food safety and development of technical information, are crucial to ensure the improvement of fish and meat market.

The aim of this chapter was to compile relevant information concerning biological hazards in seafood and meat products, as well as risk assessment and provides processing knowledge that allows an appropriate control of food poisoning and food spoilage during the whole shelf life.

This chapter includes a compilation of pathogenic agents, preventive measures of control and the effect of processing on seafood and meat and poultry.

8.1 INTRODUCTION

Food-borne disease has been defined by the World Health Organization (WHO) as "a disease of an infectious or toxic nature caused by, or thought to be caused by, the consumption of food or water" and is increasing throughout the all world. Although large efforts in research, according to European Food Safety Authority (EFSA) and the European Centre for Disease Prevention and Control (ECDC) food poisoning is increasing throughout the European Union and continue to present a major problem of both health and economic significance. Consequently, the microbial spoilage of foods and the growth of pathogens are of major importance to community safety and to food process industries. Outbreaks are associated with meat, fresh fruits and vegetable salads and the microorganisms responsible are bacteria, viruses and fungi, commonly called "germs" and parasites. Pathogenic bacteria are defined as those bacteria that may cause illness in humans. Pathogens, such as *Campylobacter*, *Listeria*, *E. coli* and *Salmonella* are the main responsible of food-borne illness in humans. The leading causes of those occurrences are:

- Inadequate handling practices;
- Insufficient thermal treatments;
- Inadequate cooling/refrigeration;
- Absence of hygienic standards;
- Cross-contamination from raw to high/ready to eat foods;
- Contaminated raw foods and ingredients;
- Improper cleaning of equipment and utensils.

Besides this, a relatively significant amount of microorganisms is naturally present on the surface or is transferred to the surfaces during slaughter/catching and processing.

Specific properties of some microorganisms, such as ubiquity and capacity to survive under chilling temperatures (for example, *Listeria*) and even under freezing temperatures (for example, *E. coli*), makes them dangerous contaminants. Moreover, the presence of such microorganisms in food products is of major risk for the consumers' health. So, it is of great importance, the application of adequate processes that allow the conveniently inactivation of bacteria.

Microbial behaviour in foods over time is influenced by food processing conditions (e.g. temperature and pressure), by intrinsic media factors (e.g. pH and water activity) and by products' characteristics (e.g. physical state, type of surfaces when in solid phase and nutrients' content). The binomial microorganism-food is also important (the microbial cell morphology and type of adhesion of the microbe to food surfaces is one example of the possible interaction that may affect kinetic microbial responses).

The application of physical stress to microorganisms is the most widely used method to induce cell death. Knowledge of the factors that most influence the microbial behaviour and the proper quantification of these effects (and their interactions) are fundamental for a convenient design and efficient control of safe food processes.

In this chapter an overview on important microorganisms present in seafood and meat is presented.

Underlying concepts related to processing knowledge that allows producers to attain the appropriate control of food poisoning and food spoilage during the whole shelf life, are also highlighted and discussed.

8.2 ASSESSMENT OF SEAFOOD AND MEAT PRODUCTS SAFETY

Most of the food microorganisms are originate from the microflora of the raw material, introduced at the time of catching, or transferred during slaughter, processing, storage and distribution. The level of contamination depends on the properties of the food, characteristics of contaminant microorganisms, conditions of processing and storage environment, which makes predicting the microbial behaviour in foods a difficult task.

Microorganisms require water, nutrients, oxygen and a suitable temperature for optimal growth and reproduction. High risk foods are those perishable foods which have high water content, proteins and abundant nutrients, making them an excellent medium to support microbial growth. As referred before, foods such as seafood, produced dishes, poultry and beef are the major cause of outbreaks.

It is well known that fruit and vegetables are naturally resistant to microorganisms. Contrarily, meat and fish, which are high perishable products, are not resistant to microbial decomposition. The main factors that great contribute to the seafood and meat food-borne diseases, are: (i) all healthy animals carry a complex microbial flora which can adapted to growth and survival on its host and may be transient; (ii) the animal surfaces are exposed to air, soil, water, food handling and equipment contact; (iii) improper handling after catching or slaughtering; (iv) the post-mortem biochemical and microbiological changes.

8.3 IDENTIFICATION/CHARACTERIZATION OF BIOLOGICAL HAZARDS IN SEAFOOD

The aspects related to safety and spoilage of seafood will be discussed. The evaluation of potential hazards, as well as the risk associated and control options will be listed and an overview of growth limiting factors and resistance of pathogens will be presented. Information on control options for disease agents will be also summarized.

The muscle and internal organs of healthy seafood are normally exempt of microorganisms. However, skin, gills and alimentary tract normally carry a great number of those, being the level of contamination different. For example, in seafood the level of contamination is normally: (i) 10^2 to 10^6 cfu cm^{-2} in the skin; 10^3 to 10^6 cfu g^{-1} in the gills; 10^3 to 10^6 cfu ml^{-1} in the gut. This contamination depends on the aquatic environment than on the fish species. Marine cold water fish carry lower numbers while tropical warm water fish have considerably higher counts. According to the US Center for Science in the Public Interest (CSPI), seafood is more likely to cause food-borne illness than any other category of food product. This is mainly caused by biological hazards contamination (which includes pathogenic bacteria, viruses, parasites and biotoxins) due to improper handling and storage, and by naturally occurring chemical toxins.

The evaluation of those potential hazards, as well as the risk associated and control options will be following presented.

8.3.1 Spoilage and pathogenic bacteria

The marine ecosystem is recognized as natural habitat of pathogenic bacteria and infectious diseases caused by these microorganisms can be dangerous to marine animal's health and could potentially affect several species as well those are endangered. Besides this, the level of natural contamination is normally quite low and insufficient to cause human disease. Nevertheless, growth conditions must be avoided.

Aquatic environment is the natural habitat of *Plesiomonas shigelloides*, *Vibrio* and *Aeromonas* species (indigenous bacteria) which are pathogenic to human and animals. Due to the selective effect of water temperature, the microflora of tropical waters differs from the temperate waters. In temperate waters psychotropic bacteria, Gram – (genera *Pseudomonas*, *Moraxella*, *Acinetobacter*, *Shewanella* and *Flavobacterium*) are the typical microorganisms present. In the warmer temperate waters mesofilic bacteria can be isolated. Gram + and enteric bacteria are characteristic of tropical waters. Those microorganisms should be the natural microflora of seafood. In Table 8.1 an overview of specific spoilage bacteria of fresh fish are presented.

Table 8.1. Specific spoilage bacteria of fresh fish [1].

Storage temperature (°C)	Packaging atmosphere	Specific spoilage microorganisms
0	Aerobic	*S. putrefaciens* *Pseudomonas* spp.
	Vacuum	*S. putrefaciens* *P. phosphoreum*
	Modified Atmosphere Packaging (CO_2 containing)	*P. phosphoreum*
5	Aerobic	*Aeromonas* spp. *S. putrefaciens*
	Vacuum	*Aeromonas* spp. *S. putrefaciens*
	Modified Atmosphere Packaging	*Aeromonas* spp.
20-30	Aerobic	*Aeromonas* spp. (motile)

Besides the low level of initial contamination, bacteria can grow and multiply easily, producing compounds responsible for "fishy" odours and flavours, and discolorations associated with stale seafood. In addition, if pathogen bacteria exist, they can multiply and cause illness to the consumers. In Table 8.2, compounds formed by bacteria growth, during spoilage of fish are presented.

Table 8.2. Spoilage compounds during spoilage of fresh fish stored aerobically or packed in ice or at room temperature.

Specific spoilage organism	Typical spoilage compounds
Shewanella putrefaciens	TMA, H_2S, CH_3SH, $(CH_3)_2S$, Hx
Photobacterium phosphoreum	TMA, Hx
Pseudomonas spp.	ketones, aldehydes, esters, non-H_2S sulphides
Vibrionaceae	TMA, H_2S
anaerobic spoilers	NH_3, acetic, butyric and propionic acid

where:

TMA-Trimethylamine; H_2S – Hydrogen sulphide; CH_3SH – Methyl mercaptan; $(CH_3)_2S$ – Dimethyl sulphide; Hx – Hypoxantine; NH_3 – Ammonia.

Seafood spoilage is affected by temperature. High temperatures speed spoilage and low temperatures slow spoilage. For many seafood species, increasing the temperature from 0 °C to 4.5 °C doubles the rate of spoilage and cuts the shelf life in half. Moreover, bacteria on temperate water native fish enter the exponential growth just after the fish have died. Contrarily, a lag phase of 1 to 2 weeks can be observed, in native tropical waters fish, if stored in ice.

There are three main ways of breaking the fish spoilage chain. Some general points are considered as follows:

1. Protecting fish from contamination (healthy fish are normally sterile):

As healthy fish are normally sterile, the main preventive measure is to avoid contamination by applying good hygiene practices (GHP) and good manufacturing practices (GMP). Short handling time should also be regarded, as well as, avoiding fish that have been catch in contaminated areas.

2. Preventing any bacteria present in the food from multiplying:

- Time for catch handling (time from catch to chilling) should be limited to maximum of 3 hours.
- Keep fish at low temperature (i.e. out of the danger zone). Fish should be kept below 1 °C in a refrigerated unit (the critical limit for fish temperature is 1°C).
- Using various packing methods like modified atmosphere or vacuum packing.

Table 8.3. Growth limiting factors and heat resistance of seafood pathogenic bacteria [2].

Organism	Temperature (°C)		pH	a_w	NaCl (%)	Heat Resistance
	minimum	optimum	minimum	minimum	maximum	
Clostridium botulinum non-proteolytic	3.3	25-28	5.0	0.97	3-5	D_{100} (spores) (0.1 min; $D_{82.2}$ = 0.5-2.0 min (broth); D_{80} (spores) = 4.5-10.5 min in products with high fat content
Vibrio spp.						
V. cholerae	10	37	5.0	0.97	<8	D_{55} = 0.24 min
V. parahaemolyticus	5	37	4.8	0.93	8-10	D_{60} = 0.71 min
V. vulnificus	8	37	5.0	0.96	5	D_{50} = 1.15 min (buffer) ; 0.66 min (oysters)
Aeromonas hydrophila	0-4	28-35	4.0	0.97	4-5	D_{55} = 0.17 min
Plesiomonas shigelloides	8	37	4.0	-	4-5	All cells killed after 30 min at 60 °C
Clostridium perfringens	12	43-47	5.5	0.93	10	D_{90} (spores) = 0.015-4.93 min (buffer) D_{100} (spores) = 0.31-13.0 min (broth)
Listeria monocytogenes	0-2	30-37	4.6	0.92	10	D_{60} = 1.95-4.48 min in fish
Bacillus cereus	4	30-40	5.0	0.93	10	D_{121} (spores) = 0.03-2.35 min (buffer)
Salmonella spp.	5	37	4.0	0.95	3	D_{60} = 0.2-6.5 min.
Shigella spp.	7	37	5.5	-	4-5	D_{60} = 5 min.
Staphylococcus aureus	6.5	37	4.7	0.86/0.90	18-20	D_{60} = 0.43-7.9 min.

Table 8.4. Pathogenic bacteria in seafood [3].

Organism		Primary habitat	Quantitative levels presented	Mode of action disease	Preventive measure and control
Indigenous bacteria	*Clostridium botulinum*	Temperate and Arctic aquatic environment; multiplication in aquatic carrion (type E)	Generally low (<0.1 spores/g fish) but up to 5.3 spores/g fish has been recorded	Intoxication	Time and temperature storage adequate, i.e., storage at all times at <3.3 °C; storage at 5-10 °C and a shelf life of <5 days. A heat treatment of 90 °C for 10 min combined with chill storage (<10 °C). 3% NaCl in the water phase is sufficient to inhibit growth of type E. A pH (5.0 throughout the food combined with chilled storage (<10 °C)
	Vibrio spp. *V. cholerae* *V. parahaemolyticus* *V. vulnificus*	Ubiquitous in warm (>15 °C) seawater environment	Up to 10^2-10^3 cfu/g in shellfish; up to 10^4-10^8 cfu/g in intestines of shellfish-eating fish	Infection-High MID	Low temperature storage has been proposed as a means of eliminating pathogenic vibrios from food. Rapid and efficient cooling (time x temperature control) is one of the most important control parameters in prevention of *V. parahaemolyticus* gastroenteritis. Cooling to 5 °C will prevent growth. Proper cooking is sufficient to eliminate most vibrios.

Table 8.4. (Continued).

Organism	Primary habitat	Quantitative levels presented	Mode of action disease	Preventive measure and control
				V. cholera survived boiling for up to 8 min and steaming for up to 25 min in naturally contaminated crabs. High NaCl-concentrations (>10% NaCl in water phase) or acidification as used in several semi-preserved products can prevent growth. Good Hygienic Practices (GHP) programmes should ensure that cooked products are not cross-contaminated. Depuration of molluscan shellfish has no significant effect on the level of Vibrio that may even multiply in depurating shellfish.
Aeromonas hydrophila	Aquatic environment	Generally low, but up to 10^4 cfu/ml in seawater; 10^7 cfu/ml in sewage and 10^6 cfu/g in raw seafood	Infection-High MID	*Aeromonas* is very sensitive to acid conditions and to salt and growth is unlikely to be a problem in foods where pH is less than 6.5 and the NaCl content greater than 3.0%.
Plesiomonas shigelloides	Warm aquatic environment; Freshwater fish (animals)		Infection-High MID	Time and temperature storage adequate, i.e., storage at maximum of 4 °C or moderate salting/acidifying conditions will prevent growth of the organism.

Organism		Primary habitat	Quantitative levels presented	Mode of action disease	Preventive measure and control
	Clostridium perfringens	Soil (type A); animals (type B, C, D and E)	10^3-10^4 cfu/g soil	Intoxication	Prompt refrigeration of unconsumed cooked fish. Proper refrigeration and sanitation.
	Listeria monocytogenes	Soil, decaying vegetation ubiquitous in general (temperate) environments	<100 cfu/g in freshly produced fish products	Infection-High or low MID	Proper cooking or pasteurization. Proper GMP and factory hygiene.
	Bacillus spp.	Ubiquitous in general environment (soil, natural waters, vegetation)	10^1-10^3 cfu/g or ml raw, processed food	Intoxication	Spore germination can be greatly reduced by unfavourable conditions such as low temperature, low a_w and low pH.
Nonindigenous bacteria	*Salmonella* spp.	Intestines of warm blooded animals/humans	Levels in symptomatic and asymptomatic carriers vary; levels in seafood assumed to be sporadic and low. May accumulate in molluscan shellfish	Infection-High MID	Proper pasteurization (hot-smoking) temperatures. Although *Salmonella* does not grow well at low water activity, it has been found to survive well in dry environments and (re)-contaminate products such as fish meal. *Salmonella* does not grow below pH 4.5. Proper GMP and factory hygiene.

Table 8.4. (Continued).

Organism	Primary habitat	Quantitative levels presented	Mode of action disease	Preventive measure and control
Shigella spp.	Intestines of warm blooded animals/humans	Levels in symptomatic and asymptomatic carriers vary; levels in seafood assumed to be sporadic and low. May accumulate in molluscan shellfish	Infection-Low MID	Storage at temperatures lower then 6-7 °C. Proper cooking, as they are sensitive to heating, as well as, sensitive to salting. As mentioned, they may survive for long periods of time in bivalves. Proper GHP and GMP.
Staphylococcus aureus	Outer surface (skin) and mucus membranes (nose)	Transient, but present on 50% of population. Generally <100 cfu/cm^2 skin	Intoxication	Time and temperature storage adequate. Avoidance of cross contamination of heat treated (cooked) products is also important. Also, they are resistant to heat and will resist boiling for some time. Proper GHP and GMP.

3. Destroying the existing bacteria:

Control of seafood decomposition is simple since low temperature will retard spoilage. A combination of a suitable temperature and sufficient time is always required to destroy bacteria. The time and temperature required will depend on the particular organism, (e.g. spores of *Clostridium perfringens* are much more heat resistant than *Salmonella* bacteria) and will be discussed later.

The application of physical stress to microorganisms is the most widely used method to induce cell death and promote stability. Knowledge of the environmental factors that most influence the microbial behaviour in foods is essential for the development, as well as for the practical use, of processing conditions. In Table 8.3 an overview of growth limiting factors and resistance of considered microorganisms is presented. Under favourable conditions bacteria multiply exponentially and consequently, seafood becomes spoiled and unfit for human consumption.

Because inactivation treatments can effectively eliminate pathogens, information on control options for disease agents is required for any processing schedules and is summarized in Table 8.4.

Different prevention and control measures have diverse impact in reducing the bacterial level. Depended on the food product and bacterial contamination, the appropriate control process should be carefully chosen. As an example, the comparative effectiveness of a number of mitigation strategies in reducing *Vibrio* spp. is presented in Table 8.5.

Table 8.5. Comparative effectiveness of mitigation strategies in reducing *Vibrio* spp. [4].

Mitigation	Comparative effectiveness in reducing *Vibrio* spp.
Hydrostatic pressure	Significant reductions
Rapid cooling	Some reduction/moderate reduction
Irradiation	Significant reductions
Pasteurization	Significant reductions
Freezing and thawing	Moderate reduction
Depuration	No effect/some reduction
Relay at high salinity for 2 weeks (for *V. vulnificus*)	Moderate reduction
Comercial heat treatment	Significant reductions

Fish should not contain microorganisms or their toxins or metabolites in quantities that present an unacceptable risk for human health. Regulation (EC) n° 1441/2007 laying down the microbiological criteria for foodstuffs, giving guidance on the acceptability of foodstuffs and their manufacturing, handling and distribution processes. CAC (1997) also gives guidance on the establishment and application of microbiological criteria for food at any point in the food chain from primary production to final consumption.

8.3.2 Viruses

Food-borne and water-borne viral infections are increasingly recognized as causes of illness in humans. This increase is partly explained by changes in food processing and consumption patterns that lead to the worldwide availability of high-risk food. As a result, vast outbreaks may occur due to contamination of food by a single food-handler or at a single source. Although there are numerous fecal-orally transmitted viruses, most reports of food-borne transmission describe infections with Norwalk-like caliciviruses (NLV) and hepatitis A virus (HAV), suggesting that these viruses are associated with the greatest risk of food-borne transmission, especially associated with fishery products [5].

NLV has been reported as the second causal agent of food-borne outbreaks after *Salmonella* in Spain, and NLV outbreaks were larger than bacterial outbreaks, suggesting greater underreporting and, consequently, draw-backs in the investigation of NLV outbreaks [5].

Assuming that the NLV has a human reservoir and a very low infective dose, along with prolonged persistence in the environment, food handlers should be aware of their relevance in this infection and receive appropriate preventive training. In order to avoid secondary cases, when a food-borne outbreak of viral gastroenteritis in closed or partially-closed institutions is suspected, rapid control measures should be adopted with an emphasis on hand washing and correct disinfection of environmental surfaces [5, 6].

Hepatitis A viruses are readily identified in shellfish harvested in certain European regions, and have caused significant outbreaks of human food-borne disease. As a result of its physiological filtration process, bivalve shellfish biocentrate microbial pathogens from marine and estuarine waters. While fecal coliforms and other pathogenic bacteria from human and animal wastes do not persist within shellfish tissues beyond a few days, enteric viruses such as HAV can persist in estuarine waters and within shellfish tissues for periods of several weeks or more [7].

HAV is partially resistant to heat, but still infectious after 10 to 12 hours at 60 ºC. Proper inactivation of HAV suspended in buffered saline occurs after 4 minutes at 70 ºC and immediately at 85 ºC and is inactivated by radiation [8].

Various studies also confirm that HAV could be inactivated by high hydrostatic pressure.

8.3.3 Parasites

Food-borne parasitic diseases are a global public health problem that affect millions of people and cause enormous suffering, particularly amongst poor and vulnerable groups in developing countries. While 700 million people around the world are at risk for food-borne trematode infections, 40 to 50 million people are infected with one or more of the parasites. The majority of these infections occur in Asia as a direct result of consuming raw fish.

Parasites (in the larval stage) consumed in uncooked, or undercooked, unfrozen seafood can represent a human health hazard. Among parasites, the nematodes or roundworms (*Anisakis* spp., *Pseudoterranova* spp., *Eustrongylides* spp. and *Gnathostoma* spp.), cestodes or tapeworms (*Diphyllobothrium* spp.) and trematodes or flukes (*Chlonorchis sinensis, Opisthorchis* spp., *Heterophyes* spp., *Metagonimus* spp., *Nanophyetes salminicola and Paragonimus* spp.) are of most concern in seafood. Some products that have been implicated in human infection are: ceviche (fish and spices marinated in lime juice); sashimi (slices of raw fish); sushi (pieces of raw fish with rice and other ingredients); green herring (lightly brined herring); cold-smoked fish; and, undercooked grilled fish [9].

The effectiveness of freezing to kill parasites depends on several factors, including the temperature of the freezing process, the length of time needed to freeze the fish tissue, the length of time the fish is held frozen, the fat content of the fish, and the type of parasite present [9].

The FDA's Food Code recommends freezing conditions to retailers who provide fish intended for raw consumption, in order to kill fish parasites, although these conditions may not be suitable for large fish (e.g. thicker than six inches). These recommended conditions are:

Freezing and storing at -4 °F (-20 °C) or below, for 7 days (total time);

Freezing at -31 °F (-35 °C) or below, until solid and storing at this temperature or below for at least 15 hours;

Freezing at -31 °F (-35 °C) or below until solid and storing at -4 °F (-20 °C) or below for 24 hours [9].

In conclusion, cconsumers can help protect themselves by avoiding consumption of tropical or subtropical reef fish, by refrigerating all seafood, and by eating only cooked shellfish or raw shellfish that have been treated to eliminate hazardous bacteria. To avoid parasitical contamination by fish consumption, proper cooking or proper freezing are required.

8.4 Identification/Characterization of Biological Hazards in Meat Products

8.4.1 Pathogenic bacteria

Meat is contaminated by a variety of microorganisms before, during slaughter and subsequent processing. The bacterial flora can be broadly divided into two groups, namely, pathogenic and spoilage bacteria. The pathogenic bacteria are of public health significance, whereas spoilage bacteria are primarily responsible for quality losses to the food industry [10].

Most important pathogenic bacteria are elucidated in Table 8.6, as well as theirs control measures and preventive strategies.

The most common bacteria typically associated with spoilage of aerobically stored refrigerated meat include *Pseudomonas fluorescens*, *Ps. putida*, *Ps. fragi*, *Ps. aureofaciens*, *Acinetobacter calcoaceticus*, *Enterobacter liquifaciens*, *Flavobacterium* spp., *Moraxella* spp. and *Brochothrix thermosphacta*. Under aerobic conditions, there is no perceptible change in the sensory properties of the meat until bacterial numbers exceed counts of 10^7-10^8 cfu/cm^2. About this time, spoilage odors can be detected and bacteria begin to form a visible slime layer [10, 11].

The complete pathway for ground meat contamination by food-borne bacteria, such as *Escherichia coli* O157:H7 and *Salmonella* beef contamination, it's to be elucidated but the general model begins with cattle shedding the bacteria in their feces, occasionally at levels exceeding 10^6 cfu/g [11].

Table 8.6. Meat pathogenic bacteria, growth limiting factors and preventive control strategies for meat and meat products [12].

Pathogen	Growth temperature (ºC)	pH	a_w	Control
Campylobacter jejuni	30-47	4.7-7.5	> 0.97	Proper cooking, chilling and freezing. Avoid cross contamination. Hygiene procedures from farm to consumers' house.
Clostridium botulinum Group I (toxins: A. B. F) *Group II* (toxins: B. E. F)	10-48 3.3-45	> 4.6	0.95 0.97	Salt and nitrats addition. Acidification at pH < 4.6 and a_w reduction under 0.93.
Clostridium perfringens	15-50	5.5-8.0	0.95	Proper cooking, boiling and chilling.
Escherichia coli O157:H7	10-42	4.5-9.0	0.95	Proper cooking, boiling and chilling. Avoid contamination during hide-removal process.
Listeria monocytogenes	2.5-44	5.2-9.6	0.92	Proper temperature and hygiene procedures. Avoid cross contamination during food handling.
Salmonella	5-46	4.0-9.0	0.95	Proper temperature and hygiene procedures. Avoid cross contamination during food handling.
Staphylococcus aureus	6.5-46	5.2-9.0	Aerobics: 0.86 Anaerobics: 0.91	pH control, proper treatment, a_w reduction. Good manufacturing practices and hygiene procedures.
Yersinia enterocolitica	2-45	4.6-9.6	0.95	Proper cooking, salt control and acidification, proper chilling and preventing cross contamination.

The hide of the individual animal as well as those of the other animals in the same feedlot pen or pasture become laden with the fecal-bacteria mixture including *E. coli* O157:H7 and other pathogens. At slaughter, a proportion of the bacterial inhabitants of the hide is transferred to the carcass during the hide-removal process [11].

In 1994, the United States Food Safety and Inspection Service (FSIS) declared *E. coli* O157:H7 to be an adulterant in ground beef, making it the first microorganism given such status under the Federal Meat Inspection Act. The FSIS has now extended that declaration to include ''non-intact'' beef, such as mechanically tenderized or reconstructed products [12].

Salmonella contamination of raw beef has also garnered attention recently, primarily because of the establishment of *Salmonella* performance standards by the FSIS. Several official control inspections have been applied to primary sector, such as poultry and swine keeping, by European member states [12].

Besides the limiting growth factors explained in table 8.6, garlic, clove and cinnamon, at 1% have been described as having an inhibitory action to *E. coli* O157:H7 in sausages. Other herbs and spices have been described for its inhibitory effect to *L. monocytogenes*, *Clostridium botulinum*, *C. perfringens*, among others.

8.4.2 Viruses

Avian influenza viruses (in poultry) and hepatitis E virus (pork, boar and deer meat), stand out from other viruses related to meat-borne illness. Highly pathogenic avian influenza (HPAI) H5N1 cause systemic infections and have been detected in blood, bone, and breast and thigh meat of chickens.

This virus spread rapidly among chickens and turkeys and typically causes death within 48 hours. HPAI H5N1 viruses also infect ducks, but ducks may remain healthy even when virus is present in muscles and internal organs. This virus has also been detected in over 40 species of wild birds, and there is concern that migrating birds may facilitate the spread of this virulent avian flu strain [13].

Currently, deer, pig, and wild boar are suspected sources of foodborne zoonotic transmission of Hepatitis E Virus (HEV) in Japan, and genotypes 3 and 4 of HEV are believed to be indigenous. Direct evidence for transmission of genotype 3 HEV from animals to humans was observed in acute hepatitis in 4 persons who had eaten uncooked deer meat that contained copies of HEV RNA [14]. High antibody-positive rates in domestic pig and wild boar, including HEV genotypes 3 and 4, have been frequently detected, suggesting that persons who eat uncooked meat are at risk for infection with HEV [13, 14].

Proper cooking and good manufacturing practices are the major strategies to, respectively, destroy and prevent those agents.

8.4.3 Parasites

Cysticercosis, Toxoplasmosis and Sarcocystosis are important parasites present in meat.

Cysticercus bovis is the larval stage of *Taenia saginata*, the bovine tapeworm. Humans are the final host and bovines the intermediate host to this infection. Human taeniosis, or infection with the adult *T. saginata*, is characterized by the presence of up to 30-meter-long worm in the small intestine of the infected person, who may pass millions of eggs daily. Estimates based on computations from incidence indicate that 2% of the human population in Europe is infested with *T. saginata* [15].

In the meat industry, economic losses are closely associated with the status of infection. If a heavy infestation or generalized cysticercosis is found in a carcass, it must be totally condemned. Light infection or localized cysticercosis leads to condemnation of the infected parts; furthermore, the carcass must be kept in cold storage at a temperature not exceeding −7 °C for up to 3 weeks to inactivate the parasites. In England alone, the costs of refrigeration, handling, and transport are estimated at £100 per carcass, or £4.0 million annually. Africa suffers great losses due to bovine cysticercosis estimated to be $1.8 billion annually [15].

With regard to cysticercosis, the integration of serological methods in the inspection procedure could result in a tenfold improvement of diagnostically sensitivity [15].

Toxoplasma gondii is an intracellular parasite that affects one third of the world's population. It can affect all mammals who serve as the intermediate host, but cats are the definitive hosts. Toxoplasma may be transmitted via hand-to-mouth contact from improper handling of, or ingestion, of raw or undercooked meat containing cysts from cat excrements; by congenital transmission from mother to fetus; and rarely by transplantation of infected organs [16].

Acute infection in an immune-competent host usually causes a self-limited, flulike illness. The situation is very different for the fetus or for an individual who is immune compromised due to human immunodeficiency virus (HIV), immune-suppressant drugs, or other illness. Fulminant infection with significant morbidity and mortality or ongoing low-grade symptoms is possible in these populations [17].

With regard to Toxoplasmosis, it is obvious that meat containing Toxoplasma-cysts may reach the consumer, as animals infected with Toxoplasma gondii can neither be recognised in the ante-mortem inspection nor in the meat inspection. Systematically serological investigations on farm level would allow an appropriate judgement during meat inspection and minimize the consumer exposure to this parasite. Further, reliable methods for the detection of Toxoplasma-cysts have to be developed [18, 19].

With regard to Sarcocystosis, there is a need for suitable microscopic, serological and molecular biological methods for the detection of Sarcocystis-species and reliable information's on the seroprevalence of the parasite in slaughtering animals [20, 21].

8.5 Issues in Fish Processing

As previously referred, bacteria are the most important cause of seafood spoilage. Spoilage begins as soon as seafood species die. Their normal protection mechanisms stops and a series of changes begin causing spoilage. Those changes are caused by bacteria, enzymes and chemical action. Seafood processing is a mean to control spoilage, making sure that products are safe to consumers. Products are prepared and processed for sale to consumers, retail and wholesale businesses, for export, for restaurants and catering industry. However those processes may compromise the final quality of the products and consumers demand for high quality foods, which are less processed and preserved. Thus, preservation techniques are becoming more important in modern food industries.

Traditional food processes relies heavily on chill and frozen storage, reduction in water activity by addition of NaCl, air vacuum or modified atmosphere packaging, heat treatment, smoking and addition of preservatives. They mainly inhibit the growth of microorganisms in food rather than inactivate cells. However, lethal effect and pathogen control can be obtained if adequate heat treatments are applied (temperature is probably the major environmental factor affecting the growth and survival of microorganisms in foods. This makes effective heat treatments critical to produce safe food. Consequently, they must be designed to provide an adequate margin of safety against food-borne pathogens. However, it is difficult to determine the exact amount of microbial inactivation when these treatments are applied [22].

The severity of a thermal process obviously depends on the extension of the heat treatment considered and on the pair-wise microorganism/food. Generally, to extended the microbial inactivation, and consequently extended the shelf life of the product, other contributing preservative factors, such as pH and water activity, should be taken into account and controlled.

Due to a spectrum of heat resistances in the microbial population, some organisms or spores are destroyed sooner, or later, than others. Thus, the survival curves are only reflections of heat resistance distributions having a different mode, variance, and skewness. While overestimating the heat resistance negatively impacts on product quality, underestimating increases the likelihood that the contaminating pathogen persists after heat treatment [23].

The negatively impacts of heat treatments on seafood quality, which are inevitable, includes undesirable changes in flavour, texture and colour. The use of poor quality raw material can also influence the final quality, since undesirable changes are more pronounced. Conventional canning processes can also compromise the flavour and texture resulting in unmarketable products. In order to minimise such losses, pre-processes such as curing, pickling, smoking and cooking should be applied. However, losses of soluble proteins may be observed.

Low temperatures slow down deteriorative processes and, when temperature is low enough, spoilage can almost be stopped. However, freezing and frozen storage of seafood products may change the flavour and texture of the final cooked products. Changes in quality depend on the freezing method and rate and on the store conditions. The faster the freezing rate, the more nucleation of ice crystal is promoted, and the greater number of small crystal size. Therefore, changes in water-holding capacity are minimized, as well as, a decrease in drip-loss, consequently small quality attributes tend to occur. Even if rapid freezing is applied, advantages gained initially may be lost due ice crystal growth of subsequent frozen storage.

The development of rancid off-flavours, essentially the oxidation of unsaturated fats, also occurs during frozen storage. If frozen store is applied for only a short period of time, the changes in sensory attributes on subsequent thawing and cooking are quite small. Besides this, previous chilling conditions can contribute to improve the final quality. Losada et al. concluded that the application of the slurry ice technology as a preliminary processing step prior to freezing and frozen storage is considered a promising strategy to achieve frozen fish products of higher quality [24]. Esaiassen et al. studied the effect of the addition of cryoprotectants in fresh fish to preserve texture during frozen storage (helps to control the ice crystal size, reducing drip loss and maintaining textural quality [25]. NaCl, polyphosphates, glucose, sodium ascorbate, starch, functional proteins and antifreeze proteins are some examples of cryoprotectants that can be used.

Reducing the moisture content through drying, smoking or curing, can also be considered to preserve seafood products, resulting in a stable product. This processes has been practised either as a means of prolonging shelf-life, or to produce desired flavours and texture.

Water activity (a_w) is defined as the ratio of the vapour pressure of water in a material (p) to the vapour pressure of pure water (p_0), at the same temperature. The water activity of a solution depends upon the number of "particles" (molecules or ions). An increase in solute concentration results in a decrease of a_w, but in different ways depending on the solute used.

The microorganisms' cell membrane is selectively permeable (e.g. glycerol penetrates the membrane readily, glucose penetrates poorly, sucrose very poorly and NaCl is almost non-penetrating [26]).

The difference in osmotic pressure (osmolarity) between intracellular medium and the surroundings, determines how water moves. The osmotic pressure is independent of the type of solute particles contained in a solution, being only affected by its concentration. Changing a_w of the environment directly affects osmolarity. When an organism is exposed to low a_w conditions, the cells may accumulate solutes and loose water, which cause dehydration and, as a consequence, damage to the cell membranes may occur. For this reason, water activity also greatly affects bacterial inactivation behaviour. Several authors studied the influence of those effects on microbial behaviour. The majority of research works reported that maximum thermal resistance is obtained for lower water activity values (for temperature values in the range used in thermal processes). This tendency is particularly important for higher temperature values. Therefore, it can be concluded that the protective effect of low water activity depends on the temperature range considered (in drying, smoking or curing processes).

The quality of smoked, cured and dried products can be assessed using a range of physical, chemical and organoleptic methods.

The ability of modified atmosphere packaging (MAP) or vacuum packaging to extend the shelf-life of foods has been recognised for many years. The use of MAP for fish has been reviewed by some authors [27]. Recently, interest by fish industries has resurged. Shelf-life of modified atmosphere fish packaging can increase a hundred percent. The three major gases used commercially in MAP are carbon dioxide, nitrogen and oxygen. The replacement of air by a gas mixture allows the inhibition of microbial growth (the overall effect on microorganisms is an extension of the lag phase and a decrease in the growth rate). Besides this, no organoleptic characteristics changes are normally observed and a decrease of drip-loss can be obtained (if oxygen is used in the mixture).

In the last few decades, a variety of novel non-thermal techniques has been studied. The new methods, such as irradiation, high pressure and pulse electric fields, cause minimal changes in the appearance and properties of the foods. However, they provide insufficient microbial reduction to prevent spoilage or to sterilise the sample.

An emerging and challenging technology, currently receiving attention, is the application of ozone in aqueous solution, as a possible alternative to chloride, with a better balance between quality and safety. The organoleptic characteristics are maintained, while microbiological and chemical parameters are improved. Some authors concluded that ozonation greatly reduced the growth of mesophilic, spoilage and acid lactic bacteria, as well as, *Pseudomonas* spp., *Brochothrix thermosphacta* and enterobacteriacea in trout [28].

8.6 Issues in Meat Processing

As in fish processing, traditional food processes applied on meat products rely heavily on chill and frozen storage, reduction in water activity, modified atmosphere packaging, heat treatment, smoking and addition of preservatives, in order to control microorganisms.

Several superior techniques such as the emergent irradiation, ozonation and hyperbaric treatment, or more commonly applied automated washing cabinets, steam pasteurization, and steam vacuuming, are major investments, that limited space, manpower, and financial resources often make difficult or impossible to implement, especially in small plants.

Meat and poultry processing attempts to avoid, restrict growth, or to destroy bacteria and other contamination agents.

8.6.1 Prevention of contamination

This includes contamination from the live animal, equipment, man, and the environment. Appropriate sanitation procedures as good manufacturing practices, are essential at this level.

Cross-contamination of ready-to-eat foods with not-ready-to-eat (raw or partially cooked) meat or poultry is of great concern. It is particularly important to ensure complete separation of not-ready-to-eat and ready-to-eat products.

8.6.2 Restriction of growth

Temperature, acidity, salt and drying, and combinations of these can be used to restrict growth of pathogens.

The growth of most bacteria can be slowed (controlled) by maintaining the product at refrigeration temperatures (less than 5 °C), or by freezing. Some bacteria survive freezing, so freezing cannot be considered a method to eliminate bacteria. Holding products at higher temperatures (greater than 55 °C) also restricts the growth of the bacteria.

Fermentation restricts the growth of bacteria of public health concern by increasing the acidity (lowering the pH) of the product. Generally a pH of less than 5 will severely restrict or completely stop the growth of harmful bacteria. Some bacteria can survive in acidic conditions, so fermentation alone cannot be relied upon to completely eliminate all harmful bacteria. Rinsing red meat carcasses with a 1.5-2.5% acetic acid solution, is an USDA approved treatment, which reduces microbial growth as in food contact surfaces [29, 30].

Some products contain high levels of salt. Salt and low moisture content in a product can be effective in controlling growth of some harmful bacteria, but some organisms (e.g., *Staphylococcus aureus*) survive in high salt environments.

8.6.3 Destruction of bacteria

Most pathogenic bacteria can be fairly easily destroyed using a rather mild cooking process, as maintaining a minimum temperature within the range of 55 °C to 75 °C for a specific amount of time. However, cooking at this temperature range and for the specified dwell time will not destroy the heat resistant forms (spores) of certain bacteria, nor will some types of toxins be destroyed if they have already been formed in the product. Thermal processing (canning) at a minimum retort temperature of greater than 115 ºC for a specific amount of time is necessary to destroy most spores and toxins.

The pathogen of concern in thermally processed is *Clostridium botulinum*, the cause of botulism in humans. In proper heat processing, the heat resistant spores of *Clostridium botulinum* cannot survive and grow into the vegetative form that produces toxin in product. In this category, it is essential that the heat process be adequate to destroy *Clostridium botulinum* spores [30].

As relevant as meat processing, meticulous sanitation procedures are currently applied. Some bacteria, such as *Listeria* (including *Listeria monocytogenes*), can be found in the processing environment, what emphasizes the need for adequate sanitation, not only of the equipment, but also the floors. Employee hygiene, air flow, and traffic flow of people and equipment between areas used for not-ready-to-eat processing and ready-to-eat processing is very important and should be strictly controlled [30].

8.7 Predictive Methods

The use of accurate microbial inactivation models, that describe the behaviour of microorganisms in foods, would be of considerable help to the food industry in the development of process systems. This will in turn lead to safer foodstuffs with improved quality and shelf-life.

The development of accurate and precise models, able to predict the behaviour of pathogens or spoiling microorganisms populations under stress factors, such as high temperature, particular ranges of pH and a_w, is a research field commonly referred as *predictive food microbiology*. It is a field of study that combines elements of microbiology, mathematics, and statistics to develop models that describe and predict the growth or decline of microbes under specified environmental conditions. With the ability to describe growth and inactivation behaviour emerges the ability to predict under different environmental conditions, where no experimental data exist.

In the last years, microbiologists jointly with food engineers have been applying sophisticated mathematical approaches to predict microbial loads on foods. Consequently, several software applications (commercially available, or down-loaded free of charge from the internet) in the field of predictive microbiology were developed. Examples, such as Pathogen Modelling Program [31], Food Micromodel [32], Food Spoilage Predictor, Seafood Spoilage Predictor [33], Pseudomonas Predictor [34], ComBase (http://wyndmoor.arserrc.gov/ombase/), Sym´Previus [35], Bugdeath 1.0 [36] and GInaFiT [37] illustrate the potential of predictive microbiology to users with lack of comprehensive skills in mathematics. These applications focus essentially on kinetic models for microbial growth and on shelf life prediction, while the (currently still available) versions 6.1 and 7.0 of the Pathogen Modelling Program do include a range of pathogen (non-) thermal inactivation and irradiation models. Further development of accurate and versatile mathematical software dealing with the microbial inactivation on the surface of food product is needed.

Conclusion

Studies on the bacterial spoilage of foods and on the survival of microbial pathogens are extremely important for the food process industry. The major incidence of food contamination by microorganisms occurs on food surfaces during slaughter/catching of animals and further processing. As the surface of foods is the interface for environmental contamination, every effort must be made to minimize contamination at every stage of processing. Good processing techniques and high standards of personal and plant hygiene, as well as, the development of suitable treatments, are important to reduce microbial content, thus leading to safer products with improved shelf life and quality.

The rapid reduction of temperature is the most important method to avoid microbial growth and the maintenance of this temperature ensures that, independently on further processes or packaging is applied, the microorganisms has the minimum chance of proliferate.

References

[1] Gram, L., & Huss, H.H. (1996). Microbiological spoilage of fish and fish products. *International Journal of Food Microbiology*, *33*, 121-137.

[2] Huss, H.H., Ababouch, L., & Gram, L. (2003). Assessment and management of seafood safety and quality. FAO Fisheries Technical paper 444, Rome, pp. 230.

[3] FAO/WHO Expert Consultation. (2002). Risk assessment of *Campylobacter* spp. in broiler chickens and *Vibrio* spp. in seafood. http://www.fao.org/docrep/008/y8145e/y8145e00.htm.

[4] Koopmans, M., & Doornum, G. (2009). Norovirus in a Dutch tertiatry care hospital. National institute of public health and the environment, Research Laboratory for infectious diseases. *The Journal of Hospital Inspection*, 71, 199-205.

[5] Martinez, A., Dominguez, A., Turner, N., Ruiz, L., Camps, N., Barrabeig, I., Arias, C., Alvarez, J., Godoy, P., Balaña, P., Pumares, A., Bartolome, R., Ferrer, D., Perez, U., Pinto, R., & Buesa, J. (2008). Epidemiology of foodborne Norovirus outbreaks in Catalonia, Spain. *BioMed Central Infectious Diseases*, *8* (47).

[6] Calci, K.R., Meade, G.K., Tezzloff, R.C., & Kingsley, D.H. (2005). High-Pressure inactivation of hepatitis A virus within oysters. *Applied and Environmental Microbiology*, *71*, 339-343.

[7] Anonymous, (2001). Hepatitis a virus. Available from: http://www.nzfsa.ovt.nz/science/data-sheets/hepatitis-a-virus.pdf

[8] U.S. Food & Drug Administration (2001). Fish and fisheries products hazards and controls guidance (3rd edition). Available from: http://www.fda.gov/Food/GuidanceComplianceRegulatoryInformation/GuidanceDocuments/Seafood/FishandFisheriesProductsHazardsandControlsGuide/default.htm.

[9] Arthur, T.M., Bosilevac, J.M., Brictha.Harhay, D., Kalchayanand, N., King, D.A., Shackelford, S.D., Wheeler, T.L., & Koohmaraie, M. (2008). Source tracking of *Escherichia coli* O157:H7 and *Salmonella* contamination in the lairage environment at commercial U.S. beef processing plants and identification of an effective intervention. *Journal of Food Protection*, *71*, 1752-1760.

[10] Barkocy-Gallagher, G.A., Arthur, T.M., Rivera-Betancourt, M., Nou, X., Shackelford S.D., Wheeler, T.L., & Koohmaraie, M. (2003). Seasonal prevalence of Shiga toxin-producing *Escherichia coli*, including O157:H7 and Non-O157 Serotypes, and *Salmonella* in commercial beef processing Plants. *Journal of Food Protection*, *66*, 1978-1986.

[11] Direcção Geral de Veterinária. (2008). http://www.dgv.min-agricultura.pt.

[12] Doyle, M.E., Schultz-Cherry, S., Robach, M., & Weiss, R. (2007). Destruction of H5N1 Avian Influenza Virus in Meat and Poultry Products. Food research institute briefings. Available from: http://www.wisc.edu/fri/briefs/FRI_Brief_H5N1_Avian_Influenza_8_07.pdf.

[13] Sonoda, H., Abe, M., Sugimoto, T., Sato, Y., Bando, M., & Fukui, E. (2004). Prevalence of hepatitis E virus (HEV) infection in wild boars and deer and genetic identification of a genotype 3 HEV from a boar in Japan. *Journal of Clinical Microbiology*, *42*, 5371-5374.

[14] Meng, X., Halbur, P., Shapiro, M., Govindarajan, S., Bruna, J., & Mushahwar, I. (1998). Genetic and experimental evidence for cross-species infection by swine hepatitis E virus. 1998. *Journal of Virology*, *72*, 9714-9721.

[15] Abuseir, S., Epe, C., Schnieder, K., & Kühne, G. (2006). Visual diagnosis of *Taenia saginata* cysticercosis during meat inspection: is it unequivocal? *Parasitology Research*, *99*, 405-409.

[16] Singh, D., & Sinert, R. (2007). Toxoplasmosis. Available from: http://emedicine.medscape.com/article/787505-overview.

[17] Takahashi, M., Nishizawa, T., Miyajima, H., Gotanda, Y., Lita, T., & Tsuda, F. (2003). Swine hepatitis E virus strains in Japan form four phylogenetic clusters comparable with those of Japanese isolates of human hepatitis E virus. *Journal of Genetics and Virology*, *84*, 851-862.

[18] Bliss, R.M. (2007). Retail Meat analyzed for Parasites. *Agricultural Research Magazine*, *55*, 22.

[19] Campillo, M.C. & Vasquez, F.A. (2001). *Parasitologia veterinária*. Madrid. McGraw-Hill Interamericana.

[20] Gil, J.I. (2000). *Manual de Inspecção Sanitária de Carnes*. Lisboa, Fundação Calouste Gulbenkian.

[21] James, C., & James, S. (1997). *Food Process Engineering*. UK, MAFF.

[22] Peleg, M. & Cole, M.B. (1998). Reinterpretation of Microbial Survival Curves. *Critical Reviews in Food Science*, *38*, 353-380.

[23] Losada, V., Barros-Velázquez, J., & Aubourg, S.P. (2007). Rancidity development in frozen pelagic fish: Influence of slurry ice as preliminary chilling treatment. *Food Science and Technology*, *40*, 991-999.

[24] Esaiassen, M., Østli, J., Joensen, S., Prytz, K., Olsen, J.V., Carlehog, Elvevoll, E.O., & Richardsen, R. (2005). Brining of cod fillets: effects of phosphate, salt, glucose, ascorbate and starch on yield, sensory quality and consumers liking. *Food Science and Technology*, *38*, 641-649.

[25] Gibbs, P., & Gekas, V. (2003). Water activity and microbiological aspects of foods: A knowledge base. Available from: www.nelfood.com/help/library/nelfood-kb02.pdf.

[26] Gibson, D.M., & Davis, H.K. (1995). Principles of modified-atmosphere and sous vide product packaging: Fish and shellfish products in sous vide and modified atmosphere packs. Pennsylvania, CRC Press.

[27] Nerantzaki, A., Tsiotsias, A., Paleologos E.K., Savvaidis, I.N., Bezirtzoglou E., & Kontominas M.G. (2005). Effects of ozonation on microbiological, chemical and sensory attributes of vacuum-packaged rainbow trout stored at 4±0.5°C. *Euro Food Research Technology*, *221*, 675-683.

[28] Anonymous, (2006). Guidelines for submitting requests for acetic and citric acid sprays to the CFIA for meat products for export from the United States to Canada. Available from: http://www.fsis.usda.gov/ofo/export/gu_acid.htm.

[29] Anonymous, (2009). Microbiological Hazard Identification Guide For Meat And Poultry Components Of Products Produced By Very Small Plants.from:http://www.fsis.usda.gov/Frame/FrameRedirect.asp?main=http://www.fsis.usda.gov/oa/haccp/hidguide.htm.

[30] International Commission of Microbiological Specifications for Foods (ICMSF) (1996). *Microorganisms in Foods 5: Microbiological Specifications of Food Pathogens*. New York, Blackie Academic and Professional.

[31] Buchanan, R.L. (1993). Developing and distributing user-friendly application software. *Journal of Industrial Microbiology*, *12*, 251-255.

[32] McClure, P.J., Blackburn, C.D., Cole, M.B., Curtis, P.S., Jones, J.E., Legan, J.D., Ogden, I.D., Peck, M.W., Roberts, T.A., Sutherland, J.P., & Walker, S.J. (1994). Modelling the growth, survival and death of microorganisms in foods: the UK food micromodel approach. *International Journal of Food Microbiology*, *23*, 265-275.

[33] Dalgaard, P., Buch, P., & Silberg, S. (2002). Seafood Spoilage Predictor – development and distribution of a product specific application software. *International Journal of Food Microbiology*, *73*, 343-349.

[34] Neumeyer, K., Ross, T., & McMeekin, T.A. (1997). Development of *Pseudomonas* predictor. *Australian Journal of Dairy Technology*, *52*, 120-122.

[35] Leporq, B., Membre, J.M. Dervin, C. Buche, P., & Guyonnet, J.P. (2005). The "Sym′Previus" software, a tool to support decisions to the foodstuff safety. *International Journal of Food Microbiology*, *100*, 231-237.

[36] Gil, M.M., Pereira, P.M., Brandão, T.R.S., Silva, C.L.M., Kondjoyan, A., Valdramidis, V.P., Geeraerd, A.H., Van Impe, J.F.M., & James, S. (2006). Integrated approach on heat transfer and inactivation kinetics of microorganisms on the surface of foods during heat treatments – software development. *Journal of Food Engineering*, *76*, 95-103.

[37] Geeraerd, A.H., Valdramidis, V.P., & Van Impe, J.F. (2005). GInaFiT, a freeware tool to assess non-log-linear microbial survivor curves. *International Journal of Food Microbiology*, *102*, 95-105.

In: Practical Food and Research
Editor: Rui M. S. Cruz, pp. 219-245

ISBN: 978-1-61728-506-6

Chapter IX

TEXTURE AND MICROSTRUCTURE

Purificación García-Segovia, Amparo Andrés-Bello and Javier Martínez-Monzó

Grupo CUINA, Departamento de Tecnología de Alimentos, Universidad Politécnica de Valencia, Camino de Vera s/n 46022 Valencia, Spain, pugarse@tal.upv.es

ABSTRACT

Understand the structural and physical properties of meat and fish is decisive to obtaining the product with the highest quality as well as to contribute in development of new technologies. Microscopic methods can be able to characterize the physical and textural properties of meat and fish products and their changes after technological processes (hydrothermal or enzymatic treatment, cooking process, drying or fermentation).

The underlying concept is that texture is not a simple property of food but is extremely complex, being composed of several interrelated parameters.

The objective measurement of texture (hardness, toughness, stringiness, brittleness, viscosity, and other characteristics, which go by various names) is important in itself, since such texture is subjectively sensed by the consumer; or it may be important because the texture has been found to be strongly correlated with flavour, maturity, colour, or other factors which determine quality in raw and cooked foods.

The purpose of this chapter is to review the concept of food texture, the characteristics of texture, the relationship between texture and food and the different methodologies used to measure this property in meat and fish as well as illustrate the application of microscope methods to understand the changes in the structure of these products.

9.1 INTRODUCTION

Texture is one of the most important attributes used by consumers to assess food quality. The quality and acceptability of foods are influenced considerably by textural properties normally associated with them. Texture can also be used as a criterion for selection of raw materials or as a factor in ascertaining market grades of many food products. The underlying concept is that texture is not a simple property of food but is extremely complex, being composed of several interrelated parameters. Rheological studies of food and objective measurement of textural properties of foods continue to be a problem to the food technologists. The objective measurement of texture (hardness, toughness, stringiness, brittleness, viscosity, meatiness, and other characteristics, which go by various names) is important in itself, since such texture is subjectively sensed by the consumer; or it may be important because the texture has been found to be strongly correlated with flavour, maturity, colour, or other factors which determine quality in raw and cooked foods. The foods, for which a satisfactory method of objective evaluation has been found, represent an important portion of the foods that are consumed. Because food texture is composed of too many variables, it is not possible to obtain an overall index in a single measurement. In general, therefore, only those properties which have the greatest influence on consumer acceptance are measured. The mechanical properties are the most critical for many foods.

Works on texture dates back to the late 19th and early 20th centuries [1]. It was not until the late 1950s that texture began to be looked at as a subject in itself (the way flavour had been studied for some time) mainly owing to a group of forward thinking technical research managers at the General Corporation in the USA.

A general agreement has been reached on the definition of texture which evolved from the efforts of a number of researches. It states that "texture is the sensory and functional manifestation of the structural, mechanical and surface properties of foods detected through the senses of vision, hearing, touch and kinesthetics". This definition conveys important concepts such as:

1. *texture is a sensory property* and, thus, only a human being (or an animal in the case of animal food) can perceive and describe it. The so called texture testing instruments can detect and quantify only certain physical parameters which then must be interpreted in terms of sensory perception.
2. *it is a multi-parameter attribute*, not just tenderness or chewiness, but a gamut of characteristics.
3. *it derives from the structure of the food* (molecular, microscopic and macroscopic); and
4. *it is detected by several senses*, the most important ones being the senses of touch and pressure [2].

In the other hand, the organization of structural elements and the interactions into a food is studied by the microstructure. Therefore the knowledge of the course and mechanisms of structural changes in food components is important for designing and development of new technologies and products. Changes in food microstructure affect product properties and microscopic techniques are necessary to understand the following relationship: microstructure-physical properties-food quality [3]. It is worth mentioning that both identity and quality of food products and raw material may be well reached by microscopy observation of such structural elements as: meat fibre, food colloids, and restructured proteins, among others.

However, even very sophisticated microscopic techniques should be supported by other analytical and physico-chemical methods since only such combination of all these methods provides complete and useful information [4].

The purpose of this chapter is to review the concept of food texture, the characteristics of texture, the relationship between texture and food and the different methodologies used to measure this property in meat and fish as well as illustrate the application of microscope methods to understand the changes in the structure of these products.

9.2. Evaluation of Changes in Meat and Fish Texture

9.2.1 Changes in textural properties in meat and fish due to manufacturing process.

The manufacturing procedures (i.e. cooking and cooling) are among the principal determinants of the product quality. During processing, significant changes occur in the composition and structure of such meats, influencing the quality accordingly.

Processing procedures vary widely and are influenced by rigor development which is influenced by post-mortem temperature and pH. There are several processes that can alter meat texture such as hot boning, prerigor cooking, prerigor cooling, cooking, smoking or marinating, etc. All of them are important in the meat tenderness but the one that produces the most important changes in the structure of meat is cooking process. Cooking is one of the most important factors that affects the quality of cooked meats, due to a series of chemical and physical changes as cooking produces certain textures and flavors while simultaneously killing pathogens and keeping the food safe [5].

Davey and Gilbert [6] defined cooking as the heating of meat to a sufficiently high temperature to denature proteins. Temperature and cooking time have a large effect on physical properties of meat and eating quality. The components of muscle that control toughness are the myofibrillar proteins and the connective tissue proteins, collagen and elastin. The structural changes that cooking brings about in the different proteins in their structural environment of the meat have been investigated by many authors [7-9]. During heating, the different meat proteins denature and they cause structural changes in the meat, such as the destruction of cell membranes, shrinkage of meat fibers, the aggregation and gel formation of myofibrilar and sarcoplasmic proteins and shrinkage and solubilisation of the connective tissue [10].

In the other hand, the porosity and pore sizes of samples tend to decrease with cooking [11], which can be attributed to the physicochemical changes that trigger certain viscoelastic behavioural characteristics of proteins [12]. The action of cooking causes loss of water, and consequently decreases the water content and increases the porosity of cooked meats. It has been demonstrated that both cooking and cooling can lead to the increase in porosity of cooked meats due to water loss: greater porosity indicates a higher water loss in cooked meats.

In fish, process technology is also influenced by rigor development, post-mortem temperature and pH [13]. Unlike terrestrial animal, fish muscle has higher levels of indigenous proteases, which immediately start to break down the proteins after the harvesting, during processing, improper handling storage, and cooking [14, 15].

The fish fillet is divided into blocks of muscle known as myotomes which are separated by collagenous myocomata. On cooking the collagen of the myocomata is denatured and the myotomes may separate, forming the flakes of the cooked fish. The fish muscle fibres run between adjacent myocomata. Within the muscle fibres are the myofibrils, the contractile elements of the muscle. The myofibrils are made up of thick filaments of myosin and thin filaments of actin. The proteins of fish muscle are usually considered as three main groups: the sarcoplasmic proteins, the myofibrillar proteins and the connective tissue. The sarcoplasmic proteins are water-soluble and precipitate on cooking and do not contribute significantly to the texture of the fish. The connective tissue comprises mainly collagen. In mammalian muscle, the collagens are chemically crosslinked to varying degrees, sometimes requiring extensive cooking to tenderise the muscle. In contrast the collagens of fish muscle have lower melting temperatures and are easily converted to gelatine on cooking [16].

Another process that alters so much the textural properties in fish is the frozen process.

Although frozen storage of fish can inhibit microbiological spoilage, the muscle proteins undergo a number of changes which modify their structural and functional properties [17, 18]. The reasons for changes in fish muscle proteins causing insolubility and formation of aggregates during frozen storage [19] is not clear. However, it is widely acknowledged that the formation of ice crystals, presence of formaldehyde as well as lipid oxidation products may be involved [20, 21]. Protein aggregation in frozen fish depends on the fish species, storage temperatures, temperature fluctuation, storage time and enzymatic degradation. The susceptibility of fish species to changes induced by frozen storage is significantly different. In frozen fatty fish, oxidative changes in lipids [22] and pigments affect the odour and colour as well as proteins [23], while in lean fish, the main changes are reported to involve aggregation in proteins which alter muscle texture [24]. Frozen muscle inevitably loses some of the special quality attributes of fresh fish, usually observed as a loss in juiciness and increase in toughness [18].

The smoking, marinating or pressurization processes are other techniques that bring about changes in quality parameters such as flavour, colour and texture in meat and fish.

9.2.2 Texture analysis in meat

The measurement of meat texture has been studied for many years because texture is an important quality characteristic of this product. Some of the most important sensory attributes of meat are appearance, juiciness, flavour and texture [25]. Texture values in meat depend on zootechnical characteristics of the animal such as breed, age and sex [26, 27], on anatomical characteristics such as type of muscle [28], on factors external to the animal, as handling and feeding characteristics [29], or on technological characteristics such as electrical stimulation [29] or meat cooking method [30-32]. Texture involves such attributes as hardness, tenderness, fibrousness, guminess, elasticity, firmness, juiciness and many others. In the evaluation of meat product quality, texture takes usually the second place after flavour [33, 34]. Among texture attributes, hardness is the most important to the consumer, as it decides the commercial values of a meat [35, 33]. Texture, as already aforementioned, is by definition a sensory parameter that only a human being can perceive, describe and quantify [36].

During the mastication of meat, deformation and fracture of the samples takes place. The mechanical forces acting on meat can include *shear, compressive and tensile forces* and they should be defined in the mechanical test in use. As meat is a composite, it is important to study in which structural elements failure takes place, and where cracks propagate, to be able to understand its mechanical properties.

Instrumental texture assessment on meat can be made by means of a texturometer (Instron, TA.XT2i SMS Stable Microsystems Texture Analyser). This system attempts to achieve universality by providing a number of different cells that allows tissue resistance both to shearing and to compression to be measured. These methods are of empirical nature involving the application of force and deformation in an arbitrary until the meat samples ruptures. Some of the tests most commonly used for characterize textural properties in meat are described in this chapter.

9.2.2.1 Warner-Bratzler test (WB)

There are many different tests to determine the texture of meat but a survey of the literature indicates that the most widely used for the assessment of the texture of whole meat is the empirical method of Warner-Bratzler shear device [37-39]. The most commonly used configuration is the one in which the shearing plane is perpendicular to the muscle fibres. Tensile, shear and compression forces operate in this type of test. The WB technique is, also, the instrumental technique that usually yields the best correlation with sensory panel scores for meat toughness [10]. This test measures the force (Newtons) necessary to shear a piece of meat. The test cell is constituted with a 3 mm-thick steel blade which has a 73° V cut into its lower edge (Figure 9.1a), and is fitted through a 4 mm wide slit in a small table (like a guillotine with a V cut into the blade).

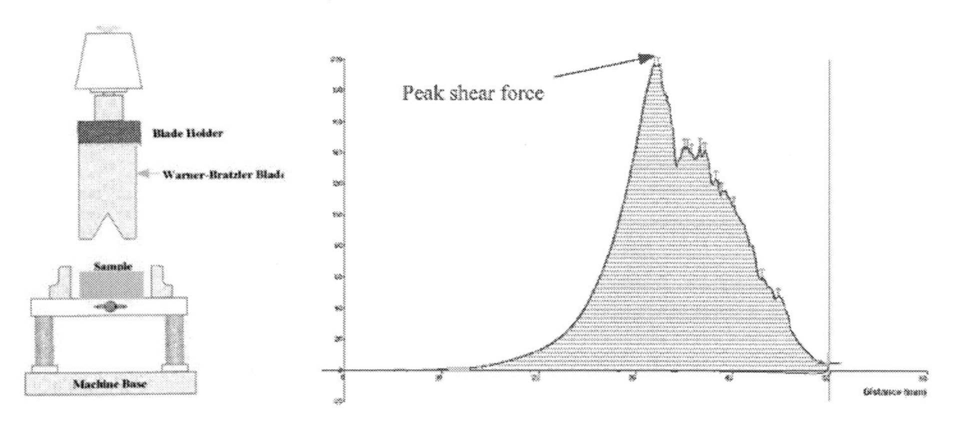

Figure 9.1. (a) Warner-Bratzler cell test and (b) typical W-B response trace.

Blade edge is not sharpened and is fitted loosely into the slit in the table. The meat sample to test is placed on the table under the V blade, and is cut through as the blade moved down with a constant speed through the slit of the table. The parameter recorded is the maximum shear force (Figure 9.1b), which is the highest peak of curve, and therefore, is the maximum resistance of the sample to shearing [40].

9.2.2.2 Texture Profile analysis (TPA)

Another commonly test for measure textural properties in meat is the Texture Profile Analysis (TPA), which simulates the conditions to which the material is subjected throughout the mastication process [1, 41]. This analysis was developed by A. S. Szczesniak at the General Foods Technical Center, Tarrytown, New York in the mid-1960s. The method has the advantage of being imitative and capable of subjecting the food to continuous chewing operations. In this test sample is subjected to a sinusoidal compressive deformation and a series of textural parameters are calculated from the readings from two successive cycles. The following parameters are quantified [42]: hardness (N), the maximum force required to compress the sample, springiness (m), the ability of the sample to recover its original form after the deforming force is removed; adhesiveness (N*s), the area under the abscissa after the first compression; and cohesiveness, the extent to which the sample can be deformed prior to rupture.

9.2.2.3 Tensile test

The tensile test is the most logical choice for studying meat texture because this type of stress predominates in resistance to chewing and in empirical test methods, particularly at the point of rupture (i.e. the maximum force region which may be the major factor in sensory reaction) [43].

The tensile test has been used to study the mechanical properties of whole meat, single muscle fibres and perimysial connective tissue [8, 44-48]. Recently, the tensile test has been successfully used on meat products to obtain more textural property information on fermented sausages [49] and meat spaghetti [50].

This test has been developed and is based on resistance of the sample to force deformation and has been developed [51]. Several tensile parameters can be obtained such as the maximum rupture force (maximum peak height resisted by the material), breaking strength (maximum rupture force by the cross-sectional area of the product) and energy to fracture (area under the deformation curve) [51, 52].

9.2.2.4 Puncture Tests

Puncture test is popular in texture testing because of its simplicity. All that is required to determine the force required to push a punch into or through the sample. Those types of tests are more used for fruits and vegetables. Standard penetration techniques have been used in which various shaped probes have been forced into the meat using a cone penetrometer [53-55], an Instron or a similar machine, or a hand held instrument similar to a fruit pressure tester [56].

9.2.2.5 Bite Testers

The use of two blunt edges to compress and cut meat samples has been used to measure meat texture. The method was proposed by Volodkevich [57] who devised a mechanical deformation mechanism and force-deformation recording system to show the force used to cut a meat sample between two rounded parallel edges having a wedge-shaped form. The objective of this method is to simulate the action of the teeth during chewing.

9.2.2.6 Grinding and Extrusion

Grinding is a method of meat texture evaluation which has been little reported in the literature. Miyada and Tappel [58] were the first to use the method. They equipped a food mixer with a power unit, a grinder and a grinder plate containing 36 holes of 5 mm in diameter. The motor was wired in series with an A.C. ammeter. A plot was manually made of the ampere readings as a function of time, so that the energy required per unit weight of sample could be computed. Using parawax as homogeneous test material, they obtained a 2% coefficient of variation. Emerson and Palmer [59] improved the equipment used by Miyada and Tappel, by adding a recorder so that the curve 'current intensity versus time' could be automatically drawn. However Schoman et al. [60] produced a critical analysis of the variability of tenderness measurement by electric grinder, pointing out the weaknesses of the method. With the progress of electronics, Voisey and Deman [61] reported a more precise electronic system using a torque transducer. In grinding tests, sample preparation is less tedious than in the classical, it would be of interest to check how far grinding parameters relate to other texture parameters.

Extruders do not appear to have been applied to testing meat tenderness but, in theory, this could be done employing the same recording methods as applied to grinders.

9.2.2.7 Compression Testing

The compression test has been utilized for intact frankfurters [62]. Voisey et al. [63] found that a compression test on cores cut from wieners was useful in predicting firmness and chewiness considered to be the major sensory characteristics of this product. First, a non-destructive compression cycle provided readings of firmness (force to achieve a given deformation) and the energy absorbed or resilence. A second rupture cycle was then used to establish cohesiveness in shear since failure occurred at shear planes. Segars et al. [64] used an unaxial compression to test raw and cooked meat and found it necessary to devise fixtures to prevent the samples from the slipping laterally relative to the compression surfaces. The results showed that elastic properties of raw and cooked meat were related.

9.2.2.8 Kramer/Ottawa test

The Kramer shear cell is a multi-bladed fixture designed to produce shear stresses in a specimen that relates to firmness. The Kramer cell is used mostly for fruits and vegetables but can also be used to evaluate meat cubes, nuggets, pork and poultry.

The Kramer Type Shear Cell comprises five parallel steel blades, which are driven down through guide slots into a rectangular container with corresponding slots in the base. The sample is sheared, compressed and extruded through the bottom openings. Since the blades are set further apart on the 10-blade version, there is a reduction in the force of bulk shearing or compression.

As the blades of the Kramer cell move down the food specimen is compressed. As deformation continues the food is extruded upwards between the blades and down through slots in the bottom of the cell. As the blades reach the bottom slots the specimen is sheared. The force required to move the blades relates to texture.

The force at various stages of the test (compression, extrusion, shear) provide additional information about texture properties.

9.2.2.9 Other techniques to measure texture properties in meat:

9.2.2.9.1 New probe

Recent research [65] has developed a means for rapidly estimating tenderness from subsample slices from cooked meat.

An instrument is proposed [66], designed, and built at Meat Industry Research Institute of New Zealand (MIRINZ) to predict the tenderness of meat. The instrument consists of two sets of pins on which meat samples are impaled. Tension is applied to the muscle fibbers by one set of pins which rotate relative to a static set of pins.

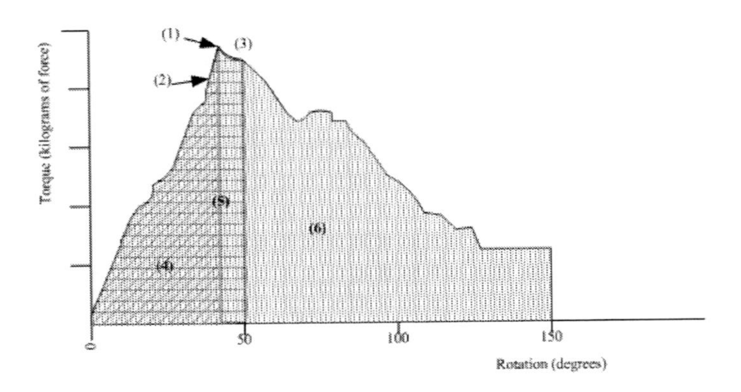

Figure 9.2. Typical torque/rotation response trace.

The torque required to rotate the inner (rotating) pin set is measured by use of a torque arm pressing against a load cell, and the torque signal is recorded against the angle of rotation (time, since the driving motor is an alternating current synchronous type).

Two configurations have been developed for the two sets of pins. One configuration is designed to provide a shearing action, while the other applies tension only. The tension head has a significant gap between the rotating and static pins, while the shear head has only about 0.5 mm of clearance between the pin sets.

The torque/rotation angle trace (Figure 9.2) of individual instrument readings was analyzed to produce characteristics to describe the trace. These were: (1) peak torque value (peak), (2) maximum slope before the peak (slope), (3) torque at 50 of rotation (D50), (4) area under the trace before the peak (area 1), (5) area under the trace before 50 of rotation (area 2), and (6) area under the whole trace (area 3) [67].

Results of this series of experiments indicate the MIRINZ tenderness probe provides a quick, viable alternative to the Warner-Bratzler shear for obtaining an objective estimate of cooked meat tenderness without the possibility of introducing additional variation through the coring process.

9.2.2.9.2 Ultrasonic techniques

In a study with sobrassada (cured sausage), Llull et al. [68] evaluated ultrasonic technique like non-destructive method to measure textural properties of a meat-based products and with the following experimental conditions. The influence of temperature on the ultrasonic velocity measurements is important to determine and also its relationship with the chemical composition of the meat based product.

The experimental set up for the velocity measurements consisted of a couple of narrow-band ultrasonic transducers (1 MHz, 0.75 in. crystal diameter, A314SSU Model, Panametrics, Waltham, MA), a pulserreceiver (Toneburst Computer Controlled, Model PR5000-HP, Matec Instruments, Northborough, MA), and a digital storage oscilloscope (Tektronix TM TDS 420, Tektronix, Wilsonville, OR) linked to a personal computer using a GPIBinterface. Proprietary software is developed to capture the signal from the oscilloscope and calculate the velocity. For each velocity measurement five signal acquisitions were performed and averaged in the computer. The time of passage of the sound wave is computed from the average signal [68].

The distance is measured with a digital height gage (Electronic Height Gage, Model 752A, Athol, MA), and sent to the computer through an RS232 interface. In order to calculate the system delay, pulse transit time measurements are performed with a set of calibration cylinders of different thickness. The delay time is obtained from the plot of time versus thickness. This delay is introduced into the velocity computation. The samples are placed between the transducers and the ultrasonic velocity is determined over three points previously marked on the pieces. In order to improve the acoustical coupling, thin layer of olive oil is placed between the samples and the transducers.

The results seem to indicate that ultrasonic technique would be a reliable and non-destructive method to determine the textural properties of meat-based products. Although, since ultrasonic velocity measurements are affected by the temperature, these must be taken into account for the determination of the textural parameters [68].

9.2.3. Texture analysis in fish

The texture of fish can have a big influence on its acceptability. Texture is an important attribute of the quality of fresh and frozen fish and fish products. After reviewing the scenarios for measurements of texture and types of texture measurement relating to the fish processing industry, the discussion focuses on physical instrumental methods of rapid non-destructive measurements of mechanical properties of fish muscle tissue.

Quality of any product is judged by its firmness for a particular purpose. In the fish production and processing path there are various critical points where quality needs to be quantified. For example a fish buyer at fish market needs to assess the quality of fresh fish. A fish processor assesses the quality of imported frozen fish. A fish product developer monitors the various attributes of a newly formulated product to ensure its acceptance by the consumer. The consumer judges the quality of the fish product by several attributes such as appearance, colour, smell and feel. The fish product is ultimately eaten and therefore the sensations during chewing and swallowing are important as well. These sensory attributes depend on the textural properties of the product. Freshness makes a major contribution to the quality of fish or fishery products. Texture of raw fish can be measured by different methods using mechanical food testing equipment. The main techniques applied for fish are puncture, compression, shear and tensile stress. Among them, the shearing force methods are recommended to be used for fresh fish [69]. When the texture of raw fish is measured, hardness and springiness are often the major variables [70]. Measurement of hardness in fish can be used as a quality indicator and shows good correlation with sensory and chemical results. Texture of fish is commonly tested in the industry by the "finger method". A finger is pressed on the skin and firmness is evaluated as a combination of the hardness when pressed on fish and the mark or hole in the fish after pressing. This technique is not desirable in the retail situation. This method depends to a large extent upon subjective evaluation of the person who is performing the measurements [71]. Borderias et al. [72] found that comparison of texture measurements of fish with sensory analysis have shown good correlation in some cases. The most reliable method is still a sensory panel in conjunction with an appropriate set of criteria for judging the quality.

The instrumental measurements in turn are affected by the inhomogeneity and variability of fish as any biological material. Therefore achieving the goal of obtaining a complete correlation between mouth-feel and instrumental is very difficult.

There are many instrumental methods that have been developed for measuring the mechanical properties of foods in general [1, 73, 74] and fish products in particular [75, 76]. These reviews classify the tests according to whether they are destructive or non-destructive, fundamental or empirical (penetration, extrusion, cutting, flow or mixing), sensory or instrumental or imitative (attempt to simulate the action of the mouth), test performed on large immobile laboratory instruments or portable hand-held instruments [70]. Many tests have been developed specifically for quantifying the quality or assisting in product development of specific products such as surimi gels and monitoring their gelling and storage performance. These kinds of test can be helpful to understand how texture is affected by the protein fibre matrix. This matrix is modified by the biochemical reactions of the autolysis and externally driven spoilage processes.

The tests are the same that the ones for measure meat texture (Warner-Bratzler shear test, Kramer Shear Compression Cell, Texture Profile Analysis, puncture test, etc). In fish there are two particulars methods: tooth and finger method. This chapter describes some of the specific cells for measuring textural properties in fish. In fish is important to standardize the sampling method.

The novel sampling technique for fillet texture test is developed by sampling certain thickness points on contour plots without reforming the fillet into regular shapes [77]. To sample a certain thickness level, the height of the probe is set at a specific distance first. Then the equivalent thickness level on the fillet can be located easily by sliding the fillet under the probe until contact is made between the probe tip and the fillet surface.

9.2.3.1 "Tooth" method

The destructive ''tooth'' method used an incisor blade (length ¼ 34.04 mm, width ¼ 9.90 mm, thickness ¼ 1.50 mm) to simulate a human tooth. The probe is attached to the Texture Analyzer with its longitude side perpendicular to the test platform and blade side (9.90×1.50 mm) facing the fillet surface. Each cut is made at directions that perpendicular to the middle horizontal septum on the fillet. After a touch is achieved between the blade tip and fillet surface, the ''tooth'' probe is allowed to cut the muscle fiber perpendicularly with the trigger type of ''button'' and then returned to the starting position. The fillet is compressed first and then cut through by the blade. The shear force (g) is measured at the maximum force required to cut through the samples (failure point) [77].

9.2.3.2 "Finger" method

The non-destructive ''finger'' method used an aluminium cylinder probe (length ¼ 123.0 mm, diameter ¼ 7.9 mm) with a ball tip (diameter ¼ 6.6 mm) to simulate compression with a human finger. The probe is attached to the Texture Analyzer with its longitude side perpendicular to the test platform. After the contact between the ball tip and fillet surface a test is performed at a speed of 1 mm/s with the trigger type ''button''. The probe is allowed to press the muscle fiber to a certain depth and then returns to the starting position. Fillet thickness must be taken into consideration when press depths are selected. In a study of Sigurgisladottir et al. [69] no difference in indentation force was observed on the head and middle parts of salmon (*Salmo salar*) fillet when a 5 mm indention was made by a ball probe (Diameter ¼ 25.4 mm) on the fillet samples of natural thickness. Only the tail part showed higher indentation force than that of the head and middle parts. The non-linear relationship between the indentation force and thickness if measured by fixed press depth method, contributes to a complex results interpretation. The press depths can be determined by the equation 9.1:

$$\text{Press depth (mm)} = [\text{Thickness (mm)}/2]-1\text{(mm)} \quad (9.1)$$

such that indentation force of the fillet is measured without breaking the muscle fiber. The indentation force (g) at the deepest press depth is recorded [77]. Prior to statistical analysis, all results collected at different thicknesses are converted to a 10 mm thickness using the following equation 9.2:

$$\text{Coverted Indentation Force (g)} = [\text{Raw Indentation Force (g)}/\text{Press Depth (mm)}] *10 \text{ (mm)} \quad (9.2)$$

9.2.4 New techniques for measure textural properties in meat and fish

In the last decade, the interest in establishing the most adequate methods to assess the quality in food has increased. These interests are associated with new technological advances, the greater interest in quality of food products, the increasing of R&D laboratories in the industry and the establishment of more regulations and standards for food in general and for meat and fish in particular. The term quality is often used in different senses and meanings. Quality may be defined as the totality of features and characteristic of a product that bear on its ability to satisfy a given need. Once the meat or fish has been processed, quality will relate to other characteristics: nutritional aspects, such as good balance in their amino acid composition, sanitary aspects, such as the numbers and types of bacteria, occurrence of parasites and presence of preservatives, processing aspects such as size, texture, etc. and finally those aspects involved in consumer acceptance, mainly organoleptic attributes such as colour, odor, flavour, etc. [74, 78, 79].

The loss of these qualities in muscle food like meat or fish is very fast since this muscle is highly perishable due to the influence of numerous ante- and post-mortem factors, and unless it is preserved in some way it easy becomes inedible. Various methods described in this chapter have been developed to measure textural modifications such as Texture Profile analysis, Kramer shear-compression cell method, puncture test, Warner–Bratzler cell method, stress-relaxation test, tensile test, etc. [51]. *Raman spectroscopy* could be individually applied for the evaluation of muscle food quality in substitution of traditional methods by evaluating the changes that originate in the components of muscle food (fish and meat) due to the effect of different processes operations such as handling, processing and storage, which are responsible for the loss of quality of the product.

The use of spectroscopic techniques has several advantages compared to the traditional methods due to the fact that they are direct, non-destructive or non-invasive and their application in situ is possible [80]. That is the reason why they have increased their importance in the determination of food quality.

Raman spectra exhibit well-resolved bands of fundamental vibrational transitions, thus providing a high content of molecular structure information of several compounds. Raman spectroscopy provides information mainly about secondary and tertiary structure of proteins [81-84] and has proven to be a powerful technique for investigating the structure of water and lipids [85-87]. Raman spectroscopy not require any pre-treatment of samples, moreover, only small portions of sample are needed. This technique provides, at the same time, information about different food compounds and not only does it offer qualitative and quantitative analysis of food components, but also it allows their structural analysis.

Raman spectroscopy provides great information compared to the traditional methods of muscle food quality evaluation, since it allows identifying structural changes in situ about various muscle food components (proteins, lipids and water) which are implicated in the loss of quality of the meat and fish due to handling, processing and storage. Understanding the structural changes involved in the loss of muscle food quality could help to optimize the conditions of application of this process with the purpose of obtaining meat and fish products of greater quality.

All these possibilities and advantages offered by Raman spectroscopy open a wide range of possibilities to apply this spectroscopic technique as an adequate on-line tool in the industry of muscle food, meat and fish [81, 88].

Other important technique is the *MRI relaxometry*. There is a growing awareness that magnetic resonance imaging (MRI) of water can provide useful insight into many structural aspects of raw and processed foods.

As a result of the latter dependencies, MRI can be used to measure the molecular environment of the water and hence could be used to optimise the quality of fruit ripening, or to detect changes in the molecular dynamics of water associated with inappropriate or illicit treatment of meat, for example, selling as fresh, meat which has been previously frozen and thawed [89].

MRI is relevant to authentication of fresh meat, fish and poultry, stemmed from the substantial seasonal variation of the cost of producing meat, fish and poultry, which makes it profitable to freeze-store tissue produced during the low cost season and then to sell it as "fresh" after thawing. MRI can be used to study the molecular interactions that are responsible for the mechanical and organoleptic properties of food, and thereby help to produce specifically structured products. MRI parameters of meats, fish, fruit and vegetables should be sensitive to the "freshness" or "ripeness" of the produce and infer that MRI will have a diverse array of applications in food processing and authentication [89].

Finally texture is an important image feature and has been applied greatly in the food industry for quality evaluation and inspection. There are four different types of texture, i.e. statistical texture, structural texture, model-based texture, and transform- based texture. Statistical texture can be obtained using statistical approach from the higher-order of pixel grey values of images while structural texture is acquired through some structural primitives constructed from grey values of pixels. Transform-based texture can be obtained by using statistical measurements from the images transformed with certain methods.

The first study of image texture can be dated back to the 1950s when the autocorrelation functions method was firstly employed as an image texture analysis technique [90]. Texture of images reflects changes of intensity values of pixels, which might contain information of geometric structure of objects since usually a great change of intensity values might indicate a change in geometric structure. In food images, texture can in some extent reflect cellular structure of foodstuffs and thus can be used as an indicator of food quality [91]. For example, texture can be adopted to reveal the tenderness of beef when colour and size features are not capable to perform well [92]. However, applications of structural texture are very limited in the food industry due to that the structural primitives used in the method can only describe very regular textures [93], thus be incapable of characterising texture of food images. Texture properties are one of the most important features in images and have been increasingly used for evaluating and inspecting food attributes. Results from previous studies show that texture features from these methods are competent for indicating food qualities. Moreover, in different applications, the predicting or classifying ability of each method might vary due to the limited understanding of image texture, which makes it difficult to discriminate which methods are the most important one for certain applications. Therefore, it is hoped that by applying several kinds of methods together, clearer relationship between image texture and food properties or qualities can be established [94].

9.3 Evaluation of Microstructure in Meat and Fish Products

9.3.1 Microstructure properties in meat and fish

The study of the microstructure of foods and food products is of great importance to understand their properties and develop new product concepts. Nature offers food in large varieties, but the structure as found in nature is often unsuitable for direct consumption by man. Therefore, food needs to be processed and consequently their molecular structure can be modified [95].

In foods, the most important blocks are lipids, proteins, carbohydrates, water and air. It's the art of food industry processing to use these building blocks to obtain the desired food products. Knowledge at the molecular level has to be combined with knowledge at the structural level to control the properties of the foods products. Microscopic studies offer this information supplemented by compositional analysis and rheological experiments [95, 96]. The wide range of microstructure elements necessitates the use of a wide range of microscopic techniques, ranging from high-resolution transmission electron microscopy and scanning probe microscopy to a variety of light microscopy techniques. Although less powerful in resolution, light microscopy offers many ways to study the microstructure under environmental conditions [95].

Food products are becoming more complex. This is directly reflected in the complexity of microstructure. Visualization of the structure and interpretation of what is observed in relation to product properties is often based on qualitative criteria. Although high-quality images contain a huge amount of information about of the microstructure, food researchers like to extract quantitative data from the images that can be correlated with relevant physical characteristics of the product. This will offer tools to modify the production process or composition to obtain the required structure [97]. Combined observation of the microstructure, extraction of quantitative data and modelling of the microstructure based on characteristics of the structure elements should point to the optimal product. 2-D image analysis is a well developed technique to extract quantitative data. The power of today's computer systems allows 3-D image analysis to be performed on large 3-D image sets. These can be acquired from tilt series of scanning electron microscopy (SEM) images or high-resolution replicas in the transmission electron microscopy (TEM) [97]. Understanding the microstructure of complex food product in relation to composition and processing conditions needs a strong cooperation and interaction between different microscopy techniques, 3-D images, spectroscopy analysis, rheological and textural measurements. Since the consumer will finally judge whether food manufacturers have been successful in preparing a high-quality product, it is the art of food researcher to discover the relationship between consumer acceptance, microstructure of the product, and the optimal ways of processing the food [97]. It is a known fact that the quality of meat and fish and meat and fish products is intimately related to their microstructure [98-103]. The main objective of microstructure studies have been to find evidence of the effect of endogenous proteolytic enzymes and to explain the loss of elasticity and increased tenderness found during the post-mortem storage of diverse range of meats [104].

Meat and fish are foods physically, biologically and biochemically active, changes at the molecular levels continue to take place when the food is stored, and they may be accelerated if it is allowed to absorb or lose moisture. The cell size, cell geometry, distribution and cell walls thickness and strength, determine the mechanical properties of the product. The biochemical microstructural changes that occur during rigor mortis and cooking are very important. Instead of the expected stress relaxation, a most common phenomenon in viscoelastic materials, the force actually increases in proportion to the material's tendency to shrink [105]. In fact, the phenomenon of strain hardening in flesh can also be traced to combined microstructural and molecular changes [105]. Microscopy has been widely used to control food and food products structure. It can be split into two fields: ''optical microscopy'' and ''electron microscopy''.

9.3.1.1 Optical microscopy

Optical microscopy offers the simplest way to obtain magnified images of biological tissues. Microscopy techniques have been used for years to characterize meat and fish structures. Techniques can be classed simply depending on whether samples must be prepared in thin cuts or not.

Chéret et al. [106], studied the effect of high pressure on texture and microstructure of sea bass (Dicentrarchus labrax L.) fillets. To corroborate their hypotheses they used microstructure observations with optical microscopy. Samples of 10 x 5 x 5 mm are cut transversally to the muscles fibers from the core of the fillet in the fleshiest part, then fixed in Carnoy's solutions (60% absolute ethanol, 30% chloroform, and 10% glacial acetic acid, v/v) at 4 °C. After 24 h, the samples are brought to room temperature and dehydrated with several alcohol solutions, first in absolute ethanol for 2 h and then in 1-butanol for 2 h (repeated 3 times). Dehydrated samples are then cleared with toluene for 30 min (repeated 3 times) and embedded in paraffin at 56 °C to 58 °C. Samples are cut with a microtome in 10μm thick slices. The slices are stained for 5 min in Orange G (0.5 g of Orange G, 1 ml acetic acid dissolved in 99 ml distilled water and filtered at 0.45 μm). The sections are washed with distilled water and stained for 5 min in Aniline blue (0.01 g of Aniline blue following the same procedure that the orange G stain). The samples are again washed with distilled water before mounting with Eukitt. This staining method stained the muscle proteins orange and collagen blue. The samples are examined in a optical microscope at 400x magnification.

Non-thin-cuts samples were used for the very early phase contrast measurement [107] allowing the detection of A and I bands in muscle and for the new confocal laser scanning microscopy [108].

Confocal laser scanning microscopy is a fluorescence technique for obtaining high-longitudinal resolution optical images. It is an evolution of the more traditional fluorescence microscopy, its key feature being the ability to produce point-by-point in-focus images of thick specimens, allowing 3D reconstructions of complex tissues [108].

Because this technique depends on fluorescence, samples usually need to be treated with fluorescent dyes to make objects visible, but contrary to the histological techniques, there is no need for thin cuts. In meat science, Straadt, Rasmussen, Andersen, and Bertram [109] recently applied confocal laser scanning microscopy to monitorize changes in fresh and cooked pork muscle during ageing. Two different magnifications, i.e., at ×10 and ×200, give spectacular views of myofibers and myofilaments, respectively. Nakamura et al. [110] successfully studied changes in bovine connective tissue according to animals feeding (concentrate- and roughage-fed groups) using confocal laser scanning microscopy coupled with immunohistochemical typing of collagen and protein structure evaluation in perimysium and endomysium. Immunohistochemical/confocal laser-scanning microscopy is a useful tool for studying structural relationships among connective tissue components in skeletal muscle.

9.3.1.2 Electron microscopy

The use of electrons beam to illuminate a specimen and create an enlarged image leads to image observation that has a much greater resolving power than with optical microscopes. The greater resolution and magnification of electron microscopy comes from the electron wavelength which is much smaller than light photon wavelength.

Biological investigations use scanning (reflection method) or transmission electron microscopy according to the application. We can also cite here the coupling of an X-ray probe to the electronic microscope to perform X-microanalysis, i.e., the local measurement, at microscopic level, of the X-ray spectrum [108].

9.3.1.3 Scanning electron microscopy

Scanning electron microscopy (SEM) gives images with great depth-of-field yielding a characteristic 3D display that provides greater insight into the surface structure of a biological sample [108].

For SEM, samples require preparation, such as cryofixation, dehydration, embedding (in resins), or sputter-coating (with conductive materials). SEM is a high-performance tool for investigating process-related changes in meat ultrastructure. In combination with histological analyses, SEM is a powerful tool to better understand relations between proteins structure and meat quality [10]. Palka and Daun [111] have also worked on structural changes during heating in bovine muscle. In their work for SEM procedure, authors cut raw and cooked semitendinosus muscles from steer and fijxed in 2.5% glutaraldehyde in 0.1 M cacodylate buffer pH 7.3 for 2 h at room temperature. The samples are rinsed with distilled water and dehydrated in 25, 50, 70, 95% absolute ethanol solution (twice), 1 h in each solution. The samples are then cut in liquid nitrogen and critical point dried using liquid carbon dioxide. After the fragments of dried tissue are mounted on holders with silver cement and coated twice with AuPd, 20 nm each time.

Structural changes in intramuscular connective tissue during tenderization of bovine meat by marinating in a solution containing proteolytic enzyme [112] is also of interest in meat process control and can be accessed with SEM.

Larrea et al. [101, 113] outlined the comparisons and complementarities of SEM, cryo-SEM and transmission electron microscopy (TEM) for the measurement of process-related ultrastructural changes in ham. For SEM observation, the samples obtained are fixed (primary fixation with 2.5% glutaraldehyde and secondary fixation with 2% osmium tetroxide) and dehydrated in a series of 10%, 20%, 40%, 60%, 80% and 100% ethanol/water solution, every 20 min. They are then rinsed in acetone and ultra-dehydrated by the critical point method with CO2 (1100 psi, 31.5 °C) in a POLARON E3000 instrument. They are then coated with gold using POLARON E6100 Equipment (10^{-4} mbar, 20 mA, 80 s) and observed in a Jeol JSM 6300 Scanning Electron Microscope at 15 kV and at a working distance of 15 mm.

Lastly, Yang and Froning [114] reported structural differences in washed or unwashed mechanically deboned chicken meat, while Nishimura et al. [115] focusing on the ante-mortem stage, studied structural changes in intramuscular connective tissue during fattening.

9.3.1.4 Cryo-scanning electron microscopy

Cryo-scanning electron microscopy consists in SEM observation after cryofixation, and is well adapted to biological tissues that are relatively unaffected by this specific sample preparation. More specifically, cryofixation is a solution for visualizing the electrolytes that are suppressed in a classical dehydration preparation. Garcia-Segovia et al. [30] studied the effects of cooking temperature and cooking time on losses, colour and texture of beef steaks using cryo-SEM to assess the endomysium and perimysium microstructure (Figure 9.3). In this work, the samples are immersed in slush Nitrogen (-210 °C) and transferred to a cryo-trans (CT 15000 C) linked to a scanning electron microscope JEOL JSM 5410 (Jeol Tokio, Japan), operating at a temperature below -130 °C. Samples are cryo-fractured at -180 °C and etched at-90 °C. The observations are carried out at 15 kV and at a working distance of 15 mm.

Pig muscle cell ultrastructure versus freezing rate and storage time has also been investigated with this technique [116]. Cryofixation is also used in transmission electronic microscopy [108].

9.3.1.5 Environmental scanning electron microscopy

Environmental scanning electron microscopy (ESEM) is a new development in the field of electron microscopy. It opens up the possibility of observing samples at almost normal atmospheric pressures (unlike classical SEM) without having to dehydrate or freeze them. Yarmand and Baumgartner [117] used ESEM to study the structure of semimembranosus veal muscle. Even though ESEM offer promising possibilities for observing intact samples, contrast is often less good than in traditional SEM when the sample is stained with conductive materials [108].

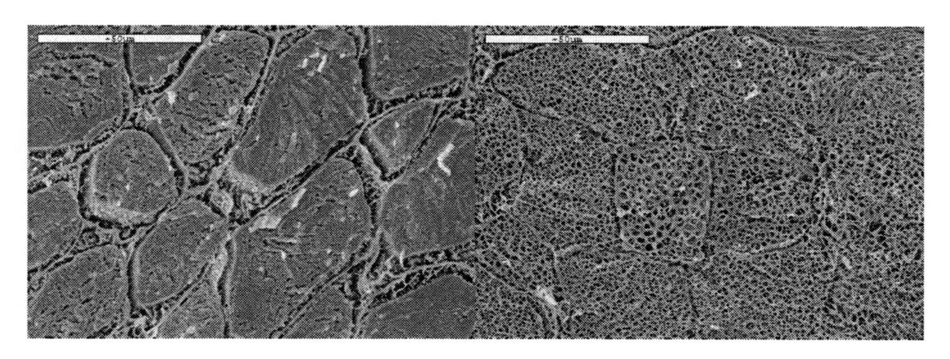

Figure 9.3. Cryo-SEM micrographs of beef meat cooked at 70 °C for 45 min: raw (a), cooked (b) (1000x).

9.3.1.6 Transmission electron microscopy

In transmission electron microscopy (TEM), electrons are passed through the sample. Resolution is higher than in SEM and the sample can be sputter-coating with conductive materials, such as gold, gold/palladium alloy or platinum, to improve image quality. Like SEM, samples may be cryofixated, dehydrated, embedded in resins and sputter-coating (with conductive materials). Unfortunately, samples have to be prepared in very thin slices and put on a grid for the observation, making the technique difficult to implement [108].

This technique has been used to observe Z line removal during culled cow meat tenderization by proteolytic enzymes [118] and by calcium chloride [119] following osmotic dehydration.

Kong et al. [120] used TEM and SEM in raw and cooked samples of chicken and salmon to study the thermal effects on tenderness, cook loss, area shrinkage, collagen solubility and microstructure on chicken and salmon muscles. In this work, for TEM analysis, pieces 2x2x2 from raw and cooked samples was fixed with 2.5% glutaraldehyde/2% paraformaldehyde in a 0.1 mol/l phosphate buffer overnight at 4 ªC. The specimens were then rinsed in phosphate buffer, post-fixed in 1% osmium tetroxide in phosphate buffer and dehydrated with a mixture of acetone and ethanol (1:1) followed by 100% acetone. The samples were infiltrated with a mixture of acetone and Spur resin (1 h) followed by 100% Spurs overnight before being polymerized at 70 °C for 24-48 h. The embedded material was sectioned (gold interference colour) to around 80–100nm using a Reichert–Jung ultratome, mounted on copper grids and stained using a solution of 4% uranyl acetate in ethanol for 10 min followed by an aqueous solution of Reynolds' lead (7 min). The material was observed in a JEOL JEM1200EX-II TEM using an accelerating voltage of 100 k.

Nakamura et al. [121] used TEM to study changes in the ultrastructural properties of tuna muscle during chilled storage. Ho et al. [122] used the same technique to study the effects of electrical stimulation of bovine carcasses on post mortem change in skeletal muscles. Nakamura et al. [121] and Sen and Sharma [123] also used TEM to show that freezing/thawing cycles do not significantly change the ultrastructural properties of buffalo muscle if freezing is applied in good conditions [108].

Table 9.1. Summary of microscopy techniques.

Microscopy techniques		Type of information	Advantages	Drawbacks	References
Optical microscopy		Fat and collagen organization; type, diameter and number of myofibers; characterize PSE zones; myofillaments organization; Z line degradation.	Selective analysis using different staining protocols. Is possible combine with several techniques like as confocal laser scaning, immunofluorescence or immunohistochemical labeling.	It needs thin cuts and staining protocols	[107, 109, 110]
Electron microscopy (Greater resolution and magnification)	SEM	Structural changes in intramuscular connective tissue; changes in meat ultrastructure.	Greater 3 D reconstruction.	Require preparation: fixation, dehydration, embedding, or sputter-coating with conductive materials.	[10, 78, 112-115]
	Cryo-SEM	Cell ultrastructure; endomysium and perimysium microstructure; ultrastructural changes	Tissues are relatively unaffected. Allow to visualize the electrolytes suppressed in a classical dehydration preparation.		[30, 113, 116]
	ESEM	Structure muscle	Observation of samples at normal atmospheric pressures without dehydration or freeze process.		[117]
	TEM	Observe Z line; changes in the ultrastructural properties; breaks in myofiber-to-myofiber; structural weakening of myofibrils and collagen.	Resolution is higher than in SEM. TEM technique may be conjugated with specific antibodies labelled with conductive materials.	Require preparation: fixation, dehydration, embedding, or sputter-coating with conductive materials. Samples have to be prepared in very thin slices and put on a grid for the observation	[118-124, 126]

Larrea et al. [113] study the microstructural changes in Biceps femoris and Semimembranosus muscles, during the processing of ''PDO Teruel'' dry-cured ham, using TEM technique. Samples are cut into 1-mm^3 cubes, fixed (primary fixation with 2.5% glutaraldehyde and secondary fixation with 2% osmium tetroxide) and dehydrated with 30%,

50% and 70% ethanol for 10 min. They are then embedded in epoxy resin and the blocks thus obtained are cut in an ultramicrotome. The sections obtained (≅100Å) are collected on copper grills and stained with 4% lead citrate to allow for observation in the Philips EM 400 Transmission Microscope at 100 kV.

In game meat, TEM has also been used to quantify ageing-related ultrastructural changes, in band breaks and sarcolemma attachment to myofibrils [124]. Myofiber attachment in salmon fillets [125] was also investigated with TEM. Authors have demonstrated that the structural change associated with loss of muscle hardness is breaks in myofiber-to-myofiber attachments, and that loss of rigor stiffness is associated with breaks in myofiber- to-myocommata attachments. These results imply that ultimate fillet texture occurs by a combination of these distinct structural changes [108].

As for histological methods, immunoelectron microscopy allows the detection of specific proteins in ultra-thin tissue sections. This technique is actually a TEM technique conjugated with specific antibodies labelled with conductive materials particles (often gold). An original application studying ageing and structural weakening of myofibrils and collagen is given in Takahashi [126].

Table 9.1 shows a summary of these microscopy methods with the type of information they give and their main advantages and drawbacks. Moreover it appears several references where to search more information.

CONCLUSION

Texture is one of the most important attributes used by consumers to evaluate food quality. There are several kinds of test to measure objectively meat and fish texture. All of these tests try to simulate the conditions to which the material is subjected throughout the mastication process, to imitate compression with a human finger, to assess the force necessary to shear a piece of meat or fish, etc. Thus, it is necessary to recognize previously the objective measurement of sensorial properties of food to an adequate selection of the texture test.

The microscopy provides information about changes in the microstructure of raw or processed food material, which allows a better understanding and controlling of the processes and their parameters.

REFERENCES

[1] Bourne, M.C. (1982). *Food Texture and Viscosity*. New York: Academic Press. (pp. 19-22).

[2] Szczesniak, A.S. (2002). Texture is a sensory property. *Food Quality and Preference*, *13*, 215-225.

[3] Aguilera, J.M., Stanley, D.W., & Baker, K.W., (2000). New dimensions in microstructure of food products. *Trends in Food Science & Technology*, *11*, 3-9.

[4] Blaszczak, W., & Fornal, J. (2008). Application of microscopy methods in food analysis. *Polish Journal of Food and Nutrition Sciences*, *58*(2), 183-198.

[5] Du C-J., & Sun D-W. (2007). Quality measurement of cooked meats. *Computer Vision Technology for Food Quality Evaluation.* (Chapter 6, 139-156), Ed. Academic Press.

[6] Davey, C.L., & Gilbert, K.V. (1974). Temperature-dependent toughness in beef. *Journal of the Science of Food and Agriculture, 25*, 931-938.

[7] Bendall, J.R., & Restall, DJ. (1983). The cooking of single myofibres, small myofibre bundles and muscle strips from beef *M. psoas* and *M. sternomandibularis* muscles at varying heating rates and temperatures. *Meat Science, 8*, 93-117.

[8] Christensen, M., Purslow, P.P., & Larsen, L.M. (2000). The effect of cooking temperature on mechanical properties of whole meat, single muscle fibres and perimysial connective tissue. *Meat Science, 55*, 301-307.

[9] Jones, S.B., Carroll, R.J., & Cavanaugh, J.R. (1977). Structural changes in heated bovine muscle. A scanning electron microscope study. *Journal of Food Science, 42*, 125-131.

[10] Tornberg, E. (2005). Effect of heat on meat proteins- Implications on structure and quality of meat products: A review. *Meat Science, 70*, 493-508.

[11] Kassama L.S., & Ngadi M O. (2005). Pore structure characterization of deep-fat-fried chicken meat. *Journal of Food Engineering, 66*(3), 369-375.

[12] Rao, M.V.N., & Skinner, G.E. (1986). Rheological properties of solid foods. In M.A., Rao, & S.S.H., Rizvi (Eds.). *Engineering Properties of Foods.* (pp 215-254). Marcel Dekker, New York.

[13] Greaser, M.L., & Pearson, A.M. (1999). Flesh foods and their analogues. In: Rosenthal, A.J. (Ed.), *Food texture measurement and perception* (pp. 236-246). Maryland: Aspen Publication.

[14] Aksnes, A. (1989). Effect of proteinase inhibitors from potato on the quality of stored herring. *Journal of the Science of Food and Agriculture, 49*, 225-234.

[15] Toyohara, H., Sakata, T., Yamashita, K., Kimoshita M., & Shimizu, Y. (1990). Degradation of oval-filefish meat gel caused by myofibrillar proteinases. *Journal of Food Science, 55*, 364-368.

[16] Goodband, R. Functional properties of fish proteins. (2002). In C., Alasalvar, & T., Taylor (Eds.). *Seafoods- Quality, Technology and Nutraceutical Applications* (pp. 73-82). Heildelberg: Springer.

[17] Conell, J. (1960). Changes in the actin of cod flesh during storage at -14 °C. *Journal of the Science of Food and Agriculture, 11,* 515-519.

[18] Mackie, I.M. (1993). The effects of freezing on flesh proteins. *Food Review International, 9*, 575-610.

[19] Badii, F., & Howell, N.K. (2002). A comparison of biochemical changes in cod (*Gadus morhua*) and haddock (*Melanogrammus aeglefinus*) fillets during frozen storage. *Journal of Science Food Agriculture, 82*, 87-97.

[20] German, J.B., & Kinsella, J.E. (1985). Lipid oxidation in fish tissue. Enzymatic initiation via lipoxygenase. *Journal of Agricultural and Food Chemistry, 33*, 680-683.

[21] Howell, N.K. (2000). Gelation properties and interactions of fish proteins. In K. Nishinari, (Ed). *Food hydrocolloids I: physical chemistry and industrial applications of gels, polysaccharides and proteins* (pp. 399-406). Elsevier: Amsterdam.

[22] Saeed, S. & Howell, N.K. (1999). High performance liquid chromatography (HPLC) and spectroscopic studies on fish oil oxidation products extracted from frozen Atlantic mackerel. *Journal of the American Oil Chemists' Society, 76*, 391-397.

[23] Saeed, S., & Howell, N.K. (2002). Effect of lipid oxidation and frozen storage on muscle proteins of Atlantic mackerel (Scomber scombrus). *Journal of the Science of Food and Agriculture*, *82*(5), 579-586.

[24] Sikorski, Z., Oley, J., & Kostuch, S. (1976). Protein changes in frozen fish. *Critical Reviews in Food Science and Nutrition*, *8*, 97-129.

[25] Barton-Gade, P.A., Cross, H.R., Jones, J.M., & Winger, R.J. (1988). In H.R., Cross, & A.J., Overby (Eds.), *Meat Science, Milk Science and Technology* (pp. 141-171). Amsterdam: Elsevier Science Publishers B.V.

[26] Huff, E.J., & Parrish, F.C.Jr., (1993). Bovine longissimus muscle tenderness as affected by postmortem aging time, animal age and sex. *Journal of Food Science*, *58*(4), 713-716.

[27] Ouali, A. (1990). La maturation des viandes: facteurs biologiques et technologiques de variation. [Meat ageing: biological and technological factors of variation]. *Viandes et produits carnés, 11(6, 6 bis, 6 ter)*, 281-290.

[28] Zamora, F. (1997). Variabilité biologique de l'attendrissage de la viande bovine. – Prédiction en fonction du facteur animal et du facteur type de muscle [Biological variability of bovine meat tenderisation. Prediction from animal and type of muscle factors]. Doctoral Thesis: Docteur d'Université, Spécialité Sciences des Aliments (Biochimie). Université d'Auvergne, pp. 44-45.

[29] Aalhus, J.L., Jones, S.D. M., Tong, A.K.W., Jeremiah, L.E., Robertson, W.M., & Gibson, L.L. (1992). The combined effects of time on feed, electrical stimulation and aging on beef quality. *Canadian Journal of Animal Science*, *72*, 525-535.

[30] García-Segovia, P., Andrés-Bello, A., & Martínez-Monzó, J. (2007). Effect of cooking method on mechanical properties, color and structure of beef muscle (*M. pectoralis*). *Journal of Food Engineering*, *80*(3), 813-821.

[31] Panea, B., Monsón, F., Olleta, J.L., Martínez-Cerezo, S., Pardos, J.J., & Sañudo, C. (2003). Estudio textural de la carne de vacuno. II. Análisis sensorial. [A texture study of bovine meat. II. Sensory analysis]. *Información Técnico-Económica Agraria*, *Vol. extra nº 24*, pp. 31-33.

[32] Sañudo, C., Monsón, F., Panea, B., Pardos, J.J., & Olleta, J.L. (2003). Estudio textural de la carne de vacuno. I. Análisis instrumental. [A textural study of bovine meat. Instrumental analysis]. *Información Técnico-Económica Agraria, vol. extra no. 24*, pp. 28-30.

[33] Cheng, C.S., & Parrish, F.C.Jr., (1976). Scanning electron microscopy of bovine muscle. Effect of heating on ultrastructure. *Journal of Food Science*, *41*, 1449-1454.

[34] Cierach, M., & Majewska, K. (1997). Comparison of instrumental and sensory evaluation of texture of cured and cooked beef meat. *Nahrung*, *41*, 366-369.

[35] Chambers, E.N., & Bowers, J.R. (1993). Consumer perception of sensory qualities in muscle foods. *Food Technology*, *47*(11), 116-120.

[36] Hyldig, G., & Nielsen, D. (2001). A review of sensory and instrumental methods used to evaluate the texture of fish muscle. *Journal of Texture Studies*, *32*, 219-242.

[37] Bratzler, L.J. (1932). *Measuring the tenderness of meat by mechanical shear*. M. Sc. Thesis, Kansas State University, Manhattan, U.S.A.

[38] Bratzler, L.J. (1949). Determining the tenderness of meat by use of the Warner-Bratzler method. *Proc. 2nd Reciprocal Meat Conference*, pp. 117-121.

[39] Warner, K.F. (1928). Progress report of the mechanical test for tenderness of meat. *Proceeding of the American Society Animal Production*, *21*, 114-118.

[40] Ruiz de Huidobro, F., Miguel, E., Blázquez, B., & Onega, E. (2005). A comparison between two methods (Warner–Bratzler and texture profile analysis) for testing either raw meat or cooked meat. *Meat Science*, *69*, 527-536.

[41] Scott-Blair, G.W. (1958). Rheology in food research. *Advances in Food Research*, *8*, 1-61.

[42] Bourne, M.C. (1978). Texture profile analysis. *Food Technology*, *32*, 62-66.

[43] Voisey, P.W. (1975). Engineering assessment and critique of instruments used for meat tenderness evaluation. Engineering Research Service, Research Branch, Agriculture Canadá, Ottawa, Ontario, Nº 520.

[44] Christensen, M., Young, R.D., Lawson, M.A., Larsen, L.M., & Purslow, P.P. (2003). Effect of added μ-calpain and post-mortem storage on the mechanical properties of bovine single muscle fibres extended to fracture. *Meat Science*, *66*, 105-112.

[45] Lepetit, J., & Culioli, J. (1994). Mechanical properties of meat. *Meat Science*, *36*, 203-237.

[46] Lewis, G.J., & Purslow, P.P. (1989). The strength and stiffness of perimysial connective tissue isolated from cooked beef muscle. *Meat Science*, *26*, 255-269.

[47] Mutungi, G., Purslow, P., & Warkup, C. (1995). Structural and mechanical changes in raw and cooked single porcine muscle fibres extended to fracture. *Meat Science*, *40*, 217-234.

[48] Willems, M.E.T., & Purslow, P.P. (1996). Effect of postrigor sarcomere length on mechanical and structural characteristics of raw and heat-denatured single porcine muscle fibres. *Journal of Texture Studies*, *27*, 217-233.

[49] Herrero, A.M., Ordóñez, J.A., Romero de Ávila, M.D., Herranz, B., de la Hoz, L., & Cambero, M.I. (2007). Breaking strength of dry fermented sausages and their correlation with Texture Profile Analysis (TPA) and physico-chemical characteristics. *Meat Science*, *77*, 331-338.

[50] Farouk, M.M., Zhang, S.X., & Waller, J. (2005). Meat spaghetti tensile strength and extensibility as indicators of the manufacturing quality of thawed beef. *Journal of Food Quality*, *28*, 452-466.

[51] Bourne, M.C. (2002). Principles of objective texture measurement. In M. C. Bourne (Ed.), *Food texture and viscosity: Concept and measurement* (pp. 107-188). San Diego, USA.

[52] Honikel, K.O. (1998). Reference methods for the assessment of physical characteristics of meat. *Meat Science*, *49*, 447-457.

[53] Birmingham, E., Naumann, H.D., & Hedrick, H.B. (1966). Relation between muscle firmness and fresh meat stability. *Food Technology*, *20*, 1222.

[54] Bouton, P.E., Harris, P.V., & Shorthose, W.R. (1971). Effect of ultimate pH upon the water holding capacity and tenderness of mutton. *Journal of Food Science*, *36*, 435-439.

[55] Paul, P.C., Mandigo, R.W., & Arthaud, V.H. (1970). Textural and histological differences among 3 muscles in the same cut of beef. *Journal of Food Science*, *35*, 505-510.

[56] Haughey, D.P., & Marer, J.M. (1971). The softening of frozen meat: criteria for transportation in insulated containers without refrigeration. *International Journal of Food Science and Technology*, *6*, 119-130.

[57] Volodkevich, N.N. (1938). Apparatus for measurement of chewing resistance or tenderness of foodstuffs. *Food Research*, *3*, 221-225.

[58] Miyada, D.S., & Tappel, A. L. (1956). Meat Tenderization. I. Two mechanical devices for measuring textures. *Food Technology*, *10*, 142.

[59] Emerson, J.A. & Palmer, A.Z. (1960). A food grinder-recording ammeter method for measuring beef tenderness. *Food Technology*, *14*, 214-216.

[60] Schoman, C.M., Bell, J., & Bell, C.O. (1960). Variations and their causes in the measurements of beef tenderness by the electric meat grinder method. *Food Technology*, *14*, 581.

[61] Voisey, P.W., & Deman, J.M. (1970). An electronic recording viscometer for food products. *Canadian Institute of Food Science Technology Journal*, *3*, 14.

[62] Townsend, W.E., Ackerman, S.A., Witnauer, L.P., Palm, W.E., & Swift C.E. (1971). Effects of types and levels of fat and rates and temperatures of comminution on the processing and characteristics of frankfurters. *Journal of Food Science*, *36*, 261-265.

[63] Voisey, P.W., Randall, C.J., & Larmond, E. (1975). Selection of an objective test of wiener texture by sensory analysis. *Canadian Institute of Food Science Technology Journal*, *8*, 23.

[64] Segars, R.A., Nordstrom, H.A., & Kapsalis, J.G. (1974). Textural characteristics of beef muscles. *Journal Texture Studies*, *5*, 283-297.

[65] Shackelford, S.D., Wheeler, T.L., & Koohmaraie, M. (1999). Tenderness classification of beef. II. Design and analysis of a system to measure beef Longissimus shear force under commercial processing conditions. *Journal of Animal Science*, *77*(6), 1474-1481.

[66] Phillips, D.M. (1992). A new technique for measuring meat texture and tenderness. *Proceeding 38th International Congress Meat Science Technology*, *38*(5), 959-962.

[67] Jeremiah, L.E, & Phillips, D.M. (2000). Evaluation of a probe for predicting beef tenderness. *Meat Science*, *55*, 493-502.

[68] Llull, P., Simal, S., Benedito, J., & Rosselló, C. (2002). Evaluation of textural properties of a meat-based product (sobrassada) using ultrasonic techniques. *Journal of Food Engineering*, *53*, 279-285.

[69] Sigurgisladottir, S., Hafsteinsson, H., Jonsson, A., Lie, Ø., Nortvedt, R., Thomassen, M., & Torrisen, O. (1999). Textural properties of raw Salmon fillets as related to sampling method. *Journal of Food Science*, *64*, 99-104.

[70] Botta, J. R. (1991). Instrument for non-destructive texture measurement of raw Atlantic cod (*Gadus morhua*) fillets. *Journal of Food Science*, *56*, 962-964, 968.

[71] Sigurgisladottir, S., Torrissen, O., Lie, Ø., Thomassen, M., & Hafsteinsson, H. (1997). Salmon quality: methods to determine quality parameters. *Reviews in Fisheries Science*, *5*, 223-252.

[72] Borderías, A.J., Lamua, M., & Tejada, M. (1983). Texture analysis of fish fillets and minced fish by both sensory and instrumental methods. *International Journal of Food Science and Technology*, *18*, 85-95.

[73] Lewis, M.J. (1996). *Physical properties of foods and food processing systems*. Woodhead Publishing, Cambrige, UK.

[74] Röhr, A., Lüddecke, K., Drusch, S., Müller, M.J., & Alvensleben, R. V. (2005). Food quality and safety – consumer perception and public health concern. *Food Control*, *16*, 649-655.

[75] Barroso, M., Careche, M., & Borderías, A.J. (1998). Quality control of frozen fish using rheological techniques. *Trends in Food Science and Technology*, *9*, 223-229.

[76] Ölafsdóttir, G., Verres-Bagnis, V., Luten, J.B., Dalgaard, P., Careche, M., Martinsdóttir, E., & Heia, K. The need for methods to evaluate fish freshness. (1998). In G., Ölafsdóttir, J., Luten, P., Dalgaard, M., Careche, V., Verrez-Bagnis, E., Martinsdóttir, & K., Heia (Eds.), *Methods to determine the freshness of fish in research and industry. Paris: International Institute of Refrigeration* (pp. 17-19).

[77] Jiang, M., Wang, Y., van Santen, E., & Chappell, J.A. (2008). Evaluation of textural properties of channel catfish (*Ictalurus punctatus Rafinesque*) fillet with the natural contour method. *LWT - Food Science and Technology*, *41*, 1548-1554.

[78] Anklam, E., & Battaglia, R. (2001). Food analysis and consumer protection. *Trends in Food Science and Technology*, *12*, 197-202.

[79] Grunert, K. G., Bredahl, L., & Bunsø, K. (2004). Consumer perception of meat quality and implications for product development in the meat sector – a review. *Meat Science*, *66*, 259-272.

[80] Li-Chan, E.C.Y. (1996). The applications of Raman spectroscopy in food science. *Trends in Food Science and Technology*, *7*, 361-370.

[81] Herrero, A.M. (2008). Raman spectroscopy for monitoring protein structure in muscle food systems. *Critical Reviews in Food Science and Nutrition, 48*(6), 512-523.

[82] Li-Chan, E.C.Y., Nakai, S., & Hirotsuka, M. (1994). Raman spectroscopy as a probe of protein structure in food system. In R.Y., Yada, R.L., Jackman, & J.L., Smith (Eds.), *Protein Structure-function Relationships in Foods* (pp. 163-197). London: Blackie Academic and Professional, Chapman and Hall.

[83] Pelton, J.T., & McLean, L.R. (2000). Spectroscopy methods for analysis of protein secondary structure. *Analytical Biochemistry*, *277*, 167-176.

[84] Tuma, R. (2005). Raman spectroscopy of proteins: From peptides to large assemblies. *Journal of Raman Spectroscopy*, *36*, 307-319.

[85] Colaianni, S.E.M., & Nielsen, O.F. (1995). Low-frequency Raman spectroscopy. *Journal of Molecular Structure*, *347*, 267-283.

[86] Maeda, Y., & Kitano, H. (1995). The structure of water in polymer systems as revealed by Raman spectroscopy. *Spectrochimica Acta*, *51*, 2433-2446.

[87] Ozaki, Y., Cho, R., Ikegaya, K., Muraishi, S., & Kawauchi, K. (1992). Potential of near-infrared Fourier transform Raman spectroscopy in food analysis. *Applied Spectroscopy*, *46*, 1503-1507.

[88] Herrero, A.M. (2008). Raman spectroscopy a promising technique for quality assessment of meat and fish: A review. *Food Chemistry*, *107*, 1642-1651.

[89] Hall, L.D., Evans, S.D., & Nott, K.P. (1998). Measurement of textural changes of food by MRI Relaxometry. *Magnetic Resonance Imaging*, *16*, Numbers 5/6, 485-492.

[90] Kaizer, H. (1955). A quantification of textures on aerial photographs. Boston, Massachusetts: Boston University Research Laboratory (Technology note 121, AD 69484).

[91] Gao, X., & Tan, J. (1996a). Analysis of expanded-food texture by image processing part I: Geometric properties. *Journal of Food Processing Engineering*, *19*, 425-444.

[92] Li, J., Tan, J., Martz, F. A., & Heymann, H. (1999). Image texture features as indicators of beef tenderness. *Meat Science*, *53*, 17-22.

[93] Bharati, M.H., Liu, J.J., & MacGregor, J.F. (2004). Image texture analysis: methods and comparisons. *Chemometrics and Intelligent Laboratory Systems*, *72*, 57-71.

[94] Zheng, C., Sun, D-W., & Zheng, L. (2006). Recent applications of image texture for evaluation of food qualities—a review. *Trends in Food Science & Technology*, *17*, 113

[95] Blonk, J.C.G. (1998). New imaging techniques and future developments. *Journal of Food Nutrition Science*, *7/48*(2): 19(S)-30(S).

[96] Heertje, I., & Pâques, M. (1995). Advances in electron microscopy. In: E. Dickinson, (Ed). *New Physico-Chemical techniques for the characterization of complex food systems* (pp. 1-52). London: Blackie academic and professional, Chapman and Hall.

[97] Blonk, J.C.G. (2002). Viewing food microstructure. In J., Welti-Chanes, G., Barbosa-Cánovas, & J.M., Aguilera (Eds.). *Engineering and Foods for the 21st Century*, pp 513-528. Boca Raton, FL: CRC Press.

[98] Chiung, Y., Stromer, M., & Robson, R. (1996). Effect of electrical stimulation on postmortem titin, nebulina, desmin, and troponin-T degradation and ultrastructural changes in bovine longissimus muscle. *Journal of Animal Science*, *74*, 1563-1575.

[99] Gordon, A., & Barbut, S. (1992). Effect of chemical modifications on the stability, texture and microstructure of cooked meat batters. *Food Structure*, *11*, 133-146.

[100] Katsaras, K., & Budras, K.D. (1992). Microstructure of fermented sausage. *Meat Science*, *31*(2), 121-134.

[101] Larrea, V., Pérez-Munuera, I., Hernando, I., Quiles, A., & Lluch, M. A. (2007). Chemical and structural changes in lipids during the ripening of Teruel dry-cured ham. *Food Chemistry*, *101*, 1327-1336.

[102] Silva, T.J.P., Orcutt, M., Forrest, J., Bracker, C.E., & Judge, M.D. (1993). Effect of heating rate on shortening, ultrastructure and fracture behavior of prerigor beef muscle. *Meat Science*, *33*, 1-24.

[103] Taylor, R., Geesink, G., Thompson, V., Koohmaraie, M., & Goll, D. (1995). Is z disk degradation responsible for postmortem tenderization? *Journal of Animal Science*, *73*, 1351-1367.

[104] Aguilera, J.M., & Stanley, D. (1999). *Microstructural principles of food processing and engineering* (2nd edition). Springer - Verlag.

[105] Peleg, M. (2002). Levels of structure and mechanical properties of solid foods. In J., Welti-Chanes, G., Barbosa-Cánovas, & J.M., Aguilera (Eds.). *Engineering and Foods for the 21st Century*, pp 465-476. Boca Raton, FL: CRC Press.

[106] Chéret, R., Chapleau, N., Delbarre-Ladrat, C., Verrez-Bagnis, V., & De Lamballerie, M. (2005). Effects of high pressure on texture and microstructure of Sea bass (*Dicentrarchus labrax* L.) fillets. *Journal of Food Science*, *70* (8), E477-E483.

[107] Ranvier, L. (1889). *Traité technique d'histologie*. Paris: Savy.

[108] Damez, J-L., & Clerjon, S. (2008). Meat quality assessment using biophysical methods related to meat structure. *Meat Science*, *80*, 132-149.

[109] Straadt, I.K., Rasmussen, M., Andersen, H.J., & Bertram, H.C. (2007). Aging-induced changes in microstructure and water distribution in fresh and cooked pork in relation to water-holding capacity and cooking loss – A combined confocal laser scanning microscopy (CLSM) and low-field nuclear magnetic resonance relaxation study. *Meat Science*, *75*(4), 687-695.

[110] Nakamura, Y.N., Iwamoto, H., Etoh, T., Shiotsuka, Y., Yamaguchi, T., Ono, Y., Tabata, S., Nishimura, S., & Gotoh, T. (2007). Three-dimensional observation of connective tissue of bovine masseter muscle under concentrate–and roughage-fed conditions by using immunohistochemical/confocal laser-scanning microscopic methods. *Journal of Food Science*, *72*(6), E375-E381.

[111] Palka, K., & Daun, H. (1999). Changes in texture, cooking losses, and myofibrillar structure of bovine M-semitendinosus during heating. *Meat Science*, *51*(3), 237-243.

[112] Chen, Q.H., He, G.Q., Jiao, Y.C., & Ni, H. (2006). Effects of elastase from a Bacillus strain on the tenderization of beef meat. *Food Chemistry*, *98*(4), 624-629.

[113] Larrea, V., Pérez-Munuera, I., Hernando, I., Quiles, A., Llorca, E., & Lluch, M.A. (2007). Microstructural changes in Teruel dry-cured ham during processing. *Meat Science*, *76*(3), 574-582.

[114] Yang, T.S., & Froning, G.W. (1992). Selected washing processes affect thermal gelation properties and microstructure of mechanically deboned chicken meat. *Journal of Food Science*, *57*(2), 325-329.

[115] Nishimura, T., Hattori, A., & Takahashi, K. (1999). Structural changes in intramuscular connective tissue during the fattening of Japanese Black cattle: effect of marbling on beef tenderization. *Journal of Animal Science*, *77*(1), 93-104.

[116] Ngapo, T.M., Babare, I.H., Reynolds, J., & Mawson, R.F. (1999). Freezing rate and frozen storage effects on the ultrastructure of samples of pork. *Meat Science*, *53*(3), 159-168.

[117] Yarmand, M., & Baumgartner, P. (2000). Environmental scanning electron microscopy of raw and heated veal semimembranosus muscle. *Journal of Agricultural Science and Technology*, *2*(3), 217-224.

[118] Gerelt, B., Ikeuchi, Y., & Suzuki, A. (2000). Meat tenderization by proteolytic enzymes after osmotic dehydration. *Meat Science*, *56*(3), 311-318.

[119] Gerelt, B., Ikeuchi, Y., Nishiumi, T., & Suzuki, A. (2002). Meat tenderization by calcium chloride after osmotic dehydration. *Meat Science*, *60*(3), 237-244.

[120] Kong, F., Tang, J., Lin, M., & Rasco, B. (2008). Thermal effects on chicken and salmon muscles: Tenderness, cook loss, area shrinkage, collagen solubility and microstructure. *LWT - Food Science and Technology*, *41*, 1210-1222.

[121] Nakamura, Y. N., Ando, M., Seoka, M., Kawasaki, K., & Tsukamasa, Y. (2006). Changes in physical/chemical composition and histological properties of dorsal and ventral ordinary muscles of full-cycle cultured Pacific bluefin tuna, *Thunnus orientalis*, during chilled storage. *Journal of Food Science*, *71*(2), E45-E51.

[122] Ho, C.Y., Stromer, M.H., & Robson, R.M. (1996). Effect of electrical stimulation on postmortem titin, nebulin, desmin, and troponin-T degradation and ultrastructural changes in bovine longissimus muscle. *Journal of Animal Science*, *74*(7), 1563-1575.

[123] Sen, A. R., & Sharma, N. (2004). Effect of freezing and thawing on the histology and ultrastructure of buffalo muscle. *Asian–Australasian Journal of Animal Sciences*, *17*(9), 1291-1295.

[124] Taylor, R.G., Fjaera, S.O., & Skjervold, P.O. (2002). Salmon fillet texture is determined by myofiber-myofiber and myofiber-myocommata attachment. *Journal of Food Science*, *67*(6), 2067-2071.

[125] Taylor, R.G., Labas, R., Smulders, F.J.M., & Wiklund, E. (2002). Ultrastructural changes during aging in M-longissimus thoracis from moose and reindeer. *Meat Science*, *60*(4), 321-326.

[126] Takahashi, K. (1996). Structural weakening of skeletal muscle tissue during post-mortem ageing of meat: The non-enzymatic mechanism of meat tenderization. *Meat Science*, *43*, S67-S80.

In: Practical Food and Research
Editor: Rui M. S. Cruz, pp. 247-279

ISBN: 978-1-61728-506-6

Chapter X

VITAMINS

Janka Porubská[1] and Martin Polovka[2]

[1]Compiler of the Slovak Food Composition Database, Department of Risk Assessment, Food Composition Databank and Consumer´s Survey , VUP Food Research Institute, Priemyselná 4, 824 75 Bratislava 26, Slovakia, janka.porubska@mail.t-com.sk
[2]Department of Chemistry and Food Analysis, VUP Food Research Institute, Priemyselná 4, 824 75 Bratislava 26, Slovakia, polovka@vup.sk

ABSTRACT

Vitamins are considered to be an essential non-caloric part of human nutrition. Meat and fish and also shellfish are one of the most important sources of vitamins. In this chapter, the classification of vitamins and vitamers as well as a short overview of their function is reviewed from the nutritional point of view. In addition, their role in meat is discussed, taking into account both, their antioxidant action, preventing e.g., the oxidation of unsaturated lipids, and their effect on the meat products stability and shelf life prolongation. The degradation of vitamins during meat processing, e.g., during cooking or freezing as well as resulting from γ-irradiation of meat and meat products is presented. In the last part, the most commonly analytical methods used for vitamins detection are presented.

10.1 INTRODUCTION

Several definitions of vitamins are to be found elsewhere. Generally, it is a non-caloric organic compound indispensable in very small amounts in diet. It cannot be synthesized in sufficient quantities by an organism, and thus must be obtained from diet [1]. Vitamins form a heterogeneous group of substances. The word vitamin is rather physiological term than a chemical term, and expresses a certain physiological activity that relates to the chemical substances responsible for this activity. These substances are usually related structurally one to another and are called vitamers [2]. Thus, each vitamin may refer to several vitamer compounds that all show the biological activity associated with a particular vitamin [3].

Vitamins regulate metabolic processes, control cellular functions and prevent diseases such as scurvy and rickets. They have specific and individual functions to promote growth or reproduction, or to maintain health and life.

Table 10.1. Vitamins and vitamers and some commonly used synonyms [3, 4].

Vitamin group	Vitamins	Synonyms	Vitamers	Meat and fish as sources
Fat-soluble vitamins				
Vitamin A	Retinol	Vitamin A_1, all-*trans*-retinol, Vitamin A_1 alcohol, axerophtol or axerol		Marine fish Freshwater fish Fish liver oils (cod liver oil and halibut liver oil) Birds Mammals Animal livers and kidneys
	Retinal	Vitamin A_1 aldehyde, retinene, retinaldehyde		
	Retinoic acid	Tretinoin, vitamin A_1 acid		
	Vitamin A_2	3-dehydroretinol, 3,4-didehydroretinol		Fresh water fish
	Provitamin A	Carotenoids	α-carotene β-carotene γ-carotene β-cryptoxanthin Echinenon	Not present
Vitamin D	Ergocalciferol	Vitamin D_2		Synthesized in the skin when exposed to sunlight At low concentration in fish oils
	Cholecarciferol	Vitamin D_3		Liver Fish liver oils Fatty fish tissues
	Provitamin D_3	7-dehydrocholesterol, 7-procholesterol		Not present
	Provitamin D_2	Ergosterol		Not present

Table 10.1 (Continued).

Vitamin group	Vitamins	Synonyms	Vitamers	Meat and fish as sources
Vitamin E	Tocopherols		α-tocopherols β-tocopherols γ-tocopherols δ-tocopherols	Not present
	Tocotrienols		α-tocotrienol β-tocotrienol γ-tocotrienol δ-tocotrienol	
Vitamin K	Phylloquinone	Vitamin K_1, vitamin $K_{1(20)}$, phytomenadione		Not present
	Menaquinone	Vitamin K_2, vitamin $K_{2(n)}$, farnoquinone		Not significant amounts
	Menadione	Vitamin K_3		Not present
Water-soluble vitamins				
B-complex	Thiamin	Vitamin B_1, aneurin		Pork and beef liver
	Riboflavin	Vitamin B_2		Liver and kidney
	Folates	Vitamin B_9, folacin	Folic acid Folate	Liver and liver products
	Pantothenic acid	Vitamin B_5		Meat and organ meat
	Niacin	Vitamin PP, vitamin B_3	Nicotinic acid Nicotinamide	Organ meat: liver, kidney and heart Chicken Beef Tuna Salmon
	Pyridoxine	Vitamin B_6	Pyridoxal Pyridoxol Pyridoxamine	Mammals' meat Fish Poultry
	Cobalamin	Vitamin B_{12}	Cyanocobalamin Hydroxocobalamin	Liver Shellfish
	Biotin	Vitamin B_7		Liver and kidney
Vitamin C			Ascorbic acid Dehydroascorbic acid	Organ meats

Nutritional science now recognises 13 groups of vitamins. They are classified according to their biological and chemical activity, not their structure [3]. Most commonly, two categories of vitamins, the water-soluble (9) and the fat-soluble (4) are to be found, as listed in Table 10.1 [3]. The majority of water-soluble vitamins act as catalysts and coenzymes in metabolic processes, energy transfer and excrete rapidly. On the other hand, fat-soluble vitamins are necessary for the provision of the proper function or structural integrity of specific body tissues and membranes and are retained in the body.

10.2 SUBSTANTIAL VITAMINS IN MEAT AND FISH

Meat is a great source of water-soluble B complex vitamins (Table 10.1), amount of which is largely influenced by the fatness of meat. Due to their lipid insolubility, these vitamins are principally found in lean portions and parts of meat (e.g., offal vs. muscle). Age of the animal also has an influence on the water-soluble vitamin content. Primarily, vitamins of the B-group are synthesized by microorganisms (e.g., intestinal microflora) and plants, but animals must obtain these vitamins mostly through their diet [5].

The biosynthesis of vitamin C occurs by different pathways in both plants and animals. Vitamin C is present in liver and other offal of animals (Table 10.1).

Retinoids, i.e., vitamin A_1 and A_2 and their biologically active metabolites, are found only in animal products. Regarding meat, animal livers and kidneys, and fish liver oils (e.g., cod liver oil and halibut liver oil) are particularly their rich sources [5].

Vitamin D is not abundant in meat, however some amounts can be found in liver and large amounts are present in fish liver oils [5]. There are only negligible amounts of vitamins E and K in meat and fish [6].

10.2.1 Natural abundance of vitamins

B-vitamins or B-complex comprises thiamin, riboflavin, folates, pantothenic acid, niacin, pyridoxine, cobalamine and biotin.

Vitamin B_1 (thiamin, Figure 10.1) is a cofactor utilised in the reactions catalysed in branch-chain amino acid metabolism, the pentose phosphate pathway, and the citric acid cycle [5]. It is released by the action of phosphatase and pyrophosphatase in the upper small intestine. Animals can carry out only the phosphorylation reactions on the pre-synthesized thiamin molecule, while many microorganisms and plants synthesize thiamin *de novo* [5].

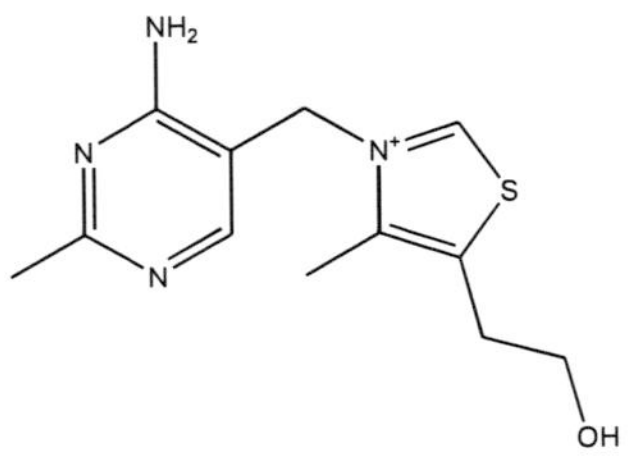

NH2 N N N+ S OH

Figure 10.1. Thiamin structure.

This vitamin is at low concentrations present in a plenty of foods. Pork and beef liver are its most concentrated sources [7]. It is sensitive to heat and alkali, thus its stability is dependent on the extent of heating and on the food matrix properties [3].

Vitamin B_2 (riboflavin, Figure 10.2) appears predominantly in the form of riboflavin 5'-phosphate (flavin mononucleotide, FMN) and flavinadenin dinucleotide (FAD). About 50 mammalian enzymes contain riboflavin in the form of FMN or FAD. Its metabolism is controlled by different hormones, which regulate its conversion into FAD and FMN. Vitamin B_2 plays a crucial role in the biosynthesis of vitamin B_{12}. It is synthesized in various microorganisms and plants, often in large amounts in the former [5].

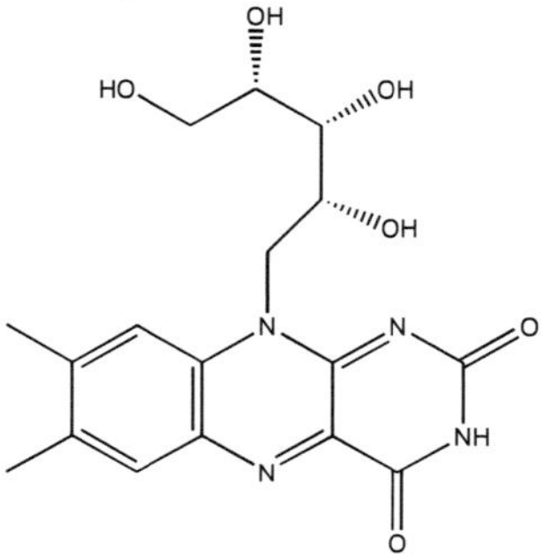

Figure 10.2. Riboflavin structure.

Liver, kidney and meat are rich sources of vitamin B_2. It is relatively resistant to dry heat, acid solutions and air (oxygen), but very sensitive to light and UV radiation, especially at high temperatures and pH values. During cooking, the vitamin leaches into water [3].

Folates comprise a group of biologically active derivates of folic acid (Figure 10.3), formerly also known as vitamin B_9 or B_c. Folic acid as such does not occur naturally in foods, but is widely used in food fortification or as a supplement [2, 8].

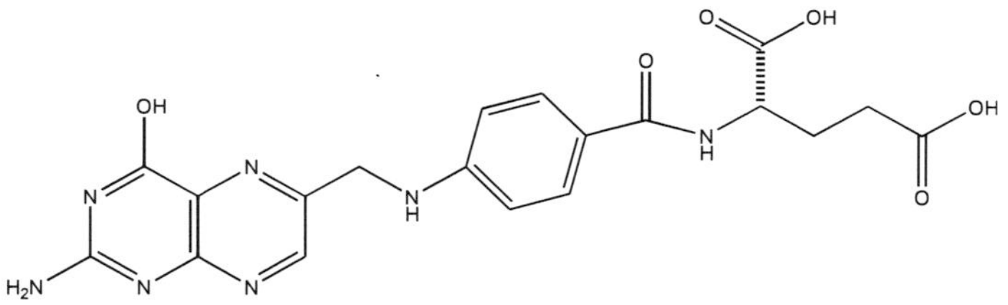

Figure 10.3. Folic acid structure.

Folic acid stimulates the formation of gastric juice and is important for well functioning of liver, folates are necessary for the metabolism of DNA and RNA. The most important animal sources are liver and liver products. Folates are sensitive to sunlight, air, and light when being heated in acid solutions [3].

Pantothenic acid (known also as vitamin B_5, Figure 10.4) is in the free form unstable and extremely hygroscopic. It is usually bounded to proteins or present in the form of salts. It naturally occurs in both L- and the D- form; but only the D- form possesses the biological activity [2, 5, 9].

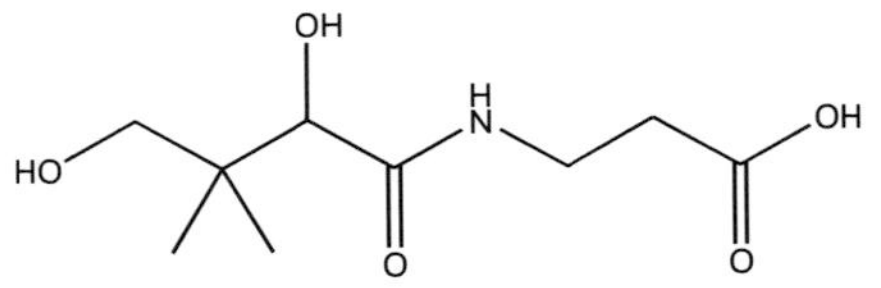

Figure 10.4. Pantothenic acid structure.

Animals cannot synthesize pantothenic acid by their own, however, similarly like microorganisms, they are capable to convert the exogenous vitamin derived from the diet to coenzyme A (HS-CoA) and acryl-carrier protein (ACP, enzyme-bound 4'-phospho-pantetheine), the two metabolically active forms of pantothenic acid [5].

Pantothenic acid is present in meat and organ meat. It is the most stable vitamin during thermal processing at pH ranging from 5 to 7. Large losses can occur via leaching into cooking water during vegetable preparation. It is also resistant to light and air, but sensitive to long-lasting cooking in water [3].

Niacin consists of two active forms, nicotinic acid (Figure 10.5), also known as vitamin PP; and nicotinamide known as vitamin B_3 [5]. As tryptophan is also metabolised to niacin, the total niacin activity must include the contribution from tryptophan, too [2].

This vitamin helps in the energy production and promotes a well working of the nervous system. It is also involved in both DNA repair, and the production of steroid hormones in the adrenal gland.

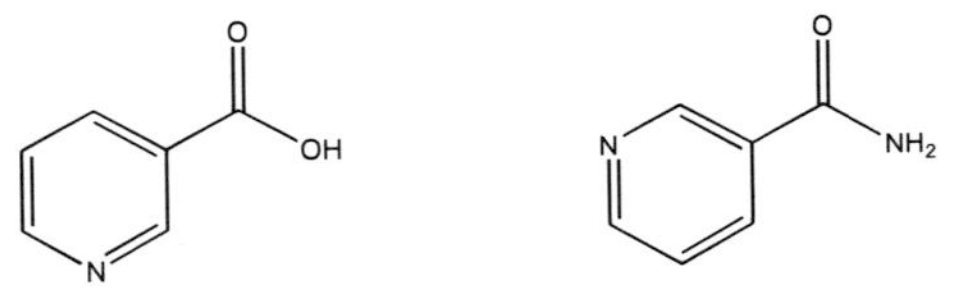

Figure 10.5. Nicotinic acid and nicotinamide structures.

Its important animal sources are meat and fish. Niacin is assumed to be the most stable water-soluble vitamin. Processing and cooking procedures do not deactivate it; it is resistant to heat, air and oxidants, but it is hydrolysed in strong acids and alkaline solutions. Leaching is usually the primary pathway of its loss during food preparation [3].

In the case of vitamin B_6 (pyridoxine, Figure 10.6), six compounds are known to exhibit its activity, i.e., pyridoxamine, pyridoxine, pyridoxal and their corresponding phosphate esters. Pyridoxine (also called pyridoxol) and its phosphate are the predominant forms occurring in plant materials, while pyridoxal, pyridoxamine and their phosphates are the main forms present in animal tissues. Animals cannot synthesize vitamin B_6 *de novo*, but they are capable to interconvert all six forms of this vitamin [5].

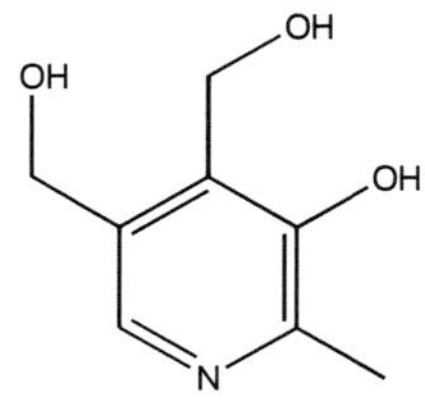

Figure 10.6. Vitamin B_6 vitamer-pyridoxine structure.

Together with folates and vitamin B_{12}, vitamin B_6 regulates the uptake of iron into body and is active in the formation of red blood cells. These three vitamins are also important for a good functioning of the nervous system and they are active in the amino acid metabolism. Foodstuffs rich in vitamin B_6 are among others fish and meat.

Thermal degradation of vitamin B_6 is pH-dependent. It is resistant to heat, acids, and alkalis, but sensitive to light in neutral and alkaline solutions. When compared one to each other, pyridoxal and pyridoxamine are more heat, oxygen and light labile than pyridoxine. Losses of this vitamin in foods of animal origin result mainly from heat degradation [3].

Vitamin B_{12} is from the chemical point of view a complex structure of cyanocobalamin (Figure 10.7), the intermediate form of the cofactors, i.e., aquacobalamin (also known as vitamin B_{12a}), and hydroxocobalamin (vitamin B_{12b}), and two metabolically active coenzyme forms, methycobalamin (methylvitamin B_{12}) and 5′-deoxy-5′-adenosylcobalamin (adenosylvitamin B_{12}).

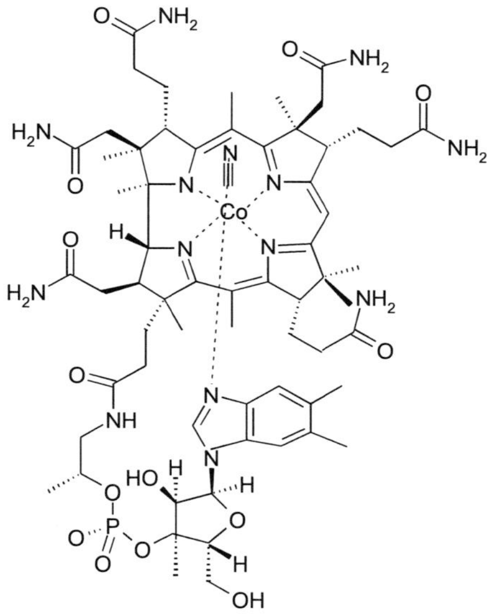

Figure 10.7. Vitamin B_{12} cyanocobalamin structure.

Cyanocobalamin does not appear in nature, but due to its good stability and lower cost, its synthetic form is frequently added into many pharmaceuticals and food supplements. It is used as food additive, as well [8].

Vitamin B_{12} helps to protect nerves and nerve tissues, and is involved in the formation of red blood cells in the bone marrow. Animal products including fish and meat are the only foods rich in vitamin B_{12}.

Vitamin B_{12} is considered stable under the most food processing operations, but like all water-soluble vitamins, it is subjected to large losses via leaching into cooking water - the higher the quantity of water used for cooking, the higher the vitamin loss during heat treatment of food. The most important associated compounds of vitamin B_{12}, cyanocobalamin is decomposed by both, oxidizing and reducing agents. Another important factor affecting the stability of this vitamin is pH level - only weak stability in alkaline solutions and strong acids was found. On the other hand, it is relatively stable to both atmospheric oxygen and heat. The stability of vitamin B_{12} is also significantly influenced by the presence of other vitamins [3].

Biotin (vitamin B_7, Figure 10.8) is found in foods as both, free vitamin and bounded to protein [2]. The biologically active isomer of biotin is required by all living cells [8].

Biotin is at low concentrations widely distributed in a variety of foods. A good animal biotin source is offal, preferably liver and kidney.

Biotin is resistant to heat and light, neutral or even strong acid solutions, but labile in alkaline solutions. In general, its retention during heat treatment is relatively high [3].

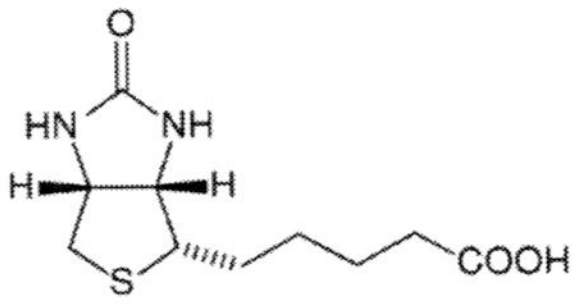

O
HN NH
H H
S COOH

Figure 10.8. Vitamin B_7 structure.

Two substances reveal vitamin C activity, i.e., L-ascorbic acid and its first-step oxidation product, L-dehydroascorbic acid [2].

In animals, L-ascorbic acid (Figure 10.9) is essential for the formation of collagen, the principal structural protein in skin, bone, tendons, and ligaments, being cofactor in hydroxylation of L-proline. It is an effective scavenger of reactive products of oxygen metabolism, e.g., superoxide radicals, hydrogen peroxide, and singlet oxygen [8].

Most animals can synthesize ascorbic acid themselves. Among others, humans as well as few avian and mammalian species lost this ability during the evolution. They ingest exogenous ascorbic acid in their food [8].

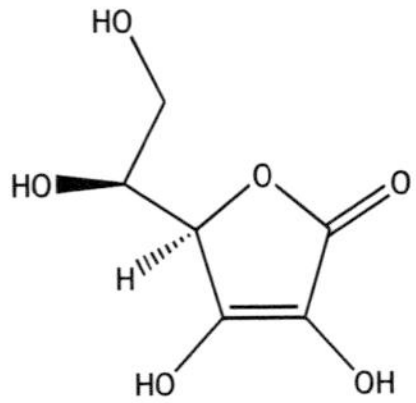

HO
HO O O
H
HO OH

Figure 10.9. L-ascorbic acid structure.

Vitamin C is besides its other sources, present in organ meats. Ascorbic acid is strong reducing agent, which is oxidised very quickly to dehydroascorbic acid and subsequent oxidation products, especially at raised temperatures and in alkaline solutions. It is one of the most sensitive vitamins [2]. Cooking losses of L-ascorbic acid depend on the way and extent of heating, leaching into the cooking medium, surface area exposed to water and oxygen, pH, presence of transition metals, and many other factors that facilitate the oxidation [3, 10].

Vitamin A (retinol, Figure 10.10) is a group of fat-soluble vitamins, which includes vitamin A_1 (all-trans-retinol, known as retinol) and vitamin A_2 (3,4-didehydro-retinol, dehydroretinol). Vitamin A_2 reveals about 40% of the activity of vitamin A_1. These vitamins and their biologically active metabolites are recognized as retinoids. Since mammals cannot synthesize retinoids *de novo*, their intake into human body depends exclusively on alimentary supplementation. Provitamins A (β-carotenes and other carotenes) are widely distributed in plants, but not present in animals. In mammals, provitamins A are transformed into vitamin A via the oxidative cleavage of the β-type carotenoids from diet [4].

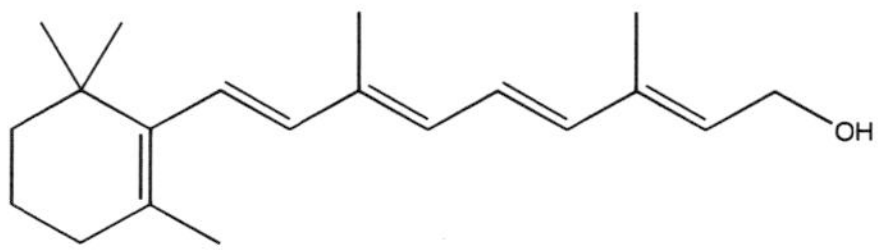

OH

Figure 10.10.Vitamin A_1 structure.

Retinoids are found only in animal products. Eggs, dairy products (butter, margarines), animal livers and kidneys, and fish liver oils are their particular rich sources. In contradiction to marine fish, birds and mammals, which appear to have only vitamin A_1, fresh water fish contain considerable amounts of both, vitamin A_1 and vitamin A_2 [4].

Free retinol is very sensitive to air oxidation. Retinyl esters are more stable than the parent compound. The entire compounds are unstable in the presence of light, oxygen and acids [3].

Vitamin D consists of several forms, but only two of them are the most important, i.e., vitamin D_2 (ergocalciferol), and vitamin D_3 (cholecalciferol) – see Figure 10.11. Vitamin D_2 appears naturally in low concentrations in fish oils. On the other hand, vitamin D_3 is much more distributed in foods, e.g., in liver, fish oil and fatty fish tissues. For humans, sunlight is the best source of this vitamin - by the sun irradiation of skin, vitamin D_3 is synthesized photochemically from the immediate precursor of cholesterol, 7-dehydrocholesterol. Some meats contain also 25-hydroxy-cholecalciferol in concentrations that contribute to vitamin D activity and must be taken into consideration [2, 4, 9].

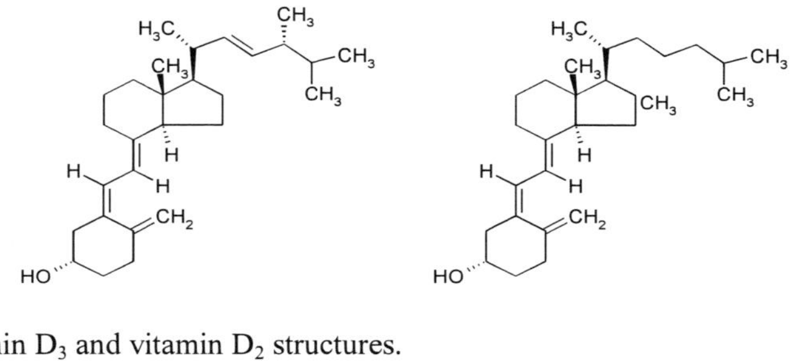

H_3C CH_3 CH_3 CH_3 CH_3 H H H CH_2 HO

H_3C CH_3 CH_3 CH_3 CH_3 H H H CH_2 HO

Figure 10.11. Vitamin D_3 and vitamin D_2 structures.

Vitamin D is susceptible to alkaline pH, light and heat [3, 11]. Although the information on the stability of vitamin D in foods is quite limited, the general assumption is that the stability is high. However, fat content is probably the crucial factor affecting the retention during culinary treatment. A high-fat content usually results in high vitamin D losses due to dripping off, while low-fat content might probably disrupt the thermal isolation and vitamin D is easily accessible to other factors (e.g., light). The stability of this vitamin is strongly dependent on the processing technique used, as well [3, 12, 13].

As mentioned above, vitamin E and K are naturally not widely present in animal products; however, they are essential for animals and humans.

Vitamin E comprises 8 biologically active lipid-soluble forms. From the chemical point of view, they are classified as chroman-6-ols (tocochromanols), i.e., 4 tocopherols (Figure 10.12): α-tocopherol, β-tocopherol, δ-tocopherol and γ-tocopherol, and 4 tocotrienols: α-tocotrienol, β-tocotrienol, δ-tocotrienol and γ-tocotrienol. Only plants and some cyanobacteria are able to synthesize vitamin E by their own [4].

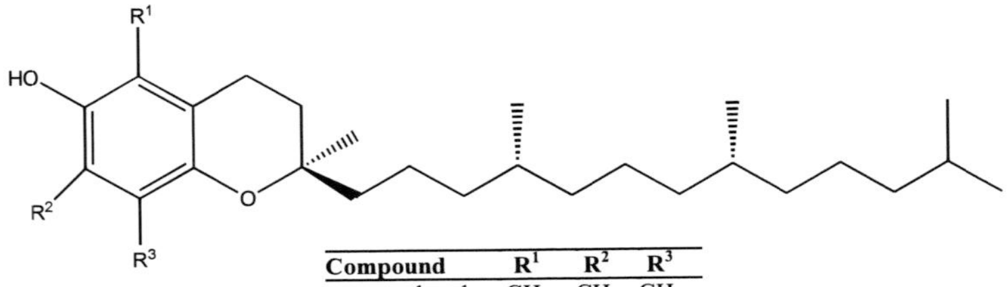

Compound	R^1	R^2	R^3
α - tocopherol	CH_3	CH_3	CH_3
β - tocopherol	CH_3	H	CH_3
γ - tocopherol	CH_3	CH_3	H
δ - tocopherol	CH_3	H	H

Figure 10.12. Tocopherol structure.

Generally, α-tocopherol has the preventive effect on ageing and ageing conditions. In mammals, it acts as a lipophilic free radical scavenger inhibiting lipid oxidation. Similarly, like vitamin C, it is a strong antioxidant [4, 14, 15].

The most abundant form of vitamin E group is α-tocopherol; while the concentration of tocotrienols is significantly lower [4, 14, 15].

Vitamin K (known also as coagulation vitamin) includes a number of fat-soluble naphto-1,4-quinone derivatives, including vitamin K_1 (phylloquinone, Figure 10.13) being of plant origin, and vitamin K_2 (menaquinone-4 and menaquinone-7) produced by microorganisms. Vitamin K is involved e.g., in normal blood clotting processes in mammals, its deficiency would lead to haemorrhage.

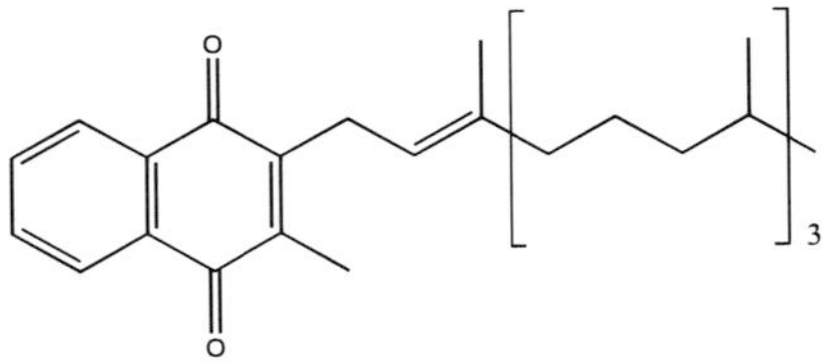

Figure 10.13.Vitamin K_1 structure.

The intestinal microflora produces a significant amount of vitamin K_2 [4]. It is quite stable to oxidation and heat and is retained after most cooking processes; on the contrary, it is sensitive to light and alkaline conditions and UV radiation [2]. Meat is the source of vitamin K_2, but not in significant amounts [16].

10.2.2 Role of vitamins as antioxidants in meat

10.2.2.1 Way of feeding and antioxidant level content in meat

Antioxidants can be incorporated in muscle through a dietary delivery. Meat derived from pasture feeding is associated with a high level of natural antioxidants and typically, it has the higher levels of α-tocopherol, β-carotene, ascorbic acid and glutathione than feedlot samples. Antioxidants including vitamins are incorporated within the cell membranes and protect tissues (muscles) against the oxidation from reactive oxygen species, thus maintaining the overall quality of meat and secondary meat products. Antioxidants retard the lipid and protein oxidation in fresh and stored meat, and preserve the colour and odour quality of meat, as well.

There are several works indicating the direct influence of the diet or feeding and meat quality [14, 17-26].

According to Driskell et al. the significant variation exists between the individual pigs with regard to the carcass quality including the nutrient content of raw meat cuts. This variation results from both, genetic and environmental factors. Influential factors include strain, sex, and age of the animal; endogenous hormone levels, stress, environmental temperature; function and training of the various muscles; type and composition of feeds; and composition of drinking water. Slaughtering, ageing, and packaging conditions may also influence carcass quality [26]. In addition, vitamin E and C supplementation of pig diets had a protective effect on riboflavin retention in liver during heating [14].

An intensive research is focused on the improvement of the myoglobin and lipid stability of fresh beef via the supplementation of animal feed with α-tocopherol, as well as on the effect of its supplementation on extended storage of meat or on its retail display quality. On the other hand, only few studies are aimed on direct meat treatment with α-tocopherol [21-25].

Currently, α-tocopherol is used as an antioxidant in the meat and poultry industry, it participates on suppression of the development of pale, soft and exudative chicken breast meat, thus improving the functional properties of meat [27, 28].

Mitsumoto et al. compared the dietary versus post-mortem meat supplementation with vitamin E dissolved in mineral oil. Endogenous vitamin E improved the pigment and lipid stability much more effectively than the exogenous vitamin E [29].

Supplementation of diets with vitamin E has shown to be effective in lipid oxidation retardation, improving also meat colour, and the consequent obtaining of meat products with extended shelf life [17]. Grass-feeding supplies natural antioxidants that are efficiently incorporated into the muscle and results in elevated concentrations of α-tocopherol within the cell membranes, as well [30, 31]. At the same time, α-tocopherol could preserve the integrity of muscle cell membranes by preventing the oxidation of membrane phospholipids during storage, which inhibits the transfer of sarcoplasmic fluid through the muscle cell membranes

[32]. Experiments carried out in Uruguay and Argentina confirmed that beef fatten up on grass has the higher vitamin E, β-carotene and ascorbic acid content and better lipid stability than those feed up on concentrates [18-20]. Cattle grazed on good-quality pasture have the higher content of α-tocopherol in skeletal muscle than cattle feed on an unsupplemented high concentrate diet [17].

Indeed, antioxidants incorporated within cell membranes are more efficient that those added post-mortem to preserve meat from oxidative damage [17].

10.2.2.2 Extending antioxidant stability and shelf life of meat

10.2.2.2.1 Natural antioxidants

A bright red colour of meat is perceived by consumers as indicator of its freshness; in fact, colour is the most important single characteristic on which they decide meat purchase [33, 34]. Upon the extended storage, the oxymyoglobin is oxidised to metmyoglobin, resulting in the development of an unattractive brown colour of meat. To extend the shelf life of fresh red meats, the modified atmosphere packaging is usually preferred. However, besides its positive effect on meat colour, oxygen presence also favours the oxidation reactions and therefore, discolouration of so-stored meat [35, 36]. In addition, oxidation of membrane phospholipids leads to the formation of unpleasant flavours. Oxymyoglobin and lipid oxidation seems to be interrelated in meat. Usually, the balance between pro-oxidative factors and antioxidative capacity of naturally present antioxidants is in favour of oxidation in stored meat [21, 37, 38]. Thus, antioxidants have been used for more than 50 years to avoid, or at least delay, the autooxidation process occurring in meat [39].

The combined use of antioxidants and modified atmosphere packaging for meat represents a realistic and attractive strategy to increase the shelf life of fresh meat [40]. However, due to concerns about the toxicological safety of synthetic antioxidants such as butylated hydroxytoluene (BHT) and butylated hydroxyanisole (BHA), it is desirable to replace these conventional synthetic antioxidants with natural ones. Thus, the interest on the application of naturally occurring antioxidants has increased over recent years [41-43].

The effect of α-tocopherol addition into feed and/or meat post-mortem is partially discussed in the previous part of this chapter. In addition, it is widely accepted that the oxidative stability of muscle lipids is dependent on α-tocopherol concentration in tissue, which, in turn, is dependent on the concentration of α-tocopheryl acetate in feed [44]. The susceptibility of meat types to lipid oxidation has been found to be in the following order: pork > beef > sheep [45]. Pork, because of its relatively high content of unsaturated fatty acids undergoes the oxidation process much more rapidly than either beef or lamb. Pre-cooked pork is even more susceptible to lipid oxidation with warmed-over flavour (WOF) developing in a matter of hours when the cooked meat is stored at 4 °C. Dietary supplementation of pig diets with α-tocopheryl acetate stabilises also cholesterol in muscle against the oxidative deterioration. This is primarily due to the assimilation of the vitamin into the subcellular membranes, where it maximises the antioxidant capacity of the system and possibly increases the physical stability of meat, as well [44].

In the contradiction, the direct addition of α-tocopherol has only non-significant effect on the oxidative stability of porcine *Muscularis longissimus dorsi* (muscle along spinal) roasts or cooked turkey breast patties. In addition, post-slaughter addition of α-tocopherol to pork offers no significant advantage in inhibiting lipid oxidation when compared to either dietary supplementation or control diets [44].

Ascorbic acid plays a crucial role in the extension of the retail display quality and shelf life of meat [46, 47]. Its synergism, when used in the combination with other antioxidants, especially vitamin E, and the promotion of their antioxidant effects is well documented. It is also known, that vitamin C regenerates vitamin E *in vitro* [48]. Okayama et al. indicated that post-mortem dipping with vitamin C and vitamin E solutions before the modified atmosphere packaging is very suitable for storage of beef steaks [49]. Mitsumoto et al. concluded that the simultaneous addition of vitamin E (at the concentration of 6 ppm) and vitamin C (at the concentration of 500 ppm) results in a lower pigment and lipid oxidation than when any of the antioxidants is used alone, confirming thus their synergistic effect [50].

Post-mortem addition of vitamin C to ground beef effectively retards the red colour deterioration in grain or grass produced meat [51].

Regarding the vitamin C, it is also important to consider that ascorbyl radicals, formed after the reaction of ascorbic acid with higher reactive radicals, are strong metal-reducing agents. The reduced forms of these metals (especially iron) are able to decompose the peroxides into radicals that can further promote a lipid and protein oxidation [17]. The role of ascorbic acid is thus ambivalent, it can acts either as promoter or as an inhibitor of lipid oxidation reactions in meat products, in dependence on its concentration. The net antioxidant capacity of vitamin C is a balance between its radical scavenging capacity and its influence on muscle pro-oxidants [17, 46, 47].

Addition of chitosan glucose complex to lamb meat can increase its shelf life by more than 2 weeks during chilled storage. It also significantly enhances the shelf life of pork cocktail salami when stored at 0-3 °C [52].

Butler and Larick reported that rosemary oleoresin improves both the sensory characteristics and oxidative stability of aseptically processed beef gels. In addition, in minced meat produced from dietary vitamin E supplemented *Muscularis semimembranosus* muscles from cattle, the oxidative stability is improved by adding the antioxidants under the aerobic packaging conditions or under the modified atmospheres packaging containing elevated oxygen levels, during refrigerated and illuminated storage at 4 °C for 8 days. Rosemary extracts demonstrated similar antioxidative efficacy as a BHA/BHT mixture [53, 54].

10.2.2.2.2 Synthetic antioxidants in meats

Addition of antioxidants during meat processing is considered to be traditional and effective technique used to control lipid oxidation and/or to delay the autooxidation of unsaturated lipids. Some of the commonly added compounds are found in meat at low levels (e.g., ascorbic acid or carnosine), while the others are usually derived from plants, i.e., phenolics/polyphenolics. The most frequently used phenolic antioxidants include BHA, BHT, propyl gallate (PG) and tertiary butylhydroquinone (TBHQ) [53-56].

Several studies revealed also the antimicrobial activity of these compounds. The activity of *Escherichia coli, Salmonella typhimurium* or *Staphylococcus aureus* is in nutrient broth meaningfully inhibited by BHA [55]. Another survey demonstrated that BHA inhibits the species of *Aspergillus flavus*, *A. parasiticus*, *Penicillium*, *Geotrichum*, *Byssochlamys* and *Saccharomyces cerevisiae* in microbiological media and inhibits also the production of mycotoxins [55]. As proven unambiguously, Gram-positive bacteria are more susceptible to BHA than Gram-negative ones. The similar susceptibility was confirmed also for TBHQ, which is extremely effective inhibitor of e.g., *Staphylococcus aureus* and *Listeria monocytogenes* at concentrations generally lower than 64 μg/ml. On the other hand, BHT is less effective than other phenolic antioxidants [55].

A number of studies have been carried out to determine the antimicrobial effectiveness of phenolic antioxidants in BHA containing foods [55]. In almost all studies, the concentration of phenolic antioxidants required for inhibition in a food, especially a meat product, is significantly higher than that needed for *in vitro* inhibition. This is probably because the presence of lipid or protein dramatically decreases the activity of phenolic antioxidants due to binding [55]. BHA and BHT are stable to heat (having the property referred to as 'carry-through') and are often used for stabilisation of fats in baked and fried products. In addition, some antioxidants, including BHA and BHT are used in combination, thus possessing the resulting synergistic effects. The disadvantage of gallates lie in their tendency to form dark coagulates with the iron ions and also on their heat sensitivity.

Very often sulphite is added to foods to act as an anti-oxidant and prevent browning. However, this will increase the loss of thiamine. Use of sulphite on meat is restricted in some countries and there is also a limit to the amount that is used in the preservation of seafood, as this may result in allergic reactions in some consumers [57].

Toxicological studies are crucial in the determination of the safety of antioxidants and their acceptable daily intake (ADI) levels. ADIs for widely used antioxidants such as BHA, BHT and gallates have changed over the years mainly because of their varying toxicological effects in various species, as presented in Table 10.2 [55, 56, 58-60]. Thus, the synthetic antioxidants have been very thoroughly tested for their toxicological behaviours, but some of them are coming, after a long period of use, under heavy pressure as new toxicological data impose some caution in their use. In this context, natural products appear healthier and safer than synthetic antioxidants. Since about 1980, natural antioxidants have appeared as an alternative to synthetic antioxidants [53, 54, 56].

The use of antioxidants in food products is governed by regulatory laws of the individual country or by internal standards. Even though many natural and synthetic compounds have antioxidant properties, only a few of them have been accepted as 'generally recognised as safe (GRAS)' substances for use in food products by international bodies such as the Joint FAO/WHO Expert Committee on Food Additives (JECFA) and the European Community's Scientific Committee for Food (SCF) [60].

There are various maximal limits for addition of synthetic antioxidants set by individual responsible competent authorities. For example, according to US Drug Administration (USDA), the addition of synthetic antioxidants is permitted in dry sausage, fresh sausage, specific cooked fresh meats, poultry products and rendered fats [61]. As follows from data presented in Table 10.3, permitted amounts are product-dependent. Different limits can be found in the EU legislation [62].

Table 10.2. Acceptable daily intakes (ADIs) of some antioxidants permitted in foods [60].

Antioxidant	ADI (mg/kg bw)
Propyl gallate	0-2.5
BHA	0-0.5
BHT	0-0.125
TBHQ	0-0.2
Tocopherols	0.15-2.0
Gum guaiac	0-2.5
Ethoxyquin	0-0.06
Phosphates	0-70.0
EDTA	2.5
Tartaric acid	0-30.0
Citric acid	Not limited
Lecithin	Not limited
L-Ascorbic acid	Not limited
Sulphites (as sulphur dioxide)	0-0.7
Ascorbyl palmitate or ascorbyl stearate (or the sum of both)	0-1.25

bw = body weight

Table 10.3. USDA approved synthetic antioxidants [61].

Product	Antioxidant/label	Limit usage
Dry sausage	BHA, BHT, PG, TBHQ*	0.003% total weight 0.006% combination
Rendered fats	BHA, BHT, PG, glycerin, resin guaiac, TBHQ*	0.01% 0.02% combination
	Tocopherols	0.03%
Fresh sausage, Italian Pizza toppings, meat balls, pre-grilled beef patties	BHA, BHT, PG, TBHQ*	0.01% fat content 0.02% combination
Dried meats	BHA, BHT, PG, TBHQ*	0.01% fat content 0.01% combination
Poultry	BHA, BHT, PG, TBHQ*	0.01% fat content 0.02% combination
	Tocopherols	0.03% fat content 0.02% combination except TBHQ

* TBHQ only in combination with BHA and BHT

10.3 Thermal Degradation of Vitamins

Meat processing results in losses of vitamins, which vary with technological treatment, processing method and type of food [3]. Several factors will influence the nutritional content of the food and the type and level of losses due to processing. These include the genetic make-up of the animal, growing conditions, packaging, storage conditions and method of preparation for processing [57]. Degradation of vitamins depends also on specific parameters during processing, e.g., temperature, oxygen, light, moisture, pH and obviously, time of treatment (exposure) [3]. The storage conditions and handling after processing are also important to the vitamin content in meat and fish. The nutrient retention may vary with a combination of conditions, such as the characteristics of the food being processed, and the concentration of the nutrient in the food. In considering the effects of processing on nutrient content of specific foods, it should be considered whether the food is one that serves as a worthwhile source of a particular nutrient [57].

Many experimental studies were carried out in order to generalise the retention of food constituents including vitamins, particularly during culinary treatment or processing of foods. However, conditions under which individual investigations were conducted as well as food properties affected meaningfully the variability of published results.

As follows from recent studies, retention factors of food constituents taking into account the effect of cooking of composite dishes (dishes consisting of several ingredients) are practically the same as these reflecting the effect of cooking of single food dish. Retention factors for cooking of individual food items are therefore transferred to recipes (dishes) containing several ingredients. Nevertheless, cooking method can influence nutrient retention considerably. Therefore, the retention factors typical for the cooking method have to be applied. For boiling, steaming and frying, nutrient retention with or without the consideration of boiling medium and dripping juice may also be of importance [63].

Moreover, cooking of dishes, especially of breaded meat and fish, by using fat as heating medium (e.g., frying in pan, deep-frying) can lead to fat uptake in food. Quantities of fat uptake are case-sensitive. As experimentally proved by Bognár, an averaged fat uptake may represent 6 g per 100 g of food ingredients.

Average retention factors of fat-soluble vitamins by cooking of fish depend on total fat content of fish. For low-fat fish (fat content <5%), such as cod, plaice, sole or trout, retention of fat-soluble vitamins is a bit higher than for high-fat fish (fat content >5%), e.g., carp, herring, mackerel, or sprout [63].

The retention of water-soluble vitamins in cooked ground beef, lamb and pork at different fat levels of raw product was investigated by Rhee. The overall vitamin retention ranged from 66% for thiamin to 78% for niacin, with the vitamin B_{12} retention (70%) being intermediate. Samples of lower initial fat levels tended to have lower vitamin retention. This is in accord with the observation that ground meats of lower initial fat content lost more of their weight during cooking through water loss than fat loss, resulting in a higher amounts of the water-soluble vitamins leached out into the drip [64].

Besides cooking, riboflavin retention ranged from 100% to 139% in other applied culinary processing methods (frying, stewing and roasting), probably because of releasing of the bounded forms of riboflavin by heat processing [65]. In the contradiction, Bognár reported on riboflavin as a stable vitamin with the average retention of 70-100% for meat, poultry and fish (for lamb, mutton and game meat, the least retention of 55% was noticed) during cooking by different methods. The increase of riboflavin content upon cooking has never been observed [63].

Al-Khalifa and Dawood studied the effect of cooking method on light and dark muscles of chicken meat. The true retention of thiamin and riboflavin in the cooked light and dark meats of broilers, when calculated on the basis of raw and cooked weights, were between 39.9% and 71.5% and 70.1% and 126%, respectively [66].

Showel et al. observed that the thiamin retention for baked bacon was lower than that one for pan-fried or microwaved, in spite of the similarities in cooking yields. Probably, the thiamin in baked bacon was more labile due to a higher cooking temperature [67].

Lassen et al. studied the thiamin retention in pork meat cooked at 72 °C. At this relatively low internal temperature, no thermal degradation products of the vitamin were observed, as most of the lost vitamins were retained in the juice (on average, 26% of thiamin) [68].

Oseredczuk et al. found that both, marinade and pasteurisation of fish are the most detrimental processes to thiamin. Because of its high sensitivity to heat, the significant losses of thiamin during pasteurisation were observed [69, 70].

As reported by Fillion, the high temperature and short transit time of the frying process cause the lower losses of heat labile vitamins than the other types of cooking. As an illustration, thiamin is well retained in fried potato products as well as fried pork meat [71].

According to Uherova et al. heat treatment of pork and chicken in microwave ovens leads to smaller losses of vitamin B_1 and B_6 than that in conventional electric ovens. Moreover, similar results were obtained with microwave ovens that differed in power and design [72].

The investigations carried out by Maskova focused on the traditional methods of meat (i.e., beef, pork, chicken breast, chicken meat) processing (i.e., boiling, roasting, stewing and frying). As follows from the results obtained, thiamin loss was observed in all the methods of culinary processing and in all kinds of meat. Its retention ranged from 20% to 85%. Regarding the thiamin losses, frying is the most favourable method (even higher in breaded fried meat), while cooking revealed the lowest thiamin retention, in compliance with experiments performed by Bognár [63, 65].

As mentioned in previous parts, thiamin is highly unstable at alkaline pH. Its stability depends on the extent of heating and on the food matrix properties. Thermal degradation occurs even under slightly acid conditions [10].

Maskova also proved that vitamin B_6 content remained constant during frying, while in cooking, the lowest retention of 23% was found. On the contrary, Bognár found that the lowest vitamin B_6 retention for frying is 65%. Comparing these two reports it can be assumed that parameters and conditions of processing as well as cooking medium, meat part and initial content of vitamin in meat altogether strongly affect the final retention level of these nutrients [63, 65].

Experiments of Driskell et al. with pork meat grilled at different temperatures showed that the higher the temperature, the higher the loss of vitamin B_6; however, those differences were not significant. On the contrary, thiamin retention values grew with temperature until the highest temperature (204 °C) when the retention significantly decreased [26].

During cooking, the content of folates in food decreases significantly, as they are broken down by heating and leaching into the cooking water. The presence of reducing agents in the food can increase the folates retention during thermal processing. Folates of animal origin are quite stable during boiling and frying. Their highest retention was observed in the case of fish and shellfish (70-100%). Cooking methods that minimise the direct contact of food with the cooking water, such as pressure or microwave cooking or stir-frying, should be preferable used to maximize the folates retention [3, 73]. As proved by McKillop et al. the cooking of beef by direct heat even after a period of extended grilling up to 16 min, results only in negligible folate losses [74]. On the other hand, Aramouni and Godber pointed on the 41% and 50% decrease in folate contents in beef liver as a result of broiling (grilling) and frying, respectively [75]. The inconsistency of published results followed from fact, that different cofactor forms of folate are to be found in beef liver and in muscle, and some of them are probably more sensitive to cooking than the others. In addition, the reported folate content in beef muscle is considerably lower than in liver; therefore, the losses are more noticeable in the latter. The rate of folates loss can be influenced by environmental factors, including pH, O_2 content, metal ions and antioxidants concentrations, heat treatment duration and product:water ratio. It is well established that reported food folate values are affected by the method of analysis employed, as well [3, 74, 76-79].

In addition, Bognár reported that the losses of folates during meat culinary treatment vary between 50% and 90% of retention, while the retention of folate in different cooking methods (stew, braise, frying in pan and deep frying) of red meat (veal, beef, mutton, lamb) ranged from 65-85%, and the retention in fried breaded red meat was 90%. Similar results were obtained for poultry, where the same cooking methods were applied and the retention was 50-70%, while for frying breaded poultry, the retention was 70%. Moreover, for the same cooking methods, the lower retention for non-breaded (65%) was found, in comparison to fried breaded organ meats with the folates retention of 85%. The same phenomenon was also found for fish regardless fish is low or high fat [63]. Frozen storage did not affect the 5-methyl-tetrahydrofolate content of raw liver [76].

As mentioned in the previous part of this chapter, vitamin, B_{12} is relatively stable to both atmospheric oxygen and heat. According to Ruan et al. in microwave heating (cooking in a minimum of water), the significant amount of vitamin B_{12} was retained [3].

Similarly, like in the case of folates and thiamin, Bognár reported on the higher retention of vitamin B_{12}, pantothenic acid and biotin in fried breaded meat than when single meat is cooked. This relates to all types of meat – red meat, poultry, offal and fish. The stability of biotin is slightly higher in comparison to all of the above-mentioned vitamins, and usually not less than 70% of biotin is retained in cooked meat [63]. On the other hand, as vitamin C is known to be very labile, it is not expected to be of as high retention as presented by Bognár for heat-treated meat, i.e., not less than 70% [63].

The documentation of processing effects on retinol is less abundant. Losses of up to 40% in fish sources rich in vitamin A have been reported following boiling [80]. In a study on the traditional food system of the Sahtú Dene/Métis, there were no consistent trends in retinol levels between raw and cooked forms of various food samples probably due to biological variation [81]. Smoking fish and mammal meat did not appear to reduce retinol levels [82].

Salting is most common method used for meat and fish preservation. Salting results in liquid exuding from the flesh, taking with it some of the water-soluble proteins, vitamins and minerals [57].

Smoking usually follows salting or curing. In addition to being bactericidal, the process has an antioxidative function. It reduces the oxidative changes that take place in fats, proteins and vitamins. However, smoking causes nutrient losses due to the associated heat, flow of gases and interaction of the smoke components with proteins [57].

The results of Sungpuag et al. indicated that boiling of intact chicken liver results in 5% loss of retinol, while the boiling of small pieces cuttings and grilling resulted in losses of 8% and 16%, respectively, what is in accord with experiments of Bognár. On the other hand, the more significant losses were identified for retinol during stewing and frying of non-breaded poultry meat at temperature higher than 80 °C, when its retention decreased to 55% [62, 83].

As mentioned in the previous part, regarding the vitamin D stability, individual studies provide quite contradictory data.

Bhuiyan et al. found that cholecalciferol is quite stable during smoking of Atlantic mackerel [3, 13]. Also baking has a positive effect on the cholecalciferol content in fish, as its losses according to Mattila reached <10% (calculated on a dry matter basis) and 78-104% for true retention, respectively. In the case of one lot of Baltic herring in which the fat content was high, the loss of cholecalciferol was exceptionally great (23%). This might be due to loss of fat during baking [84].

As investigated by Scott, commercial processing has adverse effect on vitamin content in fish oils. The samples of Menhaden oil which undergone different rate of processing (crude oil, a bleached oil, a refined oil, a refined then bleached oil, and a refined-bleached-deodorized-stabilized oil) differed in vitamin content. The most processed oil had only about one-fifth as much vitamin A and vitamin D_3 as the crude fish oil. Vitamin E levels were as a result of processing decreased to half. Bleaching the oil with Fuller's earth caused the major loss of retinols. Treating the fish oil with steam for several hours caused the major loss of vitamin D_3 [85].

Clausen et al. studied the vitamin D_3 and 25-hydroxycholecalciferol (25OHD$_3$) retention in raw and cooked pork cuts. Cooking caused the significant increases in vitamin D_3 and 25OHD$_3$ calculated per 100 g of both, individual cuts and mixed meat. Thus, although the fat lost during cooking may contain vitamin D_3 and 25OHD$_3$, the major determinant for the increased concentrations of vitamin D_3 and 25OHD$_3$ seems to be water loss, i.e., the increased dry matter content. Considering this, cooking increases the contents of vitamin D_3 and 25OHD$_3$ on weight basis, but it remain unchanged when expressed in relation to dry matter content [3, 86].

Dal Bosco et al. observed a 39% reduction of α-tocopherol content in boiled, 12% in fried and 14% in roasted rabbit meat, respectively. A bit higher retention was reached in rabbit meat supplemented by vitamin E (different DL α-tocopheryl acetate levels) under the same experimental conditions as follows: 41% reduction for boiling, 21% for frying and 22% for roasting [15].

Driskell et al. revealed that pork chops grilled at 204 °C have significantly lower true retention values for vitamin E than those prepared at lower grill temperatures. Summing up these findings, even at lower temperatures only around 80% of the vitamin E is retained after cooking [26]. Although some unsaturated fatty acids and antioxidant vitamins are lost due to oxidation, fried foods are generally good sources of vitamin E [71].

Moreover, frying can increase the nutritive value of foods since many frying oils are rich in vitamin E [71].

10.4 Degradation of Vitamins by γ-Irradiation

In order to extend the shelf life of meat and fish and to eliminate pathogenic bacteria it may contain, different methods are used. The most commonly, salting, canning, freezing and modified atmosphere packaging is used, as partially mentioned in the previous parts of this chapter. Besides them, gaseous decontamination treatments, such as ozonization, ultra-high temperature, or ultra-short time pasteurization have been effectively used for the microbial decontamination and disinfection of meat and fish [87-89].

In 1981, the FAO/IAEA/WHO Joint Committee on the wholesomeness of irradiated food approved the use of irradiation treatment for food preservation. The Committee stated that irradiation of a food at doses up to 10 kGy introduces no special nutritional problem [90].

It is well known, that ionizing radiation affects the living organisms in ways dependent on the absorbed dose of radiation, dose rate and on environmental conditions during irradiation. As the ionizing radiation application on meat and poultry has only insignificant influence on physical, chemical or sensory characteristics, it is considered an ideal sterilisation tool [88, 89].

In the case of vitamin radiolysis, types of possible free radical reactions are determined by the medium in which the vitamins are present. The fat-soluble vitamins would thus be exposed to radicals produced by the direct action of radiation on lipids and the water-soluble vitamins to radicals formed by water irradiation. In the case of fat-soluble vitamins, the free radical-mediated reactions are irrelevant since they mostly recombine with positive lipid ions. For water-soluble vitamins, some may react with hydrated electrons directly or acquire an electron from the other radicals produced in the aqueous medium. The fate of the reaction is determined by the electron reduction potential of the vitamin and the weakness of its hydrogen bonds [91].

As vitamins are in quite low amounts in most foods, the hydroxyl radicals mostly react with other major food components like lipids, proteins and carbohydrates, before reacting with vitamins. The vitamins are thus more affected by the secondary radicals formed by the interactions with the major components, e.g., by hydroperoxides [12].

10.4.1 Fat-soluble vitamins and γ-irradiation

Since these vitamins are not naturally present in meat (except for vitamin A in offal and vitamin D in fish), they are of minor concern in relation to the effect of ionizing application on meat and fish. In spite of this, only little information is available on the irradiation effect to this group of vitamins and thus on meat and fish [12, 92].

Vitamin E (α-tocopherol) is the most irradiation sensitive fat-soluble vitamin. Thus, it is a good indicator of the effect of irradiation on this class of vitamins. It was revealed, that the losses of this vitamin are in relation with fat content and the products of water radiolysis upon the ionizing radiation treatment do not influence them.

Vitamin A and carotenoids are in dry state relatively radiation-stable. In solutions, the dramatic decrease of the β-carotene was found, in dependence on the solvent used. In foods, vitamin A is sensitive to the effect of radiation, whereas the stability of β-carotene only slightly decreases even at doses of around 20 kGy [92]. The effect of meat or fish irradiation on the losses of vitamin A is only poorly discussed in literature; however, Janeve et al. described some negligible decrease of this vitamin content in liver [93].

Although vitamin K_1 is considered the most stable of all the fat-soluble vitamins, it revealed the lowest radiation stability of K group vitamins. The small quantity of vitamin K_3 present in beef is destroyed by the irradiation at doses of 28-56 kGy. Generally, the whole group of K vitamin, which is present in meat, is sensitive to high irradiation doses [12, 92].

10.4.2 Water-soluble vitamins and γ-irradiation

Regarding the vitamins of B group, it should be noted here, that the sensitivity of many B vitamins seems to vary between meat cuts and from meat to meat [6].

Thiamin is the most irradiation labile water-soluble vitamin from this group. However, its sensitivity to heat is even higher [87]. Similarly, like vitamin E in the case of fat-soluble vitamins, it is used as a marker of general vitamin loses resulting from irradiation [92]. A biologically inactive form of thiamine, dihydrothiamin, is formed as a result of its γ-irradiation in aqueous systems [92, 94]. When a beef sample containing 0.24 mg of thiamin per g was submitted to irradiation doses of 28 and 56 kGy, the thiamin content decreased to 0.057 and 0.037 mg, respectively [94]. These results are consistent with those of Wilson, who showed that the destruction of thiamin by irradiation correlates with the absorbed dose of irradiation. It has also been shown that the temperature of the beef sample during irradiation has a major effect on the rate of thiamin loss. Graham, et al, found the identical effect of γ-irradiation on thiamin content in chicken also. In addition, they revealed, that the colder the meat during the treatment, the lower is the thiamin destruction in the sample [92, 95, 96]. Hanis et al. and Fox et al. have obtained similar results [97, 98]. Gallien et al. found that thiamin content is only non-significantly affected by irradiation. Oxidative damage to thiamin is responsible for its loss. When thiamin is irradiated, a decrease in spectrophotometric absorbency indicates the destruction of its pyrimidine ring [99].

Dry riboflavin is relatively stable to irradiation. Anyway, it contains a number of hetero double bonds and when exposed to γ-irradiation in an aqueous environment, these groups can interact with the radiolytic products of water [92].

No riboflavin losses were found in pork chops and chicken breasts irradiated at doses up to 6.6 kGy at temperatures between –20 °C and +20 °C. In some irradiated samples, even the increase of its concentration of up to 25% was detected [12]. As a result of γ-irradiation of a beef sample at doses of 28 and 56 kGy, the concentration of riboflavin, 1.86 mg/g (on fresh weight basis) decreased to 1.76 and 1.79 mg, respectively [94]. In fact, it is only a minor loss and, noteworthy, an increase of riboflavin concentration is noted when the radiation dose is increased. A considerable radiation resistance of riboflavin content was also proved by Fox et al. [92, 98].

Niacin itself is, similarly like riboflavin, sensitive to ionizing radiation when treated in solutions, but it is relatively resistant to γ-irradiation when present in food. In aqueous solutions, it is even more sensitive to irradiation than thiamin or riboflavin, as it is primarily attacked by hydroxyl radicals. It was also shown, that the presence of glucose has a positive effect on niacin protection against the radiation damage [92]. In pork chops irradiated at different temperatures at doses up to 5 kGy, practically no losses of niacin content was observed by Fox et al. [92, 98]. On the other hand, regarding the effect of irradiation on niacin, some other results are inconsistent. As a result of γ-irradiation of a beef sample at doses of 28 and 56 kGy, the concentration of niacin decreased from 30 μg/g (on fresh weight basis) irradiated at 28 and 56 kGy was analysed for its content after treatment. Niacin contents after irradiation were 28.90 and 29.29 μg/g, respectively [94]. Analogously with riboflavin, no major difference in niacin content after beef irradiation was detected.

Sensitivity of pyridoxine to γ-irradiation is lower than that of thiamin. The sensitivity of pyridoxine is closer to that of riboflavin at doses higher than 10 kGy [100]. As proved by Kennedy, its losses at doses <10 kGy are only minor [101]. Practically Gallien et al. obtained the same results [99].

Although the changes in folates concentration during the γ-irradiation of e.g., poultry even at extremely high radiation doses (~70 kGy) are only slight, there are some indications that some components of folates are sensitive to the γ-radiation at doses only of 25 kGy [12, 92].

Sterilization of pork by γ-irradiation at doses ranging from 20 up to 40 kGy has only insignificant influence on biotin losses. Identically, no changes in cyanocobalamin content of pork γ-irradiated at dose of 7 kGy at 0 °C were found, although this molecule consist of ring structure that is composed of four pyrrole residues, and contain six conjugated double bonds as well as sequestered cobalt [92, 98]. The same effect was also proved in radiation-sterilised chicken. Similarly, no losses resulting from the irradiation have been reported also for choline (in some cases considered to belong to B-complex vitamin) and for pantothenic acid γ-irradiated at doses of ≥10 kGy [92, 102, 103].

A study performed by the Office of the Army Surgeon General focused on the effect of different processing treatments on thiamin, riboflavin, niacin and pyridoxine content of enzyme-inactivated beef revealed, that heat sterilization reduces the vitamin content of beef more significantly than any other sterilisation method including γ-irradiation, electron treatment and frozen storage [104].

De Groot et al. concluded that, with the possible exception of a slight decrease in vitamin E and thiamin contents after irradiation at a dose ≥6 kGy, there was no indication that irradiation caused any vitamin destruction. Fox et al. demonstrated that only thiamin loss due to irradiation process is relevant [98, 105].

Vitamin C exhibits a high sensitivity to ionizing radiation. As a result of γ-irradiation even at low doses, it is easily oxidised to dehydroascorbic acid, as in the case of heat treatment or oxidation. Dehydroascorbic acid is in fact, a biologically active form of vitamin C. Thus, when monitoring the effect γ-irradiation, it is necessary to consider the degree of ascorbic acid conversion, resulting directly from radiation itself [92]. In spite of fact, that it is widely distributed in e.g., organ meats as liver, data on its radiation-induced degradation in meat are not available.

Regarding fish and shellfish products, aquatic or fishery products are an important source of proteins, but have a relatively short shelf life unless they are frozen onboard or very shortly after harvesting. The shelf life of fresh, frozen and processed marine and fresh fish and shellfish can also be extended by the irradiation, as the irradiation of these products led to the effective elimination of potentially harmful microbiological pathogens. According to the character of product and its intended use, the effectiveness of radiation application can be enhanced using the combination of treatments, such as heat, freezing or other GRAS methods

[89]. It should be noted here, that the data on vitamin losses during the exposure of fish and seashell products to γ-irradiation are not available, with exception for vitamin A, discussed in the above part of this sub-chapter.

10.5 Analysis of Vtamins in Meat and Fish

In the last decades, interest and effort in establishing the most adequate methods to assess the quality of foods has increased. These interests are associated with new technological advances, needs for good quality of foods, the increasing possibilities of research and development laboratories in the industry and the establishment of more regulations and standards for foods in general and for meat and fish in particular [106].

Critical points for the selection of proper analytical method and evaluation of data are sampling procedure, sample pre-treatment (saponification) and sensitivity of individual compound to light, oxygen or other factors. To ensure the high quality of obtained analytical data, it is necessary to provide the matrix specific validation, employing the quality control samples and certified reference materials as well as proficiency/ring tests.

Table 10.4. Qualitative and cost aspects of analytical methods used for vitamin and vitamers analysis [2].

Method	Data quality	Costs		
		Equipment	Chemicals	Infrastructure
Spectrophotometry, colorimetry	±	low	low	medium
ELISA	±	low	high	low
Microbiological assay	++	low	low	high
RIA	++	high	high	high
HPLC	++	high	medium	medium
LC-MS (/MS)	++	(very) high	(medium) high	very high

± sufficient quality; ++ high quality data

Taking all the limitations into account several methods of choice for vitamins analysis are available: spectroscopy, GC, titration, RIA, ELISA, microbiological assay, fluorometry, ESI-LC-MS (electron-spray ionization), biological method, LC-MS/MS, SIDA, enzymatic method, colorimetric method, capillary electrophoresis, IP-RP-HPLC-FLD and other. In general, selection of method must fit to purpose of analysis and is the compromise between the qualitative aspects and price of equipment, used reagents, reference materials and other infrastructure [2, 106, 107].

The most commonly applied analytical methods for vitamins and vitamers analysis from the point of the data quality, costs of equipment, chemicals and other indispensable infrastructure are summarized in Table 10.4. Standardized methods of analysis for individual vitamins are to be found in Table 10.5

Table 10.5. Standardised analytical methods applicable for analysis of vitamins and vitamers [2, 106-120].

Vitamin	Sensitivity	Analytical method	Reference standard
Fat-soluble vitamins			
Vitamin A and provitamin carotenoids	Heat, oxygen, light, acids	HPLC with UVD (FLD); spectrometry, colorimetry	EN 12823-1:2000 (Vit. A, HPLC) EN 12823-2:2000 (carotenoids, HPLC)
Vitamin E	Light, oxygen	HPLC with FLD (UV)	EN 12822:2000 (HPLC)
Vitamin D	Air, heat	HPLC with UVD, LC-MS, RIA	EN 12821:2000 (HPLC)
Vitamin K	Light, alkali, reducing agents	HPLC with FLD or ECD, LC-MS: APCI	EN 14148:2003 (HPLC)
Water-soluble vitamins			
Vitamin B_1	Heat, alkali, oxygen, radiation	HPLC with FLD, microbiological assay, fluorometry	EN 14122:2003 (HPLC)
Vitamin B_2	Light	HPLC with FLD, microbiological assay, fluorometry	EN 14152:2003 (HPLC)
Niacin	–	HPLC with FLD, microbiological assay	prEN 15652:2007 (HPLC)
Vitamin B_6	Light and cooking	HPLC with FLD, microbiological assay	EN 14166:2001 (microb. assay) EN 14164:2008 (HPLC) EN 14663:2005 (HPLC)
Vitamin B_{12}	Alkaline, light	Microbiological assay, biospecific methods	
Folic acid (folates)	Alkaline, heat, oxygen, light	HPLC with UVD and/or FLD microbiological assay	EN 14131:2003 (microb. assay) prEN 15607:2007
Biotin		Microbiological assay, radiolabeled protein-binding assay, HPLC	prEN 15607:2007 (HPLC)
Pantothenic acid	Leaching in cooking	Microbiological assay, (HPLC) LC-MS	Under discussion in CEN
Vitamin C	Alkaline, heat, oxygen, light labile, metal ions	HPLC with UVD or FLD, colorimetry/spectrophotometry, titrimetry	EN 14130:2003 (HPLC)

CONCLUSION

Meat and fish is an important and fundamental nutritional source of different vitamins and vitamers, playing thus outstanding role for human health in providing these inevitable essential nutrients, but can be used also for improvement of nutritional status of population via adjusted diet (meat and fish) composition. The knowledge on the sensitivity aspects as well as on the degradation routes of individual vitamins during meat and fish processing or sterilisation e.g., by γ-irradiation, is of crucial importance, especially in terms of well-balanced diet. Having information about relation/coherence between processing and vitamin degradation is of essential importance also for meat and fish producers. Experience in increasing nutritional value of these commodities by different methods of fortification (e.g. post mortem vs. feed fortification) is very important for regions with poor nutritional status, limited climate conditions or vitamin deficiency. Last but not least an increasing need of consumers for long-life foods is going against the nutritional retention and meat and fish quality. This issue actually emphasis necessity to apply results of experimental studies in food industry. As we know that the consumers′ needs are above the food producers′ interests, it is very demanding in present time to ensure food quality in line with user requirements considering all aspects of production process, i.e. good primary source, processing, food distribution and storage conditions.

LIST OF ABBREVIATIONS

ACP	Enzyme-bound 4'-phosphopantetheine
ADI	Acceptable daily intake
APCI	Atmospheric pressure chemical ionization
AECD	Electron capture detector
BHA	Butylated hydroxyanisole
BHT	Butylated hydroxytoluene
ESI	Electrospray ionization
FAD	Flavinadenin dinucleotide
FLD	Fluorescence detector
FMN	Flavin mononucleotide
GRAS	Generally recognised as safe

HS-CoA	Coenzyme
IAEA	International Atomic Energy Agency
JECFA	Joint FAO/WHO Expert Committee on Food Additives
PG	Propyl gallate
pr	Draft version of standard
RIA	Radioimmunoassay
SCF	European Community's Scientific Committee for Food
TBHQ	Tertiary butylhydroquinone
USDA	US Drug Administration
UVD	UV detector
WOF	Warmed-over flavour

REFERENCES

[1] Lieberman, S., & Bruning, N. (1990). *The Real Vitamin & Mineral Book.* 2nd edition. New York, USA: Avery Publishing Group.

[2] Greenfield, H., & Southgate, D.A.T. (2003). *Food Composition data: Production, Management and Use.* 2nd edition. Rome, Italy: Food and Agriculture Organisation (FAO). Available from: http://www.fao.org/infoods/publications_en.stm.

[3] Lešková, E., Kubíková, J., Kováčiková, E., Košická, M., Porubská, J., & Holčíková, K. (2006). Vitamin losses: Retention during heat treatment and continual changes expressed by mathematical models - a Review. *Journal of Food Composition and Analysis*, *19*, 252-276.

[4] Velíšek, J., & Cejpek, K. (2007). Biosynthesis of Food Constituents: Vitamins. 1. Fat-Soluble Vitamins - a Review. *Czech Journal of Food Science*, *25*, 1-16.

[5] Velíšek, J., & Cejpek, K. (2007). Biosynthesis of Food Constituents: Vitamins. 2. Water-Soluble Vitamins: Part 1 - a Review. *Czech Journal of Food Science*, *25*, 49-64.

[6] Schweigert, BS. (1987). The nutritional content and value of meat and meat products. In: Price JF Schweigert BS, editors. *The Science of Meat and Meat Products.* Westport, Connecticut, USA: Food and Nutrition Press, 275.

[7] Combs, G.F. (2008). *The Vitamins: Fundamental Aspects in Nutrition and Health.* 3rd edition. Boston, Massachusetts, USA: Elsevier Science; 2008.

[8] Velíšek, J., & Cejpek, K. (2007). Biosynthesis of Food Constituents: Vitamins. 2. Water-Soluble Vitamins: Part 2 - a Review. *Czech Journal of Food Science*, *25*, 101-118.

[9] Friedrich, W. (1988). *Vitamins*. 1st edition. Berlin, Germany: Walter de Gruyter.

[10] Eitenmiller, R.R., & Laden, W.O. (1999). Vitamin A and Carotenoids. Ascorbic acid: Vitamin C. Thiamin. Vitamin B-6. Folate. In: Eitenmiller RR, Laden WO, editors. *Vitamin Analysis for the Health and Food Science*. Boca Raton, Florida, USA: CRC Press, 3-66, 223-264, 271-293, 369-404, 411-459.

[11] Harris, RS. (1987). General discussion on the stability of nutrients. In: Karmas E, Harris RS, editors. *Nutritional Evaluation of Food Processing*. New York, USA: Van Nostrand Reinhold Company; 4.

[12] Kilcast, D. (1994). Effect of irradiation on vitamins. *Food Chemistry*, *49*, 157-164.

[13] Bhuiyan, A.K.M.A., Ratnayake, W.M.N., & Ackman, R.G. (1993). Nutritional composition of raw and smoked Atlantic mackerel (*Scomber scombrus*): oil and water-soluble vitamins. *Journal of Food Composition and Analysis*, *6*, 172-184.

[14] Leonhardt, M., Gebert, S., & Wenk, C. (1996). Stability of α-tocopherol, thiamin, riboflavin and retinol in pork muscle and liver during heating as affected by dietary supplementation. *Journal of Food Science*, *61*, 1048-1067.

[15] Dal Bosco, A., Castellini, C., & Bernardini, M. (2001). Nutritional quality of rabbit meat as affected by cooking procedure and dietary vitamin E. *Journal of Food Science*, *66*, 1047-1051.

[16] Elder, S.J., Haytowitz, D.B., Howe, J., Peterson, J.W., & Booth, S.L. (2006). Vitamin K contents of meat, dairy, and fast food in the U.S. Diet. *Journal of Agricultural and Food Chemistry*, *54*, 463-467.

[17] Descalzo, A.M., & Sancho, A.M. (2008). A review of natural antioxidants and their effects on oxidative status, odor and quality of fresh beef produced in Argentina. *Meat Science*, *79*, 423-436.

[18] Descalzo, A.M., Insani, E.M., Biolatto, A., Sancho, A.M., Garcia, P.T., & Pensel, N.A. (2005). Influence of pasture or grain-based diets supplemented with vitamin E on antioxidant/oxidative balance of Argentine beef. *Meat Science*, *70*, 35-44.

[19] Insani, E.M., Eyherabide, A., Grigioni, G., Sancho, A.M., Pensel, N.A., & Descalzo, A.M. (2008). Oxidative stability and its relationship with natural antioxidants during refrigerated retail display of beef produced in Argentina. *Meat Science*, *79*, 444-452.

[20] Realini, C.E., Duckett, S.K., Brito, Q.W., Dalla Rizza, M., & De Mattos, D. (2004). Effect of pasture vs. concentrate feeding with or without antioxidants on carcass characteristics, fatty acid composition, and quality of Uruguayan beef. *Meat Science*, *66*, 567-577.

[21] Morrissey, P.A., Sheehy, P.J.A., Galvin, K., Kerry, J.P., & Buckley, D.J. (1998). Lipid stability in meat and meat products. *Meat Science*, *49*, S73-S86.

[22] Arnold, R.N., Arp, S.C., Scheller, K.K., Williams, S.N., & Schaefer, D.M. (1993). Tissue equilibration and subcellular distribution of vitamin E relative to myoglobin and lipid oxidation in displayed beef. *Journal of Animal Science*, *71*, 105-118.

[23] Eikelenboom, G., Hoving-Bolink, A.H., Kluitman, I., Houben, J.H., & Klont, R.E. (2000). Effect of dietary vitamin E supplementation on beef colour stability. *Meat Science*, *54*, 17-22.

[24] O'Grady, M.N., Monahan, F.J., Burke, R.M., & Allen, P. (2000). The effect of oxygen level and exogenous alpha tocopherol on the oxidative stability of minced beef in modified atmosphere packs. *Meat Science*, *55*, 39-45.

[25] Okayama, T. (1987). Effect of modified gas atmosphere after dip treatment on myoglobin and lipid oxidation of beef steaks. *Meat Science*, *19*, 179-185.

[26] Driskell, J.A., Giraud, D.W., Sun, J., Joo, S., Hamouz, F.L., & Davis, S.L. (1998). Retention of vitamin B6, thiamin, vitamin E and selenium in grilled boneless pork chops prepared at five grill temperatures. *Journal of Food Quality*, *21*, 201-210.

[27] Olivo, R., Soares, A.L., Ida, E.I., & Shimokomaki, M. (2001). Dietary Vitamin E inhibits poultry PSE and improves meat functional properties. *Journal of Food Biochemistry*, *25*, 271-283.

[28] Barretto, A.C.S., Ida, E.I., Silva, R.S.F., Torres, E.A.F.S., & Shimokomaki, M. (2003). Empirical models for describing poultry meat lipid oxidation inhibition by natural antioxidants. *Journal of Food Composition and Analysis*, *16*, 587-594.

[29] Mitsumoto, M., Arnold, R.N., Schaefer, D.M., & Cassens, R.G. (1993). Dietary versus postmortem supplementation of vitamin E on pigment and lipid stability in ground beef. *Journal of Animal Science*, *71*, 1812-1816.

[30] Ashgar, A., Gray, J.I., Booren, A.M., Gomaa, E.A., Abouzied, M.M., & Miller, E.R. (1991). Effects of supranutritional dietary vitamin E levels on subcellular deposition of α-tocopherol in the muscle and on pork quality. *Journal of the Science of Food and Agriculture*, *57*, 31-41.

[31] Monahan, F.J., Buckley, D.J., Morrissey, P.A., Lynch, P.B., & Gray, J.I. (1990). Effect of dietary α-tocopherol supplementation on tocopherol levels in porcine tissues and on susceptibility to lipid peroxidation. *Food Science and Nutrition*, *42*, 203-212.

[32] Gray, J.I., Gomaa, E.A., & Buckley, D.J. (1996). Oxidative quality and shelf life of meats. *Meat Science*, *43*, 111-123.

[33] Faustman, C., Cassens, R.G., Schaefer, D.M., Beuge, D.R., Williams, S.N., & Scheller, K.K. (1989). Improvement of pigment and lipid stability in Holstein steer beef by dietary supplementation with vitamin E. *Journal of Food Science*, *54*, 858-862.

[34] Jeremiah, L.E., Carpenter, Z.L., & Smith, G.C. (1972). Beef color as related to consumer acceptance and palatability. *Journal of Food Science*, *37*, 476-479.

[35] Faustman, C., & Cassens, R.G. (1990).The biochemical basis for discoloration in fresh meat: a review. *Journal of Muscle Foods*, *1*, 217-243.

[36] O'Grady, M.N., Monahan, F.J., Bailey, J., Allen, P., Buckley, D.J., & Keane, M.G. (1998). Colour-stabilising effect of muscle vitamin E in minced beef stored in high oxygen packs. *Meat Science*, *50*, 73-80.

[37] Gandemer, G. (1997). Lipides du muscle et qualite de la viande. Phospholipides et flaveur. *Oleagineux, Corps Gras et Lipides*, *4*, 19-25.

[38] Anton, M., Salgues, C., Gatellier, P., & Renerre, M. (1993). Etude des relations oxydatives entre les lipides membranaires et la myoglobine in vitro. *Science des Aliments*, *13*, 261-274.

[39] Cuvelier, M.E., Berset, C., & Richard, H. (1994). Antioxidant constituents in sage (*Salvia officinalis*). *Journal of Agricultural and Food Chemistry*, *42*, 665-669.

[40] Giese, J. (1996). Antioxidants: tools for preventing lipid oxidation. *Food Technology*, *50*, 73-80.

[41] Branen, A.L. (1975). Toxicology and biochemistry of butylated hydroxyanisole and butylated hydroxytoluene. *Journal of the American Oil Chemists Society*, *52*, 59-63.

[42] Mielche, M.M., & Bertelsen, G. (1994). Approaches to the prevention of warmed-over flavour. *Trends in Food Science and Technology*, *5*, 322-327.

[43] Djenane, D., Sánchez-Escalante, A., Beltrán, J.A., & Roncalés, P. (2002). Ability of α-tocopherol, taurine and rosemary, in combination with vitamin C, to increase the oxidative stability of beef steaks packaged in modified atmosphere. *Food Chemistry*, *76*, 407-415.

[44] McCarthy, T.L., Kerry, J.P., Kerry, J.F., Lynch, P.B., & Buckley, D.J. (2001). Evaluation of the antioxidant potential of natural food/plant extracts as compared with synthetic antioxidants and vitamin E in raw and cooked pork patties. *Meat Science*, *57*, 45-52.

[45] Jayathilakan, K., Sharma, G.K., Radhakrishna, K., & Bawa, A.S. (2007). Antioxidant potential of synthetic and natural antioxidants and its effect on warmed-over-flavour in different species of meat. *Food Chemistry*, *105*, 908-916.

[46] Wheeler, T.L., Koohmaraie, M., & Shackelford, S.D. (1996). Effect of vitamin C concentration and co-injection with calcium chloride on beef retail display color. *Journal of Animal Science*, *74*, 1846-1853.

[47] Decker, E.A., & Xu, Z. (1998). Minimizing rancidity in muscle foods. *Food Technology*, *52*, 54-59.

[48] Kinsella, J.E., Frankel, E., German, B., & Kanner, J. (1993). Possible mechanisms for the protective role of antioxidants in wine and plant foods. *Food Technology*, *47*, 85-89.

[49] Okayama, T., Imai, T., & Yamanoue, M. (1987). Effect of ascorbic acid and α-tocopherol on storage stability of beef steaks. *Meat Science*, *21*, 267-273.

[50] Mitsumoto, M., Faustman, C., Cassens, R.G., Arnold, R.N., Schaefer, D.M., & Scheller, K.K. (1991). Vitamins E and C improve pigment and lipid stability in ground beef. *Journal of Food Science*, *56*, 194-197.

[51] Realini, C.E., Duckett, S.K., & Windham, W.R. (2004). Effect of vitamin C addition to ground beef from grass-fed or grain-fed sources on color and lipid stability, and prediction of fatty acid composition by near-infrared reflectance analysis. *Meat Science*, *68*, 35-43.

[52] Kanatt, S.R., Chander, R., & Sharma, A. (2008). Chitosan glucose complex - A novel food preservative. *Food Chemistry*, *106*, 521-528.

[53] Butler, A.J., & Larick, D.K. (1993). Effect of antioxidants on the sensory characteristics and storage stability of aseptically processed low-fat beef gels. *Meat Science*, *35*, 355-369.

[54] Formanek, Z., Kerry, J.P., Higgins, F.M., Buckley, D.J. Morrissey, P.A., & Farkas, J. (2001). Addition of synthetic and natural antioxidants to α-tocopheryl acetate supplemented beef patties: effects of antioxidants and packaging on lipid oxidation. *Meat Science*, *58*, 337-341.

[55] Davidson, P.M. (2000). Antimicrobial compounds. In: Francis FJ, editor. *Wiley Encyclopedia of Food Science and Technology*. New York, USA: John Wiley & Sons, 63-75.

[56] Yanishlieva-Maslarova, N.V. (2001). Inhibiting oxidation. In: Pokorny J, Yanishlieva N, Gordon M, editors. *Antioxidants in foods - Practical applications*. Cambridge, England, UK: Woodhead Publishing, 20-70.
[57] Morris, A., Barnett, A., & Burrows, O.J. (2004). Effect of processing on nutrient content of foods. *Cajanus*, *37*, 160-164.
[58] Jadhav, S.J., Nimbalkar, S.S., Kulkarni, A.D., & Madhavi, D.L. (1996). Lipid oxidation in biological and food systems. In: Madhavi DL, Deshpande SS, Salunkhe DK, editors. *Food Antioxidants*. New York, USA: Marcel Dekker, 5-63.
[59] Madhavi, D.L., Singhal, R.S., & Kulkarni, P.R. (1996). Toxicological aspects of food antioxidants. In: Madhavi DL, Deshpande SS, Salunkhe DK, editors. *Food antioxidants. Technological, toxicological and health perspective.* New York, USA: Marcel Dekker, 159-266.
[60] Miková, K. (2001). The regulation of antioxidants in food. In: Pokorny J, Yanishlieva N, Gordon M, editors. *Antioxidants in foods - Practical applications*. Cambridge, England, UK: Woodhead Publishing, 268-284.
[61] Code of Federal Regulations. (2003) Animals and animal products. USDA, Title 9, Volume 2, Revised as of January 1. Available from: http://www.cfsan.fda.gov/~lrd/9CF424.html.
[62] Directive 1999/3/EC. (1995). The European Parliament and Council Directive 95/2/EC of 20 February 1995 on food additives other than colours and sweeteners, 1-53.
[63] Bognár, A. (2002). Tables on weight yield of food and retention factors of food constituents for the calculation of nutrient composition of cooked foods (dishes). 1st edition. Karlsruhe, Germany: Berichte der Bundesforschungsanstalt für Ernährung.
[64] Rhee, K.S., Griffith-Bradle, H.A., & Ziprin, Y.A. (1993). Nutrient composition and retention in browned ground beef, lamb and pork. *Journal of Food Composition and Analysis*, *6*, 268-277.
[65] Maskova, E., Rysova, J., Fiedlerova, V., & Holasova, M. (1994). Vitamin and mineral retention in meat in various cooking methods. *Potravinarske vedy*, *12*, 407-416.
[66] Al-Khalifa, A.S., & Dawood, A.A. (1993). Effects of cooking methods on thiamin and riboflavin contents of chicken meat. *Food Chemistry*, *48*, 69-74.
[67] Showell, B.A., Howe, J.C., & Buege, D.R. (2003). Cooking yields and nutrient retention factors of bacon, liver, and sausages. *Poster, 5th International Food Data Conference and the 27th US National Nutrient Databank Conference*, Washington, DC, USA.
[68] Lassen, A., Kall, M., Hansen, K., & Ovesen, L. (2002). A comparison of the retention of vitamins B1, B2 and B6, and cooking yield in pork loin with conventional and enhanced meal-service systems. *European Food Research and Technology*, *215*, 194-199.
[69] Oseredczuk, M., Du Chaffaut, L., Ireland, J., & Collet-Ribbing, Ch. (2003). Effect of preservation and transformation processes on the composition of fishes. *Poster, 5th International Food Data Conference and the 27th US National Nutrient Databank Conference*, Washington, DC, USA.
[70] Bourgeois, CF. (2003). *Les Vitamines dans les Industries Agroalimentaires*. 1st ed. Paris, France: Tec & Doc Lavoisier.
[71] Fillion, L., & Henry, C.J.K. (1998). Nutrient losses and gains during frying: a review. *International Journal of Food Sciences and Nutrition*, *49*, 157-168.

[72] Uherová, R., Hozová, B., & Smirnov, V. (1993). The effect of microwave heating on retention of some B vitamins. *Food Chemistry*, *46*, 293-295.

[73] Ball, G.F.M. (1998). Folate. In: Chapman & Hall, editors. *Bioavailability and Analysis of Vitamins in Foods*. London, UK: Jones & Bartlett, 439-496.

[74] McKillop, D.J., Pentieva, K., Daly, D., McPartlin, J.M., Hughes, J., Strain, J.J., Scott, J.M., & McNulty, H. (2002). The effect of different cooking methods on folate retention in various foods that are amongst the major contributors to folate intake in the UK diet. *British Journal of Nutrition*, *88*, 681-688.

[75] Aramouni, F.M., & Godber, J.S. (1991). Folate losses in beef liver due to cooking and frozen storage. *Journal of Food Quality*, *14*, 357-365.

[76] Vahteristo, L.T., Lehikoinen, K.E., Ollilainen, V., Koivistoinen, P.E., & Varo, P. (1998). Oven-baking and frozen storage affect folate vitamer retention. *Lebensmittel Wissenschaft und Technology*, *31*, 329-333.

[77] Keagy, P.M. (1985). Folacin: Microbiological and animal assays. In: Augustin J, Klein BP, Becker D, Venugopal PB, editors. *Methods of Vitamin Assay*. New York, USA: John Wiley and Sons, 445-471.

[78] Gregory, J.F. (1989). Chemical and nutritional aspects of folate research: Analytical procedures, methods of folate synthesis, stability and bioavailability of dietary folates. *Advances in Food and Nutrition Research*, *33*, 1-101.

[79] Hawkes, J.G., & Villota, R. (1989). Folates in food: reactivity, stability during processing and nutritional implications. *Critical Reviews in Food Science and Nutrition*, *28*, 439-538.

[80] Burt, J. (1988). Fish Smoking and Drying: The Effect of Smoking and Drying on the Nutritional Properties of Fish. 1st edition. New York: Elsevier Science Publishers.

[81] Morrison, N., & Kuhnlein, H.V. (1993). Retinol content of wild foods consumed by the Sahtu (Hareskin) Dene/Metis. *Journal of Food Composition and Analysis*, *6*, 10-23.

[82] Booth, S.L., Johns, T.A., & Kuhnlein, H.V. (1997). Part I. Vitamin A in food and diets: 2. The complexities of understanding Vitamin A in food and diets: The problem. In: Kuhnlein HV, Pelto GH, editors. *Culture, Environment, and Food to Prevent Vitamin A Deficiency*. Boston, USA: International Nutrition Foundation for Developing Countries, United Nations University, 19.

[83] Sungpuag, P., Tangchitpianvit, S., Chittchang, U., & Wasantwisut, U. (1999). Retinol and β-carotene content of indigenous raw and home - prepared foods in Northeast Thailand. *Food Chemistry*, *64*, 163-167.

[84] Mattila, P., Ronkainen, R., Lehikoinen, K., & Piironen, V. (1999). Effect of household cooking on the vitamin D content in fish, eggs, and wild mushrooms. *Journal of Food Composition and Analysis*, *12*, 153-160.

[85] Scott, K.C., & Latshaw, J.D. (1991). Effects of commercial processing on the fat-soluble vitamin content of menhaden fish oil. *Journal of the American Oil Chemists'Society*, *68*, 234-236.

[86] Clausen, I., Jakobsen, J., Leth, T., & Ovesen, L. (2003). Vitamin D3 and 25-hydroxyvitamin D3 in raw and cooked pork cuts. *Journal of Food Composition and Analysis*, *16*, 575-585.

[87] Stevenson, M.H. (1994). Nutritional and other implications of irradiating meat. *Proceedings of the Nutrition Society*, *53*, 317-325.

[88] Molins, R.A. (2001). Irradiation of meats and poultry. In: Molins RA, editor. *Food irradiation: Principles and applications*. New York, USA: John Willey & Sons, 131-135.

[89] Kilgen, M.B. (2001). Irradiation of fish and shellfish products. In: Molins RA, editor. *Food irradiation: Principles and applications*. New York, USA: John Willey & Sons, 193-212.

[90] Directive 1999/3/EC. (1999). The European Parliament and of the Council of 22 February 1999 on the establishment of a Community list of foods and food ingredients treated with ionising radiation. *Official Journal of the European Communities*, L 66:24-L 66:25.

[91] Singh, H., Lacroix, M., & Gagnon, M. (1991). Post-irradiation chemical analyses of poultry: a review. *Literature Review done for Health and Welfare Canada*.

[92] Stewart, E.M. (2001). Food irradiation chemistry. In: Molins RA, editor. *Food irradiation: Principles and applications*. New York, USA: John Willey & Sons, 37-76.

[93] Janave, M.T., & Thomas, P. (1979). Influence of post-harvest storage temperature and gamma irradiation on potato carotenoids. *Potato Research*, *22*, 365-369.

[94] Ziporin, Z.Z., Kraybill, H.F., & Thach, H.J. (1957). Vitamin content of foods exposed to ionizing radiation. *Journal of Nutrition*, *63*, 201-209.

[95] Wilson, G.M. (1959). The treatment of meats with ionizing radiations. 2. Observations on the destruction of thiamine. *Journal of the Science of Food and Agriculture*, *10*, 295-300.

[96] Graham, W.D., Stevenson, M.H., & Stewart, E.M. (1998). Effect of irradiation dose and irradiation temperature on the thiamin content of raw and cooked chicken breast meat. *Journal of the Science in Food and Agriculture*, *78*, 559-564.

[97] Hanis, T., Jelen, P., Klir, P., Mnukova, J., Perez, B., & Pesek, M. (1989). Poultry meat irradiation. Effect of temperature on chemical changes and inactivation of microorganisms. *Journal of Food Protection*, *52*, 26-29.

[98] Fox, J.B., Thayer, D.W., Jenkins, R.K., Phillips, J.G., Ackerman, S.A., Beecher, G.R., Holden, J.M., Morrow, F.D., & Quirbach, D.M. (1989). Effect of gamma irradiation on the B vitamins of pork chops and chicken breasts. *International Journal of Radiation Biology*, *55*, 689-703.

[99] Gallien, C.L., Paquin, J., Ferradini, C., & Sadat, T. (1985). Electron beam processing in food industry-technology and costs. *Radiation Physics and Chemistry*, *25*, 81-96.

[100] Richardson, L.R., Wilkes, S., & Ritchey, S.J. (1961). Comparative vitamin B6 activity of frozen, irradiated and heat-processed foods. *Journal of Nutrition*, *73*, 363-368.

[101] Kennedy, T.S. (1965). Studies on the nutritional value of foods treated with gamma irradiation. 1. Effects on some B-complex vitamins in egg and wheat. *Journal of Food Science and Agriculture*, *16*, 81-84.

[102] Diehl, J.F., Hasselmann, C., & Kilcast, D. (1991). Regulation of food irradiation, in the European Community: is nutrition an issue? *Food Control*, *2*, 212-219.

[103] Thayer, D.W., Fox, J.B., & Lakritz, L. (1991). Effects of ionizing radiation on vitamins. In: Thorne S, editor. *Food Irradiation*. London, UK: Elsevier Applied Science, 285-325.

[104] Josephson, E.S., Thomas, M.H., & Calhoun, W.K. (1978). Nutritional aspects of food irradiation: an overview. *Journal of Food Processing and Preservation*, *2*, 299-313.

[105] De Groot, A.P., van der Mijll, D., Slump, P., Vos, H.J., & Willems, J.L. (1972). Composition and nutritive value of radiation-pasteurized chicken. *Research report no. R-3787.* Central Institute for Nutrition and Food Research.
[106] Herrero, A.M. (2008). Raman spectroscopy a promising technique for quality assessment of meat and fish: A review. *Food Chemistry*, *107*, 1642-1651.
[107] Blake, C.J. (2007). Analytical procedures for water-soluble vitamins in foods and dietary supplements: a review. *Analytical and Bioanalytical Chemistry*, *389*, 63-76.
[108] EN 12821:2000 Foodstuffs - Determination of vitamin D by high performance liquid chromatography - Measurement of cholecalciferol (D3) and ergocalciferol (D2).
[109] EN 12822:2000 Foodstuffs - Determination of vitamin E by high performance liquid chromatography - Measurement of alpha-, beta-, gamma-, and delta-tocopherols.
[110] EN 12823-1:2000 Foodstuffs - Determination of vitamin A by high performance liquid chromatography - Part 1: Measurements of all-trans-retinol and 13-cis-retinol.
[111] EN 12823-2:2000 Foodstuffs - Determination of vitamin A by high performance liquid chromatography - Part 2: Measurements of Beta-carotene.
[112] EN 14122:2003 Foodstuffs - Determination of vitamin B1 by HPLC.
[113] EN 14122:2003/AC:2005 Foodstuffs - Determination of vitamin B1 by HPLC.
[114] EN 14130:2003 Foodstuffs - Determination of vitamin C by HPLC.
[115] EN 14131:2003 Foodstuffs - Determination of folate by microbiological assay.
[116] EN 14148:2003 Foodstuffs - Determination of vitamin K1 by HPLC.
[117] EN 14152:2003 Foodstuffs - Determination of vitamin B2 by HPLC.
[118] EN 14152:2003/AC:2005 Foodstuffs - Determination of vitamin B2 by HPLC.
[119] EN 14164:2008 Foodstuffs - Determination of vitamin B6 by HPLC.
[120] EN 14663:2005 Foodstuffs - Determination of vitamin B6 (including its glycosylated forms) by HPLC.

PART III

MILK AND DAIRY PRODUCTS

In: Practical Food and Research
Editor: Rui M. S. Cruz, pp. 283-297

ISBN: 978-1-61728-506-6

Chapter XI

COLOUR AND PIGMENTS

***Gabriela Grigioni*[1,2,3]*, Andrea Biolatto*[4]*, Leandro Langman*[1,3]*, Adriana Descalzo*[1,3]*, Martín Irurueta*[1]*, Roxana Páez*[5] *and Miguel Taverna*[5]**

[1]Instituto Tecnología de Alimentos, Centro de Investigación de Agroindustria, Instituto Nacional de Tecnología Agropecuaria INTA, CC 77 (B1708WAB) Morón, Buenos Aires, Argentina, ggrigioni@cnia.inta.gov.ar

[2]Consejo Nacional de Investigaciones Científicas y Técnicas CONICET, Av. Rivadavia 1917 (C1053AAY) Buenos Aires, Argentina

[3]Facultad de Agronomía y Ciencias Agroalimentarias, Universidad de Morón, Cabildo 134 (B1708JPD) Morón, Argentina

[4]EEA Concepción del Uruguay, Instituto Nacional de Tecnología Agropecuaria INTA Ruta Provincial 39 Km 143,5 (3260) Concepción del Uruguay, Entre Ríos, Argentina

[5]EEA Rafaela, Instituto Nacional de Tecnología Agropecuaria INTA, Ruta Nacional 34 Km 27, CC 22 (2300), Rafaela, Santa Fé, Argentina

ABSTRACT

This chapter surveyed different aspects related to milk and dairy colour characteristics.

Appearance of dairy products is a complex topic involving many factors related to primary and transformation processing and storage conditions. As an aspect of the appearance of food, consumers are sensitive to product colour. In this context, colour measurement has become a useful tool for quality product and process management in the dairy industry.

Several dietary factors have been identified as being responsible for the characteristics of the raw milk obtained. Milk contains different amounts of carotenoids that contribute to the nutritional and sensory properties of dairy products. Therefore, carotenoids are relevant in determining the colour of dairy products.

In processed milk, raw milk characteristics are overlapped with processing parameters affecting colour stability as consequences of several reactions that occurred during the transformation process. In dehydrated diary products, the storage conditions determine the stage reached by the Maillard reaction and, therefore, the induced variations in colour.

Emerging processing technology in food production is a growing field. Within these technologies, high pressure is the most developed and different types of devices are commercially available. Even though, high pressure technology is still not widely adopted by the industry sector. Dairy products treated with high pressure exhibit changes in colour characteristics that depend on both the pressure exerted and the time of exposure.

11.1 INTRODUCTION

Colour is one of the attributes that affect consumer perception of quality. As well as flavour and texture, they are considered to be major attributes that contribute to the overall quality products. Hence, in the food industry, the assessment of the colour has become an important part of quality product and process management [1].

In some foods, colour is the first criterion to be perceived by the consumer. As stated by Burrows [1], the repeated recognition of a particular brand of a food commodity largely depends on its typical colour.

As an aspect of the appearance of food, consumers are sensitive to product colour. Even though, preferences differ among and within countries. In many European countries, a yellow colour in milk is associated with pasture, bringing connotations of "natural" feeding [2]. In contrast, it is considered negative for certain colour-sensitive markets of the Middle-East.

Food colour is the result of natural products associated with the raw material from which it is processed and/or coloured compounds generated as a result of processing [3]. It is influenced by how the food matrix interacts with light, regarding as its reflecting, absorbing or transmitting characteristics, which in turn is related to its physical structure and chemical nature [4].

Colour measurements to characterize dairy products have been employed by Rhim et al. [5], Pagliarini et al. [6], Kneifel et al. [7], Nielsen et al. [8, 9], Celestino et al. [10, 11], Morales and van Boekel [3], Priolo et al. [12] and Grigioni et al. [13] to cite few examples.

In the following sections several aspects of colour characteristics in milk and dairy product are screened. In the first one, some basic definitions are given. In the second section, the relevance of colour pigment is presented. In the third, the colour stability due to processing by conventional and emerging technology is discussed. Finally, some conclusions are pointed out.

11.2 MEASUREMENT TECHNIQUES

The procedures used to describe colour are based on the specification of the three stimuli. Due to the phenomenon of trichromacy, any colour stimulus can be matched by a mixture of three primary stimuli in adequate amounts [14]. This involves a process of integration. Clarity, tone and saturation can be discriminated by an observer when seeing a colour. But in contrast, the observer can not detail the spectral composition of the stimulus.

A colorimeter is used to evaluate in physical terms psychological feelings.When a colour is described, the observer usually refers to attributes of chromatic sensation as hue, lightness and saturation. In colorimetry these three aspects are considered psychological correlates of the physical dimensions of the stimulus.

Besides the reflectance spectrum, each colour can be identified by certain independent coordinates. Using these coordinates it is possible to build colour-spaces where each colour is represented by a point in that space.

CIE (Commission Internationale de l'Eclairage) is devoted to the world wide cooperation and the exchange of information relating to the science and art of light and lighting, colour and vision, photobiology and image technology (http://www.cie.co.at). CIE derived the most used systems for colour determination which are based on the use of standard illuminate and observer.

Among the several existing colour scales (Hunter Lab, XYZ system, etc.) CIE recommended the CIELAB colour space that is a three-dimensional spherical system defined by three colorimetric coordinates. The coordinate L* is called the lightness. The coordinates a* and b* form a plane perpendicular to the lightness. The coordinate a* defines the deviation from the achromatic point corresponding to lightness, to red when it is positive and toward the green if negative. Similarly, the coordinate b* defines the turning to yellow if positive and to blue if negative [15].

11.3 Important Pigments in Milk and Dairy Products

Several dietary factors have been identified as being responsible for the obtained raw milk characteristics. These factors include those associated with the diet fed to animals. Among these, the nature and stage of maturity of forage, the pasture system in use, the supplements given, the adaptation periods and the energy balance have special significance.

Carotenoids are a family of more than 600 molecules that are synthesized by higher plants and algae. They form the main group of natural pigments and are natural pigment precursors of the yellow to red colour range in vegetal and animal tissues. Plant carotenoids are transferred into animal products. As stated by Noziére et al. [16], carotenoids are involved in the nutritional and sensory characteristics of dairy products, either indirectly through their antioxidant properties or directly through their yellowing properties. Several articles in the literature considered their potential as biomarkers for traceability of products associated to feeding conditions. As a result, the colour of dairy products highly depends on their carotenoid concentration.

Forages represent the main source of carotenoids for ruminants, where they develop several functions including provitamin A function, antioxidant function, cell communication, enhancement of immune function, and UV skin and macula protection. Nearly 10 carotenoids have been identified in forages: lutein, epilutein, antheraxanthin, zeaxanthin, neoxanthin and violaxanthin for xanthophylls, all-trans ß-carotene, 13-cis ß-carotene and α-carotene [17], being the most quantitatively important ß-carotene and lutein. Differences in the numbers of carotenoids described in forages could arise from the variety of molecules in natural grasslands [16].

In cows' milk, carotenoids principally consist of all-trans- ß-carotene and, to a lesser extent, lutein, zeaxanthin, ß-cryptoxanthin [16]. Since the amount of ß-carotene deposited in adipose tissue and/or secreted in milk fat varies widely according to the carotenoid content in the feed, it plays a key role in the sensorial and nutritional value of dairy products.

Carotenoids are found in higher concentrations in milk produced through grass-based diets, specially pasture. In grazing systems, a change in carotenoids in milk in the course of time may depend on both the amount of carotenoid intake and milk yield. [17]. In this context, diets based on grass, mainly pastures, lead to a higher concentration of β-carotene in milk as compared to diets rich in corn silage or concentrates [18, 19], since processing greatly reduces their concentration [20, 21].

11.3.1 Milk colour

As reviewed by Chatelain et al. [22], milk colour characterization is mainly applied to identify technological parameters such as homogenisation, thermal treatment (including Maillard reactions), fat concentration, photo-degradation, storage conditions or additives.

The white appearance of milk is the result of its physical structure. The casein micelles and fat globules disperse the incident light and, consequently, milk exhibit a high value of parameter L* (lightness). Technological treatments that influence the physical structure of milk also have an effect on L*. The other colour components (parameters a* and b*) are influenced by factors related to natural pigment concentration of milk.

Several studies concerning the instrumental measurement of colour in milk and dairy products have been conducted. Some of them [13, 23] focused on the evaluation of the effect of the milking season, obtaining a seasonal effect that was evidenced by the variations observed in the levels of L* (lightness) and b* (blue-yellow component) colour parameters.

Through a study by Prache and Theriez [24], based on the spectrophotometric properties of carotenoids accumulated in sheep milk and plasma, the effect of diet on the concentration of β-carotene was shown. Thus, it was possible to differentiate milk obtained from animals fed on diets containing different levels of carotenoids.

As compared to diets consisting of silage, alfalfa-based diets provide a substantially higher supply of β-carotene, an effect that is entirely attributed to the contribution of alfalfa pasture. Among carotenoid pigments, the β-carotene and lutein provide the yellow colour. On this account, carotenoids could be used as indicators of pasture production systems [24].

Langman et al. [25] reported changes in the colour of raw milk in a trial which compared alfalfa- and silage-based diets fed to Holstein milk cows.

Briefly, the experiment was conducted during spring (October to December) at the National Institute of Agricultural Technology in Rafaela (province of Santa Fe, Argentina: 31°11'S; 61°30'W). During a first four-week pre-experimental period, ten Holstein cows were fed on a silage-only diet with at least 50% of forage; this diet also contained soy expeller and sunflower pellets (3.5 and 1.1 kg/day per cow, respectively) and hay (1.5 kg/day per cow). Thereafter, five cows were randomly assigned to an alfalfa diet (at least 60% alfalfa of dry matter on dietary basis) while another group remained as control, during 60 experimental days. The evaluation of the colour was carried out using a reflectance spectrophotometer (BYK Gardner Colour View model 9000) according to CIELab scale. The instrumental settings were large port area (5 cm diameter) and D65-artifitial daylight.

A significant increase of the β-carotene content was observed after 20 days, a tendency that was maintained at least up to 60 days after the change of diet (Table 11.1), with dissimilar values in the range of 5-6 μg/g of fat (in milk obtained from animals fed on alfalfa) vs. 1-1.4 μg/g of fat (in milk obtained from animals fed on silage). These data are consistent with the results reported by Calderón et al. [2] where similar values of β-carotene concentration were observed in milk obtained from Montbéliarde dairy cows fed on diets rich in carotenoids (67% on dry basis of grass-based silage).

Table 11.1- Effect of day of pasture in raw milk β-carotene content. Mean plus standard deviation.

	β-carotene (μg/g milk fat)			
	Days			
	0	20	40	60
Silage diet	2.4±0.4	1.1±0.2	0.9±0.3	1.4±0.4
Alfalfa diet	2.9±1.1	5.8±0.9	5.0±0.8	5.6±0.9

As regards the study of the b* colour component, raw milk corresponding to the alfalfa-based diet showed significant differences only 60 days after implementing the diet. These samples presented higher b* values, which indicates a more yellow colour.

In order to summarize spectral results obtained from milk samples, several authors use indexes that involve the study of the main pigments of milk. The Integral Value (IV), as proposed by Prache and Theriez [24], is a widely used index that allows characterizing milk according to the diet fed to animals based on the study of spectra in the wavelength range between 450 and 530 nm, corresponding to carotenoids absorption.

This index allowed a clear differentiation between milk obtained from cows fed on silage and those fed on alfalfa pasture 20 days after changing their diet. In contrast, the differentiation of milk obtained from animals fed on different types of silage – when comparing, for example, sorghum-based and corn-based diets- was not possible using IV. Figure 11.1 illustrates percentage distributions after applying defined ranges. In the range of integral values between 450 and 550, milk obtained from silage-fed cows accounted for 100% of the cases. Also, milk obtained from fresh alfalfa-fed cows accounted for 100% of the cases in the range between 651 and 850. The only range of IV values within which there was an overlapping between milk obtained from both diets was that between 551 and 650.

Thus, it was observed that milk samples obtained from silage-fed cows presented lower IV (450-550), while those from the diet based on alfalfa pasture presented higher values (651-850).

The results of this research match those obtained by Prache and Theriez [24]. These authors established that the IV was useful to differentiate milk obtained from animals fed on diets with various carotenoid levels. In this case, the IV index proved to be useful to distinguish between milk from a hay- and concentrate-based diet (low in carotenoids) and milk from a pasture-based diet (rich in carotenoids) 36 days after implementing the diet. In contrast, after replacing a diet rich in carotenoids (grass-based silage) with a diet low in carotenoids (hay-based diet), both b* component and IV of milk did not allow to differentiate milk, even after 50 days. In addition, a significant correlation of the IV was established with milk β-carotene content. Thus, this index also allowed differentiating milk rich in carotenoids, such as β-carotene, from milk with a low carotenoid content.

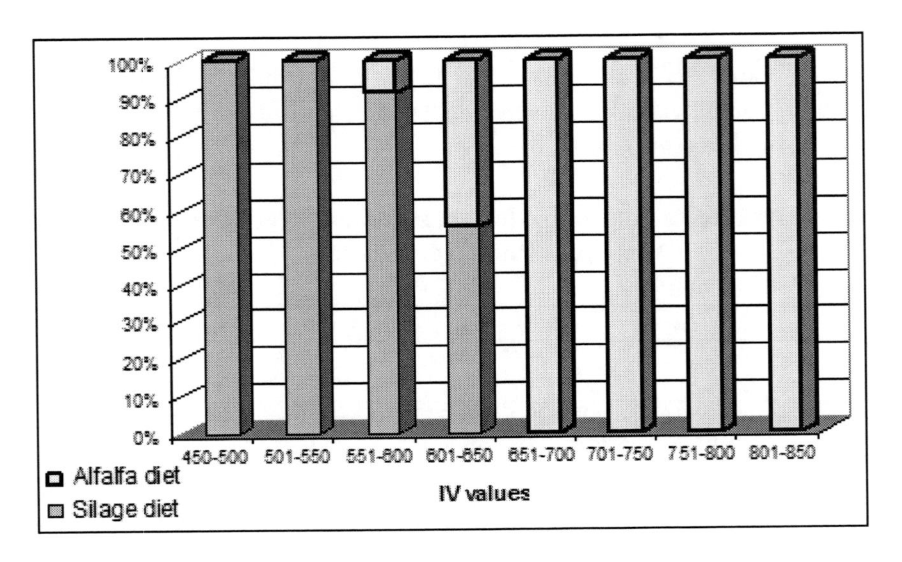

Figure 11.1. Cumulative relative histogram of Integral Value (IV) determined in raw milk under different feeding systems.

Integral value emerges like a promising tool in feed system traceability. Even though, more research is needed in order to explore and explain its response under different feeding conditions.

11.3.2 Dairy products colour

Milk carotenoids are transferred to butter and cheese with minimal losses and hence contribute to their yellow colour. Since carotenoids and retinol are soluble in fat, they mainly behave as milk fat. Nevertheless, a small proportion of retinol and carotenoids are related with whey protein and/or concentrated in the membrane of fat globules in milk. As a result, a certain amount of these micronutrients may be lost to whey during cheese and butter processing [16].

In addition, process-related factors are involved in the colour of butter and cheese, such as ripening and storage environment (length, temperature, display conditions, etc) or contamination by pigment-producing microorganisms.

Hurtaud et al. [26] characterized the effect of hay and maize silage on the sensory properties of butter. Physical measurements were done 14 days after manufacture and the colour of the butter was measured with a Minolta chromameter. The assay consisted of two trials in order to compare hay and maize silages during winter of 2001 and 2002. Authors reported that hay diet had little effect on butter colour and they observed that the dietary effect differed between trials. This difference was detected instrumentally and assessed by a sensory panel. Authors related the differences between the butters from hay, elaborated in successive winter, with a greater loss of pigments in one of the trial resulting from prolonged weather exposure and greater loss of carotenoid hay pigments.

Kneifel et al. [7] studied colour characteristics in butter samples that were purchased from local retail outlets. These authors used a microcolor tristimulus colorimeter with a 10° standard observer and D65 standard illuminant (Dr Bruno Lange GmbH, Berlin, Germany). Colour differences between summer and winter butter were reported and authors associated these variations to differences in β-carotene contents. Also, it was pointed out that L*, a*, b* parameters were strongly influenced by the sample temperature due to the temperature-dependent extent of fat crystallization.

In the literature several articles investigate the relationships between milk characteristics and the sensory properties of cheese, especially those related to animal feeding. As pointed out by Verdier-Metz et al. [27], the effect of upstream factors on cheese sensory properties depends on cheese varieties and cheese making parameters such as partial skimming, pasteurization, acidification kinetics or ripening time.

Cheese colour is dependent on forage composition. Milk contains variable amounts of pigments and the variability in milk explains a high proportion of the variability in dairy products. Feeding system has a marked effect on carotene content in milk and therefore on the colour of cheeses [28].

Carpino et al. [29] studied whether inclusion of native pastures in the diet of dairy cows changes the colour, odour, taste, consistency, or mouth structure of Ragusano cheese. Ragusano is a Protected Denomination of Origin cheese produced in the Hyblean area of Sicily. Authors compared two different feeding systems (pasture and total mixed ration, TMR) after 4 and 7 moths of aging. A Macbeth Color-Eye Spectrophotometer (model 2020; Kollmorgen Instruments Corp., Newburgh, NY) was used to measure the colour of cheeses using Hunter Lab scale and illuminate A. As result of the assay, cheeses made from milk produced by cows that consumed native pasture were more yellow than cheeses produced from TMR fed cows. Authors related this difference to compounds transfer from pasture plants to cheese. Also, the reflectance spectra were recorded. Significant differences in the amount of reflected light at 460, 480, and 500 nm were observed between pasture and TMR samples. The absorbance maximum for β-carotene and related carotenoids are in this range.

11.4 Colour Stability During Milk and Dairy Products Processing

11.4.1 Conventional treatments

Thermal treatments of milk have been successfully applied in industrial practice, ranging from mild to severe ones. As expected, the more severe the heat treatment, the more extensive the damage [30].

Dairy powders are sensitive to the Maillard reaction as they contain high concentration of lactose and proteins with high lysine level [31]. In addition, relatively high temperature and water content during processing and prolonged storage are the major factors involved in the high susceptibility of dehydrated dairy products, as they are favourable conditions for the Maillard reaction [32].

During the manufacture process of milk powder, the heat treatment applied prior to the concentration is very important because many of the physical, chemical and functional properties of the powder for particular end-uses are determined by the processing conditions used. Many physical and chemical reactions occur at this stage [33]. For example, lactose can interact with many components of milk during heat treatments, and most of the changes associated with lactose involve the Maillard reaction.

Changes in the colour of whole milk powder (WMP), as regards the type of thermal treatment applied to raw milk before drying, were objectively evaluated by Grigioni et al. [13]. Colour measurements were carried out using a reflectance spectrophotometer (BYK Gardner Colour View model 9000) according to CIELab scale, with a 5 cm port area and D65 illuminate. In this study, it was reported that WMP obtained under indirect heat treatment (90-93 °C; 180 s = IHT) had significantly lower L* values as compared to the values of WMP obtained under direct heat treatment (105 °C; 30 s = DHT). The L* parameter indicates the lightness of samples and depicts the capacity of an object to reflect or transmit light. Furthermore, the authors showed that reflectance, the amount of light reflected by a surface [14] measured at 450 nm (r450), followed the same behaviour as L*. The researchers pointed out that the decrease in lightness may be due to the formation of brown pigments in casein-sugar mixtures as a consequence of the Maillard reaction. In contrast, as regards the application of thermal treatments, the b* parameter was found to have an opposite behaviour to that of L* and r450. It was observed that WMP obtained under IHT was prone to show higher b* values than WMP obtained under DHT. The authors concluded that the combined action of temperature and time during IHT contributed to the most noticeable changes in colour since the Maillard reaction had reached a more advanced stage.

Ordóñez et al. [34] reported that the greater the intensity of the thermal treatment applied, the greater the chemical changes occurred, mainly regarding the sensory quality and nutritive value. It is known that direct heating processes usually result in less browning than indirect heating. This occurs because indirect processes need a longer residence time in order to reach working temperature, being the amount of heat applied on milk higher in this type of treatments [11].

The storage conditions of dehydrated diary products determine the stage reached by the Maillard reaction and consequently the shift of colour in products [32, 35].

Biolatto [36] studied the shift of colour in WMP processed under indirect heat treatment (90-93 °C; 180 s; denatured whey protein nitrogen index WPNI = 0.72 mg/g), packed in 400 g polyethylene bags contained in cardboard boxes, and stored at 20 °C ± 0.5 °C during 6 months. The evaluation of the colour was done with a reflectance spectrophotometer (BYK Gardner Colour View model 9000) with large port area (5 cm diameter) and D65 illuminate. Data were expressed in CIELab system. This study showed that, in general, the L* parameter of WMP processed from raw milk produced in different seasons was not significantly affected during storage. The exception was whole milk powder manufactured in spring which presented a significant decrease (92.7 vs. 91.4; $P < 0.05$) of L* parameter during storage. In addition, the b* parameter of WMP manufactured in autumn and winter showed a decrease during storage (autumn: 21.5 vs. 19.6; $P < 0.05$ – winter: 21.4 vs. 18.4; $P < 0.05$). Contrary to autumn and winter, WMPs manufactured in spring and summer were prone to show an increase in b* value between the start and end of storage (spring: 18.3 vs. 21.2; $P < 0.05$ – summer: 20.0 vs. 20.8). As regards the value of r450, autumn and winter WMPs showed a statistically higher value by the end of storage as compared to the start of storage. While for

WMPs manufactured in spring and summer, the r450 value at the end of storage was lower than that at the start of storage, the difference between the start and end of storage was statistically significant in spring WMPs. According to Kwok et al. [37], the measure of r450 may be used to indicate brown pigment concentration. Renner [38] pointed out that during prolonged storage, the Maillard reaction occurs in a detectable degree only when storage temperature is above 20 °C.

In the same study, Biolatto [36] evaluated the shift of colour in WMP obtained under direct heat treatment (105 °C; 30 s; WPNI = 1.16 mg/g), packed in 800 g aluminium cans under an inert nitrogen atmosphere and stored at 20 °C ± 0.5 °C during 12 months. The results showed that although the value of L* parameter at the end of storage was lower than the value at the start of storage, the differences were not statistically significant (autumn: 93.3 vs. 93.2; winter: 92.7 vs. 92.6; spring: 93.5 vs. 93.1; summer: 92.3 vs. 91.7). As regards the b* parameter, although values corresponding to WMPs manufactured in autumn (18.4 vs. 19.3), winter (18.7 vs. 19.3) and spring (18.5 vs. 18.6) by the end of storage were higher than those at the start of storage, such differences were not statistically significant. Summer WMP, in turn, showed a significant decrease in b* parameter (21.6 vs. 16.6) by the end of storage. The value of r450 showed a decrease when storage times (start and end) were compared in each season (autumn: 59.8 vs. 59.1; winter: 58.0 vs. 57.6; spring: 60.2 vs. 59.4; summer: 59.1 vs. 58.6); however, this decrease was not statistically significant. Considering the general evolution of colour and reflectance parameters, WMP processed under DHT, packed under a nitrogen atmosphere and stored during 12 months, seems to have undergone no substantial change in colour, which indicates that the combination of a less intense thermal treatment and a nitrogen atmosphere might have contributed to preserve the colour characteristics of WMP during the storage period. In addition, according to Celestino et al. [10], whole milk powder packed in a vacuum or inert atmosphere, such as nitrogen, may extend its shelf life over 12 months.

As describe by Schebor et al. [39], Maillard reaction is one of the reactions that might be affected by the glass transition phenomenon, since it can be diffusion-limited. Milk powders may contain amorphous solids that can suffer glassy-to-rubbery transition when they are stored at temperatures higher than glass transition temperature (Tg).

Fernandez et al. [40] analysed the glass transition temperatures of milk powder with hydrolyzed lactose and regular milk powder. As part of their study, authors analysed browning development in milk stored during 1 month at 37 °C at different relative humidity using a spectrophotometer Minolta 508-d (illuminant D65 and an observation angle of 2°). Authors concluded that Tg values of the main carbohydrates do not account completely for the behaviour of flowing characteristics and the development of nonenzymatic browning. Other components in milk powder, like proteins and fat, may also be important in the physical and the chemical stability.

Consumer perception of cheese is strongly related to its appearance and texture, which in turn depends on microbiological, biochemical and technological parameters that affect microstructure directly or indirectly [30].

Processed cheese is produced by heating a mixture of cheese, water, emulsifying salts and optional ingredients, being its structure and sensory characteristics determined by ingredients and processing conditions. Changes with age can be influenced mainly by product composition, processing, packaging and storage conditions like temperature and duration. In general, this product is considered to have reasonable shelf life. However, shelf life should be reduced by non-enzymatic browning or lipid oxidation during storage at ambient temperature for long periods [41, 42]. The relevance of nonenzymatic browning in processed cheese was discussed by Berger et al. (1989), as cited by Schär et al. [41]. The extent of the reaction is reduced by lower lactose content, less severe heating conditions and a lower storage temperature.

Bley et al. [43] investigate how manufacturing practises and composition of stirred-curd Cheddar cheese affect nonenzymatic browning. These authors found a high correlation between galactose content and the brown colour intensity and pointed out that faster cooling of processed cheese reduced the intensity of brown colour.

Kristensen et al. [42] studied the effect of temperature and light exposure on the colour stability and lipid oxidation in processed cheese in simulated retail (dark or under a light intensity of about 2000 lx) in a hot climate (one-year storage period at 5 °C, 20 °C and 37 °C). Colour was measured by a Minolta Tristimulus Chromometer CR-300 (Minolta Camera Co. Ltd., Osaka 541, Japan). Authors pointed out that the most prominent change was browning of the product, which depend strongly on storage temperature but not on exposure to light. Each colour parameter changes linearly with time, indicating a zero-order browning reaction. Lightness decreased while a* and b* parameters increased during the storage.

11.4.2 Emerging technologies

High pressure (HP), power ultrasonics, and pulsed electric field are non-thermal processing technologies with promising impact in food processing [44]. Among these technologies, HP devices are commercial available since the 90's while others are still at laboratory or prototype scale.

Non thermal technologies were though for preservation purposes, enhancing quality product due to the absence (or low) of thermal stress applied. A range of other application is appearing in dairy systems, such improving process effectiveness, ingredients differentiation, preservation of heat-labile bioactive compounds, improving microstructure through component interaction, between others [44]

As stated by Pereira et al. [30], high pressure (HP) has been proposed as a suitable technology for milk treatment to substitute or in addition to thermal treatment. This procedure has the advantage to produce minimal food quality deterioration. However, it must be considered the effect upon milk constituents and consequently on the final characteristics of dairy product.

In milk, the applied of HP treatment induced changes related to the physicochemical properties of casein micelles and whey proteins: HP affects intra molecular bonds either reinforcing or weakening them. HP produces casein micelles disintegration into casein particles of smaller diameters that increase the translucence of the milk, consequently a decrease in lightness, reduce milk turbidity and increase viscosity [45].

Several authors suggest potential benefits to the dairy industry by HP technology, especially in cheese manufacture and ripening; even though this technology it is slowly adopted by the food industry.

Sheehan et al. [46] investigate the effects of HP treatment on the appearance, rheological and cooking properties of reduced-fat Mozzarella cheese. Samples were treated at 400 MPa for 5 min at 21 °C, with rates of compression and decompression of 300 MPa min^{-1}. Once the

HP treatment was complete, samples were stored at 4 °C. Colour was measured at room temperature on 2-cm thick, freshly cut cheese slices using a Minolta Colorimeter CR-300 (Minolta Camera, Osaka, Japan), immediately after HP treatment and after 75 days storage. Authors reported that HP treatment resulted in a marked decrease in the L*-value (whiteness of cheese), in small but significant increase in greenness and reduction in yellowness at the beginning of storage. While, after 75 days no differences between treated and no treated cheese were observed.

Rynne et al. [47] examine the effect of HP treatment (400 MPa for 10 min at room temperature), applied at 1 day post-manufacture, on a range of ripening characteristics of full-fat Cheddar cheese. Colour was evaluated on the Hunter Lab scale using a portable Minolta Colorimeter CR-300 (Minolta Camera Co., Osaka, Japan), immediately after HP treatment and after 42, 90 and 180 days of ripening. Authors reported that HP treatment significantly decreased *a* (green-red component) and increased *b* (blue-yellow component) values of the cheese; while it had no effect on lightness. These changes in colour parameters are consistent with the higher pH and lower levels of expressible serum in the HP-treated cheese observed in the trail, both of which could contribute to reduce casein hydration and thus altered light-scattering properties of treated cheese.

Okpala et al. [48] examine the effects of HP treatment on the physico-chemical characteristics (colour, pH, fat, lipid oxidation, moisture, protein, and texture) of a rennet-coagulated fresh Scottish cheese treated at a constant temperature of 25 °C. Colour measurements were carried out immediately after high pressurisation of fresh cheese samples using a LUCI™ 100 colorimeter Version 01-08-92 (Dr. Bruno Lange GmbH, D-14163 Berlin, Germany). As a conclusion of the assay, authors advised that both pressure applied and exposures influenced colour significantly in the HP fresh cheese. The b*-parameter increased appreciably with increased pressure, and its pronounced effect was significantly higher than the effect on the a*-parameter.

CONCLUSION

Colour of milk and dairy products depend on factors related either to primary and transformation processes.

Carotenoids play a major role in colour of dairy products. The concentration of carotenoids and retinol in milk depends on several dietary and non-dietary factors, like animal breed and feeding management. In this context, feeding systems are relevant in the nutritional and sensory quality of milk and dairy products. Spectral characterization of milk using visible reflectance spectroscopy appears to be a promising tool as a global approach in feed system traceability.

Process-related factors are relevant in the colour of butter and cheese, such as ripening and storage. The storage conditions of dehydrated diary products determine the stage reached by the Maillard reaction, being the development of brown colour an evident indicator of the extent of the reaction. Among the emerging technologies, dairy products treated with high pressure exhibit changes in colour characteristics that depend on both the pressure exerted and the time of exposure.

Colour assessment yields objective and well-defined physical data on milk and dairy products, being an important part of the product quality and process management.

REFERENCES

[1] Burrows, A. (2009). Palette of our palates: a brief history of food coloring and its regulation. *Comprehensive Reviews in Food Science and Food Safety*, *8*, 394-408.

[2] Calderón, F., Chauveau-Duriot, B., Pradel, P., Martin, B., Graulet, B., Doreau, M., & Nozière, P. (2007). Variations in carotenoids, vitamins A and E, and color in cow's plasma and milk following a shift from hay diet to diets containing increasing levels of carotenoids and vitamin E. *Journal of Dairy Science*, *90*, 5651-5664.

[3] Morales, F.J., & van Boekel, M.A. (1998). A study on advanced maillard reaction in heated casein/sugar solutions: colour formation. *International Dairy Journal*, *8*, 907-915.

[4] Kaya, S. (2002). Effect of salt on hardness and whiteness of Gaziantep cheese during short-term brining. *Journal of Food Engineering*, *52*, 155-159.

[5] Rhim J.W., Jones V.A., & Swartzel K.R. (1988). Kinetic studies on the colour changes of skim milk. *Lebensmittel-Wissenschaft und-Technologie*, *21*, 334-338

[6] Pagliarini E., Vernile M., & Peri C. (1990). Kinetics study on color changes in milk due to heat. *Journal of Food Science*, *55*, 1766-1767.

[7] Kneifel, W., Ulberth, F., & Schaffer, E. (1992). Tristimulus colour reflectance measurement of milk and dairy products. *Lait*, *72*, 383-391.

[8] Nielsen, B.R., Stapelfeldt, H., & Skibsted, L.H. (1997). Early prediction of the shelf-life of medium-heat whole milk powders using stepwise multiple regression and principal component analysis. *International Dairy Journal*, *7*, 341-348.

[9] Nielsen, B.R., Stapelfeldt, H., & Skibsted, L.H. (1997). Differentiation between 15 whole milk powders in relation to oxidative stability during accelerated storage: analysis of variance and canonical variable analysis. *International Dairy Journal*, *7*, 589-599.

[10] Celestino, E.L., Iyer, M., & Roginski, H. (1997). The effects of refrigerated storage of raw milk on the quality of whole milk powder stored for different periods. *International Dairy Journal*, *7*, 119-127.

[11] Celestino, E.L., Iyer, M., & Roginski, H. (1997). Reconstituted UHT-treated milk: effects of raw milk, powder quality and storage conditions of UHT milk on its physico-chemical attributes and flavor. *International Dairy Journal*, *7*, 129-140.

[12] Priolo, A., Lanza, M., Barbagallo, D., Finocchiaro, L., & Biondi, L. (2003). Can the reflectance spectrum be used to trace grass feeding in ewe milk?. *Small Ruminant Research*, *48*, 103-107.

[13] Grigioni, G., Biolatto, A., Irurueta, M., Sancho, A.M., Páez, R., & Pensel, N. (2007) Color changes of milk powder due to heat treatments and season of manufacture. *Ciencia y Tecnología Alimentaria*, *5*(5), 335-339.

[14] Guirao, M. (1980). El sistema visual. In Guirao, M., Editor, Los sentidos. Bases de la percepción (235-280). Madrid, Alambra Universidad: Publisher.

[15] Pérez Alvarez, J. (2006). Color. In Y., Hui, I., Guerrero, & M., Rosmini (Eds.), *Ciencia y Tecnología de Carnes* (199-228). Mexico DF, Limusa: Publisher.

[16] Nozière, P., Graulet, B., Lucas, A., Martin, B., Grolier, P., & Doreau, M. (2006). Carotenoids for rumiants: From forages to dairy products. *Animal Feed Science and Technology*, *131*, 418-450.

[17] Calderón, F., Tornanbé, G., Martin, B., Pradel, P., Chauveau-Duriot, B., & Nozière, P. (2006). Effects of mountain grassland maturity stage and grazing management on carotenoids in sward and cow's milk. *Animal Research*, *55*, 533-544.

[18] Havemose, M.S., Weisbjerg, M.R., Bredie, W.L., & Nielsen, J.H. (2004). Influence of feeding different types of roughage on the oxidative stability of milk. *International Dairy Journal*, *14*, 563-570.

[19] Martin, B., Fedele, V., Ferlay, A., Grolier, P., Rock, E., Gruffat, D., & Chilliard, Y. (2004). Effects of grass-based diets on the content of micronutrients and fatty acids in bovine and caprine dairy products. *Grassland Science Europe*, *9*, 876-886.

[20] Reynoso, C.R., Mora, O., Nieves, V., Shimada, A., & De Mejia, E.G. (2004). Beta-carotene and lutein in forage and bovine adipose tissue in two tropical regions of Mexico. *Animal Feed Science and Technology*, *113*, 183-190.

[21] Park, Y.M., Anderson, M.J., Walters, J.L., & Mahoney, A.W. (1983). Effects of processing methods and agronomic variables on carotene contents in forages and predicting carotene in alfalfa hay with near-infrared-reflectance spectroscopy. *Journal of Dairy Science*, *66*, 235-245.

[22] Chatelain, Y., Aloui, J., Guggisberg, D., & Bosset, J.O. (2003). La couleur du lait et des produits laitiers et samesure—un article de synthèse (1972-2002). *Millelungen aus Lebensmitteluntersuchung und Hygiene*, *94*, 461-488.

[23] Biolatto, A., Grigioni, G., Irurueta, M., Sancho, A.M., Taverna, M., & Pensel, N. (2007). Seasonal variation in the odour characteristics of whole milk powder. *Food Chemistry*, *103*(3), 960-967.

[24] Prache, S., & Theriez, M. (1999). Traceability of lamb production systems: carotenoids in plasma and adipose tissue. *Animal Science*, *69*, 29-36.

[25] Langman L. (2009). Calidad organolética en leche expresada en su color y perfil de olor. Relación de estos parámetros con la incorporación de antioxidantes naturales en la dieta implementada en las vacas. MSc Thesis, La Plata National University, Argentina.

[26] Hurtaud, C., Delaby, L., & Peyraud, J.L. (2007). The nature of preserved forage changes butter organoleptic properties. *Lait*, *87*, 505-519.

[27] Verdier-Metz, I., Martin, B., Pradel, P., Albouy, H., Hulin, S., Montel, M.C., & Coulon, J.B. (2005). Effect of grass-silage vs. hay diet on the characteristics of cheese: interactions with the cheese model. *Lait*, *85*, 469-480.

[28] Coulon, J.B., Delacroix-Buchet, A., & Martin, B. (2004). Relationships between ruminant management and sensory characteristics of cheeses: a review. *Lait*, *84*, 221-241.

[29] Carpino, S., Horne, J., Melilli, C., Licitra, G., Barbano, D.M., & Van Soest, P.J. (2004). Contribution of Native Pasture to the Sensory Properties of Ragusano Cheese. *Journal of Dairy Science*, *87*, 308-315.

[30] Pereira, C., Gomes, A.M.P., & Malcata, F.X. (2009). Microstructure of cheese: Processing, technological and microbiological considerations. *Trends in Food Science & Technology*, *20*, 213-219.

[31] Palombo, R., Gertler, A., & Saguy, I.A. (1984). Simplified method for determination of browning in dairy powders. *Journal of Food Science*, *49*, 1609-1613.

[32] Labuza, T.P. (1972). Nutrient losses during drying and storage of dehydrated food. *Critical Review in Food Technology*, *3*, 217-240.

[33] Singh, H., & Newstead, F. (1998). Aspects of proteins in milk powder manufacture. In P. F. Fox (ed.), Advanced Dairy Chemistry, Aspects of proteins in milk powder manufacture (735-765). London, Elsevier Applied Science: Publisher.

[34] Ordónez, J.A., Cambero Leónidez-Fernández, M.I., García, M.L., García de Fernando, G., de la Hoz, L., & Selgas, M. D. (1998). Leche de consumo. In Ordónez, J. A., Editor, Tecnología de Alimentos. Vol II: Alimentos de Origen Animal (64-87). España, Síntesis: Publisher.

[35] Stapelfeldt, H., Nielsen B.R., & Skibsted, L.H. (1997). Effect of heat treatment, water activity and storage temperature on the oxidative stability of whole milk powder. *International Dairy Journal*, *7*, 331-339.

[36] Biolatto, A. (2005). Incidencia de la estación del año sobre el perfil de olor y la evolución de color en leche entera en polvo bajo prácticas comerciales de procesamiento y almacenamiento. PhD Tesis, La Plata National University, Argentina.

[37] Kwok, K.C., MacDougall, D.B., & Niranjan, K. (1999). Reaction kinetics of heat-induced colour in soymilk. *Journal of Food Engineering*, *40*, 15-20.

[38] Renner, E. (1988). Storage stability and some nutritional aspects of milk powders and ultra high ambient temperatures. *Journal of Dairy Research*, *55*, 125-142.

[39] Schebor, C., Buera, M.P, Karel, M., & Chirife, J. (1999). Color formation due to non-enzymatic browning in amorphous, glassy, anhydrous, model systems. *Food Chemistry*, *65*, 427-432.

[40] Fernández, E., Schebor, C., & Chirife, J. (2003). Glass transition temperature of regular and lactose hydrolyzed milk powders. Research note. *Lebensm Wiss u Technol*, *36*, 547-551.

[41] Schär, W., & Bosset, J.O. (2002). Chemical and physico-chemical changes in processed cheese and ready-made fondue during storage. A Review. *Lebensm. Wiss u Technol*, *35*, 15-20.

[42] Kristensen, D., Hansen, E., Arndal, A., Appelgren Trinderup, R., & Skibsted, L.H. (2001). Influence of light and temperature on the colour and oxidative stability of processed cheese. *International Dairy Journal*, *11*, 837-843.

[43] Bley, M.E., Johnson, M.E., & Olson, N.F. (1985). Factors affecting nonenzymatic browning of processed cheese. *Journal of Dairy Science*, *68*, 555-561.

[44] Smithers, G., Versteeg, C., & Sellahewa, J. (2008). Introduction to non-thermal processing technologies and dairy systems. In Symposium Dairy Foods: Emerging Non-Thermal Food Processing Technologies-Their Potential in Dairy Systems. *J. Anim. Sci. Vol. 86, E-Suppl. 2/J. Dairy Sci. Vol. 91, E-Suppl. 1*, 553-556.

[45] Trujillo, A.J., Capellas, M., Saldo, J., Gervilla, R., & Guamis, B. (2002). Applications of high-hydrostatic pressure on milk and dairy products: a review. *Innovative Food Science and Emerging Technologies*, *3*, 295-307.

[46] Sheehan, J.J., Huppertz, T., Hayes, M.G., Kelly, A.L., Beresford, T.P., & Guinee, T.P. (2005). High pressure treatment of reduced-fat Mozzarella cheese: Effects on functional and rheological properties. *Innovative Food Science and Emerging Technologies*, *6*, 73-81.

[47] Rynne, N.M., Beresford, T.P., Guinee T.P., Sheehan, E., Delahunty, C.M., & Kelly, A.L. (2008). Effect of high-pressure treatment of 1 day-old full-fat Cheddar cheese on subsequent quality and ripening. *Innovative Food Science and Emerging Technologies*, *9*, 429-440.

[48] Okpala, C.O.R., Piggott, J.R., & Schaschke, C.J. (2010). Influence of high-pressure processing (HPP) on physico-chemical properties of fresh cheese. *Innovative Food Science and Emerging Technologies*, *11*, 61-67.

In: Practical Food and Research
Editor: Rui M. S. Cruz, pp. 299-335

ISBN: 978-1-61728-506-6

Chapter XII

ENZYMES

Golfo Moatsou
Laboratory of Dairy Research, Department of Food Science and Technology, Agricultural University of Athens, Iera Odos 75, 118 55 Athens, Greece, mg@aua.gr

ABSTRACT

A great variety of enzymatic activities is present in milk and dairy products that originate from the milk as secreted (indigenous enzymes), from the native milk microflora or the starters and from rennets used in cheesemaking. The enzymes can affect the manufacture of dairy products and their stability during storage. A variety of processes are applied to raw bovine milk with the aim to eliminate microbial hazards, to increase shelf-life and finally to formulate dairy products with particular characteristics. The purpose of this chapter is the presentation of the characteristics and the behaviour of the enzymes of raw milk, in particular of the major indigenous ones, under heat treatments or under various processes alternative to heat treatments, i.e. high hydrostatic pressure, ultra-high pressure homogenization, high-intensity pulsed electric fields and membrane processes, which are applied to milk. In addition, issues related to enzymes of psychrotrophic microorganisms that are the main spoilage microorganisms related to refrigerated storage of raw milk are presented.

12.1 INTRODUCTION

A great variety of enzymatic activities is present in milk and dairy products that can affect both their production and stability during storage. They can be grouped according to their origin:

- Indigenous enzymes of milk; the milk is a source of numerous indigenous enzymatic activities of various types.
- Enzymes from microorganisms, mainly bacteria of the native milk microflora or starters.
- Enzymes from the rennets used in cheesemaking; rennets contain milk clotting proteinases and in some cases show also lipolytic activity.

The purpose of this chapter is to present the effect of heat treatments and of processes alternatine to heat treatments on the behaviour of the dairy enzymes, in particular of the indigenous ones.

12.2 Dairy Enzymes

12.2.1 Indigenous milk enzymes

Among the constituents of bovine milk as secreted are included more than 60 indigenous enzymes that belong to the fraction of minor proteins. About 20 of them have been characterized in detail and they are for the most part very significant with regard to dairy technology. This is especially true for proteinases, taking into consideration the processes and the stability of dairy products. Due to the significance of milk enzymes a lot of research has been carried out and excellent reviews have been presented very recently [1-4]. As summarized by Fox [1] and Fox and Kelly [2] indigenous milk enzymes come from:

- Blood, entering milk through leaky junctions between secretory cells (e.g. lipoprotein lipase, plasmin).
- Cytoplasm of secretory cells, a part of which may be enclosed in milk fat globule membrane (MFGM) during the excretion process.
- Milk fat globule membrane (MFGM), which is formed from the apical membrane of the mammary secretory cell. As the apical membrane originates from the Golgi membranes, this mechanism may be the source of the majority of milk indigenous enzymes.
- Somatic cells, which are blood leucocytes fighting against mammary gland infection; their enzymes can enter milk (e.g. cathepsin D, other lysosomal proteinases).

Thus, the majority of indigenous milk enzymes are located in the milk fat globule membrane, whereas enzymes of great technological importance such as lipoprotein lipase, plasmin and plasminogen are associated with casein micelles. Finally, several enzymes are dispersed in the milk serum or are associated with leucocytes.

The present review presents data regarding bovine milk, since it is the most studied milk kind with the highest production world wide (Table 12.1). From the limited literature information regarding other milk kinds, it is concluded that their indigenous enzymatic activities are similar to that of bovine milk, although there is a great variation in terms of concentration.

The level of the indigenous enzymatic activities in milk is affected by genetic, physiological and nutritional factors. They are species-specific, and are related to the metabolic activity of cells, stage of lactation, whether the enzyme is secreted in constitutive or inductive manner and the hormonal, nutritional and metabolic status of the producing animal [5]. Therefore:

- There are differences related to the animal species, to the breed and to the individuality of the animal. Alkaline phosphatase activity is much greater in ovine milk than in bovine milk, whereas in the bovine is greater than in caprine milk [6-8]. Ribonuclease activity is very low in ovine and caprine milks and the same is true for lysozyme and xanthine oxidoreductase, whereas glutathione peroxidase activity is higher in caprine milk [9-11]. Plasmin activity varies among breeds of cow and it is related to the β-lactogloboulin genotype [12].
- The enzymatic activities in the milk of a particular animal may vary due its age, and mainly due to the stage of lactation. In general, enzymatic activities are highest during early and late lactation, especially those coming from blood as plasmin and xanthine oxidoreductase, due to changes in the mammary gland associated with involution. Regarding plasmin, in early lactation there is an increased influx of plasminogen from blood whereas in late lactation there is an increased activation of plasminogen to plasmin [13, 14].
- Mastitis is associated with an influx of leucocytes and other blood constituents including enzymes into milk. As a result, an increase in the activities of many indigenous enzymes in milk is observed, e.g. plasmin [15, 16], catalase [17], lysosomal proteinases [18], acid phosphatase [19], β-N-acetylglucosaminidase [20], lactate dehydrogenase [10] and lipoprotein lipase [21].
- Feed can influence the levels of lactoperoxidase, xanthine oxidoreductase and plasmin in bovine milk [2].

The biological role of many indigenous enzymes is not fully elucidated. However, it is well documented that lysozyme and lactoperoxidase are involved in the antimicrobial system of milk and that lactoperoxidase, superoxide dismutase and sulhydryl oxidase protect the milk constituents from undesirable changes [5]. Factors correlated to the physicochemical composition of milk such as substrate limitations, pH and redox potential, restrain enzymatic activities, which could affect dramatically milk nutritional value and technological behavior. As an example, lipoprotein lipase and plasmin can hydrolyse milk fat and casein respectively, causing among others loss of dairy products yield or undesirable changes in their constituents or loss of stability during storage. However, indigenous enzymes can play a key role in the biochemical changes that occur during cheese ripening process. Finally, milk enzymes are a very significant topic of Dairy Science and Technology, because many of them are used as indices for either the thermal treatment of milk (alkaline phosphatase, lactoperoxidase, γ-glutamyl-transferase) or for the detection of subclinical mastitis (catalase, lactate dehydrogenase).

Table 12.1. Important indigenous enzymes in bovine milk (based on references 2, 3, 10, 12, 22-26) SM: skim milk; MFGM: milk fat globule membrane; MMSM: membrane material in skim milk.

Enzyme	EC number	Optimum pH	Optimum T (°C)	Source	Distribution in milk phases
Plasmin (serine proteinase)[1]	3.4.21.7	8	37	Blood[2]	Casein micelles
Lysosomal proteinases					
Cathepsin D (aspartic proteinase)	3.4.23.5	3-4	37	Somatic cells	Acid whey
Cathepsin B (cysteine proteinase)	3.4.22.1	5-6		Somatic cells	
Others (probably cathepsins G, S, K, H, L; elastase; thrombin)				Somatic cells	
Lipoprotein lipase (LPL)	3.1.1.34	9	33	Mammary gland	Casein micelles
Phosphohydrolases					
Alkaline phosphatase (ALP)	3.1.3.1	9[3]	37	Mammary gland	Mainly MFGM
Acid phosphatase (ACP)	3.1.3.2	4	~50		MFGM/SM
Ribonuclease (RNase)	3.1.27.5	7-7.5	37	Blood	Serum
Oxidases					
Lactoperoxidase (LPO)	1.11.1.7	6.7	20	Mammary gland	Serum
Catalase	1.11.1.6	7		Somatic cells	Cream/SM
Xanthine oxidase (oxidoredouctase, XOR)	1.1.3.22	8.3	37	Blood	MFGM
Superoxide dismutase (SOD)	1.15.1.1				Serum
γ-glutamyltransferase (transpeptidase, γ-GGT)	2.3.2.2	8.5-9	~45	Mammary gland	MMSM/MFGM
β-N-acetylglucosaminidase (NAGase)	3.2.1.30	4.2	~50	Somatic cells	SM
Amylase (diastade, mainly α-amylase)	3.2.1.1	6.5-7.5	44	Blood	Serum/SM
Lysozyme	3.2.1.17	7.5		Lysosomal enzyme	Serum
Others, e.g. Sulphydryl oxidase (SHOx, EC 1.8.3.2), L-lactate dehydrogenase (LDH, EC 1.1.1.27), Glutathione peroxidase (GSHPOx, EC 1.11.1.9), Aldolase (EC 4.1.3.13)					

[1] part of a complex system of active and inactive forms; [2] in the form of plasminogen; [3] pH 6-8 on caseinate

12.2.2 Enzymes of microbial origin in raw milk

Enzymes from starters that are deliberately added to milk during the manufacture of fermented dairy products do not affect raw milk quality. Therefore, enzymes of microbial origin in raw milk are practically enzymes derived from bacteria, which contaminate milk. They are not desirable and actually they are enzymes of psychrotrophic bacteria that are the main spoilage microorganisms related to refrigerated storage of milk. Since raw milk is kept under refrigeration usually for more than 48 h before processing, the significance of this group of enzymes that can hydrolyse milk fat and protein is obvious.

Psychrotrophs are capable of growing in milk at temperatures close to 0 °C. It is a group of both Gram-negative (e.g. *Pseudomonas*, *Achromobacter*, *Aeromonas*, *Serratia*, *Alcaligenes*, *Chromobacterium* and *Flavobacterium* spp.) and Gram-positive bacteria (e.g. *Bacillus*, *Clostridium*, *Corynebacterium*, *Streptococcus*, *Lactobacillus* and *Microbacterium* spp.). The most important of them are *Pseudomonas* spp. that can dominate the microflora of raw milk after prolonged refrigeration storage. Another problem related to psychrotrophs is the survival of spores of *Bacillus cereus* after heat treatment that can seriously affect pasteurized milk and cream. Surviving spores are activated by heat treatments in the range 65-75 °C and the vegetative cells can grow at temperatures as low as 6 °C [27].

Most of these microorganisms produce extracellular proteinases and lipases that hydrolyse milk fat and protein producing thus off flavours such as rancidity and bitterness. Proteolytic activity of psychrotrophs can result in milk coagulation without causing acidification of milk. Although psychrotrophs in general, and in particular *Pseudomonas* spp. are killed by low pasteurization, their enzymes are heat-stable and can produce off-flavours later during the shelf-life of the product. In fact, off-flavours can be evident when *Pseudomonas* spp. are present in $>10^6$ bacteria/ml of raw milk [28]. Therefore, the production and the accumulation of enzymes of psychrotrophs in milk can be avoided by using milk of low microbial counts, by decreasing the duration of the refrigeration storage of raw milk or by applying a mild heat treatment (thermalization) in raw milk after collection [26]. Regarding cheese milk, the enzymes of psychrotrophs have been related to the reduction in cheese yield and losses of protein in the whey.

Most extracellular proteinases from *Pseudomonas* are metalloenzymes containing one zinc atom and up to eight calcium atoms per molecule and have milk-clotting activity. They hydrolyse caseins but not whey proteins. In addition to the very active extacellular lipase, the production of different phospholipases from psychrotrophs has been also reported [27, 29].

The enzymes of starters are responsible for the most part of flavour compounds in fermented dairy products, especially in cheese, and their contribution has been extensively reviewed [30-34]. However, they are not an objective of the present chapter, which focus mainly on indigenous milk enzymes and to a lesser extent on the enzymes of psychrotrophs. Both these groups have a decisive effect on the technological behaviour and storage stability of milk and dairy products.

12.2.3 Rennets

The main application of exogenous enzymes in dairy technology is the use of rennets in cheesemaking. The basis for most cheese varieties manufacture is the hydrolysis of κ-casein by rennet. Hydrolysis results in the destabilization of casein micelles and in the formation of a curd consisted of paracasein micelles, which are connected through calcium bridges. Rennets are NaCl extracts of the stomachs of young calves, lambs or kids, which have milk clotting activity. This activity is mainly due to chymosin, which is an aspartyl proteinase optimally active at pH 4.4, with high specificity for the Phe105-Met106 bond of κ-casein. About 10% of the milk clotting activity comes from pepsin but this percentage increases as the age of the animal increases. Due to the shortage of young animal stomachs available for rennet production, suitable substitutes of microbial and plant origin are used for cheesemaking, whereas fermentation produced chymosin is used extensively nowadays. Rennets and rennet-induced coagulation of milk are presented by Crabbe [35] and Horne and Banks [36]. Issues related to milk coagulating enzymes that can be used as rennet substitutes have been reported by Guinee and Wilkinson [37] and Chitpinityol and Crabbe [38].

The proteolytic activity of rennets is considered to be low. However, the residual chymosin activity that remains in the cheese curd after draining is an important factor for the ripening of most cheese varieties. Processes applied to cheese curd such as heat treatment and acidification can affect its presence and activity in the cheese curd. The effect of cheese making conditions on the levels of the residual chymosin and its role in cheese ripening has been presented recently by several researchers [e.g. 39-41]. Other types of exogenous enzymes used in dairy technology are lipases or proteinases and peptidases used for the acceleration of cheese ripening [42, 43].

Similarly to starter enzymes, the detailed presentation of milk clotting and other exogenous enzymes is not within the objective of the present chapter as stated above.

12.3 Enzymes and Milk Processing

12.3.1 Milk processing

A variety of processes are applied to raw milk with the aim to eliminate microbial hazards, to increase shelf-life and finally to formulate dairy products with various characteristics. In this respect, heat treatment is applied to milk destined to be consumed as liquid milk or to be transformed to dairy products. Apart from the widely used heat treatments, efforts have been made to substitute them with other types of preservation methods like high-pressure treatment or high intensity pulsed electric fields. The reason for these efforts is that heat treatment of milk, despite its well-documented efficacy, is in some cases related to the loss of nutrients like vitamins and to undesirable changes in sensory characteristics of milk due to Maillard reaction [26, 44, 45].

12.3.1.1 Homogenization

The main effect of homogenization is the reduction of the diameter of the raw milk fat globules to <1 μ in order to counteract creaming and the partial coalescence of the cream layer. Therefore, it is a method for the stabilization of milk fat emulsion, which results in a homogenous appearance of liquid milk and in the formation of particular rheological properties in dairy products such as cream and yogurt. On the other hand, homogenization is not applied in cream for butter production. Regarding cheese milk, homogenization is undesirable because the high number of the fat globules and the characteristics of their newly formed membrane change the behavior of the paracasein matrix [26].

Homogenization is carried out by means of two-stage homogenizers. At first, milk is forced to pass through a high-pressure valve (~200 bar) to disrupt milk fat globules into smaller ones and then it is passed through the second low-pressure valve (~35 bar) to disperse the clumps of the small milk fat globules, that occur after the first-stage homogenization. Milk homogenization is carried out at 40-70 °C, because milk fat must be liquid. Furthermore, at ~65 °C lipoprotein lipase is inactivated and cannot act on the newly formed milk fat globules. The main effects of homogenization are [26, 44]:

- The reduction of the average diameter of the milk fat globules from 3-5 μm to 0,1-1 μm, depending on the pressure and the type of homogenizer and the very great increase of their number and average surface from 1,3-2,2 m^2/g to 7-34 m^2/g of fat.
- The denaturation of cryoglobulins, that results in the prevention of cold agglutination of milk fat globules.
- The change of the nature of the milk fat globule membrane. The newly formed membrane consists apart from patches of native membrane also from milk plasma proteins, mainly casein (70-90%). As a result lipoprotein lipase is transferred from the casein micelle to the milk fat globule and therefore fat becomes more susceptible to lipolysis, if not immediately pasteurized. However, the tendency to fat autoxidation is reduced, because phospholipids and pro-oxidants like copper and xanthine-oxidoreductase take a small part of the newly formed surface.

12.3.1.2 Heat treatment

Heat treatment of milk involves heating processes of different intensity [26, 44], involving the temperature and the duration of heating:

- Thermalization, 60-69 °C for about 20 s
- (Low) Pasteurizarion, 63 °C for 30 min or 72 °C for 15 s
- High pasteurization, up to 100 °C for a few seconds
- Sterilization, which is differentiated in “in-bottle sterilization”, 110 °C for 30 min and in “UHT (ultra-high temperature) treatment”, 130 °C for 30 s or 145 °C for 1 s.

The main objective of the heat treatment of milk is to provide a safe product by killing pathogens like *Mycobacterium tuberculosis*, *Coxiella burnetii*, *Staphylococcus aureus*, *Salmonella* spp., *Listeria monocytogenes* and *Campylobacter jejuni*. This objective is achieved by using at least low pasteurization treatment. Furthermore, heat treatments of various intensities are applied with the aim to suppress or kill spoilage microorganisms and their spores and to inactivate enzymes that can downgrade the quality of milk during storage. Finally, heat treatments with intensities higher than that of low pasteurization can be applied for special reasons, such as denaturation of whey proteins. Denaturation of whey proteins results in a consistent yogurt coagulum and protects evaporated milk from gelling during sterilization. Also, such heat treatments can inactivate indigenous milk bacterial inhibitors, which affect negatively the efficacy of yogurt starters.

12.3.1.3 Processes alternative to heat treatment

12.3.1.3.1 High-hydrostatic pressure (HP)

High-hydrostatic pressure (HP) processing is a non-thermal method for food preservation that aims to achieve microbial inactivation equivalent to thermal processing without any practical loss of its nutritional and sensory characteristics. In addition to the antimicrobial effect, research has focused on the potential of HP at 100-1000 MPa to cause reversible and irreversible changes in milk. Comprehensive reviews of the effects of HP on milk are available [45-49], with the following main points:

i. The effect of HP in milk microflora is variable depending on the conditions of treatment (pressure, duration, temperature), on the microbial species and strain, on the growth phase and on the characteristics of the suspended medium (composition, pH, a_w). The most baroresistant species in milk are the Gram-positive *Listeria monocytogenes* and *Staphylococcus aureus* and the Gram-negative *Escherichia coli*. In general, milk with microbiological characteristics similar to pasteurized milk can be obtained after HP treatment at 400 MPa for 15 min or at ≥ 500 MPa for 5 min. The resistance of bacterial spores to HP limits the sterilizing effect of this process. For the inactivation of bacterial spores a first HP treatment at 50-300 MPa is applied to induce their germination followed by a second treatment using heat or HP.

ii. The application of HP affects the individual components of milk, mainly by causing structural changes, which are related to the conditions of pressurization (pressure, duration, temperature). HP treatment increases the transfer of individual casein from the colloidal to the soluble phase reducing the average micelle size, due to the disruption of hydrophobic and electrostatic repulsions and colloidal calcium phosphate solubilization. The major whey proteins α-lactalbumin and β-lactoglobulin are denatured, with the latter being less resistant. The mineral equilibrium is modified due to the increase, although reversibly at 20 °C, of the ionic calcium concentration in milk and of the amount of calcium and phosphate in the serum phase.

iii. HP treatment changes some biochemical characteristics of milk such as indigenous enzyme activities, which are related to the deterioration of milk during storage and to cheese ripening. Plasmin activity is reduced at pressures ≥400 MPa at 20 °C and the reduction is enhanced in the presence of β-lactoglobulin or by pressurizing at elevated temperatures. The other enzymes that have been studied, e.g., alkaline phosphatase, lactoperoxidase, xanthine oxidoreductase, lipoprotein lipase and γ-glutamyl transferase, maintain in general their activities after treatments at pressures ≤400 MPa. The opposite is true for acid phosphatase, which is inactivated significantly at pressures ≥200 MPa.

12.3.1.3.2 Ultra High-Pressure Homogenization (UHPH)

Recently, ultra high-pressure homogenization by means of valves able to withstand pressures >100 MPa and up to 400 MPa has been proposed for milk processing as a combined process for both reduction of the size of milk fat globules and inactivation of microorganisms and enzymes [50]. According to Hayes et al. [51], this type of treatment, owing to the significant increase of temperature observed during processing and the possibility of varying holding times, may be considered as a novel liquid milk processing technique. The effect of UHPH treatment, at 300 MPa at an inlet temperature of 30 °C, has been considered as efficient as high pasteurization at 90 °C for 15 s with regard to the shelf life of bovine milk. It causes important reductions of about 3.5 log cfu/ml of psychrotrophic, lactococci and total bacteria, while coliforms, enterococci and lactobacilli are eliminated [51-53]. In general, Gram-positive bacteria have been found more resistant than Gram-negative species; however, their resistance is variable [54].

No significant effect of UHPH has been found on fatty acid composition of raw milk treated up to 350 MPa [55]. However, treatment at 200 MPa induces lipolysis and at 300 MPa promotes oxidation [56, 57]. Finally, the volatile profile of UHPH-treated has been found different from that of the heat-treated milk, which has been considered as a promising result regarding the consumer acceptance [58]. A series of structural and physicochemical changes is caused in milk constituents by this type of process:

- The size of milk fat globules is decreased and their average diameter is smaller compared to conventional homogenization. The decrease is higher as the inlet temperature and the homogenization pressure increase. Furthermore, further decrease is induced by a second homogenization pass [51, 53, 59, 60].
- A few studies have been carried out regarding the effect of UHPH on bovine milk proteins. Average casein micelle size does not change at pressures <150 MPa and a small decrease is observed at 200 MPa [59]. Extended denaturation of β-lactoglobulin occurs at pressures ≥150 MPa, while α-lactalbumin is unaffected [51, 61].
- The pH of milk treated at pressures ≤200 MPa is decreased during storage [51, 53, 59].
- Rennet coagulation time of bovine milk treated with UHPH can be enhanced, although curd firmness may be reduced [59, 61].
- Plasmin and plasminogen-derived activities are significantly decreased after UHPH treatment, depending on the applied pressure [56, 62, 63]. Lipoprotein lipase activity appears to be enhanced by this treatment [56]. The pattern of inactivation of alkaline phosphatase and lactoperoxidase are similar to thermal inactivation, the inactivation depending on the outlet temperature [56, 62].

12.3.1.3.3 High Intensity Pulsed Electric Fields (HIPEF)

High Intensity Pulsed Electric Fields (HIPEF) are considered as a very promising non-thermal technology for milk, alternative to pasteurization. As described by Mosqueda-Melgar et al. [64], HIPEF is carried out by applying short pulses (from 1 to 10 μs) with high intensity electric field (from 20 to 80 kV/cm) to fluid foods placed between two electrodes in batch or continuous flow treatments. A few studies have been carried out regarding the effect of

HIPEF on milk constituents and technological behavior [65]. Most of them present promising results about the effectiveness of HIPEF to destroy pathogenic bacteria, like *Listeria* spp. and *Staphylococcus aureus* [64]. However studies about other pathogens in milk are missing. Finally, some of the HIPEF studies focus on indigenous milk enzymes but information with regard to the effect of HIPEF on major milk constituents is in general limited. Recently, Odriozola-Serrano et al. [66] have reported that treatment of 1000 μs at 35.5 kV/cm, which ensures the microbiological stability of whole milk, causes limited denaturation of whey proteins, without affecting physicochemical characteristics such as pH, acidity and free fatty acid content. In conclusion, the levels of microorganisms inactivation by HIPEF vary depending on the microorganisms, processing conditions and treatment media, whereas spore inactivation is in general low [65].

12.3.1.3.4 Membrane processes

Membrane processes have been used in the dairy sectors as alternatives to some unit operations since 1970. Their main applications are the manufacture of dairy-based protein ingredients by ultrafiltration, microfiltration or nanofiltration, the pre-concentration of milk before cheese manufacture by ultrafiltration or microfiltration and the extension of shelf-life of milk, by removing somatic cells, spores and a great part of microorganisms by microfiltration [67]. The different types of membrane processes are presented in the book of Walstra et al. [26].

Extended shelf life (ESL) products can be manufactured by passing defatted milk through microfiltration (MF) membranes (pores >0.2 μm) that concentrate somatic cells and bacteria in the retentate. The permeate after fat standardization is pasteurized at minimal time-temperature conditions to produce an ESL milk, with shelf life of 21 to 28 d at refrigeration temperature, with high flavour quality [68]. Developments of microfiltation technology (MF) and its use for improving fluid milk quality are reported by Saboya and Maubois [69] and by Elwell and Barbano [70]. MF does not cause changes in the physicochemical and biochemical characteristiscs of milk. However, enzymatic activities originating from somatic cells (Table 12.1) are expected to be altered in the MF milk.

12.3.2 Effect of milk processing on important indigenous enzymes of milk

12.3.2.1 Plasmin

The physiological function of plasmin and its role in milk and dairy products is presented among others in the review papers of Bastian and Brown [71], Kelly and McSweeney [12] and Kelly et al. [4]. Plasmin (fibrinolysin EC 3.4.21.7) is a trypsin like serine – proteinase that dissolves blood clots. Its activity in milk is controlled by a complex enzymatic system, in which plasminogen (its zymogen), plasminogen activators and inhibitors of both plasmin and plasminogen are also included. Active enzyme, its zymogen and plasminogen activators are associated with casein micelles, whereas inhibitors are soluble in the milk serum. Plasmin and plasminogen are dissociated from casein micelles at pH <4.6. There are two types of plasminogen activators, i.e. urokinase-type associated with somatic cells and tissue-type associated with casein [72, 73]. They can convert plasminogen to plasmin, which has a molecular mass of about 81 kDa and its activity is related to 5 intramolecular disulphide-linked loops. It exhibits optimum activity at pH 7.5 and at 37 °C. It is highly specific for peptide bonds at the C-terminal side of Arg and Lys residues. The most susceptible milk protein to plasmin action is β-casein. It is hydrolysed at Lys28-Lys29, Lys105-is106 and Lys107-Glu108 bonds resulting in the production of γ-caseins that are included in the casein fraction of milk and in the production of proteose-peptone, which are in the whey fraction. Among the other caseins, αs2-casein is the most susceptible followed by αs1-casein, while κ-casein seems resistant. The major whey proteins β-lactoglobulin and α-lactalbumin are not sensitive to plasmin action. Mastitis, advanced stage of lactation and the age of cow increase plasmin activity in bovine milk [13, 15, 16]. The role of plasmin in dairy products is a subject of great importance, since it acts upon casein, which is a key-constituent for all types of milk processing. As a result, its behaviour under various processes applied to milk has been studied by various research groups.

Plasmin activity during refrigerated storage is configured by the rate of autolysis of mature enzyme and plasminogen activation, the former mechanism being more significant than the latter [74]. Both plasmin and plasminogen activators are very heat stable as presented below. This characteristic has been related to the appearance of defects in UHT milk during storage.

Plasmin is also significant for cheese technology. It has been related to inferior quality of cheese curds [75], although no relation of plasmin to milk coagulation properties has been also reported [76, 77]. However, it is an important proteolytic factor for the ripening process of many cheese varieties. This is especially true for Swiss type cheeses, in which chymosin is for the most part inactivated due to scalding conditions [39, 71].

12.3.2.1.1 Effect of heat treatment of milk on the plasmin system

Bovine plasmin and plasminogen activators survive pasteurization and they are resistant to many UHT treatments [78]. Thermal inactivation of the purified enzyme is achieved after heating at 80 °C for 10 min at pH 7.0, whereas temperature/time combinations equivalent to heat treatment at 73 °C for 40 min are necessary for its inactivation in milk [26, 79]. Plasmin is reversibly inactivated in the range of 55-65 °C and irreversible inactivation starts at temperatures >65 °C, obeying first order kinetics [80].

In fact, low pasteurization enhances plasmin activity in milk by causing inactivation of plasminogen inhibitors, allowing thus the conversion of plasminogen to plasmin [81, 82]. Nevertheless, the activation of bovine plasminogen observed at pasteurization temperatures could be attributed to its denatured form that is more readily activated by plasminogen activators than the native form [83]. Denatured β-lactoglobulin occurred in milk during heat treatment destabilizes plasmin on heating due to thiol-disulfide interactions, whereas casein increases the heat stability of plasmin [78, 80, 84]. It is considered that residual plasmin and plasminogen-derived activities contribute to gelation of UHT milk or to undesirable precipitation or to the development of bitter flavour [85-88]. The dissociation of β-lactoglobulin/κ-casein complex from the casein micelle caused by plasmin action may be an explanation for the gelation of UHT milk [89]. In conclusion, the role of plasmin as a sole agent for the deterioration of UHT milk during storage is under question [2, 74] and it is considered that an appropriate UHT treatment, i.e. 140 °C for 15 s can inactivate plasmin system [26].

12.3.2.1.2 Effect of processing alternative to heat treatment of milk on the plasmin system

The effect of high pressure (HP) on plasmin activity in bovine milk has been extensively studied [90-95]. The general conclusion is that plasmin activity is reduced at pressures ≥400 MPa at 20 °C and the reduction is enhanced in the presence of β-lactoglobuling or by pressurizing at elevated temperatures. Data presented by Moatsou et al. [95] indicate that HP affects plasmin and plasminogen activators in bovine milk by at least two different mechanisms. The first mechanism involves transfer of both enzymes from the casein to the milk serum fraction that occurs at 200, 450 and 650 MPa at room temperature. Apparently, this transfer is related to the well known disruption of casein micelles caused by HP described by Needs et al. [96] and Huppertz et al. [97]. The second mechanism, in which HP affects plasmin and plasminogen activators, involves decreases in activity of both enzymes, after treatment at pressures 450 and 650 MPa, at room temperature. The decrease of activity may be related to the increase of whey proteins denaturation induced by HP treatment. Whey proteins inhibit plasmin activity in milk serum, because the presence of β-lactoglobulin greatly destabilizes the enzyme under HP treatments, due to thiol-disulfide interactions of disulfide bonds in the plasmin molecule, similarly to heat inactivation [90, 91, 93, 94]. Moreover, it is interesting to note that HP affects the enzymes (plasmin and plasminogen activators) in a different manner [95]. Strong synergistic effects of temperature and pressure have been observed for plasmin for pressures 450 and 650 MPa. In contrast, synergistic effect, very mild in nature, has been observed regarding residual plasminogen activators activity and only for the 450 MPa.

The results of a study regarding the effect of HP on ovine milk enzymes showed that HP (200, 450 and 650 MPa) at 20 °C does not affect ovine plasmin system, whereas at 450 and 650 MPa there is a strong synergistic effect of temperature [98]. These findings indicate that the major effect of HP is the reductions of plasmin and plasminogen activators activities in ovine milk. The distribution of enzyme activities between the casein and the milk serum fractions of ovine milk is not affected by HP/temperature treatments used, opposite to bovine milk subjected to the same treatments.

The effect of ultra-high pressure homogenization (UHPH) on plasmin activity has been also investigated. Plasmin activity in conventionally-homogenized milk (18 MPa) is about 40% lower compared to untreated milk, probably as a result of the adsorption of casein to the fat/serum interface in order to coat the newly formed fat globules. UHPH at ≥100 MPa causes significant decreases in plasmin and plasminogen activities, the residual activities being about 35% of the original. The effect of two-stage processing is more intense than that of one-stage. The inactivation increases as the fat content increases from 0 to 2% for plasmin and from 0 to 4% for plasminogen indicating that the effect of fat content has to be related to transfer of casein micellar fragment on the surface of the newly formed fat globules [62]. Very extended inactivation accounting for 85% and 95% of the original plasmin activity in raw milk has been reported by Hayes et al. [51] for treatments at 150 and 250 MPa respectively. They attribute the inactivation to the higher outlet temperature observed in their latter study compared to a previous work of Hayes and Kelly [62], the inlet temperature being ~45 °C and the outlet temperature 67 and 83.6 °C respectively. Similarly, UHPH of milk at 200 MPa at outlet temperatures in the range of 65-80 °C causes inactivation of plasmin in the range 74-90% [56]. Lower plasmin inactivation (~45% and 70%) has been reported for UHPH treatment at 200 and 300 MPa despite the high outlet temperatures (78-95 °C) using another type of instrumentation [63]. Finally, according to the conclusions of Datta et al. [56] and Hayes and Kelly [62], some effect of this processing is due to thermal effect, while others are induced by the combination of physical forces and heating to which the milk is exposed during UHPH.

The effect of HIPEF on the plasmin activity has been investigated by Vega-Mercado et al. [99] using simulated milk ultrafiltrate. They report a 90% decrease of activity after treatment at 30 or 50 kV/cm at 15 °C, with 50 pulses of 0.1 Hz and 2 μs duration. Increase of the number of pulses, of the field intensity and of the temperature enhances inactivation.

Ultrafiltration of milk has been reported to cause decrease of plasmin activity in cheeses made therefrom, due to the increase of plasmin inhibitors into the UF cheeses or to inactivation of the plasminogen activator system caused by UF-concentration [100]. Also, according to Kelly et al. [4] micellar casein preparations produced using microfiltration can have enhanced specific plasmin activity, because plasmin is associated with the micelles, while inhibitors are removed in the permeate.

Finally, rennet cheese whey has lower plasmin activity than acid whey, because plasmin is associated with the micelles that form the cheesecurd. In addition, lowering of the milk pH to obtain acid whey cause dissociation of plasmin from the micelles [101].

12.3.2.2 Lysosomal somatic cells proteinases

The increase of proteolytic activity in milk as the somatic cells counts increases has been attributed apart from plasmin also to lysosomal proteinases. In fact, the proteolytic activity of lysosomal proteinases has been connected with the poor quality of dairy products, mainly cheese [102-103]. This group of enzymes in milk includes the aspartic (acid) proteinase cathepsin D, the cysteine (thiol) proteinase cathepsin B and probably cathepsin C and serine proteinases like cathepsin G and elastase. Most of these enzymes are present in milk but they are inactive due to its high redox potential [2, 4]. Among them, only cathepsins B and D have been identified in milk. An activity similar to that of cathepsin D has been firstly identified in milk by Kaminogawa and Yamauchi [104]. Cathepsin D is the most studied lysosomal proteinase and its characteristics and action on milk proteins has been reviewed by Hurley et al. [105]. It is a lysosomal aspartic proteinase with acidic pH optimum and proteolytic activity similar to that of chymosin. It is a glycoprotein synthesized as procathepsin D, which can convert itself by an autoproteolytic pathway into the active pseudocathepsin D, whereas the formation of the mature cathepsin D seems to require the involvement of other enzyme [106]. Although these three types exist in milk, procathepsin D is the major one. It is present in milk at a level of about 0.4 μg/ml and it is mainly present in the whey, about ¼ of its activity being casein-bounded [107].

As stated above, cathepsin D shows proteolytic activity in milk similar to chymosin. The degradation patterns resulted from hydrolysis of αs1- and β-casein by cathepsin D are similar to that produced by chymosin, whereas hydrolysates of αs2-casein are different from those resulted from the action of chymosin [108]. The main hydrolysis product of κ-casein by cathepsin D is para-κ-casein. However, coagulation of milk due to cathepsin D is not possible under normal conditions. Approximately 10 times the indigenous amount of milk are necessary for milk coagulation [108, 109].

Heating at 70 °C for 10 min inactivates completely the cathepsin D in bovine milk [104]. According to Larsen et al. [107] and Hayes et al. [109], cathepsin D partially survives low pasteurization and thus it may affect the stability of dairy products during storage. As expected, it can survive (45% survival) processes applied during manufacture of high-cooked cheese varieties, i.e. at 55 °C for 30 min. Therefore, it can be a proteolytic factor during the ripening of cheese varieties, in which the high-cooking has inactivated residual chymosin.

Cysteine proteinase activity in milk has been firstly reported by Susuki and Kato [110] and O'Driscoll et al. [111]. According to the findings of Magboul et al. [112] a heterogenous group of cysteine proteinases exist in the acid whey of bovine milk. One of these enzymes is probably cathepsin B, which can potentially survive in more than 20%, following typical low pasteurization (72 °C for 30 s). This enzyme has a very broad specificity against the caseins with cleavage sites located throughout β- and αs1-caseins and close to or identical to those of plasmin. Furthermore, cathepsin B shows some similarity to chymosin in its cleavage sites on both β- and αs1-caseins [113].

Information about the effect of processes alternative to heat treatment on somatic cells proteinases is very limited. Cathepsin D activity in the acid whey from HP-treated bovine milk has been found in general baroresistant at room temperature. Its residual activity decreased significantly at 650 MPa and 40 °C and at 450 and 650 MPa and 55 °C. In general, proteinase activities in the ovine whey assayed under the same conditions have been found similar to that of bovine whey in qualitative terms. Cathepsin D - like activity in ovine whey decreases significantly at 450 MPa at 40 and 55 °C and at 650 MPa at 20, 40 and 55 °C [95, 98].

12.3.2.3 Lipoprotein lipase (LPL)

Lipoprotein lipase (LPL) is an enzyme of major significance for the Dairy Technology accounting for almost all lipolytic activity in raw milk. LPL is a glycoprotein elecrtostatically and hydrophobically associated with the casein micelle, only active at oil-water interface. The detailed characterization of the enzyme is given by Olivecrona et al. [22]. It catalyses the hydrolysis of triglycerides, inducing thus lipolysis in milk that results in rancidity. Milk fat contains for the most part (~98%) triglycerides and lipolysis results in the production of free fatty acids, partial glycerides and glycerol. Also, LPL is very important for the biosynthesis of lipids in the mammary gland because it liberates fatty acids from lipoprotein and chylomicrons in the blood.

LPL liberates fatty acids from the sn-1 and sn-3 positions of triglycerides. Since short-chain fatty acids are for the most part at the sn-3 position, lipolysis results in the accumulation of free short-chain fatty acids in milk and dairy products [114]. There is no fatty acid specificity. According to an explanation given by Deckelbaum et al. [115], the greater solubility and mobility of the short chain triglycerides compared to that of long chain is the determining factor for LPL specificity. In fact, hydrolysis as little as 1-2% of milk triglycerides to free fatty acids makes the milk unacceptable for consumption due to rancid flavour [22].

Since LPL is inactivated by low pasteurization (see below) it contributes largely to the ripening process of cheeses made from raw or thermalized milk by producing free fatty acids (FFA). Short-chain FFAs can contribute directly to the flavour of some cheese varieties or indirectly by being the substrates for chemical and enzymatic modifications that result in volatile compounds. Therefore lipolysis can be desirable in cheese.

According to Deeth [117] and Deeth and Fitz-Gerald [118], lipolysis in milk has also functionality effects, such as depression of its foaming ability when injected with steam, impaired creaming ability during separation and increased churning time in the manufacture of butter. Reviews about LPL in milk and cheese have been recently published by Olivecrona et al. [22], Collins et al. [114] and Deeth et al. [117].

Actually, LPL activity is a problem of raw milk. Its potential activity in milk is enough to produce rancid flavours in less than 10 min. Since the milk triglycerides are protected by the milk fat globule membrane (MFGM) and provided that MFGM remains intact, the access of the enzyme to the milk fat is not allowed. However, agitation, foaming, temperature fluctuations and homogenization can damage MFGM. Furthermore, there are additional factors that limit the lipase activity in milk as presented by Walstra et al. [26]. First of all ionic strength, ionic composition and pH of milk are suboptimal. Then, in milk there are enzyme inhibitors. Casein micelles, to which the 90% of the enzyme is bounded, act as inhibitors. The same is true for the long-chain saturated monoglycerides that are products of lipolysis. Finally, the activity of the enzyme decreases slowly in raw milk kept under refrigeration.

There is another type of lipolysis that occurs in the milk of some cows that is called spontaneous lipolysis. Spontaneous lipolysis happens immediately soon after cooling of fresh milk and it is related to particular activating factors such as (apo)lipoproteins, which can induce adsorption of LPL onto the fat at low temperatures. However, the crucial factor for this phenomenon is the activator-inhibitor balance in milk. Spontaneous lipolysis can occur at the farm and the FFA production is 20 times as high as in normal milk after 24 h of refrigerated storage. Finally this type of lipolysis has been connected to late lactation, mastistis and feed of poor quality [116-118].

It has to be mentioned that lipolysis in milk may be also induced by bacterial lipases, if the raw milk has poor microbiological quality or it has been kept under prolonged refrigerated storage (section 12.2.2).

12.3.2.3.1 Effect of milk processing on LPL

Homogenization can induce lipolysis due to the change of the nature of the membrane that coats the newly formed small milk fat globules. As stated above (section 12.3.1.1), a part of the membrane is consisted of casein micelles, which "carry" also lipoprotein lipase activity. In practice, homogenization of milk is carried out immediately before pasteurization than inactivates LPL.

LPL is heat sensitive and it is almost inactivated by low pasteurization (63 °C for 30 min or 72 °C for 15 s), it has a D-value at 70 °C of 20 s. Therefore it does not affect pasteurized milk products [26, 117].

There are also very few reports about the effect of treatments of milk alternative to heat treatment on LPL. The enzyme is resistant to HP treatments at 400 MPa and 3 °C for 100 min [119].

According to the findings of Datta et al. [56], UHPH does not cause more inactivation of lipase than the corresponding thermal treatment and on the contrary it enhances its activity. They determined lipase activity in milk samples subjected to UHPH (two-stage at 200 MPa, using different inlet temperatures), assuming that the measured activity was due to LPL and not to bacterial lipases, due to the freshness of the milk used. They report higher lipase activity in treated samples compared to the raw milk samples. The activity reaches the 240% of the control milk activity at an outlet temperature of ~57 °C. For the total inactivation at 200 MPa outlet temperatures >71 °C are necessary.

Residual lipase activity in a buffer solution after treatment using 13-87 kV/cm field, 0.5 Hz pulse frequency and 2 μs pulse width, at 20 °C has been found 15-35% of the original [120]. HIPEF applied to milk using electric field strength in the range 15-35 kV/cm and treatment time up to 75 μs, with a pulse duration of 1 μs, induces marginal reduction of lipase activity at a maximum of 14.5%. The reduction is linearly increased with increasing electric field intensity and treatment time [121].

12.3.2.4 Phosphohydrolases

There are several phosphorohydrolase activities detected in bovine milk. Among them phosphatases (phosphomonoesterases) have been characterized in detail due to their great technological significance. Phosphatases are hydrolases that catalyze hydrolysis of phosphate monoesters using water as acceptor.

12.3.2.4.1. Alkaline phosphatase (ALP)

The enzyme is a dimmer of two identical 85 kDa subunits and it contains four Zn atoms per mole, which are required for activity. Mg^{2+}, Ca^{2+}, Co^{2+}, Mn^{2+} and Zn^{2+} are activators. Its optimum pH differs according to the substrate, e.g. pH 6.8 for caseinate, pH 10.5 for p-nitrophenylphospate. Most of ALP in milk is associated with lipoprotein particles in the milk fat globule membrane (MFGM) and about 50% is found in the cream phase. Since, MFGM material is released into buttermilk during butter production, buttermilk is a rich source of the enzyme. In fact, there are two ALPs in milk, one acquired intracellularly and concentrated on the MFGM and another that originates mycoepithelial cells of the mammary gland that is in the skim milk [122]. Its activity varies among species, individuals and throughout lactation [23]. Therefore, ALP could dephosphorylate casein under the milk pH conditions, but this does not happen; a possible explanation is its inhibition by inorganic phosphate. In acid dairy products, it is not expected to be active. However, it can dephosphorylate caseins in cheese made from raw milk but it has to be taken into consideration that in cheese there are also phosphatases of bacterial origin [3, 123].

Effect of milk processing on ALP

ALP has been studied extensively, because the determination of its activity is used to monitor the efficacity of low pasteurization of milk. Inactivation of the enzyme ensures that all the non-spore forming pathogenic microorganisms present in milk during heat treatment are killed (section 12.2.2). Under these conditions most but not all lactic acid bacteria and Gram-negative rods have also been killed. The D-value for ALP in bovine milk at 70 °C is 33 s [26]. Because of the great importance of ALP for the quality control of pasteurized milk and therefore for the safety of the consumers, several analytical methods have been developed for the determination of its activity, which were recently reviewed by Fox and Kelly [3].

A complication related to its use as a heat treatment indicator is its reactivation after UHT treatment of milk, which has been recognized since more than 50 years. Homogenization before heat treatment reduces the extent of reactivation. Reactivation of ALP is a problem of

UHT milk because no reactivation is observed in pasteurized milk. Reactivation increases in the presence of Mg^{2+} and Zn^{2+}. The most important factor for reactivation seems to be the –SH groups of the whey proteins that are denatured under UHT conditions but not under low pasteurization conditions. The role of these groups is considered to be the chelation of heavy metals, which could otherwise bind to –SH groups of the enzymes that are also inactivated on denaturation. Finally, there is an official AOAC method for the determination of renatured residual phosphatase [23, 44].

The effects of processing alternative to heat treatment on milk ALP have been studied by various researchers, due to its importance as an indicator of milk heat treatment. In general, the complete inactivation of ALP in bovine milk treated with HP results from overprocessing of the product [124]. In particular, residual ALP activity is reported after treatment of milk at 400 MPa for 60 min at 25 °C, whereas pressure treatment of raw bovine milk at 800 MPa for 20 min or at 600 MPa for 30 min at 55 °C is necessary for an almost complete inactivation of the enzyme; D-values of about 300 min have been estimated at 400 MPa [125-127]. In addition, there is an antagonistic effect between pressure up to 200 MPa and high temperatures [128]. Reactivation of ALP at low enzyme activity takes place after severe pressure treatment; however, during cold storage which is the industrial practice, reactivation is not observed. Finally, the enzyme has been found to be more resistant than *Escherichia coli* and *Listeria monocytogenes* in milk and hence its inactivation could indicate the absence of these pathogens in pressure treated milk [129].

Two stage UHPH treatment of milk at the range 50-200 MPa and outlet temperatures up to 54 °C does not decrease ALP activity, indicating that thermal conditions during treatment are not adequate for inactivation as reported by Hayes and Kelly [62] and Datta et al. [56]. The same group applied UHPH at 150, 200 or 250 MP to preheated raw milk that resulted in outlet temperatures 67, 76.8 and 83.6 °C respectively for 20 s [51]. They report that milk treated at 200 or 250 MPa were ALP negative and suggest that the thermal load applied alone would account for inactivation. Results of Picart et al. [50] are in agreement with the above-mentioned results. However, they attribute the inactivation at higher outlet temperatures to the mechanical forces in the valve and not to the short residence time ($\ll$1 s) at the outlet temperatures. Additionally, they report that after UHPH treatment at the range 100-250 MPa and outlet temperature up to 58 °C, slight ALP activation is observed.

Limited or no inactivation of ALP has been reported using various configurations of HIPEF treatments in raw milk or in buffer [99, 120, 121, 130, 131].

Inactivation of ALP using microfiltration occurs during single passage at temperatures ≥60 °C [132].

12.3.2.4.2 Acid phosphatase (ACP)

It appears that bovine milk contains three acid phosphatases, two of them located in the skim milk and one on the MFGM. It is possible that two of them originate from leucocytes and it has been reported that the ACP activity increases up to 10 fold in mastitic milk [19]. The characteristics of acid phosphatase (ACP) are presented by Shakeel-Ur-Rehman et al. [23] and Fox and Kelly [3]. Its activity in milk is only ~2% of that of ALP and variation among individuals occurs. Its pH optimum is ~4.0 using phenylphosphate as substrate and its activity increases linearly between 14 and 50 °C. In contrast to ALP, it is not activated by Mg^{2+}.

Indigenous ACP may play a role in cheese ripening, possibly by dephosphorylating peptides. Taking into consideration the low pH optimum and the heat stability of ALP along with reports about several small partially dephosphorylated peptides isolated from some cheese varieties, a possible role of this enzyme during cheese ripening is under discussion [23]. However, a similar effect can result also from the action of bacterial proteinase [133, 134].

According to the study of Andrews [135], after low pasteurization there is significant residual activity of the enzyme in milk, which increases as the pH decreases, but UHT treatment inactivate it. Although ACP is one of the most heat resistant indigenous enzymes of bovine milk, its activity is reduced after 10 min treatment at pressures $\geq$ 200 MPa, in contrast to ALP behaviour. Furthermore, the majority of ACP activity has been lost within 10 min on exposure to pressures of 500 and 550 MPa [136].

12.3.2.4.3 Ribonuclease (RNase)

Milk is a rich source of RNase activity, i.e. 11-25 mg/l in bovine milk, which is located almost entirely in the serum phase. It has no technological significance in milk; however it has potent bactericidical and anti-viral activity. The characteristics of RNases found in milk and their physiological effects are presented by Meyer et al. [137], Stepaniak et al. [24] and Fox and Kelly [3]. Substantial RNase activity survives low pasteurization, but it is almost entirely destroyed after UHT treatment, whereas it is completely stable in the acid whey [137, 138].

12.3.2.5 Oxidases

12.3.2.5.1 Lactperoxidase (LPO)

Lactoperoxidase (LPO) is the predominant enzyme responsible for the antimicrobial properties of bovine milk; these properties require sufficient concentrations of hydrogen peroxide and thiocyanate ion. It is a glycoprotein with a molecular mass of 78 kDa that contains one haem group and ~10% carbohydrate. It has a monomeric and very ordered structure stabilized by eight disulfide bonds and a calcium ion [139]. The presentation of the main points of lactoperoxidase activity and its biological significance in milk is based on the reviews of Pruitt and Kamau [140], Pruitt [25], Seifu et al. [141], Fox and Kelly [2] and Walstra et al. [26].

Its presence in milk has been firstly reported in the 19th century and an assay developed by Storch in 1898 [142] that used phenylenediamine as reducing agent is still in use as an indicator of super-pasteurization, i.e. heat treatment at ≥76 °C for 15 s. It is the second most abundant enzyme in bovine milk after xanthine oxidoreductase. Its concentration in bovine milk is 0.4 μM, i.e. 10-30 μg/ml and it makes up about 0.5% of the whey proteins. It catalyses the reaction $H_2O_2+2HA \rightarrow 2H_2O+2A$, where substrate HA is an oxidisable substrate or a hydrogen donor that can be aromatic amine, phenol, aromatic acid or leukodyes. It catalyses also the oxidation of thiocyanatae (SCN^-) by H_2O_2 to $OSCN^-$. Both SCN^- and $OSCN^-$ are harmless for the animal but inhibit most bacteria that produce H_2O_2 themselves. The reaction of the LPO system in milk, which includes H_2O_2 and thiocyanate anion, depends on SCN^- concentration and pH and it is schematically presented in Figure 12.1. The limiting factors are the concentration of both SCN^- and H_2O_2; the concentration of the former is variable and depends on the cyanoglucoside content of the animal feed. Xanthine oxidoreductase (XOR, section 12.3.2.5.3) and superoxide dismutase (SOD, section 2.2.5.4) can form hydrogen peroxide from various sources enhancing thus the activity of LPO system. Principal functions of the enzyme may be the prevention of the accumulation of toxic levels of peroxide in the bovine udder and the protection of the nursing calf through the potent antibacterial properties of the oxidised forms of thiocyanate. The antimicrobial spectrum and the antiviral effect of LPO system in bovine milk is reported in the review of Seifu et al. [141]. The activation of LPO system in milk through the addition of thiocyanate ions and sodium percarbonate can be used for the preservation of raw milk during storage or for the extension of milk shelf life, if applied prior to pasteurization [139, 140, 143].

Effect of milk processing on LPO

LPO is one of the most heat stable indigenous enzymes in milk. After low pasteurization treatment (63 °C for 30 min or 72 °C for 15 s), there is enough residual LPO activity in milk, i.e. about 60% of that in the untreated milk [144], to catalyse the reaction involving thiocyanate and hydrogen peroxide. Its D-value at 80 °C is 4 s [26]. Complete inactivation is achieved at temperatures ≥78 °C for 15 s, and its thermal inactivation in the range of 69-73 °C could be accurately described by a first order kinetic model. Its heat stability is decreased at pH ~5.0, in which Ca^{2+} is released from the molecule [139, 144-148]. Due to its heat stability, LPO activity is used as an indicator for heat treatments of milk more severe than low pasteurization, e.g. super- or high-pasteurization [138] and its detection is carried out by means of various methods presented in the reviews of Seifu et al. [141] and Fox and Kelly [2].

Regarding the effect of high pressure (HP), LPO is considered to be a very baroresistant enzyme due to its highly ordered structure. It can withstand milk treatment at 400 MPa for 60 min, at 25 °C [48]. Residual LPO activity about 80% of the original after treatment at 800 MPa at 25-40 °C for 30 min has been reported [126]. Furthermore, milk treatment at 700 MPa for 140 min, at 65 °C does not cause substantial inactivation of LPO, whereas minor inactivation of the enzyme has been observed in whey treated under the same conditions. A very pronounced antagonistic effect of high pressure and high temperature occurs at 73 °C, at which pressures from 150 to 700 MPa inhibit completely the inactivation of the enzyme. Following this finding, it is suggested that in heat treatments involving heating at temperatures >70 °C, at which LPO is strongly inactivated, the application of relatively low pressure (50-100 MPa) might have potential on the antimicrobial activity of milk [148].

UHPH at 250 MPa cause significant reductions in LPO activity that are due to outlet temperatures. At 250 MPa, at which the outlet temperature has estimated as 84 °C, with a residence time of 20 s, LPO is fully inactivated [51]. Since, loss of its activity in UHPH milk has been found greater than that in heated milk, it has been suggested that the single-chain structure of LPO is more susceptible to inactivation by the extreme forces associated with this type of processing [56].

Inactivation of LPO has not been detected after treatment with exponential decaying high voltage pulses at a field strength 21.5 kV/cm; no inactivation has been found by the same authors for ALP [130]. According to Van Loey et al. [149] enzymes are generally more resistant to electric pulses than vegetative microorganisms. They report that LPO is not inactivated after HIPEF treatment of raw milk with up to 100 pulses of 5 μs and 19 kV/cm at 1 Hz. This is in accordance with the recent study of Riener et al. [121], who report that LPO activity can withstand HIPEF treatments from 15 to 34 kV/cm up to 80 pulses of 1 μs at 15 Hz.

Finally, microfiltered bovine milk is LPO positive as happens with milk treated under low pasteurization conditions [150].

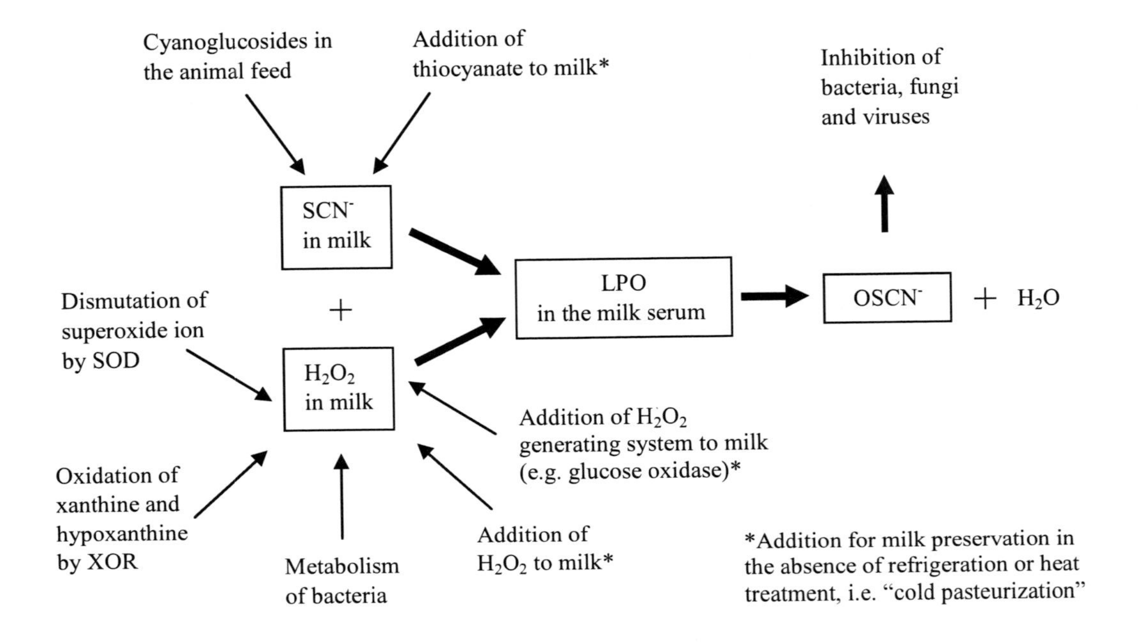

Figure 12.1. Schematic presentation of the LPO - SCN^- - H_2O_2 antimicrobial system in bovine milk.

12.3.2.5.2 Catalase

Milk catalase (H_2O_2:H_2O_2 oxidoreductase) is a haem protein that catalyses the decomposition of H_2O_2 as follows: $2H_2O_2 \rightarrow 2H_2O+O_2$. It has also peroxidase activity oxidizing acids, lower alipahatic alcohols and amines. Its properties, structure and presence in different milk phases have been reviewed by Kitchen et al. [151] and Ito and Akuzawa [152, 153]. Recent reviews about milk catalase are those of Farkye [10] and Fox and Kelly [2]. It comes from somatic cells and its activity in milk is affected by feed and by the stage of lactation. Its activity increases as the somatic cell counts increase and its determination has been proposed as an indicator of mastitis [17, 154]. However, due to the above-mentioned variations it is not considered as a reliable indicator and the somatic cell count is nowadays widely used for this purpose. Catalase activity is partitioning into the cream and skim milk fractions [151] and it is in general heat labile. About 26% of its activity is destroyed after milk thermalization at 60 °C for 16 s. After heat treatment at 72 °C for 16 s (low pasteurization), its activity is reduced by 92% [155, 156]. Its D-value at 80 °C is 2 s [26]. Reactivation of catalase observed during refrigerated storage of heat-treated milk or during cheese ripening has been attributed to heat resistant microorganisms [138, 156].

Almost zero reduction in catalase activity has been observed in buffer or in milk after treatment at 600 MPa [126].

12.3.2.5.3 Xanthine Oxidoreductase (XOR)

Xanthine oxidoreductase (XOR) of bovine milk has been extensively studied. It is a complex dimeric molybdenoflavoenzyme that contains two Mo centers, four $Fe_2S_2(Cys)_4$ groups and two flavin adenine dinucleotide (FAD) groups per molecule. It comes from blood and its activity depends on the Mo content of cow's feed. It is very abundant in bovine milk. It is the 20% of the protein content of the MFGM, which is ~700 mg/l in bovine milk; i.e. bovine milk contains ~140 mg XOR/l [157]. However, the high XOR content is a characteristic of bovine milk because its activity in the milk of other species is low.

Recent reviews about milk xanthine oxidoreductase and its physiological role have been presented by Harrison [11, 158-160] and Fox and Kelly [2]. It acts as a late enzyme of purine metabolism catalysing the oxidation of hypoxanthine to xanthin and of xanthin to uric acid. Apart from xanthine and hypoxanthine, there are many other substrates for XOR as nitrogen heterocycles and relatively simple aldehydes. The oxidation of XOR substrate yields very reactive superoxide ion and hydrogen peroxide. Therefore, it is considered as a potent prooxidant and its presence at high levels in milk is related to spontaneous oxidative rancidity. Hydrogen peroxide produced by the action of XOR can serve as a substrate for LPO in its action as antibacterial agent (section 12.3.2.5.1, Figure 12.1). XOR can also reduce nitrate (that is added to some cheese varieties) to nitrite, which prevents the growth of boutyric acid bacteria. The most important role of XOR in milk is in the secretion of milk fat, in which it does not act as an enzyme but as a protein. Furthermore, it is an important factor for the immune system of the neonate.

There are two forms of XOR in bovine milk, which are xanthine oxidase (XO) and xanthine dehydrogenase (XDH). A part of XO does not take part in the structure of MFGM and may affect milk quality [161]. Milk constituents like major milk proteins can act as activators or inhibitors of XOR activity, provided that they are present in sufficiently high concentrations in bovine milk [162].

Effect of milk processing on XOR

Treatments that damage or alter the MFGM, i.e. cooling, homogenization or heat treatment cause the release of XOR from the MFGM into the skim milk that renders the enzyme more active. However, its heat stability is higher in cream than in skim milk [2, 26]. XOR activity is enhanced by cold storage of milk; after 24 h, an increase of activity in the range of 60-100% has been reported [163, 164]. Crystallization onset of milk fat can be related to the activation of XOR [165]. Also, homogenization of fresh uncooled milk has been found to increase activity from 59 to 89% [163].

Similar effect has the heat treatment. Heating at 70 °C for 5 min almost duplicates the activity estimated in the untreated milk [163]. Heating at 60 °C for 5 min increases the activity of XOR in whole milk by 40%; the increase is more intense in the cream fraction [164]. Finally after bovine milk heat treatment at 80 °C for 120 s, residual XOR activity is detected. Heating at 91.4 °C for 4.2 s inactivates totally the enzyme [162]. Inactivation of XOR in bovine milk is accomplished at 73 °C for 7 min and its D-value at 80 °C is 17 s [26]. However, its behaviour under heat treatment depends on the previous treatments of milk (cold storage, homogenization), which can reduce its heat stability [166].

XOR is very resistant to high pressure treatment of milk at pressures lower than 400 MPa and its inactivation at higher pressures follows first order kinetics. After 12 min at 600 MPa and 25 °C its residual activity has been found 17% of the original and a pressure of 700 MPa for at least 10 min is required for the full inactivation of the enzyme [167].

12.3.2.5.4 Superoxide dismutase (SOD)

Superoxide dismutase is a superoxide oxidoreductase that catalyses the dismutation of superoxide ion to hydrogen peroxide that in turn can be reduced to H_2O and O_2 by catalase or lactoperoxidase in bovine milk (sections 12.3.2.5.1 and 12.3.2.5.2, Figure 12.1). It is an antioxidant constituent of milk that can counteract the autoxidation of milk fat. The type of enzyme present in milk serum is the CuZn-SOD type of eukaryotes, in concentrations about 150 times lower than that in blood. Its concentration is variable among individuals and different cow breeds. Mastitis does not affect its concentration. High SOD activity has been related to low oxidative rancidity in milk [3, 10, 168-170]. It is extremely heat resistant. Low pasteurization does not affect its activity; its D-value at 80 °C is 345 s and treatment at 75 °C for 65 min is necessary for its inactivation [26, 171].

12.3.2.6 Other indigenous enzymes

12.3.2.6.1 γ-glutamyl transferase (γ-GGT)

γ-glutamyl transferase (γ-GGT) catalyses the transfer of γ-glutamyl residues from γ-glutamyl containing peptides and plays a role in the transfer of γ-glutamyl residues from the blood into the mammary gland during milk biosynthesis. The characteristics of γ-GGT have been recently reviewed by Farkye [10] and Fox and Kelly [3]. In bovine milk, it is associated mainly with the membrane material; skim milk membrane material has >70% of the total activity in whole milk, the specific activity being 2 to 3 times higher in the skim milk membranes [172, 173].

Effect of milk processing on γ-GGT

Heat treatments at 65 °C with a 15 s holding time (thermalization) results in a slight decrease in activity. Residual activity more than 50% of the original is observed after low pasteurization treatment at 72 °C for 15 s, whereas less than 10% is observed after treatment at 75 °C for 15 s. This residual activity is not affected by storage at 37 or 40 °C. No activity is observed after heating at >77 °C for 15 s [173, 174]. Taking into consideration that its kinetics under heat treatments applied to milk are similar to that of lactoperoxidase, the γ-GGT activity has been proposed as an indicator of milk pasteurization at temperatures >77 °C [173, 175].

Due to its significance as a potential indicator for the high pasteurization of milk, γ-GGT behaviour under HP treatments has been studied [119, 129]. Its activity has been reported to be enhanced after short HP treatments in the range of 300-400 MPa, with lower pressure resulting in greater enhancement. The elongation of pressure times inactivates the enzyme, following a first order kinetic model [119]. Treatments of raw bovine milk at pressures 400-800 MPa for 0-128 min have showed that inactivation follows first order reaction kinetics. Complete inactivation could be achieved at 600 MPa and 20 °C for 30 min and γ-GGT is more pressure sensitive than ALP treated under the same conditions. Furthermore, the kinetics of inactivation of γ-GGT at 20 °C at HP>500 MPa have been reported to be close to the inactivation of *Listeria monocytogenes* and *Escherichia coli*; therefore it might provide a useful marker for their destruction [129].

12.3.2.6.2 N-acetyl-β-D-glucosaminidase (NAGase)

N-acetyl-β-D-glucosaminidase (NAGase) is a lysosomal enzyme, which catalyses the hydrolysis of terminal non-reducing N-acetyl-β-D-glucasamine residues from glycoproteins. Its optimum pH and temperature are 4.2 and 50 °C respectively. Most of NAGase in milk originates from leucocytes and mammary gland epithelial cells. Therefore, its activity in mastitic milk is higher than that in normal milk and for this reason it has been proposed as an indicator for bovine mastitis. During freezing-thawing of milk or during storage at temperatures >40 °C, it is released from somatic cells to the milk serum. Low pasteurization of bovine milk results in inactivation of NAGase [10, 20, 138, 176-179].

12.3.2.6.3 Amylases

In bovine milk there are α- and β-amylase the predominating type being α-amylase. The α-amylase of milk is salivary-type amylase and hydrolyse 1,4-α-D-glucosidic linkages in polysaccharides containing at least three 1,4 α-linked D-glucose units. The function of α-amylase in bovine milk is not clear because milk contains no starch and only low levels of oligosaccharides. Furthermore, the structure of milk oligosaccharides is not consistent with the catalytic specificity of the enzyme. About half of its activity is destroyed after low pasteurization [2, 10, 138].

12.3.2.6.4 Lysozyme

Lysozyme is also called muramidase (peptidoglucan-N-acetylmuranoylhydrolase) and cleaves β-1-4 linkages between N-acetylmuramic acid and N-acetyl-D-glucosamine residues in peptidoglycan, a constituent of bacterial cell wall. It is an antibacterial agent that is present in many body fluids and in the milk of many mammalian species that cause lysis of many bacteria types. A rich source of lysozyme is the egg-white of hen. Its concentration in mammalian milks varies from <0.3 mg/100 ml in bovine milk to about 79 mg/100 ml in equine milk, with the human milk having a high lysozyme content (10-20 mg/100 ml). It is a lysosomal enzyme and in milk is usually isolated from whey [3, 10]. The partial purification of lysozyme and the existence of two types of the enzyme in bovine milk have been reported [180, 181]. It has a molecular mass of 18 kDa and it is consisted of 154 amino acids. It is different from human milk lysozyme and egg-white lysozyme regarding amino acid sequence. Amino acid and three dimensional structures of lysozymes are similar to that of the protein α-lactalbumin, which takes part in the biosynthesis of lactose [182]. In the case of milk, the physiological role of lysozyme is to act as a 'spill-over' enzyme or it may have a definite protective role. However, no effect of lysozyme on the shelf-life or on physicochemical properties of milk has been reported [3, 44].

Bovine lysozyme is heat stable at pH 4.0, losing only 43% of its activity after 20 min at 100 °C. At pH 7.0, it is totally inactivated after 4 min at 100 °C [183] and it can survive heating of milk at 80 °C for 15 s [138].

12.3.3 Effect of processing on the extracellular enzymes of psychrotrophs

Psychrotrophs are a problem of tank milk that has been kept at low temperatures for more than 4-5 days. Also, milk tankers not rigorously cleaned can contaminate milk with high number of psychrotrophs [26]. Their heat stable extracellular proteinases and lipases affect negatively mainly the sensory characteristics and the shelf life of UHT milk. The inactivation of most extracellular enzymes from psychrotrophs requires very severe heat treatments that are not adequate for dairy products [184]. D-values of proteinase and lipase from *Pseudomonas* spp. at 130 °C are in the range of 160-700 s [26]. Some proteinases from psychrotrophs are susceptible to inactivation by heating at temperatures ~55 °C for some minutes. However, according to D-values estimated from their heat inactivation at temperatures >100 °C the duration of heat treatment at moderate temperatures would be several hours [27, 29, 184-188].

To avoid among others also the problems related to enzymes from psychrotrophs, thermalization, pasteurization, UHT treatment at 130-145 °C for <1 s combined with aseptic filling to avoid recontamination are used. Microfiltration and bactofugation that can remove spores can be used combined with heat treatments [26, 27]. Also, the enhancement of nature antimicrobial LPO system in milk by the addition of hydrogen peroxide and thiocyanate (Figure 12.1) and the treatment of milk with carbon dioxide or nitrogen have been proposed [29].

UHPH at 150, 200 and 250 MPa with an inlet temperature of 45 °C has been applied to raw milk, which was mixed with a culture of a strain of *Pseudomonas fluorescens* and a pressure dependant reduction of population from ~1 to 6 log cycles has been reported [51]. The treatment of a cell-free supernatant of *Ps. fluorescens* with UHPH at 50, 100 and 150 MPa at an inlet temperature 10-14 °C has no effect on proteinase activity, but 20% or 30% inactivation has been observed at 200 or 250 MPa respectively. Taking into consideration the heat stability of this proteinase, its inactivation has been attributed to the molecule stresses during UHPH [51].

The effect of various HIPEF treatments on highly heat-resistant proteinases of psychrotrophs *Bacillus subtilis* and *Pseudomonas fluorescens* in milk has been studied in a series of papers of Bendicho et al. and Vega-Mercado et al. which are presented in the reviews of Sampedro et al. [65]. According to their results, these proteinases are in general resistant to HIPEF treatments. Furthermore, the increase of electric field intensity of treatment duration and of frequency decreases proteinase activity. The results about the effect of HIPEF on enzymes of psychrotrophs are controversial and depend on the type and the origin of the enzyme and on the medium, in which it is suspended. Proteinase activity from *Bacillus subtilis* inoculated in milk can be inactivated up to 81% after applying 35.5 kV/cm for 866 μs, at 111 Hz, the highest levels of inactivation being achieved in skim milk rather than in whole milk. The decrease augments as treatment time and field strength and pulse repetition rate increase [189]. Riener et al. [121] have estimated a general proteinase activity in milk and they report about 35% inactivation at 35 kV/cm, for 80 μs.

A heat resistant lipase from *Pseudomonas fluoresescens* in milk has been proved to be quite resistant after treatment with HIPEF [190]. In particular, batch-mode HIPEF treatment at 27.4 kV/cm, 80 pulses, 2-3.5 Hz caused maximum lipase inhibition up to 62%; the continuous mode being practically ineffective.

Conclusion

Although indigenous milk enzymes are in general present in low quantities in milk, their significance in relation to processes applied to milk and during the manufacture of dairy products has been recognised since many decades. Milk processing may influence their activity and their distribution in the different milk phases. Both these changes can be critical for the quality of dairy products and their stability during storage, taking into consideration that many of them are heat and pressure resistant under the conditions applied to milk (Table 12.2).

Table 12.2. Effect of (low) pasteurization (63 °C for 30 min or 72 °C for 15 s) and of high-pressure (HP) treatment at the range of 400-500 MPa for 10-15 min on the activity of important indigenous enzymes of bovine milk.

Enzyme	Residual activity		References
	After (low) pasteurization [1]	After HP treatment [2]	
Plasmin and plasminogen activators	+++	+	[78-84, 90-95, 98]
Cathepsin D	+	++	[95, 98, 107, 109]
Lipoprotein lipase (LPL)	-	++	[26, 117, 119]
Alkaline phosphatase (ALP)	-	++	[26, 124-129]
Acid phosphatase (ACP)	+	very low	[135, 136]
Lactoperoxidase (LPO)	+	++	[48, 126, 144, 148]
Catalase	very low	++	[126, 155, 156]
Xanthine oxidoreductase (XOR)	++	++	[162-164, 167]
Superoxide dismutase (SOD)	++		[26, 171]
γ-glutamyl transferase (γ-GGT)	+	++	[119, 129, 173-175]
N-acetyl-β-D-glucosaminidase (NAGase)	-		[138, 179]
Amylases	+		[138]
Lysozyme	++		[138]

[1] 63 °C for 30 min or 72 °C for 15 s; [2] high-pressure (HP) treatment at the range of 400-500 MPa for 10-15 min

Moreover, the majority of them are relatively stable in milk processed using methods alternative to heat treatment. Therefore, enzymes are a topic of major significance for both Dairy Science and Dairy Industry. Finally, information about the enzymes of ovine and caprine milk is rather scarce, although these kinds of milk are mainly used for the production of fermented dairy products, i.e. yoghourt and cheese, in which enzymatic activities have a significant effect.

REFERENCES

[1] Fox, P.F. (2003). Indigenous enzymes in milk. In P.F., Fox, & P.L.H., McSweeney (Eds.), *Advanced Dairy Chemistry: Vol. 1 Proteins* (3rd edition, Part A, pp. 467-472). New York, USA: Kluwer Academic/Plenum.

[2] Fox, P.F., & Kelly, A.L. (2006). Indigenous enzymes in milk: Overview and historical aspects - Part 1. *International Dairy Journal*, *16*, 500-516.

[3] Fox, P.F., & Kelly, A.L. (2006). Indigenous enzymes in milk: Overview and historical aspects - Part 2. *International Dairy Journal*, *16*, 517-532.

[4] Kelly, A.L., O'Flaherty, F., & Fox, P.F. (2006). Indigenous proteolytic enzymes in milk: A brief overview of the present stage of knowledge. *International Dairy Journal*, *16,* 563-572.

[5] Silanikove, N., Merin, U., & Leitner, G. (2006). Physiological role of indigenous milk enzymes: An overview of an evolving picture. *International Dairy Journal*, *16*, 533-545.

[6] Atmani, D., Benboubetra, M., & Harrison, R. (2004). Goat's milk xanthine oxidoreductase is greatly deficient in molybdenum. *Journal of Dairy Research*, *71*, 7-13.

[7] Benboubetra, M., Baghiani, A., Atmani, D., & Harrison, R. (2004). Physicochemical and kinetic properties of purified sheep's milk xanthine oxidoreductase. *Journal of Dairy Science*, *87*, 1580-1584.

[8] Vamvakaki, A.-N., Zoidou, E., Moatsou, G., Bokari, M., & Anifantakis, E. (2006). Residual alkaline phosphatase activity after heat treatment of ovine and caprine milk. *Small Ruminant Research*, *65*, 237-241.

[9] Chandan, R.C., Parry, R.M.Jr., & Shahani, K.M. (1968). Lysozyme, lipase and ribinuclease in the milk of various species. *Journal of Dairy Science*, *51,* 606-608.

[10] Farkye, N.Y. (2003). Other Enzymes. In P.F., Fox, & P.L.H., McSweeney (Eds.), *Advanced Dairy Chemistry: Vol. 1 Proteins* (3rd edition, Part A, pp. 571-603). New York, USA: Kluwer Academic/Plenum.

[11] Harrison, R. (2006). Milk xanthine oxidase: Properties and physiological roles. *International Dairy Journal*, *16*, 546-554.

[12] Kelly, A.L., & McSweeney, P.L.H. (2003). Indigenous proteinases in milk. In P.F., Fox, & P.L.H., McSweeney (Eds.), *Advanced Dairy Chemistry: Proteins, Vol. 1* (3rd edition, Part A, pp. 495-522). New York, USA: Kluwer Academic/Plenum.

[13] Politis, I., Lachance, E., Block, E., & Turner, J. D. (1989). Plasmin and plasminogen in bovine milk: A relationship with involution. *Journal of Dairy Science*, *72*, 900-908.

[14] Baldi, A., Savoini, G., Cheli, F., Fantuz, F., Senatore, E., Bertochi, L., & Politis, I. (1996). Changes in plasmin-plasminogen-plasminogen activator system in milk from Italina Friesian herds. *International Dairy Journal*, *6*, 1045-1053.

[15] Zachos, T., Politis, I., Gorewit, R.C., & Barbano, D.M. (1992). Effect of mastitis on plasminogen activator activity of milk somatic cells. *Journal of Dairy Research*, *59*, 462-467.

[16] Leitner, G., Krifucks, O., Merin, U., Lavi, Y., & Silanikove, N. (2006). Interactions between bacteria type, proteolysis of casein and physico-chemical properties of bovine milk. *International Dairy Journal*, *16*, 648-654.

[17] Kitchen, B.J. (1981). Review of the progress of Dairy Science, bovine mastitis: compositional changes and related diagnostic tests. *Journal of Dairy Research*, *48*, 167-188.

[18] Larsen, L.B., McSweeney, P.L.H., Hayes, M.G., Andersen, J.B., Ingvartsen, K.L., & Kelly, A.L. (2006). Variation in activity and heterogeneity of bovine milk proteinases with stage of lactation and somatic cell count. *International Dairy Journal*, *16*, 1-8.

[19] Andrews, A.T., & Alichanidis, E. (1975). The acid phosphatases of bovine leucocytes, plasma and the milk of healthy and mastitic cows. *Journal of Dairy Research*, *42*, 391-400.

[20] Kaartinen, L., & Jensen, N.E. (1988). Use of N-acetyl-beta-glucosaminidase to detect teat can inflammations. *Journal of Dairy Science*, *55*, 603-607.

[21] Azzara, C.D., & Dimick, P.S. (1985). Lipoprotein lipase activity of milk from cows with prolonged subclinical mastitis. *Journal of Dairy Science*, *68*, 3171-3175.

[22] Olivecrona, T., Vilaró, S., & Olivecrona G. (2003). Lipases in milk. In P.F., Fox, & P.L.H., McSweeney (Eds.), *Advanced Dairy Chemistry: Vol. 1 Proteins* (3rd edition, Part A, pp. 473-494). New York, USA: Kluwer Academic/Plenum.

[23] Shakeel-Ur-Rehman, Farkye, N.Y., & Fox, P.F. (2003). Indigenous phosphatases of milk. In P.F., Fox, & P.L.H., McSweeney (Eds.), *Advanced Dairy Chemistry: Proteins, Vol. 1* (3rd edition, Part A, pp. 524-543). New York, USA: Kluwer Academic/Plenum.

[24] Stepaniak, L., Fleming., C.M., Gobbetti, M., Corsetti, A., & Fox, P.F. (2003). Indigenous nucleases in milk. In P.F., Fox, & P.L.H., McSweeney (Eds.), *Advanced Dairy Chemistry: Proteins, Vol. 1* (3rd edition, Part A, pp. 545-566). New York, USA: Kluwer Academic/Plenum.

[25] Pruitt, K.M. (2003). Lactoperoxidase. In P.F., Fox, & P.L.H., McSweeney (Eds.), *Advanced Dairy Chemistry: Proteins, Vol. 1* (3rd edition, Part A, pp. 563-570). New York, USA: Kluwer Academic/Plenum.

[26] Walstra, P., Wouters, J.T.M., & Guerts, T. J. (2006). *Dairy Science and Technology*, (2nd edition). Boca Raton FL, USA: CRC Press.

[27] Sørhaug, T., & Stepaniak, L. (1997). Psychroytophs and their enzymes in milk and dairy products: Quality aspects. *Trends in Food Science and Technology*, *8*, 35-42.

[28] Barbano, D.M., Ma, Y., & Sants, M.V. (2006). Influence of raw milk quality of fluid milk shelf-life. *Journal of Dairy Science*, *89*, *(E.Suppl.):* E15-E19.

[29] Shah, N.P. (1994). Psychrotrophs in milk: A Review. *Milchwissenschaft*, *49*, 432-437.

[30] McSweeney, P.L.H., & Sousa, M.J. (2000). Biochemical pathways for the production of flavour compounds in cheese during ripening. A review. *Lait*, *80*, 293-324.

[31] McSweeney, P.L.H. (2004). Biochemistry of cheese ripening. *International Journal of Dairy Technology*, *57*, 127-144.

[32] McSweeney, P.L.H., & Fox, P.F. (2004). Metabolism of residual lactase and of lactate and citrate. In P.F., Fox, P.L.H., McSweeney, T.M., Cogan, & T.P., Guinee (Eds.), *Cheese: Chemistry, Physics and Microbiology, Volume 1* (3rd ed., pp. 361-372). London, UK: Elsevier.

[33] Beresford, T., & Williams, A. (2004). The microbiology of cheese ripening. In P.F., Fox, P.L.H., McSweeney, T.M., Cogan, & T.P., Guinee (Eds.), *Cheese: Chemistry, Physics and Microbiology* (3rd edition, pp. 287-318). London, UK: Elsevier Academic Press.

[34] Tamime, A.Y., Skriver, A. & Nilsson, L.E. (2006). Starter cultures. In: A.Y. Tamime (Ed) *Fermented Milks* (pp. 11-52). Oxford, UK: Blackwell Science Ltd.

[35] Crabbe, M.J.C. (2004). Rennets: General and molecular aspects. In P.F., Fox, P.L.H., McSweeney, T.M., Cogan, & T.P., Guinee (Eds.), *Cheese: Chemistry, Physics and Microbiology* (3rd edition, pp. 19-46). London UK: Elsevier Academic Press.

[36] Horne, D.S., & Banks, J.M. (2004). Rennet Induced coagulation of milk. In P.F., Fox, P.L.H., McSweeney, T.M., Cogan, & T.P., Guinee (Eds.), *Cheese: Chemistry, Physics and Microbiology* (3rd edition, pp. 47-70). London, UK: Elsevier Academic Press.

[37] Guinee, T.P., & Wilkinson, M.G. (1992). Rennet coagulation amd coagulants in cheese manufacture. *Journal of the Society of Dairy Technology*, *45*, 94-103.

[38] Chitpinityol, S., & Crabbe, M.J.C. (1998). Review: Chymosin and aspartic acid proteinases. *Food Chemistry*, *61*, 395-418.

[39] Upadhyay, V.K., McSweeney, P.L.H., Magboul, A.A.A., & Fox, P.F. (2004). Proteolysis in cheese during ripening. In P.F., Fox, P.L.H., McSweeney, T.M., Cogan, & T.P. Guinee (Eds.), *Cheese: Chemistry, Physics and Microbiology, Vol. 1* (3rd edition, pp. 391-434). London, UK: Elsevier.

[40] McSweeney, P.L.H., Hayaloglu, A.A., O'Mahony, J.A., & Bansal, N. (2006). Prespectives on cheese ripening. *Australian Journal of Dairy Technology*, *61*, 69-77.

[41] Bansal, N., Fox, P.F., & McSweeney, P.L.H. (2007). Factors affecting the retention of rennet in cheese curd. *Journal of Agricultural and Food Chemistry*, *55*, 9219-9225.

[42] Fox, P.F., Wallace, J.M., Morgam, S., Lynch, C.M., Niland, E.J., & Tobin, J. (1996). Acceleration of cheese ripening. *Antonie van Leeuwenhoek, International Journal of General and Molecular Microbiology*, *70*, 271-279.

[43] Law, B.A. (2001). Controlled and accelerated cheese ripening: The research base for new technologies. *International Dairy Journal*, *11*, 383-398.

[44] Fox, P.F., & McSweeney, P.L.H. (1998). *Dairy Chemistry & Biochemistry*. London, UK: Blackie Academic & Professional.

[45] Balci, A.T., & Wilbey, R.A. (1999). High-pressure processing of milk – the first 100 years in the development of a new technology. *International Journal of dairy Technology*, *52*, 149-155.

[46] Patterson, M.F. (2005). Microbiology of pressure-treated foods. *Journal of Applied Microbiology*, *98*, 1400-1409.

[47] Huppertz, T., Smiddy, M.A., Upadhyay, V.K., & Kelly, A.L. (2006). High-pressure-induced changes in bovine milk: A review. *International Journal of Dairy Technology*, *59*, 58-66.

[48] Lopez-Fandiño, R. (2006). High Pressure induced changes in milk proteins and possible applications in dairy technology. *International Dairy Journal*, *16*, 1119-1131.

[49] Considine, K.M., Kelly, A.L., Fitzgerald, G.F., Hill, C., & Sleator R.D. (2008). High-pressure processing – Effects on microbiological food safety and food quality. *FEMS Microbiology Letters*, *281*, 1-9.

[50] Picart, L., Thiebaud, M., René, M., Guiraud, J.P., Chefter, J.C., & Dumay, E. (2006). Effects on high-pressure homogenization of raw bovine milk on alkaline phosphatase and microbial inactivation. A comparison with continuous short-time thermal treatments. *Journal of Dairy Research*, *73*, 454-463.

[51] Hayes, M.G., Fox, P.F., & Kelly, A.L. (2005). Potential applications of high pressure homogenization in processing of liquid milk. *Journal of Dairy Research*, *72*, 25-33.

[52] Pereda, J., Ferragut, V., Guamis, B., & Trujillo, A. J. (2006). Effect of ultra high-pressure homogenization on natural occuring microorganims in bovine milk. *Milchwissenschaft*, *61*, 245-248.

[53] Pereda, J., Ferragut, V., Quevedo, J.M., Guamis, B., & Trujillo, A. J. (2007). Effects of ultra high-pressure homogenization on microbila and physicochemical shelf-life of milk. *Journal of Dairy Science*, *90*, 1081-1093.

[54] Wuytack, E.Y., Diels, A.M.J., & Michielis, C.W. (2002). Bacterial inactivation by high pressure homogenization and high hydrostatic pressure. *International Journal of Food Microbiology*, *77*, 205-212.

[55] Rodríguez-Alcalá, L.M., Harte, F., & Fontecha, J. (2009). Fatty acid profile and CLA isomers content of cow, ewe and goat milks processed by high pressure homogenization. *Innovative Food Science and Emerging Technologies*, *10*, 32-36.

[56] Datta, N., Hayes, M.G., Deeth, H.C., & Kelly, A.L. (2005). Significance of frictional heating for effects of high-pressure homogenization on milk. *Journal of Dairy Research*, *72*, 393-399.

[57] Pereda, J., Ferraut, V., Quevedo, J.M., Guamis, B., & Trujillo, A.J. (2008). Effects of ultra-high-pressure homogenization treatment on the lipolysis and lipid oxidation of milk during refrigerated storage. *Journal of Agricultural and Food Chemistry*, *56,* 7125-7130.

[58] Pereda, J., Jaramillo, D.P., Quevedo, J.M., Ferragut, V., Guamis, B., & Trujillo A.J. (2008). Characterization of volatile compounds in ultra-high-pressure homogenized milk. *International Dairy Journal*, *18*, 826-834.

[59] Hayes, M.G., & Kelly, A.L. (2003). High pressure homogenization of raw bovine milk (a) effects on fat globule size and other properties. *Journal of Dairy Research*, *70*, 297-305.

[60] Thiebaud, M., Dumay, C., Picart, L., Guiraud, J.P., & Cheffel, J.C. (2003). High-pressure homogenization of raw bovine milk. Effects on fat globule size distribution and microbial contamination. *International Dairy Journal*, *13*, 427-439.

[61] Zamora, A., Ferragut, V., Jaramillo, P.D., Guamis, B., & Trujillo, A.J. (2007). Effects of ultra-high-pressure homogenization on the cheese making properties of milk. *Journal of Dairy Science*, *90*, 13-23.

[62] Hayes, M.G., & Kelly, A.L. (2003). High pressure homogenization of milk (b) effects on indigenous enzymatic activity. *Journal of Dairy Research*, *70*, 307-313.

[63] Pereda, J., Ferragut, V., Buffa, M., Guamis, B., & Trujillo, A.J. (2008c). Proteolysis of ultra high-pressure homogenized treated milk during refrigerated storage. *Food Chemistry*, *111*, 696-702.

[64] Mosqueda-Melgar, J., Elez-Martínez, P., Raybaudi-Massilia, R.M., & Martín-Belloso, O. (2008). Effects of pulsed electric fields on pathogenic microorganisms of major concern in fluid foods: A review. *Critical Reviews in Food Science and Nutrition*, *48*, 747-759.

[65] Sampedro, F., Rodrigo, M., Martínez, A., Rodrigo, D., & Barbosa-Cánovas, G.V. (2005). Quality and safety aspects of PEF application in Milk and Milk Products. *Critical Reviews in Food Science and Nutrition*, *45*, 25-47.

[66] Odriozola-Serrano, I., Bendicho-Porta, S., & Martin-Belloso, O. (2006). Comparative study on Shelf life of whole milk processed by high-intensity pulsed electric field or heat treatment. *Journal of Dairy Science*, *89*, 905-911.

[67] Pouliot, Y. (2008). Membrane process in the dairy technology – From a simple aidea to worldwide panacea. *International Dairy Journal, 18*, 735-740.

[68] Goff, H.D., & Griffiths, M.W. (2006). Major advances in fresh milk products: Fluid milk products and frozen desserts. *Journal of Dairy Science, 89*, 163-1173.

[69] Saboya, L.V., & Maubois, J.-C. (2000). Current developments of microfiltration technology in the dairy industry. *Le Lait, 80*, 541-553.

[70] Elwell, M.W., & Barbano, D.M. (2006). Use of microfiltration to improve fluid milk quality. *Journal of Dairy Science, 89 (E. Suppl.):* E10-E30.

[71] Bastian, E.D., & Brown, R.J. (1996). Plasmin in milk and Dairy Products: An Update. *International Dairy Journal, 6*, 435-457.

[72] Politis, I., Zhao, X., McBride, B.W., Burton, J.H., & Turner, J.D. (1991). Plasminogen activator production by bovine milk macrophages and blood monocytes. *American Journal of Veterinary Research, 52*, 1208-1213.

[73] Politis, I., Barbano, D.M., & Gorewit, R.C. (1992). Distribution of plasminogen and plasmin in fractions of bovine milk. *Journal of Dairy Science, 75*, 1402-1410.

[74] Crudden, A., Fox, P.F., & Kelly, A.L. (2005). Factors affecting the hydrolytic action of plasmin in milk. *International Dairy Journal, 15*, 305-313.

[75] McMahon, D.J., & Brown, R.J. (1984). Enzymic coagulation of casein micelles: A review. *Journal of Dairy Science, 67*, 919-929.

[76] Bastian, E.D., Brown, R.J., & Ernstrom, A. (1991). Plasmin activity and milk coagulation. *Journal of Dairy Science, 74*, 3677-3685.

[77] Considine, T., McSweeney, P.L.H., & Kelly, A.L. (2002). The effect of lysosomal proteinases and plasmin on the rennet coagulation properties of skim milk. *Milchwissenschaft, 57*, 425-428.

[78] Alichanidis, E., Wrathall, J.H., & Andrews, A.T. (1986). Heat stability of plasmin (milk proteinase) and plasminogen. *Journal of Dairy Research, 53*, 259-269.

[79] Saint-Denis, T., Humbert, G., & Gaillard, J.L. (2001). Enzymatic assays for native plasmin, plasminogen and plasminogen activators in bovine milk. *Journal of Dairy Research, 68*, 437-449.

[80] Metwalli, A.A.M., De Jong, H.H.J., & Van Boeckel, M.A.J.S. (1998). Heat inactivation of bovine plasmin. *International Dairy Journal, 8*, 47-56.

[81] Richardson, B.C. (1983). The proteinases of bovine milk and the effect of pasteurization on their activity. *New Zealand Journal of Dairy Science and Technology, 18*, 233-245.

[82] Prado, B.M., Sombers, S.E., Ismail, B., & Hayes, K.D. (2006). Effect of heat treatment on the activity of inhibitors of plasmin and plasminogen activators in milk. *International Dairy Journal, 16*, 593-599.

[83] Burbrink, C.N., & Hayes, K.D. (2006). Effect of thermal treatment on the activation of bovine plasminogen. *International Dairy Journal, 16*, 580-585.

[84] Kennedy, A., & Kelly, A. L. (1997). The influence of somatic cell count on the heat stability of bovine milk plasmin activity. *International Dairy Journal, 7*, 717-721.

[85] Kohlmann, K.L., Neilson, S.S., & Ladish, M.R. (1988). Effect of serine proteolytic enzymes (trypsin and plasmin), trypsin inhibitor and plasminogen activator addition to ultra-high temperature processed milk. *Journal of Dairy Science, 71*, 1728-1739.

[86] Harwalkar, V.R., Cholette, H., McKellar, R.C., & Emmons, D.B. (1993). Relation between proteolysis and astringent off-flavor in milk. *Journal of Dairy Science*, *76*, 2521-2527.

[87] Kelly, A.L., & Foley, J. (1997). Proteolysis and storage stability of UHT milk as influenced by milk plasmin activity, plasmin/β-lactoglobulin complexation, plasminogen activation and somatic cell count. *International Dairy Journal*, *7*, 411-420.

[88] Kelly, A.L., & McSweeney, P.L.H. (1997). Proteolysis and storage stability of UHT milk as influenced by milk plasmin activity, plasmin/β-lactoglobulin complexation, plasminogen activation and somatic cell count. *International Dairy Journal*, *7*, 411-420.

[89] Auldist, M.J., Coats, S.J., Sutherland, B.J., Hardman, J.F., McDowell, G.H., & Rogers, G.L. (1996). Effect of somatic cell count and stage of lactation on the quality and storage life of ultra high temperature milk. *Journal of Dairy Research*, *63*, 377-386.

[90] Scollard, P.G., Beresford, T.P., Murphy, P.M. & Kelly, A.L. (2000). Barostability of milk plasmin activity. *Le Lait*, *80*, 609-619.

[91] Scollard, P.G., Beresford, T.P., Needs, E.C., Murphy, P.M., & Kelly, A.L. (2000). Plasmin activity, β-lactoglobulin denaturation in high-pressure-treated milk. *International Dairy Journal*, *10*, 835-841.

[92] García-Risco, M.R., Cortes, E., Carrascosa, A.V., & López-Fandiño, R. (2003). Plasmin activity in pressurized milk. *Journal of Dairy Science*, *86*, 728-734.

[93] Borda, D., Indrawati, Smout, C., Van Loey, A., & Hendrickx, M. (2004). High pressure thermal inactivation of a plasmin system. *Journal of Dairy Science*, *87*, 2351-2358.

[94] Huppertz, T., Fox, P.F., & Kelly, A.L. (2004). Plasmin activity and proteolysis in high-pressure treated milk. *Le Lait*, *84*, 297-304.

[95] Moatsou, G., Bakopanos, C., Katharios, D., Katsaros, G., Kandarakis, I., Taoukis, P., & Politis, I. (2008). Effect of high-pressure treatment at various temperatures on indigenous proteolytic enzymes and whey protein denaturation in bovine milk. *Journal of Dairy Research*, *75*, 262-269.

[96] Needs, E.C., Stenning, R.A., Gill, A.L, Ferragut, V., & Rich, G.T. (2000). High-pressure treatment of milk: effects on casein micelle structure and on enzymatic coagulation. *Journal of Dairy Research*, *67*, 31-42.

[97] Huppertz, T., Fox, P.F., & Kelly, A.L. (2004). High-pressure treatment of bovine milk: effects on casein micelles and whey proteins. *Journal of Dairy Research*, *71*, 97-106.

[98] Moatsou G., Katsaros, G., Bakopanos, C., Kandarakis, I., Taoukis, P., & Politis, I. (2008). Effect of high-pressure treatment at various temperatures on activity of indigenous enzymes and denaturation of whey proteins in ovine milk. *International Dairy Journal*, *18,* 1119-1125.

[99] Vega-Mercado, H., Powers, J.R., Barbosa-Canovas G.V., & Swanson B.G. (1995). Plasmin inactivation with pulsed electric fields. *Journal of Food Science*, *60*, 1143-1146.

[100] Benfeldt, K. (2006). Ultrafiltration of cheese milk: Effect of plasmin activity and proteolysis during cheese ripening. *International Dairy Journal*, *16*, 600-608.

[101] Crudden, A., & Kelly, A.L. (2003). Studies of plasmin activity in the whey. *International Dairy Journal*, *13*, 987-993.

[102] Cooney, S., Tiernan, D., Soyce, P., & Kelly A.L. (2000). Effect of somatic cell count and polymorphonuclear leucocyte content on the ripening of Swiss cheese. *Journal of Dairy Research*, *67*, 301-307.

[103] Marino, R., Considine, T., Sevi, A., McSweeney, P.L.H., & Kelly A.L. (2005). Contribution of proteolytic activity associated with somatic cells in milk to cheese ripening. *International Dairy Journal*, *15*, 1026-1033.

[104] Kaminogawa, S., & Yamauchi, K. (1972). Acid proteinase of bovine milk. *Agricultural and Biological Research*, *36*, 2351-2356.

[105] Hurley, M.J., Larsen, L.B., Kelly, A.L., & McSweeney, P.L.H. (2000). The milk acid proteinase cathepsin D: A review. *International Dairy Journal*, *10*, 673-681.

[106] Larsen, L.B., Boisen, A., & Petersen, T.E. (1993). Procathepsin D cannot autoactivate to cathepsin D at acid pH. *FEBS Letters*, *319*, 54-58.

[107] Larsen, L.B., Wium, H., Benfeldt, C., Heegaard, C.W., Ardo, Y., Qvist, K.B., & Petersen, T.E. (2000). Bovine milk procathepsin D: presence and activity in heated milk and in extracts of rennet-free UF-Feta cheese. *International Dairy Journal*, *10*, 67-73.

[108] McSweeney, P.L.H., Fox, P.F., & Olson, N.F. (1995). Proteolysis of bovine caseins by cathepsin D: Preliminary observations and comparison with chymosin. *International Dairy Journal*, *5*, 321-336.

[109] Hayes, M.G., Hurley, M.J., Larsen, L.B., Heegaard, C.W., Magboul, A. A.A, Oliveira, J.C., McSweeney, P.L.H., & Kelly, A.L (2001). Thermal inactivation kinetics of bovine cathepsin D. *Journal of Dairy Research*, *68*, 267-276.

[110] Susuki, J., & Kato, N. (1990). Cysteine proteinase in bovine milk capable of hydrolyzing caseins as substrate and elevation of the activity during the course of mastistis. *Japanese Journal of Veterinary Science*, *52*, 947-954.

[111] O'Driscoll, B.M, Rattray, F.P., McSweeney, P.L.H., & Kelly A.L. (1999). Proteinase activities in raw milk determined using a synthetic heptapeptide substrate. *Journal of Food Science*, *64*, 606-611.

[112] Magboul, A.A.A., Larsen, L.B., McSweeney, P.L.H., & Kelly, A.L. (2001). Cysteine proteinase activity in bovine milk. *International Dairy Journal*, *11*, 865-872.

[113] Considine, T., Healy, A., Kelly, A.L., & McSweeney, P.L.H. (2004). Hydrolysis of bovine caseins by cathepsin B, acysteine proteinase indigenous to milk. *International Dairy Journal*, *14*, 117-124.

[114] Collins, Y.F., McSweeney, P.L.H., & Wilkinson, M.G. (2003). Lipolysis and free fatty acid catabolism in cheese: A review of the current knowledge. *International Dairy Journal*, *13,* 841-866.

[115] Deckelbaum, R.J., Hamilton, J.A., Moser, A., Bengtsson-Olivecrona, G., Butbul, E., Carpentier, Y.A., Gutman, A., & Olivecrona, T. (1990). Medium-chain vs long-chain triaglycerol emulsion hydrolysis by lipoprotein lipase and hepatic lipase: Implications for the mechanism of lipase action. *Biochemistry*, *29*, 1136-1142.

[116] Deeth, H.C., & Fitz-Gerald, C.H. (1975). Factors governing the susceptibility of milk to spontaneous lipolysis. In IDF (Ed) *Proceedings of the lipolysis symposium* (IDF Bulletin No 86, pp 24-34). Brussels, Belgium: International Dairy Federation.

[117] Deeth, H.C. (2006). Lipoprotein lipase and lipolysis in milk. *International Dairy Journal*, *16*, 555-562.

[118] Deeth, H.C., & Fitz-Gerald, C.H. (1995). Lipolytic enzymes and hydrolytic rancidity in milk and dairy products. In P.F. Fox (Ed), *Advanced Dairy Chemistry: Lipids, Vol. 2* (2nd edition, pp. 247-308). London, UK: Chapman & Hall.

[119] Padney, P.K., & Ramaswamy, H. S. (2004). Effect of high-pressure treatment of milk on lipase and γ-glutamyl transferase activity. *Journal of Food Biochemistry*, *28*, 449-462.

[120] Ho, S.Y., Mittal, G.S., & Gross J.D. (1997). Effects of high field electric pulses on the activity of selected enzymes. *Journal of Food Engineering*, *31*, 69-84.

[121] Riener, J., Noci, F., Cronin, D.A., Morgan, D.J., & Lyng, J.G. (2008). Effect of high intensity pulsed electric field on enzymes and vitamins in bovine raw milk. *International Journal of Dairy Technology*, *62*, 1-6.

[122] Bingham, E.W., Garrer, K., & Powlem, D. (1992). Purification and properties of alakaline phosphatase in the lactating mammary gland. *Journal of Dairy Science*, *75*, 3394-3401.

[123] Shakeel-Ur-Rehman, Farkye, N. Y. & Yim, B. (2006). A preliminary study on the role of alkaline phosphatase in cheese ripening. *International Dairy Journal*, *16*, 697-700.

[124] Clays, W.L., Indrawati, Van Loey, A.M., & Hendrickx M.E. (2003). Review: Are intrinsic TTIs for thermally processed milk applicable for high-pressure processing assessment? *Innovative Food Science and Emerging Technologies*, *4*, 1-14.

[125] Lopez-Fandiño, R., Carrascosa A.V., & Olano, A. (1996). The effects of high pressure on whey protein denaturation and cheese making properties of raw milk. *Journal of Dairy Science*, *79*, 929-936.

[126] Seyderhelm, I., Bogulawski, S., Michaelis, G., & Knorr, D. (1996). Pressure induced inactivation of selected food enzymes. *Journal of Food Science*, *61*, 308-313.

[127] Mussa, D.M., & Ramaswamy, H.S. (1997). Ultra high pressure pasteurization of milk: kinetics of microbial destruction and changes in physicochemical characteristics. *Lebensmittel Wissenschaft und Technologie*, *30*, 551-558.

[128] Ludikhuyze, L., Claeys, W., & Hendrickx, M. (2000). Combined pressure-temperature inactivation of alkaline phosphatase in bovine milk: A kinetic study. *Journal of Food Science*, *65*, 155-160.

[129] Radenmacher, B., & Hinrichs, J. (2006). Effects of high pressure treatment on indigenous enzymes in bovine milk: Reaction kinetics, inactivation and potential application. *International Dairy Journal*, *16*, 655-661.

[130] Grahl, T., & Märkl, H. (1996). Killing microorganisms by pulsed electric fields. *Applied Microbiology and Biotechnology, 45,* 148-157.

[131] Shamsi, K., Versteeg, C., Sherkat, F. & Wan, J. (2008). Alkaline phosphatase and microbial inactivation by pulsed electric field in bovine milk. *Innovative Food Science and Emerging Technologies*, *9*, 217-223.

[132] Bindith J.L., Cordier, J.L., & Jost, R. (1996). Cross-flow microfiltration of skim milk, germ production and effect on alkaline phosphatase and serum proteins. In IDF (Ed) *Heat treatments & Alternative methods: Proceedings of the IDF Symposium held in Vienna, Austria, 6-8 Sept. 1995* (IDF Special Issue 9602, pp. 222-231). Brussels, Belgium: International Dairy Federation.

[133] Andrews, A. T. & Alichanidis, E. (1975b). Acid phosphatase activity in cheese and starters. *Journal of Dairy Research, 42*, 327-339.

[134] Akuzawa, R. & Fox, P. F. (2004). Acid phosphatase in cheese. *Animal Science Journal, 75,* 385-391.

[135] Andrews, A.T. (1974). Bovine milk acid phosphatase. II. Binding to casein substrates and heat-inactivation studies. *Journal of Dairy Research*, *42*, 401-417.

[136] Balci, A.T., Ledward D.A., & Wilbey, R.A. (2002). Effect of high pressure on acid phosphatase in milk. *High Pressure Research*, *22*, 639-642.

[137] Meyer, D.H., & Kunin, A.S., Maddalena, J., & Meyer, W.L. (1987). Ribonuclease activity and isoenzymes in raw and processed cows' milk and infant formulas. *Journal of Dairy Science*, *70*, 1797-1803.

[138] Griffiths, M.W. (1986). Use of milk enzymes as indices of heat treatment. *Journal of Food Protection*, *49*, 696-705.

[139] Kussendrager, K.D., & Van Hooijdonk, A.C.M. (2000). Lactoperoxidase: physicochemical properties, occurrence, mechanism of action and applications. *British Journal of Nutrition*, *84*, 519-525.

[140] Pruitt, K.M., & Kamau, D.N. (1994). Quantitative analysis of bovine lactoperoxidase system on bacterial growth and survival. In IDF (Ed) *Proceedings of the International Dairy Federation Seminar: Indigenous antimicrobial agents of milk – Recent developments, Uppsala Sweeden 31 Aug-1 Sept. 1993.* (IDF, Special Issue 9404, pp. 73-87). Brussels, Belgium: International Dairy Federation.

[141] Seifu, E., Buys, E.M., & Donkin, E.F. (2005). Significance of the lactperoxidase system in dairy industry and its potential application: A Review. *Trends in Food Science and Technology*, *16*, 137-154.

[142] Storch, V. (1898). Chemical Method of ascertaining whether milk has been heated to at least 80 °C. *Centra Agrikulturchemie*, *27*, 711-714.

[143] Björck, L. (1994). Preservation of raw milk by lactoperoxidase system – legal situation. In IDF (Ed) *Indigenous antimicrobial agents of milk – Recent developments: Proceedings of the IDF Seminar held in Uppsala Sweeden, 31 Aug-1 Sept. 1993* (IDF, Special Issue 9404, pp. 211-213). Brussels, Belgium: International Dairy Federation.

[144] Villamiel, M., López-Fandiño, R., Corzo, N., & Olano, A. (1997). Denaturation of β-lactoglobulin and native enzymes in the plate exchanger and holding tube section during continous flow pasteurization of milk. *Food Chemistry*, *58*, 49-52.

[145] Marín, E., Sánchez, L., Pérez, M.D., Puyol, P., & Calvo, M. (2003). Effect of heat treatment on bovine lactoperoxidase activity in skim milk: kinetic and thermodynamic analysis. *Journal of Food Science*, *68*, 89-93.

[146] Wolfson, L.M., & Summer, S.S. (1993). Antibacterial activity of the lactoperoxidase system: A Review. *Journal of Food Protection*, *56*, 887-892.

[147] De Wit, J.N., & Van Hooydonk, A.C.M. (1996). Structure, functions and applications of lactoperoxidase in natural antimicrobial systems. *Netherlands Milk and Dairy Journal*, *50*, 227-244.

[148] Ludikhuyze, L., Claeys, W., & Hendrickx, M. (2001). Effect of temperature and/or pressure on lactoperoxidase activity in bovine milk and acid whey. *Journal of Dairy Research*, *68*, 625-637.

[149] Van Loey, A., Verachtert, B., & Hendrickx, M. (2002). Effects of high electric field pulses on enzymes. *Trends in Food Science and Technology*, *12*, 94-102.

[150] Larsen, P.H. (1996). Microfiltration for pasteurized milk. In IDF (Ed) *Heat treatment and alternative methods: Proceedings of the International Dairy Federation symposium*

held in Vienna (Austria), 6-8 Sept. 1995 (IDF, Special Issue 9602, pp. 232-235). Brussels, Belgium: International Dairy Federation.

[151] Kitchen, B.J., Taylor, G.C., & White, I.C. (1970). Milk enzymes – Their distribution and activity. *Journal of Dairy Research*, *37*, 279-288.

[152] Ito, O., & Akuzawa, R. (1983). Purification, crystallization and properties of bovine milk catalase. *Journal of Dairy Science*, *66*, 967-973.

[153] Ito, O., & Akuzawa, R. (1983). Isoenzymes of bovine milk catalase. *Journal of Dairy Science*, *66*, 2468-2473.

[154] Kitchen, B.J. (1976). Enzymatic methods for estimation of the somatic cell count in bovine milk. I. Development of assay techniques and a study of their usefulliness in evaluation and somatic cell count in milk. *Journal of Dairy Research*, *43,* 251-258.

[155] Hirvi, Y., Griffiths, M.W., McKellar, R.C., & Modler, H.W. (1996) Linear transform and non-linear modelling of bovine milk catalase inactivation in a high-temperature short-time pasteurized. *Food Research International*, *29*, 89-93.

[156] Hirvi, Y., & Griffiths, M.W. (1998). Milk catalase activity as an indicator of thermalization treatments used in the manufacture of Cheddar cheese. *Journal of Dairy Science*, *81*, 338-345.

[157] Fong, B.Y., Norris C.S., & MacGibbon, A.K.H. (2007). Protein and lipid composition of bovine milk-fat-globule-membrane. *International Dairy Journal*, *17*, 275-288.

[158] Harrison, R. (2002). Milk oxidoreductase: Hazard or benefit? *Journal of Nutrition and Environmental Nutrition*, *12*, 231-238.

[159] Harrison, R. (2002). Structure and function of xanthine oxidoreductase. Where are we now? *Free Radical Biology and Medicine*, *33*, 774-797.

[160] Harrison, R. (2004). Physiological roles of xanthine oxidoreductase. *Drug Metabolism Reviews*, *36,* 363-375.

[161] Silanikove, N., & Shapiro, F. (2007). Distribution of xanthine oxidase and xanthine dehydrogonase activity in bovine milk: Physiological and technological implications. *International Dairy Journal*, *17*, 1188-1194.

[162] Hwang, Q.S., Ramachandran, K.S., & Whitney, R. (1967). Presence of inhibitors and activators of xanthine oxidase in milk. *Journal of Dairy Science*, *50*, 1723-1737.

[163] Gudnason, G.V., & Shipe, W.F. (1962). Factors affecting the apparent activity and heat sensitivity of xanthine oxidase in milk. *Journal of Dairy Science*, *45*, 1440-1448.

[164] Bhavadasam, M.K., & Ganguli, N.G. (1980). Free and membrane-bound xanthine oxidase in bovine milk during cooling and heating. *Journal of Dairy Science*, *63*, 362-367.

[165] Steffensen, C.L., Hermausen, J.E., & Nielsen, J.H. (2004). The effect of milk fat composition on release of xanthine oxidase during cooling. *Milchwissenschaft*, *59*, 176-179.

[166] Cerbulis, J., & Farrell, H.M.Jr. (1977). Xanthine oxidase activity in dairy products. *Journal of Dairy Science*, *60*, 170-176.

[167] Olsen, K., Kristensen, D., Rasmussen, J.T., & Skibsted, L.H. (2004). Comparison of the effect high pressure and heat on the activity of bovine xanthine oxidase. *Milchwissenschaft*, *59*, 411-413.

[168] Holbrook, J., & Hicks, C.L. (1978). Variation of superoxide dismutase in bovine milk. *Journal of Dairy Science*, *61*, 1072-1077.

[169] Hicks, C.L. (1980). Occurrence and consequence of superoxide dismutase in milk products: A Review. *Journal of Dairy Science*, *63*, 1199-1204.

[170] Lindmark-Månsson, H., & Akesson, B. (2000). Antioxidative factors in milk. *British Journal of Nutrition*, *84*(Suppl. 1), pp. S103-110.

[171] Hicks, C.L., Bucy, J., & Stofer, W. (1979). Heat inactivation of superoxide dismutase in bovine milk. *Journal of Dairy Science*, *62*, 529-532.

[172] Baumrucker, C.R. (1979). γ-glutamyl transpeptidase of bovine milk membranes: Distribution and characterization. *Journal of Dairy Science*, *62*, 253-258.

[173] McKellar, R.C., & Emmons, D.B. (1991). Gamma glutamyl traspeptidase in milk and butter as indicator of heat treatment. *International Dairy Journal*, *1*, 241-251.

[174] Patel, S.S., & Wilbey, R.A. (1989). Heat exchanger performance: γ-glutamyl transpeptidase assay as a heat treatment indicator for dairy products. *Journal of the Society of Dairy Technology*, *42*, 79-80.

[175] Zehetner, G., Bareuther, C., Henle, T., & Klostermeyer, H. (1995). Inactivation kinetics of γ-glutamyltranspeptidase during the heating of milk. *Zeitscrift für Lebensmittel Untersuchung und Forschung*, *201*, 336-338.

[176] Kitchen, B.J., Middleton, G., Durward, G., Andrews, R.J., & Salmon, M.C. (1980). Mastitis diagnostic tests to estimate mammary gland epithelial cell damage. *Journal of Dairy Science*, *63*, 978-983.

[177] Kitchen, B.J., Kwee, W.S., Middleton, G., & Andrews, R.J. (1984). Relationships between the level of N-acetyl-beta-D-glucosamininidase (NAGase) in bovine milk and the presence of mastitis pathogens. *Journal of Dairy Research*, *51*, 11-16.

[178] Kitchen, B.J., Middleton, G., Kwee, W.S., & Andrews, R.J. (1984). N-acetyl-beta-D-glucosamininidase (NAGase) levels in bulk herd milk. *Journal of Dairy Research*, *51*, 227-232.

[179] Andrews, A.T., Anderson, M., & Goodenough, P.W. (1987). A study of the heat stabilities of a number of indigenous milk enzymes. *Journal of Dairy Research*, *54*, 237-246.

[180] Chandan, R.C., Parry, R.M.Jr., & Shahani, K.M. (1965). Purification and some properties of bovine milk lysozyme. *Biochimica et Biophysica Acta*, *110*, 389-398.

[181] White, F.H. Jr., McKenzie, H.A., Shaw, D.C., & Pearce, R.J. (1988). Studies on a partially purified bovine milk lysozyme. *Biochemistry International*, *16*, 521-528.

[182] McKenzie, H.A., & White, F.H.Jr. (1991). Lysozyme and α-lactalbumin: structure function and interrelationships. *Advances in Protein Chemistry*, *41*, 173-315.

[183] Eitenmiller, R.R., Friend, B.A., & Shahani, K.M. (1976). Relationship between composition and stability of bovine milk lysozyme. *Journal of Dairy Science*, *59*, 834-839.

[184] Speck, M.L., & Adams, D.M. (1976). Heat resistant proteolytic enzymes from bacterial sources. *Journal of Dairy Science, 59*, 786-789.

[185] Barach, J.T., Adams, D.M., & Speck, M.L. (1976). Low temperature inactivation in milk of heat-resistant proteinases from psychrotrophic bacteria. *Journal of Dairy Science*, *51*, 391-395.

[186] Barach, J.T., Adams, D.M., & Speck, M.L. (1978). Mechanism of low temperature inactivation of heat-resistant bacterial proteinase in milk. *Journal of Dairy Science*, *61*, 523-528.

[187] Andersson, R.E., Danielson, G., Hedlund, C.B., & Svensson, S.G. (1981). Effect of heat resistant microbial lipase on flavour of ultra-high temperature sterilized milk. *Journal of Dairy Science*, *64*, 375-379.

[188] Griffiths, M.W., Phillips, J.D., & Muir, D.D. (1981). Thermostability of proteinases and lipases from a number of species of psychrotrophic bacteria of dairy origin. *Journal of Applied Bacteriology*, *50*, 289-303.

[189] Bendicho, S., Barbosa-Cánovas, G.V., & Martín, O. (2003). Reduction of proteinase activity in milk by continuous flow high-intensity pulsed electric field treatments. *Journal of Dairy Science*, *86*, 697-703.

[190] Bendicho, S., Estela, C., Giner, J., Barbosa-Cánovas, G.V., & Martín, O. (2002). Effects of high intensity pulsed electric fields and thermal treatments on a lipase from *Pseudomonas fluorescens*. *Journal of Dairy Science*, *85*, 19-27.

In: Practical Food and Research
Editor: Rui M. S. Cruz, pp. 337-360

ISBN: 978-1-61728-506-6

Chapter XIII

MICROORGANISMS AND SAFETY

Fátima A. Miller, Cristina L. M. Silva and Teresa R. S. Brandão

CBQF- Centro de Biotecnologia e Química Fina and Escola Superior de Biotecnologia, Universidade Católica Portuguesa, Rua Dr. António Bernardino de Almeida, 4200-072 Porto, Portugal, mfmiller@mail.esb.ucp.pt, clsilva@esb.ucp.pt and tsbrandao@mail.esb.ucp.pt

ABSTRACT

Milk and dairy products are the most consumed foods. Their quality and safety have a great impact in humans' diet and consequently in their health. The objective of this chapter is to present an overview of the most relevant microorganisms of milk and dairy products. With the purpose of eliminating pathogens and inactivating spoiled related microorganisms, heat-based processes are usually applied. Pasteurization and sterilization, when conveniently designed, are efficient in ensuring food safety from a microbiological point of view. Some regulations and standards related to milk pasteurization were gathered.

Non-thermal treatments have emerged as alternatives to the conventional thermal processes, with the aim of reducing the negative impact of the heat in products' organoleptical and nutritional properties. These processes include pulsed electric fields, high pressure processing, ultrasounds and irradiation.

It is expected that this chapter will contribute with guided information regarding milk and dairy products processing and its impact in target microorganisms.

13.1 INTRODUCTION

Before the widespread of milk pasteurization, this product was one of the most important vehicles of human disease transmission. Several diseases, such as diphtheria, tuberculosis, brucellosis, typhoid fever and septic sore throat, were prevented by the application of adequate/efficient heat treatments to raw milk. This was possible due to the development of regulations that defined times and temperatures required to destroy the pathogens responsible for such outbreaks. Although thermal treatments are the most widely used processes, other non-thermal technologies can be applied to milk processing. Some of these technologies are discussed in this chapter and include pulsed electric field (PEF), high hydrostatic pressure (HHP), ultrasounds and irradiation.

In recent decades, food safety and quality are main concerns of dairy products' industry (e.g. cheese, cream, butter, yogurt, concentrated and dried milks). Thus, efforts have been done in the development or improvement of processes that assure safety and high quality retention levels, while extending the products' shelf-life.

In Figure 13.1 it is represented a scheme with the steps involved in milk and milk-derived products' processing. To assess quality and safety, the complete knowledge of every factor involved in the processes is required. These factors include target microorganisms related to the food under consideration, the raw product, the food matrix characteristics, handling, processing, distribution and storage. To obtain a safe milk or dairy product with high quality characteristics, every step must be monitored (from the raw milk sources until the final consumer). Raw milk, without processing to prevent contamination, should never be used in dairy foods' production. Several outbreaks were reported due to the consumption of dairy products manufactured with raw milk. Generally, if the milk used to produce the dairy products is of good quality and free from pathogens, the final milk-derived products are safe, once they suffer an efficient food processing.

Contamination has been observed in pasteurized milk and its derivates. The non-efficiency of the thermal process applied and/or re-contamination of the products are the most probable causes. To minimize this incidence, all the factors that can have an impact on microbial survival must be controlled.

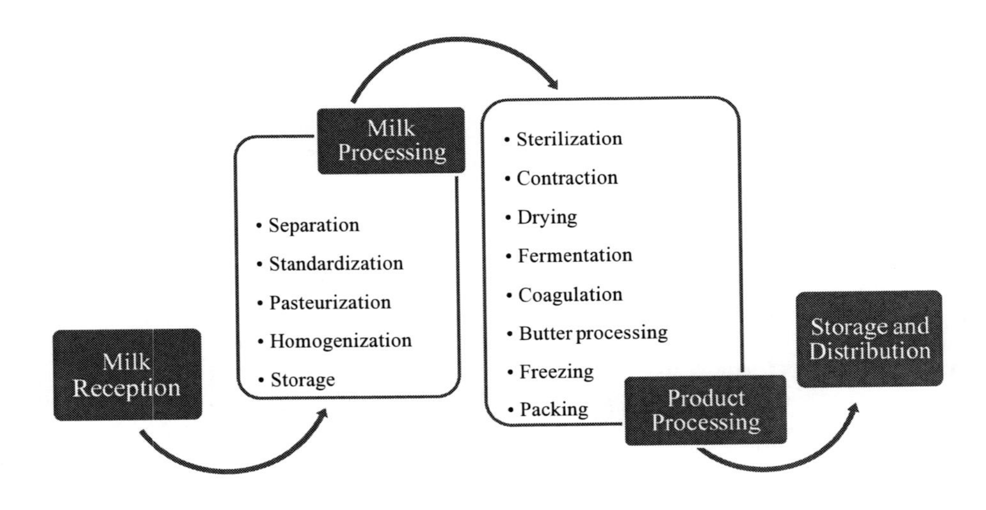

Figure 13.1. Steps involved in milk and milk-derived products' processing.

Microorganisms' activity in dairy products is important for three main reasons [1]: (i) microorganisms produce physical, chemical and organoleptical characteristics in dairy products such as in cheese and in fermented milk, (ii) they may cause spoilage, and (iii) they may represent health hazards when pathogenic bacteria or their toxins contaminate the products.

13.2 Milk

13.2.1 Composition

From a biological point of view, milk is the fluid, excluding colostrum, secreted by female mammalian, whose natural function is the newborns feeding.

From a physical-chemical approach, milk is a homogeneous mixture of many substances such as water, lactose, lipids, proteins, vitamins, salts and enzymes. Some of those substances are in emulsion (e.g. fat), some are in suspension (like caseins that are linked to mineral salts) and others are in real dissolution (e.g. lactose, water soluble vitamins, serum proteins and salts). Milk composition changes considerably between animal's species and, consequently, its microflora heat resistance may also vary significantly. This is mainly important in milks with different fat content, since this component influences heat transfer [2].

13.2.2 Microflora of raw milk

Due to its moderate pH value (around 6.6), high water content and good nutrients quality, milk is a food product with great value for the human diet. However, and for the same reason, it is an excellent substrate for the growth of a wide variety of microorganisms that, like human beings, use the nutritious power of milk.

Milk, even the one that comes from healthy animals, contains a great microbial load (varying in the range 10^2 - 10^6 cfu/ml), depending on hygienic measures that have been adopted. The International Dairy Federation stated that a total microorganism's count above 10^5 cfu/ml indicates that the milk was obtained in poor hygienic conditions, while a value below that indicates that the hygiene was appropriate.

The raw milk microbial loads are related to farms' production. The udder interior, the teat exterior and its surrounding environment, milking and milk-handling equipment are the main sources of contamination [3]. These factors have a joined effect on the number and type of bacterial flora present in raw milk.

The normal flora of the udder interior includes streptococci, staphylococci and micrococci (usually >50%) followed by *Corynebacterium* spp., *Escherichia coli* and others. Abnormal conditions due to infection, disease, unhygienic milking practices or poor milk handling may affect the microflora of milk drawn from the udder. Microbial load is frequently higher due to mastitis, an inflammatory disease of the mammary tissue. Many microorganisms can cause mastitis, being *Staphylococcus aureus*, *Escherichia coli*, *Streptococcus agalactiae*, *Strep. dysgalactiae*, *Strep. uberis* and *Actinomyces pyogenes* the most important ones. *Listeria monocytogenes*, *Salmonella*, *Staphyloccus epidermidis*, coliforms, *Pseudomonas aeruginosa*, *Mycobacterium bovis* and *Mycobacterium tuberculosis* have also been reported as critical microorganisms. The number of bacteria shed into milk changes with the stage of mastitis. Although organisms responsible for mastitis do not typically grow in refrigerated milk, they may survive and be a health concern.

The udder exterior and its direct environment is also a main source of microbial contamination, with great incidence of *E. coli*, *Campylobacter, Salmonella*, some *Bacillus* species, *Clostridia butyricum* and *C. tyrobutyricum*. Appropriate cleaning and disinfection of the animal surfaces before milking can greatly minimize the presence of these microorganisms in the raw product.

The microorganisms commonly found in milk-handling equipment are lactococci, coliforms, *Pseudomonas*, *Alcaligenes*, *Flavobacterium* and *Chromobacterium*. Milk residues left on equipment surfaces provide plenty nutrients favorable for microbial growth and biofilm formation [4, 5]. The efficient cleaning with adequate detergents will contribute to microbial removal and will prevent their occurrence.

Some other sources of contamination are the quality of the air in the milking environment, the quality of the water used for cleaning purposes and the milk-handling workers.

13.2.3 Pre-processing techniques

After milking, the product should be immediately chilled and held at low temperature before transportation. The transport must be under refrigerated conditions and, when arrived at the processing plant, the milk must be kept in chilled storage tanks. If this happens for an extended time before processing, the growth of psychrotrophic bacteria is likely to occur. The bacteria growth extent is dependent on the initial microbial load, and on the storage time and temperature. As well as the growth of psychrotrophic microorganisms, heat-stable, extracellular proteolytic and lipolytic enzymes are usually produced. These enzymes resist to pasteurization processes. Therefore, milk changes may occur and milk products' quality may be compromised [6].

The most common psychrotrophic species found in raw milk include *Pseudomonas*, *Acinetobacter*, *Alcaligenes*, *Flavobacterium* and psychrotrophic coliforms, predominantly *Aerobacter* spp. and *Bacillus* spp. A number of processes can be used to limit the growth of these psychrotrophic microorganisms during raw milk storage [7]. The most commonly used is thermization, which requires the application of a mild heat treatment (≅ 65 ºC for 15-20 seconds) followed by a quick cooling (to temperatures below 6 ºC). This technique inactivates significantly psychrotrophic bacteria and extends the milk storage time. However, vegetative pathogens like *L. monocytogenes* can survive thermization and then may grow during the chilling process [8]. Another process also applied to extend the milk storage time is the decrease of the storage temperature. A reduction from 6 to 2 ºC has been shown to give a 2-day gain in milk storage life, with good microbiological quality [9]. Adding to the milk carbon dioxide (at a concentration around 20-30 mM) also extends the storage life by inhibition of spoilage microorganisms, thus improving products' quality [10].

After storage, milk undergoes further processing with the objective of extending its commercial life. This is usually attained by application of efficient heat treatments that guarantees products' safety (from a microbiological point of view) and quality. If these standards are achieved, the raw milk can be used in the production of diverse dairy products (like butter, hard cheese and milk powders).

13.2.4 Conventional thermal processes

There are three types of thermal treatments applied to milk: thermization, pasteurization and sterilization. Thermization was already mentioned as a mitigate heat treatment for 15 - 20 seconds at 60 - 65 °C. The objective is to inactivate thermo-sensitive psychrotrophic bacteria, but the process is not severe enough to destroy the pathogenic microorganisms.

Pasteurization is applied to guarantee: (i) microbiological safety by killing pathogenic bacteria, and (ii) maximum quality retention by reducing undesirable enzymes and spoilage bacteria responsible for changes of original milk quality attributes. There are two kinds of pasteurization: the low temperature holding (LTH) and the high temperature - short time (HTST) processes. The first mentioned process, LTH, is a discontinue process adequate to small volumes of milk. The treatment time is 30 minutes and the temperature is 62 - 68 °C. The HTST process is applied in continuous flow systems during 15 to 20 seconds at 72 to 78 °C.

Milk may suffer more severe heat treatments performed either in batch systems (closed containers) or continuously with subsequent aseptic packaging. Sterilization of milk involves treatments of 120 °C for approximately 30 minutes. However, modern large scale production methods often use an initial ultra-high-temperature (UHT) treatment prior to filling the container, which reduces the time in 10 minutes and improves the product quality [7]. UHT milk requires minimal treatments for at least 1 second at 135 °C, but combinations of 138 to 142 °C for 2 to 5 seconds are usually applied. Both conventional sterilization and UHT processes, results in an extended product's shelf life without the need of refrigeration.

Another thermal process is the microwave heating. This technology is defined as the use of electromagnetic waves of certain frequencies to generate heat in a material. Being a heat-based process, the microbial inactivation is accomplished by heat effect. Several studies were conducted aiming at comparing this technology to pasteurization. However, reported tested conditions differ in terms of microwave frequency and power, and in terms of the temperature/process time binomial. All these variables are chosen according to the type of container used, and volume and milk composition. Pathogen microorganisms such as *L. monocytogenes*, *Salmonella* spp. and *E. coli* are inactivated by microwave treatment, but some authors reported that this technique is less effective than conventional heating [11, 12]. This may be explained by non uniform heating, which may lead to the occurrence of cold points and consequently less microbial inactivation. The effectiveness of this process is also dependent on the product's pH and water activity.

13.2.4.1 Target microorganisms

As mentioned in section 13.2.2 of this chapter, raw milk is a potential hazardous product that cannot be considered microbiologically safe without adequate and efficiently designed heat treatments. Pasteurization treatments, specified by governmental regulation, ensure pathogens' destruction. However, several outbreaks (campylobacteriosis, salmonellosis, yersiniosis, and listeriosis) have been linked to pasteurized milk and can be traced to inadequate pasteurization, post-pasteurization contamination or temperature abuses [13-15]. The most critical microorganisms in milk/milk processing will be following discussed.

13.2.4.1.1 *Bacillus* spp.

Bacillus cereus is a frequent contaminant of raw and pasteurized milk, reported in recent decades [16]. Post-pasteurization contamination can occur if cleaning is not fully effective. Spores of *B. cereus* can adhere to equipment surfaces and become more heat-resistant [17]. It is generally classified as a mesophile, but some psychrotrophic strains are also known. *B. cereus* spores may survive pasteurization and the vegetative cells that result from spore germination (after pasteurization) can have an adverse effect on the organoleptical and physical properties of milk products. *B. cereus* also produces extracellular toxins, being the most important ones the diarrhoeagenic enterotoxins and an emetic toxin. The first ones are inactivated by certain proteolytic enzymes present in the human gastrointestinal tract, or by thermal treatment at 56 ºC for 30 minutes. In contrast, the emetic toxin is extremely resistant to heat, exhibiting thermotolerance at 126 ºC for 90 minutes. However, it seems likely that milk spoilage would occur before sufficient toxin production had taken place to cause illness, since ingestion of small load usually is not harmful [16].

13.2.4.1.2 *Brucella* spp.

Three species of *Brucella* can cause disease: *B. abortus*, *B. melitensis* and *B. suis*. Some former studies were indicative that *B. abortus* could survive pasteurization. However, it was latter reported that this organism is destroyed by pasteurization [18]. Recent studies on the inactivation of this bacterium in milk were not found.

13.2.4.1.3 *Campylobacter* spp.

Works on the heat resistance of *C. jejuni* began to appear only in the early 1980s, when this specie was recognized as a foodborne pathogen. According to some studies, both *C. jejuni* and *C. coli* are inactivated by pasteurization, with a wide margin of safety [19].

13.2.4.1.4 *Clostridium botulinum*

Vegetative cells of *C. botulinum* are as heat sensitive as the vegetative cells of most bacteria and are readily destroyed by commonly used pasteurization temperatures [20]. However, this specie is able to produce endospores, which may germinate and produce neurotoxins in foods. As for *Bacillus* spores, pasteurization is inadequate to inactivate *C. botulinum* spores. It appears that endospores of this organism have variable degrees of natural heat resistance [21]. However, thermal processes at 121 ºC for 3 minutes are recognized to destroy *C. botulinum* toxins [22].

13.2.4.1.5 *Coxiella burnetti*

Earlier studies showed that *C. burnetti* might be present “in great numbers” in infected dairy cows, and had a great heat resistance. This was confirmed by the presence of this organism in pasteurized milk according to the recommended minimum standards for the low temperature holding (LTH) method in 1950 [23]. Due to studies carried out by Enright et al. [23], the pasteurization standard was increased from 61.7 ºC (for 30 minutes) to 62.8 ºC (for the same process time), since they concluded that the former temperature was not effective in the microorganism inactivation. Thermization at 62 ºC for 15 seconds does not inactivate *C. burnetti*.

13.2.4.1.6 *Cronobacter sakazakii*

In 2007, a group of researchers clarified the taxonomic relationship of *Enterobacter sakazakii* strains. This resulted in the proposal of an alternative classification of *E. sakazakii* as a new genus, *Cronobacter*, comprising five species [24].

Although *C. sakazakii* is considered a thermotolerant organism, it does not survive a standard pasteurization process [25]. However, caution must be taken after pasteurization to prevent re-contamination.

13.2.4.1.7 *Escherichia coli*

Diverse *E. coli* strains have been considered pathogenic, but *E. coli* O157:H7 is the best known and the most widely studied serotype. One of its natural habitats is the cattle intestines, which creates the potential for milk and dairy products contamination. However, when this organism is destroyed by pasteurization, it rarely causes problems in those products. Some studies were carried out concerning the thermal inactivation of this organism in milk [26, 27]. Different heat resistances were observed due to different *E. coli* strains used, to different initial number of microorganisms present and to different milk types and compositions. Although it was concluded that this organism is destroyed by pasteurization with a wide margin of safety, effective pasteurization and prevention of post-process contamination are crucial to ensure product's safety. This is of main importance, because of the very low infective dose of this pathogen. Epidemiological evidence suggests that only about 100 cells are required to cause human illness [28]. Thermization treatments only achieve a few log-reductions.

13.2.4.1.8 *Listeria monocytogenes*

L. monocytogenes has been implicated as the causative agent in several outbreaks of foodborne listeriosis. After an important one in Massachusetts during 1983, a series of studies on the heat resistance of the organism were conducted. However, the results of these studies are apparently contradictory. Some works showed that pasteurization conditions were effective [29, 30], some demonstrated that they were not [31, 32] while some concluded that the process is effective if the initial population of *L. monocytogenes* did not exceed 10^4 cfu/ml [33]. A possible explanation for the discrepant findings is the difference between methods procedures for thermal resistance determination. Mackey and Bratchell [34] gathered significant literature-reported data on the heat resistance of *L. monocytogenes* in milk. More recent studies proved that pasteurization kills *L. monocytogenes* in the product. However, this is not true for thermization, since *L. monocytogenes* may survive this treatment. *L. monocytogenes* is likely to be present in wet dairy processing environments and post-process contamination and therefore it is a particular hazard. Another characteristic of main concern in dairy industry is its ability to grow at temperatures below 4 °C. Therefore, effective HACCP system is of main relevance for these bacteria, mostly the cleaning and sanitizing of all milk contact surfaces.

13.2.4.1.9 *Mycobacterium* spp.

There are three *Mycobacterium* species of interest concerning their heat resistance in milk: (i) *M. tuberculosis*, (ii) *M. bovis* and (iii) *M. avium* subsp. *paratuberculosis*, also known as *M. paratuberculosis* or MAP.

(i) *M. tuberculosis*

This specie is the causative agent of human tuberculosis and can be transmitted by milk. Since 1880 that diverse studies were made on the thermal resistance of *M. tuberculosis*, due to the importance of tuberculosis on that time. All indicate that the organism is destroyed by pasteurization. Indeed, the standards for pasteurization recommended by National Milk Standards Committee in the United States were originally based on the heat resistance of *M. tuberculosis*.

(ii) *M. bovis*

Tuberculosis in cattle is caused by *M. bovis* and can be spread to humans by milk. Although kinetic data of this organism is scarce, it is considered that pasteurization destroy it with a considerable margin of safety. Thermization is not efficient.

(iii) *M. paratuberculosis*

A huge attention has been given to this organism due to its presence in milk of infected animals, and to its potential link to Crohn's disease in humans. An extended review of the controversy effect of pasteurization on this organism was published by Lund et al. [35]. Some of the reasons for the incongruent results included: the slow growth rate of *M. paratuberculosis* in culture media that can allow the overgrown of other heat resistant bacteria which may be present; the hydrophobic nature of the *M. paratuberculosis* cells; the need of sophisticated cultural techniques; the tendency of the organism to clumps formation, which may have a protective effect during the heat treatment and may lead to imprecise dilutions and, consequently, to incorrect enumeration data; and different methods of inoculums preparation.

Nevertheless, it was concluded that *M. paratuberculosis* does not survive pasteurization, but would survive thermization treatment [35, 36].

13.2.4.1.10 *Salmonella* spp.

Outbreaks of human salmonellosis have highlighted the significance of milk and milk products as vehicles of disease dissemination. Doyle and Mazzotta [37] gathered information related to thermal inactivation of salmonellae in dairy products, and particularly in milk. All salmonellae, even the most heat resistant one often used as a test indicator organism (*S. seftenberg)*, are destroyed by pasteurization. However, thermization would only reduce the *Salmonella* load. Hence, the presence of this organism in pasteurized milk indicates that the treatment was not efficient, and/or that post-process contamination happened. Therefore, it is of main importance that contamination post-pasteurization should not occur and effective measures, based on HACCP principles, should be adopted.

13.2.4.1.11 *Staphylococcus aureus*

Staphylococcal food poisoning is not caused by ingestion of the organism itself, but by an enterotoxin produced by *S. aureus*. This organism, under suitable conditions, grows in milk and releases enterotoxins. Only the absence of these toxins in the food sample can guarantee its safety. However, it is recognized that the thermal resistance and stability of the enterotoxins produced by *S. aureus* are much higher than its vegetative cells. Although vegetative cells of *S. aureus* do not survive pasteurization, if the organism has grown and has produced enterotoxin in raw milk prior to pasteurization, enterotoxins will not be destroyed.

Therefore, inadequate chilling of raw milk is one of the key factors for the increase of *Staphylococcus* enterotoxins [38].

13.2.4.1.12 *Streptococcus* spp.

Thermal inactivation data of Streptococci is scarce, but International Commission on Microbiological Specifications for Food [39] collected information available on *Strep. pyogenes* heat resistance. The majority of the studies indicate that this organism does not survive pasteurization, being thermization not severe enough to destroy it.

13.2.4.1.13 *Yersinia enterocolítica*

Although varying with strains, all *Y. enterocolítica* species have low heat resistance. Therefore, it is recognized that this organism would not survive pasteurization. However, culture growth temperature and pH, and composition of the heating medium can have a significant effect on its heat resistance [40]. *Y. enterocolítica* can grow at refrigerated temperatures, and therefore can multiply in pasteurized milk throughout storage. Measures to prevent post-pasteurization contamination must be taken to prevent its proliferation in pasteurized milk.

13.2.4.2 Regulations and standards

There are a diversity of regulations and policies in many milk-producing countries, related to raw milk production. These include the areas of milk quality and safety, milking machine performance, and worker and animal welfare.

The Food and Drug Administration (FDA) and United States Department of Agriculture (USDA) are entities that coordinate national food safety strategies and inspection services, with mission of protecting public health through food safety and defense. In Europe this is attained by agencies with similar missions.

According to FDA by establishment of Grade "A" Pasteurized Milk Ordinance [41], pasteurization is defined as

"... the process of heating every particle of milk or milk product, in properly designed and operated equipment, to one of the temperatures given in the following chart and held continuously at or above that temperature for at least the corresponding specified time:

Temperature	**Time**
*63 °C (145 °F)	30 minutes
*72 °C (161 °F)	15 seconds
89 °C (191 °F)	1.0 second
90 °C (194 °F)	0.5 second
94 °C (201 °F)	0.1 second
96 °C (204 °F)	0.05 seconds
100 °C (212 °F)	0.01 seconds

* If the fat content of the milk product is 10% or more, or if it contains added sweeteners, or if it is concentrated, the specified temperature shall be increased by 3 °C (5 °F)."

In the same document, the temperature and bacteriological limits for Grade "A" raw milk for pasteurization and Grade "A" pasteurized milk are established. Table 13.1 summarizes the information.

EEC Directive 92/46/EEC of 1992 defines the requirements for animal health and the milk quality parameters of Somatic Cell Count and Bacteria Count in the European Union. In this directive it is stated that *"pasteurized milk must have been obtained by means of a treatment involving a high temperature for a short time (at least 71.7 ºC for 15 seconds or any equivalent combination) or a pasteurization process using different time and temperature combinations to obtain an equivalent effect. Immediately after pasteurization, milk must have been cooled to a temperature not exceeding 6 ºC as soon as possible."*

Raw milk used for the production of heat-treated drinking milk, fermented milk, junket, jellied or flavored milk and cream must meet a plate count standard of 100,000 per ml (when tested at 30 ºC). This is a geometric mean over a period of two months with at least two samples a month tested.

The following standards are applied to pasteurized milk:

Pathogenic microorganisms must be absent in 25 g, with n=5, c=0, m=0, M=0

Coliforms (per ml), n=5, c=1, m=0, M=5

After incubation at 6 ºC for 5 days:

Plate count at 21 ºC (per ml), n=5, c=1, m=5x10^4, M=5x10^5

being:

n = number of sample units comprising the sample;

m = threshold value for the number of bacteria; the result is considered satisfactory if the number of bacteria in all sample units does not exceed "m";

M = Maximum value for the number of bacteria; the result is considered unsatisfactory if the number of bacteria is one or more sample units is "M" or more;

c = number of sample units where the bacteria count may be between "m" and "M", the sample being considered acceptable if the bacteria count of the other sample units is "m" or less.

Table 13.1. Temperature and bacteriological standards of milk.

Grade "A" raw milk for pasteurization	Temperature	Cooled to 10 °C (50 °F) or less within 4 hours or less, of the commencement of the first milking, and to 7 °C (45 °F) or less within 2 hours after the completion of milking. Provided, that the blend temperature after the first milking and subsequent milkings does not exceed 10 °C (50 °F).
	Bacterial limits	Individual producer milk not to exceed 100,000 per ml prior to commingling with other producer milk. Not to exceed 300,000 per ml as commingled milk prior to pasteurization.
Grade "A" pasteurized milk	Temperature	Cooled to 7 °C (45 °F) or less and maintained thereat.
	Bacterial limits	20,000 per ml
	Coliform	Not to exceed 10 per ml.

13.2.5 Non-thermal processes

Although current pasteurization processes are effective in destroying pathogenic microorganisms, the products present relatively short lives. This can be overcome by applying UHT processes, which may cause undesirable changes in the product's organoleptical and nutritional properties. Therefore, food industrials are looking for alternative technologies that can assure products' safety while improving products' quality and stability. Some of these methods, already applied to milk, include: pulsed electric field (PEF), high hydrostatic pressure (HHP), ultrasounds and irradiation. In each of these processes, the microorganisms and sometimes the enzymes were destroyed, but the product temperature did not significantly increase. As a result, changes in pigments/colour, aromatic compounds or vitamins were not significant and sensorial and nutritional characteristics only slightly degraded.

13.2.5.1 Pulsed electric field (PEF)

Pulsed electric fields (PEF) is one promising non-thermal processing, since several studies have demonstrated its effectiveness on microbial inactivation [42]. The PEF processing of food consists of a treatment with very short electric pulses (for periods of time in the magnitude of μs) at high electric field intensities at moderate temperatures. The effectiveness of PEF treatments depends on several factors, such as electric field intensity, treatment time, temperature of food, and target microorganism or enzyme [43]. PEF is known to inactivate bacteria by causing dielectric breakdown of the cell membrane, thus altering the functionality of the membrane as a semi-permeable barrier. The extent of cell membrane damage, whether visible in the form of a pore or as loss of membrane functionality, leads to the inactivation of the microorganism. Calderón-Miranda et al. [44] studied the effect of PEF on *L. innocua* in skimmed milk and observed that the bacterium exhibited an increase in the cell wall roughness, cytoplasmic clumping, leakage of cellular material, and rupture of the cell walls and cell membranes. Several researchers reported the inactivation of microorganisms such as *E. coli, Bacillus subtillis, Lactobacillus plantarum, L. monocytogenes,* and *C. sakazakii* by application of PEF [45-48]. Some studies on the effects of PEF on milk quality and composition, proved that this treatment causes fewer changes in the original food's composition than thermal treatments [43]. However, numerous critical process factors exist and carefully designed experiments aiming at optimizing PEF processing conditions are lacking.

13.2.5.2 High pressure processing (HPP)

High pressure processing (HPP), also described as high hydrostatic pressure (HHP) or ultra high pressure (UHP) processing, imposes to liquid or solid foods pressures between 100 and 800 MPa. Pressures between 300 and 600 MPa were effective in microbial inactivation, including most infectious foodborne pathogens. Many studies on the inactivation of pathogenic and spoilage microorganisms (autoctone flora or inoculated one) by HPP have been carried out in milk throughout the last years. It is possible to obtain 'raw' milk pressurized at 400 to 600 MPa with a microbiological quality comparable to the one obtained by HTST pasteurization (depending on the milk microbiological quality) but not comparable to sterilized milk due to HPP resistant spores [49-51].

The bacterial spores are more resistant than vegetative cells and they can survive at pressures of 1000 MPa. The inactivation of bacterial spores by HPP, unlike the inactivation of vegetative bacteria, occurs in two steps [52]. First, pressures between 50 and 300 MPa cause spore germination (germination is the process by which a dormant spore is converted into a vegetative cell), and then germinated forms are easily killed by mild treatments.

The resistance to pressure of microorganisms in foods is variable, depending on process conditions (i.e. pressure, time, temperature and cycles), food constituents and properties and physiological state of the microorganism [53]. HPP destroys microbial cells by inducing morphological alterations (at wall and cell membrane level) and by modifying biochemical reactions and genetic mechanisms [53, 54].

Food components influence microbial inactivation targets [55]. Raso et al. [52] reported that factors, such as pH and water activity, significantly influence the inactivation of *B. cereus*. However, no protective effect of milk fat during high-pressure pasteurization of milk was observed. Dogan and Erkmen [56] compared the HPP effects in milk and fruit juices with *L. monocytogenes*. They concluded that the treatment was not so effective in milk, due to its high nutrient content and consequent increased microbial resistance.

In addition to microbial inhibition and destruction, HPP treatment causes less deterioration of essential vitamins, phytochemicals and aroma compounds, when compared to conventional heat treatments. Another positive feature of the process is the instantaneously uniform pressure distribution attained in the product, avoiding complications such as non-stationary conditions typical of convection-type and conduction-type processes [57, 58].

The combined effects of HPP and temperature have also been investigated [59-61]. HPP can be applied either as a high pressure–cold pasteurization process (HPP–CP) or as a high pressure–temperature pasteurization process (HPP–TP). In HPP–CP, adiabatic effects produce a (small) rise in product temperature. In HPP–TP additional thermal energy is added and microbial inactivation is achieved through a combined pressure-thermal treatment. HPP–

CP processes allow higher sensory properties and nutrients (vitamins) retention, and also involve lower processing costs due to significant energy savings [62].

In addition to microbial destruction, the effects of HPP on protein structure and mineral equilibrium suggest different applications on dairy products. These include the microbiological stabilization of milk and dairy products (i.e. cream, yogurt and cheese), the processing of milk used for cheese and yogurt production, and the preparation of dairy products with novel textures [63].

13.2.5.3 Ultrasounds

Ultrasound is defined as sound waves generated by mechanical vibrations of frequencies between 20 kHz and 800 kHz. When these waves propagate into liquid media, alternate compressions and rarefactions are produced. The waves cause formation of air or vapor bubbles, which collapse releasing high temperatures and pressures, causing cellular stress. This process is known as cavitation [64]. This violent collapse generates mechanical forces, resulting in the break and shear of cell walls, leading to cell death. Cameron et al. [65] studied the destructive effect of cavitation on microbial cells, by transmission electron microscopy techniques. The cavitational forces, induced by ultrasonication, cause irreparable damages to the outer cell wall and inner cell membrane of the tested microorganism.

Some studies have been conducted in sonicated milk, in order to analyze functional properties after processing [66-68] and to assess *L. monocytogenes*, *E. coli*, *Sacharomyces cerevisiae* and *Lactobacillus acidophilus* inactivation [65, 69].

Ultrasonication is, in many situations, combined with temperature (commonly referred as thermosonication) [70, 71] and pressure [72], resulting in increased microbial inactivation due to synergistic effects.

Although it is recognized that sonication inactivates microorganisms and extends milk shelf life without significant nutritional or physicochemical changes, further investigation is still necessary for the implementation of safe food industrial processes.

13.2.5.4 Irradiation

Irradiation is the process by which food is exposed to enough radiation energy to cause ionization. Ionization can lead to microorganisms' death due to DNA damages. Ionizing radiations such as gamma rays (emitted from radioisotopes cobalt-60 or cesium-137), high energy electrons (with a maximum energy of 10 million electron volts) and x-rays (with a maximum energy of 5 million electron volts) can be used in food processes. These types of radiation are chosen because: (i) they produce the desired food preservative effects; (ii) they do not induce radioactivity in foods or packaging materials and (iii) they are available in quantities and at costs that allow commercial use of the irradiation process [73].

Lethality of irradiation depends on the target microorganism, the product and environmental factors. Food composition, moisture content, time/temperature, and oxygen presence are factors that mostly influence the antimicrobial effect of irradiation. The exposure time and dose of irradiation must be adequate to attain microbial destruction, while maintaining food quality standards.

Two terminologies have been used to classify the degree of pathogen reduction with irradiation: (i) pasteurization-radiation that refers to the destruction of vegetative pathogenic cells, and (ii) sterilization-radiation that refers to total elimination of the most resistant spore-forming foodborne bacteria.

Irradiation, being a cold process, is considered a more efficient alternative for pasteurization of solid foods without significant changes in products' quality [74]. However, irradiation should not be applied to some foods such as milk and various dairy products, since undesirable organoleptic changes occurs (e.g. rancid odors) [75]. Studies about the impact of UV irradiation [76], gamma irradiation [77, 78] and electron-beam irradiation [79-81] on *E. coli*, *S. typhimurium*, *S. enteritidis*, *L. monocytogenes*, *S. aureus*, *Enterococcus faecallis* and *B. cereus* were carried out in meat, poultry, seed sprouts and powdered weaning food.

Although the irradiation effect on microbial populations suggests its potential application in a wide range of food products, additional studies concerning process assessment are required.

13.3 Milk Products

The foodstuffs produced from milk are usually defined as dairy products. These milk-derived products are the result of several processes applied. A diagram, with related decrease or increase of temperature (T), pH and water activity (a_w) of milk to attain dairy products' production, is presented in Figure 13.2. Heat and mechanical treatments, water extraction, biochemical or microbiological fermentation are processes involved in those products' manufacture.

In this chapter, we focus on milk-derived products with greater hazard impact.

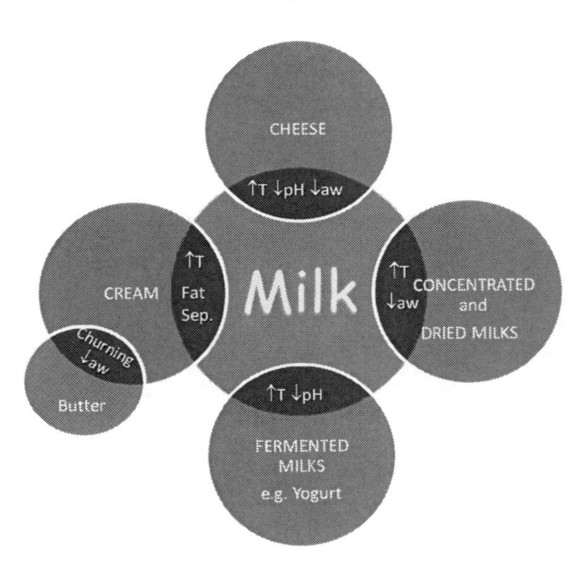

Figure 13.2. Relations between milk properties' alterations (T, pH and a_w) and milk-derived products.

13.3.1 Cheese

Cheese is made by casein coagulation which traps milk fat into a curd matrix. The water content is deeply reduced by the separation and removal of whey from the curd. There are many cheese varieties with significant variations in their flavor, texture and appearance. The basic steps of a typical cheese production are included in Figure 13.3.

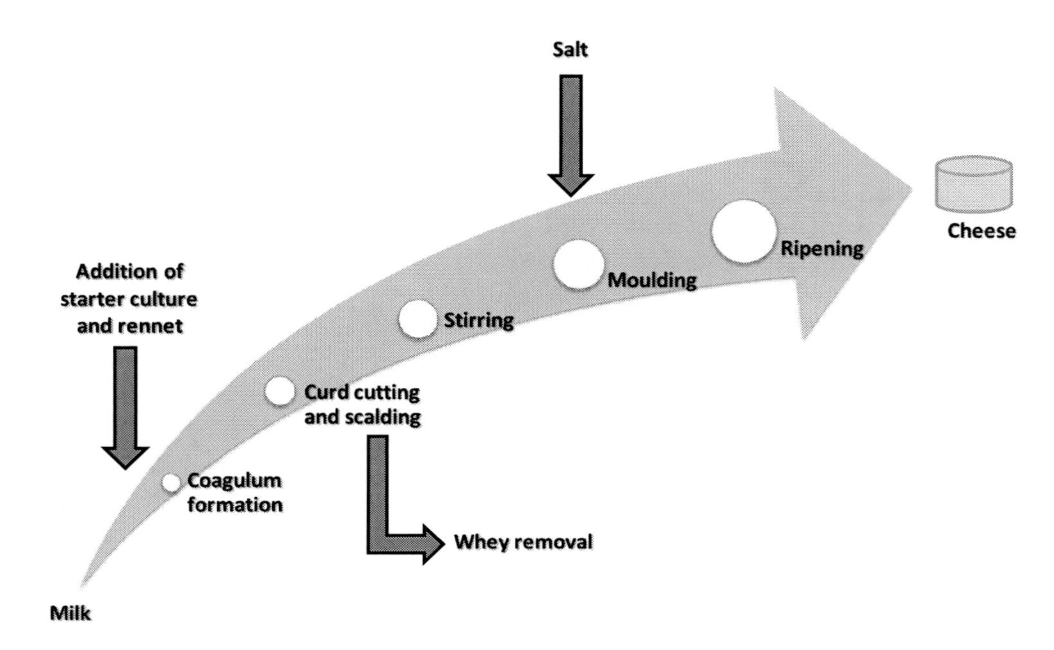

Figure 13.3. Flow diagram of a typical cheese production.

Cheese manufacture involves the combination of four mainly components (i.e. milk, rennet, microorganisms and salt), which are processed through a number of common steps such as coagulum formation, whey exclusion, acid production and salt addition, followed by a period of ripening.

The milk treatment (whether heat-based or not), is the first step of cheese production and plays an important role in the product's microbiological quality. Milk is pasteurized with the purpose of eliminating pathogens and spoilage microorganisms. However, pasteurization is known to adversely affect the development of many sensory attributes of cheese, leading to texture and flavor alterations. A full and mature flavor is only developed if cheese is made with raw milk. Nevertheless, this can lead to the presence of pathogens such as *Listeria*, *Salmonella* and *E. coli*, which have been the cause of several outbreaks associated with unpasteurized cheese [82-84].

HPP technology has been used to ensure the microbiological safety of dairy products with improved quality [63]. HPP processing of milk at room temperature causes several changes that increase the technological ability of milk in cheese production, improving the curd coagulation [85].

Acidification of milk is the key step in cheese production: it supports the development of both flavor and texture, promotes coagulation, and the pH reduction inhibits the growth of pathogens and spoilage organisms [7]. This is usually achieved by lactic acid production by bacterial starter cultures. However, direct addition of acid is also done in certain cheeses manufacture. The choice of starter depends on the type of cheese being produced as well on the temperature of scalding or cooking of the curd. *Lactococcus lactis* subsp. *lactis*, *Lc. lactis* subsp. *cremoris* and *Leuconostoc* spp. are commonly used mesophilic starters (singly or in combination), when the temperature is lower than 40 °C. Thermophilic starters, such as *Strep. thermophilus* and *Lactobacillus delbrueckii* subsp. *bulgaricus*, are applied when higher scalding temperatures (45 to 55 °C) are required. The role of bacterial starter in milk fermentation is the fast conversion of lactose into lactic acid, causing a rapid pH decrease and production of important metabolites for cheese properties [86]. However, the rate of acid production is in some situations slower than required, allowing pathogens growth (such as *S. aureus*) before an inhibitory pH is attained.

Special care must be taken in the curd formation process, since the temperature is maintained at a suitable level for microbial growth. The water used to wash the curd must be controlled, since it can be a source of bacterial spoilage.

Cheeses with high pH are very susceptive to be contaminated with gram-negative, psychrotrophic species (such as pseudomonads) and some coliforms. Moulds and yeasts, which can be controlled by effective hygiene practices, may be responsible for cheese spoilage.

Different factors can influence the presence and survival of pathogens in cheese. These include the amount of heat applied during the several steps of cheese manufacture, the rate of acid formation (by the starter bacteria), the salt levels and water availability resulting from salting and ripening/maturation, the cheese composition, the ripening and storage temperature and the post-manufacturing conditions and contamination. The bacteria survival is also dependent on the strain of the microorganism, including their initial number, their physiological state and their tolerance characteristics to low pH, salt, reduced water activity and to the cooking time/temperature conditions during cheese making.

Although each of these factors has recognized importance, their combined effects have the greatest impact on the growth and survival of pathogens.

The most important pathogens related to cheese production are: *S. seftenberg*, *L. monocytogenes*, *E. coli* and *S. aureus*. Most of them tolerate salt and acidity, and then multiply during cheese manufacture.

To prevent health hazards, the following measures must be taken into consideration:

- a milk treatment prior to cheese making must be done;
- the milk used must be from healthy animals, subjected to very severe hygienic conditions;
- very rigorous hygienic conditions must be established and maintained during milk collection, storage, handling, processing and cheese aging;
- only active fast acid producing starter cultures must be used;
- the design and maintenance of cheese making conditions that involve relatively high cooking temperature, long cooking time and a rapid acid development must be applied;
- the pH, moisture content and salt must be carefully adjusted and monitored;
- effective HACCP must be established, implemented and maintained through milk production and processing.

13.3.2 Cream

Cream is composed by the fat-rich layer skimmed from the top of milk before homogenization. In many countries, cream is sold in several grades depending on the total fat content. Fat separation is performed by centrifugation, which leads to microorganisms' concentration in the fat phase. Separation is usually made at 45 to 50 °C (or higher) to inhibit possible microorganisms growth. Cream is then standardized to the desired fat content [1].

The initial three steps present in a typical cream making process are the same used in milk processing (see Figure 13.1; separation, standardization and pasteurization). Therefore, cream is submitted to temperatures within the range of pasteurization, with the same effects on microbiological inactivation. However, contamination after the heat process may occur, the hazard depending on the product and storage conditions. Due to its high fat content, lipolytic organisms such as *Pseudomonas* spp. or yeasts may lead to cream rancidity.

13.3.2.1 Butter

Butter is produced from cream by churning (or any other equivalent process). If made from properly pasteurized cream, butter is rarely associated with outbreaks of foodborne illness. Typically, salted butter contains around 16% moisture and 2% salt. Hence, these product characteristics are very restricted for microbial growth.

13.3.3 Fermented Milks - Yogurt

Yogurt is a fermented or acidified milk that result from the transformation of lactose into lactic acid, responsible for the typical fermented taste. As it happens for cream, milk used for yogurt manufacture is submitted to the first three steps of milk production. Pasteurized treatments are applied and bacterial elimination is achieved. After cooling, starter cultures are added (usually mixtures of *Lc. delbrueckii* subsp. *bulgaricus* and *Strep. thermophilus*).

As mentioned for cheese (section 13.3.1), the growth of spore-organisms and other contaminants may occur if fast acidification is not accomplished.

The most frequent spoilage organisms in this kind of product are the acid-tolerant yeasts and moulds. Contamination is mainly due to packaging material, ingredients, stabilizing agents and poor hygienic on the processing lines [1].

Hence, particular attention must be done to the milk and ingredients used, to hygienic practices and to the proper storage and distribution temperature conditions.

13.3.4 Concentrated Milks

Concentrated milk products have reduced water content. They include evaporated milks and sweetened condensed milks, having the last ones higher sugar content. Evaporated milk is preserved by heat treatment (UHT treatment or sterilization) and sweetened condensed milk is preserved by its sugar content.

No reported outbreaks of foodborne disease have been attributed to the consumption of sweetened condensed milk or evaporated milk [1]. Generally, these products do not support microbial growth due to its low water activity or to the heat treatment suffered. However, care must be taken with osmophilic microorganisms such as micrococci, yeasts and moulds, and with thermophiles such as *B. stearothermophilus*, *B. coagulans* and *B. licheniformis*, in sweetened condensed and evaporated milks, respectively.

13.3.5 Dried Milks

Whole milk, skim milk, whey, buttermilk, cheese and cream may be prepared in dry form by the application of heat. Drying is mostly accomplished by spray-drying. The extent of microbial destruction during drying depends on the number and types of microorganisms present, on the drying temperature of the exit air in spray-drying or on the drum temperature and retention time for drum drying, and on plant hygiene [1].

As well as Enterobacteriaceae, *L. monocytogenes* can survive a typical spray-drying process in the manufacture of dried milk powders [87]. For this reason, milk for drying must have at least a heat treatment equivalent to the pasteurization prior to drying. Contamination must be prevented after the drying process. During the storage of dried milk powder, surviving organisms slowly die, but spore-formers (the most resistant) retain viability for long periods of time [1].

Conclusion

The pathogens of major concern in milk and dairy products are *Salmonella*, *Campylobacter*, *Staphylococcus*, *E. coli*, *Y. enterocolítica* and *L. monocytogenes*. Their presence in these products may be due to raw milk or by re-contamination. Although pasteurization or equivalent alternative technologies inactivate these organisms, they will not destroy some of the more heat resistant bacteria (thermodurics) or bacterial spores produced by bacteria of the Genera *Bacillus* and *Clostridium*. Thus, every attempt should be made to prevent spore germination and vegetative bacterial proliferation. This can be attained by milk preservation at low temperatures during all processing stages.

Special attention must be given to hygienic practices throughout the entire process to minimize possible re-contamination. It should also be highlighted the importance of establishing strict criteria by the design of effective HACCP systems that must be implemented and maintained in throughout milk production and processing.

References

[1] International Commission on Microbiological Specifications for Food (1998). Microorganisms in Food 6: Microbial ecology of food commodities. London, ST: Blackie Academic & Professional.

[2] Ahmed, N.M., Conner, D.E., & Huffman, D.L. (1995). Heat resistance of *Escherichia coli* O157:H7 in meat and poultry as affected by product composition. *Journal of Food Science*, *60*(3), 606-610.

[3] Adams, M.R., & Moss, M.O. (2008). Food Microbiology (3rd). Cambrigde, ST: The Royal Society of Chemistry.

[4] Sharma, M., & Anand, S.K. (2002). Characterization of constitutive microflora of biofilms in dairy processing lines. *Food Microbiology*, *19*(6), 627-636.

[5] Oulahal, N., Brice, W., Martial, A., & Degraeve, P. (2008). Quantitative analysis of survival of *Staphylococcus aureus* or *Listeria innocua* on two types of surfaces: polypropylene and stainless steel in contact with three different dairy products. *Food Control*, *19*(2), 178-185.

[6] Burdova, O., Baranova, M., Laukova, A., Rozanska, H., & Rola, J.G. (2002). Hygiene of pasteurized milk depending on psychrotrophic microorganisms. *Bulletin of the Veterinary Institute in Pulawy*, *46*(2), 325-329.

[7] Lawley, R. (2001). Microbiology Handbook: Dairy Products (2nd). Surrey, ST: Leatherhead Food RA Publishing.

[8] Gandhi, M., & Chikindas, M.L. (2007). *Listeria*: a foodborne pathogen that knows how to survive. *International Journal of Food Microbiology*, *113*(1), 1-15.

[9] Griffiths, M.W., Phillips, J.D., & Muir, D.D. (1987). Effect of low-temperature storage on the bacteriological quality of raw milk. *Food Microbiology*, *4*(4), 285-292.

[10] Hotchkiss, J.H., Werner, B.G., & Lee, E.Y.C. (2006). Addition of carbon dioxide to dairy products to improve quality: a comprehensive review. *Comprehensive Reviews in Food Science and Food Safety*, *5*(4), 158-168.

[11] Kindle, G., Busse, A., Kampa, D., Meyer-König, U., & Daschner, F.D. (1996). Killing activity of microwaves in milk. *Journal of Hospital Infection*, *33*(4), 273-278.

[12] Thompson, J.S., & Thompson, A. (1990). In-home pasteurization of raw goat's milk by microwave treatment. *International Journal of Food Microbiology*, *10*(1), 59-64.

[13] Ackers, M.L., Schoenfeld, S., Markman, J., Smith, M.G., Nicholson, M. A., DeWitt, W., Cameron, D.N., Griffin, P.M., & Slutsker, L. (1996). An outbreak of *Yersinia enterocolitica* O:8 infections associated with pasteurized milk. In: *36th Interscience Conference on Antimicrobial Agents and Chemotherapy*. New Orleans, Louisiana, Univ Chicago Press.

[14] Reij, M.W., & Den Aantrekker, E.D. (2004). Recontamination as a source of pathogens in processed foods. *International Journal of Food Microbiology*, *91*(1), 1-11.

[15] Leedom, J.M. (2006). Milk of nonhuman origin and infectious diseases in humans. *Clinical Infectious Diseases*, *43*(5), 610-615.

[16] International Commission on Microbiological Specifications for Food (1996). *Bacillus cereus*. In, Roberts, T. A. *Microorganisms in Foods 5: Characteristics of microbial pathogens* (pp. 20-35). New York, ST: Blackie Academic & Professional.

[17] Simmonds, P., Mossel, B.L., Intaraphan, T., & Deeth, H.C. (2003). Heat resistance of *Bacillus* spores when adhered to stainless steel and its relationship to spore hydrophobicity. *Journal of Food Protection*, *66*(11), 2070-2075.

[18] Kronenwett, F.R., Lear, S.A., & Metzger, H.J. (1954). Thermal death time studies of *Brucella abortus* in Milk. *Journal of Dairy Science*, *37*(11), 1291-1302.

[19] Sorqvist, S. (1989). Heat resistance of *Campylobacter* and *Yersinia* strains by three methods. *Journal of Applied Bacteriology*, *67*(5), 543-549.

[20] International Commission on Microbiological Specifications for Food (1996). *Clostridium botulinum*. In, Roberts, T. A. *Microorganisms in Foods 5: Characteristics of microbial pathogens* (pp. 66-111). New York, ST: Blackie Academic & Professional.

[21] Jay, J.M. (2000). Modern food microbiology (6th). Maryland, ST: Aspen Publication.
[22] Blackburn, C.D.W., & McClure, P.J. (2002). Foodborne pathogens: hazards, risk analysis and control. Boca Raton, ST: CRC Press.
[23] Enright, J.B., Sadler, W.W., & Thomas, R.C. (1956). Observations on the thermal inactivation of the organism of Q fever in milk. *Journal of Milk and Food Technology*, *10*, 313-318.
[24] Iversen, C., Lehner, A., Mullane, N., Bidlas, E., Cleenwerck, I., Marugg, J., Fanning, S., Stephan, R., & Joosten, H. (2007). The taxonomy of *Enterobacter sakazakii*: proposal of a new genus *Cronobacter* gen. nov and descriptions of *Cronobacter sakazakii* comb. nov *Cronobacter sakazakii* subsp *sakazakii*, comb. nov., *Cronobacter sakazakii* subsp *malonaticus* subsp nov., *Cronobacter turicensis* sp nov., *Cronobacter muytjensii* sp nov., *Cronobacter dublinensis* sp nov and *Cronobacter genomospecies* I. *Bmc Evolutionary Biology*, *7*, 11.
[25] NazarowecWhite, M., & Farber, J.M. (1997). Thermal resistance of *Enterobacter sakazakii* in reconstituted dried-infant formula. *Letters in Applied Microbiology*, *24*(1), 9-13.
[26] Daoust, J.Y., Park, C.E., Szabo, R.A., Todd, E.C.D., Emmons, D.B., & McKellar, R.C. (1988). Thermal inactivation of *Campylobacter* species, *Yersinia enterocolitica*, and Hemorrhagic *Escherichia coli* O157-H7 in fluid milk. *Journal of Dairy Science*, *71*(12), 3230-3236.
[27] Usajewicz, I., & Nalepa, B. (2006). Survival of *Escherichia coli* O157:H7 in milk exposed to high temperatures and high pressure. *Food Technology and Biotechnology*, *44*(1), 33-39.
[28] Advisory Committee on the Microbiological Safety of Food (1995). Report on verocytotoxin-producing *Escherichia coli*. London, ST: HMSO.
[29] Lovett, J., Wesley, I.V., Vandermaaten, M.J., Bradshaw, J.G., Francis, D.W., Crawford, R.G., Donnelly, C.W., & Messer, J.W. (1990). High-temperature short-time pasteurization inactivates *Listeria monocytogenes*. *Journal of Food Protection*, *53*(9), 734-738.
[30] Piyasena, P., Liou, S., & McKellar, R. C. (1998). Predictive modelling of inactivation of *Listeria* spp. in bovine milk during high-temperature short-time pasteurization. *International Journal of Food Microbiology*, *39*(3), 167-173.
[31] Doyle, M.P., Glass, K.A., Beery, J.T., Garcia, G. A., Pollard, D.J., & Schultz, R.D. (1987). Survival of *Listeria monocytogenes* in milk during high-temperature, short-time pasteurization. *Applied and Environmental Microbiology*, *53*(7), 1433-1438.
[32] Garayzabal, J.F.F., Rodriguez, L.D., Boland, J.A.V., Ferri, E.F.R., Dieste, V.B., Cancelo, J.L.B., & Fernandez, G.S. (1987). Survival of *Listeria monocytogenes* in raw-milk treated in a pilot-plant size pasteurizer. *Journal of Applied Bacteriology*, *63*(6), 533-537.
[33] Bearns, R.E., & Girard, K.F. (1958). The effect of pasteurization on *Listeria monocytogenes*. *Canadian Journal of Microbiology*, *4*(1), 55-61.
[34] Mackey, B. M., & Bratchell, N. (1989). The Heat-Resistance of *Listeria monocytogenes*. *Letters in Applied Microbiology*, *9*(3), 89-94.
[35] Lund, B.M., Gould, G.W., & Rampling, A.M. (2002). Pasteurization of milk and the heat resistance of *Mycobacterium avium* subsp paratuberculosis: a critical review of the data. *International Journal of Food Microbiology*, *77*(1-2), 135-145.

[36] Pearce, L.E., Truong, H.T., Crawford, R.A., Yates, G.F., Cavaignac, S., & de Lisle, G.W. (2001). Effect of turbulent-flow pasteurization on survival of *Mycobacterium avium* subsp paratuberculosis added to raw milk. *Applied and Environmental Microbiology*, *67*(9), 3964-3969.

[37] Doyle, M.E., & Mazzotta, A.S. (2000). Review of studies on the thermal resistance of *salmonellae*. *Journal of Food Protection*, *63*(6), 779-795.

[38] International Commission on Microbiological Specifications for Food (1998). *Staphylococcus aureus*. In, Roberts, T. A., *Microorganisms in Foods 5: Characteristics of microbial pathogens* (pp. 299-333). New York, ST: Blackie Academic & Professional.

[39] International Commission on Microbiological Specifications for Food (1996). *Streptococcus*. In, Roberts, T. A., *Microorganisms in Foods 5: Characteristics of microbial pathogens* (pp. 334-346). New York, ST: Blackie Academic & Professional.

[40] Pagan, R., Manas, P., Raso, J., & Trepat, F.J.S. (1999). Heat resistance of *Yersinia enterocolitica* grown at different temperatures and heated in different media. *International Journal of Food Microbiology*, *47*(1-2), 59-66.

[41] FDA (2003 Revision). Grade "A" pasteurized milk ordinance. ST: U. S. Department of Health and Human Services.

[42] Alkhafaji, S.R., & Farid, M. (2007). An investigation on pulsed electric fields technology using new treatment chamber design. *Innovative Food Science and Emerging Technologies*, *8*(2), 205-212.

[43] Bendicho, S., Barbosa-Cánovas, G.V., & Martín, O. (2002). Milk processing by high intensity pulsed electric fields. *Trends in Food Science and Technology*, *13*(6-7), 195-204.

[44] Calderón-Miranda, M.L., Barbosa-Cánovas, G.V., & Swanson, B.G. (1999). Transmission electron microscopy of *Listeria innocua* treated by pulsed electric fields and nisin in skimmed milk. *International Journal of Food Microbiology*, *51*(1), 31-38.

[45] Bendicho, S., Marsellés-Fontanet, A.R., Barbosa-Cánovas, G.V., & Martín-Belloso, O. (2005). High intensity pulsed electric fields and heat treatments applied to a protease from *Bacillus subtilis*. A comparison study of multiple systems. *Journal of Food Engineering*, *69*(3), 317-323.

[46] Pérez, M.C.P., Aliaga, D.R., Bernat, C.F., Enguidanos, M.R., & López, A.M. (2007). Inactivation of *Enterobacter sakazakii* by pulsed electric field in buffered peptone water and infant formula milk. *International Dairy Journal*, *17*(12), 1441-1449.

[47] Sampedro, F., Rivas, A., Rodrigo, D., Martínez, A., & Rodrigo, M. (2007). Pulsed electric fields inactivation of *Lactobacillus plantarum* in an orange juice-milk based beverage: Effect of process parameters. *Journal of Food Engineering*, *80*(3), 931-938.

[48] Alkhafaji, S., & Farid, M. (2008). Modelling the inactivation of *Escherichia coli* ATCC 25922 using pulsed electric field. *Innovative Food Science & Emerging Technologies*, *9*(4), 448-454.

[49] Mussa, D.M., & Ramaswamy, H.S. (1997). Ultra high pressure pasteurization of milk: Kinetics of microbial destruction and changes in physico-chemical characteristics. *Lebensmittel-Wissenschaft and Technologie*, *30*(6), 551-557.

[50] Linton, M., McClements, J.M.J., & Patterson, M.F. (2001). Inactivation of pathogenic *Escherichia coli* in skimmed milk using high hydrostatic pressure. *Innovative Food Science & Emerging Technologies*, *2*(2), 99-104.

[51] Erkmen, O. & Dogan, C. (2004). Kinetic analysis of *Escherichia coli* inactivation by high hydrostatic pressure in broth and foods. *Food Microbiology*, *21*(2), 181-185.

[52] Raso, J., Gongora-Nieto, M.M., Barbosa-Canovas, G.V., & Swanson, B.G. (1998). Influence of several environmental factors on the initiation of germination and inactivation of *Bacillus cereus* by high hydrostatic pressure. *International Journal of Food Microbiology*, *44*(1-2), 125-132.

[53] Smelt, J. (1998). Recent advances in the microbiology of high pressure processing. *Trends in Food Science & Technology*, *9*(4), 152-158.

[54] Abee, T. & Wouters, J.A. (1999). Microbial stress response in minimal processing. *International Journal of Food Microbiology*, *50*(1-2), 65-91.

[55] Gao, Y.L., Ju, X.R., Qiu, W.F., & Jiang, H.H. (2007). Investigation of the effects of food constituents on *Bacillus subtilis* reduction during high pressure and moderate temperature. *Food Control*, *18*(10), 1250-1257.

[56] Dogan, C., & Erkmen, O. (2004). High pressure inactivation kinetics of *Listeria monocytogenes* inactivation in broth, milk, and peach and orange juices. *Journal of Food Engineering*, *62*(1), 47-52.

[57] Matser, A.A., Krebbers, B., van den Berg, R.W., & Bartels, P.V. (2004). Advantages of high pressure sterilisation on quality of food products. *Trends in Food Science and Technology*, *15*(2), 79-85.

[58] Yuste, J., Mor-Mur, M., Capellas, M., Guamis, B., & Pla, R. (1998). Microbiological quality of mechanically recovered poultry meat treated with high hydrostatic pressure and nisin. *Food Microbiology*, *15*(4), 407-414.

[59] Van Opstal, I., Bagamboula, C.F., Vanmuysen, S.C.M., Wuytack, E.Y., & Michiels, C.W. (2004). Inactivation of *Bacillus cereus* spores in milk by mild pressure and heat treatments. *International Journal of Food Microbiology*, *92*(2), 227-234.

[60] Gao, Y.I., Ju, X.R., & Jiang, H.H. (2006). Analysis of reduction of *Geobacillus stearothermophilus* spores treated with high hydrostatic pressure and mild heat in milk buffer. *Journal of Biotechnology*, *125*(3), 351-360.

[61] Ju, X.-R., Gao, Y.-L., Yao, M.-L., & Qian, Y. (2008). Response of *Bacillus cereus* spores to high hydrostatic pressure and moderate heat. *LWT - Food Science and Technology*, *41*(10), 2104-2112.

[62] Phua, S.T.G., & Davey, K.R. (2007). Predictive modelling of high pressure (<= 700 MPa)-cold pasteurisation (<= 25 degrees C) of *Escherichia coli*, *Yersinia enterocolitica* and *Listeria monocytogenes* in three liquid foods. *Chemical Engineering and Processing*, *46*(5), 458-464.

[63] Trujillo, A.J., Capellas, M., Saldo, J., Gervilla, R., & Guamis, B. (2002). Applications of high-hydrostatic pressure on milk and dairy products: a review. *Innovative Food Science and Emerging Technologies*, *3*(4), 295-307.

[64] Wu, J.R. (2002). Theoretical study on shear stress generated by microstreaming surrounding contrast agents attached to living cells. *Ultrasound in Medicine and Biology*, *28*(1), 125-129.

[65] Cameron, M., McMaster, L.D., & Britz, T.J. (2008). Electron microscopic analysis of dairy microbes inactivated by ultrasound. *Ultrasonics Sonochemistry*, *15*(6), 960-964.

[66] Ashokkumar, M., Sunartio, D., Kentish, S., Mawson, R., Simons, L., Vilkhu, K., & Versteeg, C. (2008). Modification of food ingredients by ultrasound to improve

functionality: A preliminary study on a model system. *Innovative Food Science and Emerging Technologies*, *9*(2), 155-160.

[67] Lin, S.X.Q., & Chen, X.D. (2007). A laboratory investigation of milk fouling under the influence of ultrasound. *Food and Bioproducts Processing*, *85*(C1), 57-62.

[68] Villamiel, M., & de Jong, P. (2000). Inactivation of *Pseudomonas fluorescens* and *Streptococcus thermophilus* in Trypticase (R) Soy Broth and total bacteria in milk by continuous-flow ultrasonic treatment and conventional heating. *Journal of Food Engineering*, *45*(3), 171-179.

[69] D'Amico, D.J., Silk, T.M., Wu, J.R., & Guo, M.R. (2006). Inactivation of microorganisms in milk and apple cider treated with ultrasound. *Journal of Food Protection*, *69*(3), 556-563.

[70] Bermúdez-Aguirre, D., & Barbosa-Cánovas, G.V. (2008). Study of butter fat content in milk on the inactivation of *Listeria innocua* ATCC 51742 by thermo-sonication. *Innovative Food Science and Emerging Technologies*, *9*(2), 176-185.

[71] Villamiel, M., & de Jong, P. (2000). Influence of high-intensity ultrasound and heat treatment in continuous flow on fat, proteins, and native enzymes of milk. *Journal of Agricultural and Food Chemistry*, *48*(2), 472-478.

[72] Pagan, R., Manas, P., Alvarez, I., & Condon, S. (1999). Resistance of *Listeria monocytogenes* to ultrasonic waves under pressure at sublethal (manosonication) and lethal (manothermosonication) temperatures. *Food Microbiology*, *16*(2), 139-148.

[73] Farkas, J. (2006). Irradiation for better foods. *Trends in Food Science & Technology*, *17*(4), 148-152.

[74] Loaharanu, P. (1996). Irradiation as a cold pasteurization process of food. *Veterinary Parasitology*, *64*(1-2), 71-82.

[75] Crawford, L.M., & Ruff, E.H. (1996). A review of the safety of cold pasteurization through irradiation. *Food Control*, *7*(2), 87-97.

[76] Koivunen, J., & Heinonen-Tanski, H. (2005). Inactivation of enteric microorganisms with chemical disinfectants, UV irradiation and combined chemical/UV treatments. *Water Research*, *39*(8), 1519-1526.

[77] Borsa, J., Lacroix, M., Ouattara, B., & Chiasson, F. (2004). Radiosensitization: enhancing the radiation inactivation of foodborne bacteria. *Radiation Physics and Chemistry*, *71*(1-2), 137-141.

[78] Gumus, T., Sukru Demirci, A., Murat Velioglu, H., Velioglu, S.D., Yilmaz, I. & Sagdic, O. (2008). Application of gamma irradiation for inactivation of three pathogenic bacteria inoculated into meatballs. *Radiation Physics and Chemistry*, *77*(9), 1093-1096.

[79] Mayer-Miebach, E., Stahl, M.R., Eschrig, U., Deniaud, L., Ehlermann, D.A.E., & Schuchmann, H.P. (2005). Inactivation of a non-pathogenic strain of *E. coli* by ionising radiation. *Food Control*, *16*(8), 701-705.

[80] Hong, Y.-H., Park, J.-Y., Park, J.-H., Chung, M.-S., Kwon, K.-S., Chung, K., Won, M. & Song, K.-B. (2008). Inactivation of *Enterobacter sakazakii*, *Bacillus cereus*, and *Salmonella typhimurium* in powdered weaning food by electron-beam irradiation. *Radiation Physics and Chemistry*, *77*(9), 1097-1100.

[81] Waje, C.K., Jun, S.Y., Lee, Y.K., Kim, B.N., Han, D.H., Jo, C., & Kwon, J.H. (2009). Microbial quality assessment and pathogen inactivation by electron beam and gamma irradiation of commercial seed sprouts. *Food Control*, *20*(3), 200-204.

[82] Cody, S.H., Abbott, S.L., Marfin, A.A., Schulz, B., Wagner, P., Robbins, K., Mohle-Boetani, J.C., & Vugia, D.J. (1997).Two outbreaks of multidrug-resistant *Salmonella* serotype *typhimurium* DT104 infections linked to raw-milk cheese in northern California. In: *35th Annual Meeting of the Infectious-Diseases-Society-of-America*. San Francisco, California, Amer Medical Assoc.

[83] De Buyser, M.-L., Dufour, B., Maire, M., & Lafarge, V. (2001). Implication of milk and milk products in food-borne diseases in France and in different industrialised countries. *International Journal of Food Microbiology*, *67*(1-2), 1-17.

[84] Honish, L., Predy, G., Hislop, N., Chui, L., Kowalewska-Grochowska, K., Trottier, L., Kreplin, C., & Zazulak, I. (2005). An outbreak of *E. coli* O157:H7 hemorrhagic colitis associated with unpasteurized. *Canadian Journal of Public Health-Revue Canadienne De Sante Publique*, *96*(3), 182-184.

[85] Trujillo, A.J., Royo, C., Guamis, B., & Ferragut, V. (1999). Influence of pressurization on goat milk and cheese composition and yield. *Milchwissenschaft-Milk Science International*, *54*(4), 197-199.

[86] Delorme, C. (2008). Safety assessment of dairy microorganisms: *Streptococcus thermophilus*. *International Journal of Food Microbiology*, *126*(3), 274-277.

[87] Doyle, M.P., Meske, L.M., & Marth, E.H. (1985). Survival of *Listeria monocytogenes* during the manufacture and storage of nonfat dry milk. *Journal of Food Protection*, *48*(9), 740-742.

In: Practical Food and Research
Editor: Rui M. S. Cruz, pp. 361-391
ISBN: 978-1-61728-506-6

Chapter XIV

Texture and Microstructure

Sameh Awad
Department of Dairy Science and Technology, Faculty of Agriculture, Alexandria University, Egypt, sameh111eg@yahoo.com

Abstract

Texture is a critical characteristic for both the consumer and the manufacturer's perception of dairy products quality. The physical structure of the dairy products directly influences how the product withstands and handles within subsequent unit operations such as shredding or slicing for the manufacture of ready prepared foods. As dairy products became an important part of the diet in many countries, the dairy industry responded by manufacturing new types of dairy products with varying texture to suit varied needs and to promote these products use as table products and as food ingredients. The relationship between the textural characteristics of the dairy products and the manufacturing and compositional parameters incorporated into its production has been reviewed for a range of dairy products.

14.1 Introduction

The texture is generally defined as those elements that do not involve the senses of smell and taste. The International Organization for Standardization [1] defines texture of a food product as, "all the rheological and structural, tactile, and, where appropriate, visual and auditory receptors". The textural attributes of foods play a major role in consumer appeal, buying decisions, and eventual consumption. Texture is generally limited to the sensations experienced when masticating, suggesting the predominant role of mechanical properties. However, although force and deformation do occur as part of mastication, other processes such as manipulation of the chewed mass by the tongue and mixing with saliva also occur and result in unique sensory texture perceptions that are not currently measured by instrumentation [2, 3].

Texture is the primary quality attribute of cheeses. The overall appearance and mouthfeel of cheeses are appreciated before their flavour [4]. Cheese offers a variety of textures. For each type of cheese, there is an expected dominant texture attribute. For example, Mozzarella cheese is "stretchy" or "stringy" and Parmesan cheese is "crumbly", etc [2].

Texture clearly plays a role in consumer acceptance of cheeses. The exact role that texture plays with consumer acceptance is difficult to define because flavor cannot be uncoupled from texture when the consumer evaluates cheese. Visual appearance cues may also affect both flavor and texture perception [3].

Scanning electron microscopy (SEM) is a useful tool for providing information on microstructure of dairy products, which assists researchers in understanding factors affecting functional, sensory, and physical properties. This technique has been used to study microstructure of different types of cheese such as Cheddar [5], Mozzarella [6], soft cheese [7], cream cheese [8]), and process cheese [9].

Major structure-forming constituent in cheese and yogurt is the casein matrix in which fat globules are entrapped; water or serum is both bound to casein and fills interstices of the matrix. This network structure is critically affected by the relative content of protein, fat, and water, as well as by the biochemical activities that occur almost continually during storage. The strong interrelationship between food structure and texture is well known [2].

Important milestones in the understanding of dairy products structure are highlighted. The development of complex instrumentation, such as transmission, scanning, cry-scanning and environmental electron microscopy, dynamic oscillatory low-strain rheology, confocal laser scanning microscopy, dynamic light scattering, and nuclear magnetic resonance have facilitated the development of structural models that can be used to predict functional properties [10].

The physical properties of cheese as well as flavour are influenced by a number of factors including: milk composition; milk quality; temperature; the rate and extent of acidification by the starter bacteria; the pH history of cheese; the concentration of Ca salts (proportions of soluble and insoluble forms); extent and type of proteolysis, and other ripening reactions. These factors also control and modify the nature and strength of casein interactions [11].

This review will focus on the development of texture and microstructure during manufacturing of dairy products. Cheese and yogurt are considered soft solid material consisting of a network composed of mainly protein, water, and lipid [12]. The formation of gels during the manufacture of yogurt and cheeses is basically due to destabilization of the casein complex [11, 13, 14]. These gels are classified into different groups:

- acid gels formed by the acid fermentation of milk, for example yogurt and some soft cheeses.
- enzymic gels, which are formed as the result of milk clotting enzymes action which destabilizes the k-casein allowing aggregation of the casein in the presence of calcium ions.
- acid/heat-induced gels, which are normally produced during the manufacture of Ricotta cheese.

The main differences between acid- and enzymic-induced milk gels have been reported by Walstra and van Vliet [15]; van Vliet et al. [16]; Tamime and Robinson [13] and could be summarized as follows: (a) The permeability of the acid-induced gel does not change during the first 24 hour after gelation, whilst in an enzymic-induced gel, it increases continuously during the same period, and (b) a milk gel formed by coagulant enzymes is more robust than an acid-induced gel; the latter type of gel is fragile and shatters very easily.

14.2 Texture Analysis of Dairy Products

The texture of dairy products could be analyzed by human sensory (sensory perception of dairy products structure) and/or instrumentals mechanical (instrumental tests that measure force and deformation over time). Foegeding and Drake [3] reported that the instrumental mechanical properties of cheese and cheese texture are critical attributes. Accurate measurement of these properties requires both instrumental and sensory testing. Fundamental rheological and fracture tests provide accurate measurement of mechanical properties that can be described based on chemical and structural models. Sensory testing likewise covers a range of possible tests with selection of the specific test dependent of the specific goal desired. Sensory analysis is a scientific method to measure human responses to external stimuli. There are 2 basic groups of sensory tools: analytical tests and consumer tests, with the latter often referred to as affective tests. The results from these tests are, respectively, objective and subjective. Analytical sensory tests generally use screened or trained judges depending on the specific test, whereas affective sensory tests use consumers [3, 17].

As previously described, the exact nature of the texture changes is difficult to be determined using this sensory approach, which is one of many reasons why this technique is no longer an acceptable technique for research [3]. Texture Profile Analysis (TPA) is the key instrumental method used to correlate with sensorial textural parameters. TPA studies on cheese are generally carried out at large deformations circa (25-75%) utilising two repeat cycles (Figure 14.1).

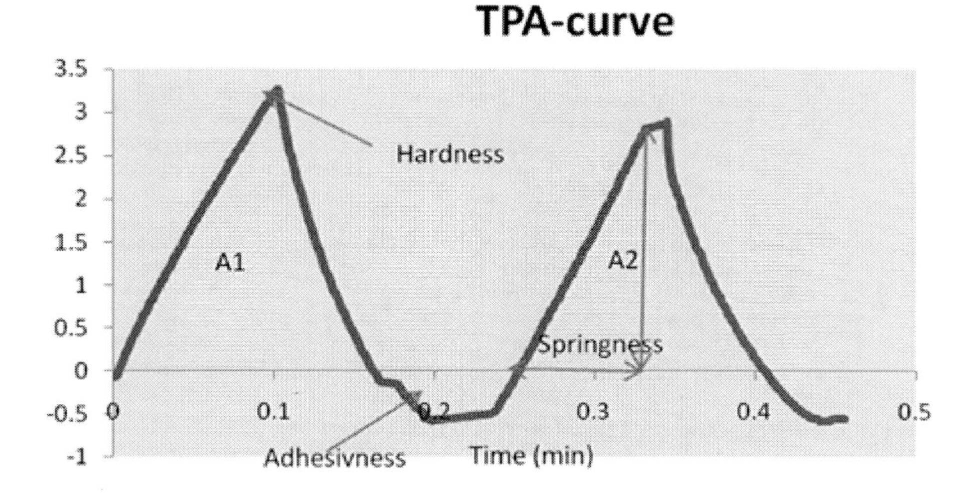

Figure 14.1. TPA curve of two deformation cycles. Calculation parameters are shown in Table 14.1.

Table 14.1. Standard textural characteristics used within sensory evaluation and TPA.

Parameter	Sensorial definition	Instrumental definition (TPA)
Hardness	Force required compressing a cheese between the molars. The extent of resistance offered by the cheese, assessed during the first 5 chews using the front teeth; ranging from soft to firm [19].	Peak force of the first compression cycle [18]
Springiness	Rate at which a deformed material goes back to its undeformed condition after the deforming force is removed. Press sample gently using thumb and 2 fingers for 1 to 2 s without breaking. Evaluate rate at which sample springs back after compression [19].	Height that the food recovers during the time that elapses between the end of the first bite and the start of the second bite [18].
Adhesiveness	The work necessary to overcome the attractive forces between the surface of the food and the surface of other materials with which the food comes into contact (e.g. tongue, teeth, palate). The degree to which the chewed mass sticks to mouth surfaces, evaluated after 5 chews [19].	The negative area for the first bite, representing the work necessary to pull the compressing plunger away from the sample [18].
Cohesiveness	The strength of internal bonds making up the body of the product. Manipulate sample using thumb and 2 fingers 5 times and then score the degree to which the sample holds together during manipulation [19].	The ratio of positive force during the second to that of the first compression cycle [18].
Viscosity	Force required to draw a liquid from a spoon over the tongue The mouthfeel associated with consuming very viscous fluids like heavy whipping cream or honey [3]	Rate of flow per unit force
Fracturability (brittleness)	Force at which a material fractures. Related to the primary parameters of hardness and cohesiveness, where brittle materials have low cohesiveness. The amount of breakdown that occurs in the sample because of mastication, evaluated after 5 chews [19].	The first significant break in the first compression cycle [18].
Gumminess	Energy required to disintegrate a semi-solid food product to a state ready for swallowing, related to foods with low hardness levels.	Calculated parameter: Product of Hardness x Cohesiveness [18]. Semi-solid products undergo permanent deformation and have no springiness.
Chewiness	Energy required to chew a solid food product to a state where it is ready for swallowing. degree of chewing needed to break up the cheese; requiring a good deal of mastication, toffee-like texture [19].	Calculated Parameter: Product of Gumminess x Springiness (essentially primary parameters of Hardness x Cohesiveness x Springiness) [18].

The first cycle utilises a trigger force from which the target test distance is commenced. The large degree of deformation generally results in rupture of the viscoelastic cheese sample from which it fails to recover i.e. the damage is beyond the elasticity of the sample. The test probe then returns to the original start distance, ready to commence its second compression cycle from the trigger point of the first compression cycle. These results in a very different textural profile being formed during the second cycle to that of the first, from which established textural characteristics can be identified. Table 14.1 discusses the standard textural characteristics utilised within sensory and TPA analysis.

14.3 Yogurt Texture and Microstructure Development

The textural properties of yogurts are critical in determining consumer preference where variation in fat content of formulation has a direct influence on the set characteristics of the product. The milk solids not fat (MSNF) content of yogurts forms strong casein-casein interaction. Homogenised fat globules are partly covered with surface-active materials, mainly proteins. The risk of whey syneresis in yogurt is reduced, and the firmness of the end product is increased giving it a better mouthfeel [13]. Texturally, fat plays a role depending on whether it acts as an active filler or not. Milk fat globules act as structure breakers in gelled dairy products. Heat treatment of a homogenised milk base leads to incorporation of the fat phase into the protein matrix. In low-fat products this can be emulated by fat mimetics such as microparticulated whey proteins [20]

Studies of casein micelle dissociation and aggregation during the acid-induced gelation of milk suggest that the mechanisms involved are pH, ion concentration and temperature-dependent. It is evident that the formation of yogurt gel is the result of both biological and physical action on the milk, such as the fortification, homogenisation and heat treatment of the milk base and the catabolism of lactose in the milk by the starter culture for its energy requirements and, as a result, the production of lactic acid and other compound [13]. These effects bring about the gelation of milk. Heertje et al. [21] and Tamime and Robinson [13] reported that, during the acidification of skimmed milk with glucono-δ-lactone (GDL) at 30 °C, the casein micelles may undergo the following changes at different pHs

- 6.6-5.9, no evidence of change in the casein micelles, size about 0.1mm and homogenously distributed in milk.
- 5.5-5.2, partial micellar disintegration occurs and at 5.2, casein particles aggregate to form structures with empty spaces between them; however, when such interaction(s) between micelles take place, the milk gel should not be disturbed.
- 5.2-4.8, contraction of casein aggregates take place, and these particles are larger in size than the native micelles.
- >4.5, rearrangement and aggregation of casein particles occurs leading to the formation of a protein matrix consisting of micellar chains and clusters.

Acid-induced gelation is an important step in making fermented milk. During gelation, casein micelles aggregate and form a 3-dimensional network that, under the microscope, looks like a sponge trapping milk serum and fat. Factors affecting the structure and microstructure of the coagulum of fermented milk determine the final product characteristics [22]. Parnell-Clunies et al. [23] concluded that acid-gel formation of milk was a multistage process consisting of an initial lag period of low viscosity, a period of rapid viscosity change and a stage of high viscosity. However, the same authors reported that dissociation of casein micelles occurred at pH 5.1 and was thought to be influenced by the conversion of colloidal

Ca to $Ca^{2+.}$ At pH 4.8 these casein subparticles reassociate to form larger casein aggregates bearing no specific shape and dimensions [13].

Overall it is reasonable to suggest that the *β*-lactoglobulin (*β*-Lg) interaction with the *k*-casein (linked by SH and SS bridges) partially protects the micelles; however, as the pH in milks is lowered, destabilisation or disruption of the micelles starts to occur. As a result, the gel network or protein matrix consists of micellar chains and/or micellar clusters and entraps within it all the other constituents of the milk base, including the water phase [13]

The microstructure of yogurt has been well studied, and some data have been published on the mechanisms of the acid induction of gels in milk by *Streptococcus thermophilus* and *Lactobacillus delbrueckii* subsp. *bulgaricus* at 30-45 °C. However, the casein micelles are composed of different protein fractions, and are associated with one another via Ca-phosphate bridges. During the fermentation of milk, the micellar or colloidal Ca^{2+} content (and possibly to a lesser extent magnesium and citrate) increases in the serum as the pH is lowered due to the solubilization of micellar Ca-phosphate [13, 24].

Scanning electron microscopy (SEM) studies on the structure of gels derived from heated and unheated milks revealed some distinctive characteristics of the casein micelles. In heated milks, the gel is formed as the casein micelles gradually increase in size and form a chain matrix. This behaviour results in an even distribution of the protein throughout the yogurt and the aqueous phase is immobilized within the network; the resultant coagulum is firm and less susceptible to syneresis. While in unheated milk, the casein micelles form aggregates or clusters in which the protein is unevenly distributed and this heterogeneity impairs the immobilization of the water; the coagulum of unheated milk was much weaker, by 50% compared with the coagulum of heated milk [13, 25-31]. Homogenisation and high-heat treatment of the milk base increases the hydrophilic properties of the coagulum and the stability of the yogurt gel due to the denaturation of whey proteins and association with k-casein [13]. It was reported that the physical properties of yogurt manufactured from milk heated at 82 °C for 30 min were better compared with that heated at149 °C for 3.3 s, and that the latter treatment is suitable only for the production of drinking yogurt or yogurt with thin consistency or low curd firmness [13]. Yogurt made by the vat process exhibited syneresis and a grainy texture; UHT treatment resulted in weak texture of the yogurt coagulum; the high pasteurization process (i.e. 98 °C for 1.87 min) represented the best process and was recommended for industrial production. However, other researchers have recommended 85 °C for 30 min for maximum starter activity. Image analysis using TEM observed that the casein aggregates were larger in the yogurt made from milk heated at high temperature [13].

Commercial yogurts are divided into three main categories, plain/natural, fruit and flavoured and these different types of yogurt are manufactured in either the set or stirred/drinking form. In the manufacture of stirred yogurts, a set-style yogurt gel is mixed at the end of fermentation, then cooled and packaged. At all stages, shearing during mixing and pumping disrupts the yogurt protein network, leading to stirred yogurt [13]. In industry, mixing and packaging are not continuous and a two-stage cooling is therefore performed. The gel is first cooled from the incubation temperature to approximately 20 °C in a heat exchanger, then stored at 20 °C prior to filling of the retail container and further cooling at 4 °C. Filling at this rather high temperature could prevent excessive structural breakdown. Viscosity of the product decreases during mixing and pumping (stirring) but increases again during cold storage [32]. This phenomenon is called "rebodying" or structure recovery [33].

Tamime and Robinson [13] reported that the method of fortification of the milk solids can affect the firmness and syneresis of the yogurt gel. Similarly, these same properties are influenced by the homogenisation pressure used. However, while the physicochemical changes in the protein components of milk could be considered to be one of the major changes influencing the quality of the manufactured yogurt, the role of the starter culture in relation to acid development should not be over looked. Currently, in order to enhance the texture and taste of set yogurt, emulsifiers and fat replacers are being used by food makers. Pectin gel is another approach that is being recently used. Milk gelation is initiated below pH 5.5. As the pH drops from 6.6 to 5.5, some strains of lactic acid bacteria (LAB) produce sufficient amounts of Exopolysaccharide (EPS) to affect structure formation [34]. Such strains produce EPS in the form of unattached slime or large capsules. The EPS can affect formation of casein gel structure by acting as filler. Therefore, the effect of EPS on protein matrix and structure formation depends on their composition, molecular properties, concentration and interactions with the protein [35]. Studying the relationship between EPS and casein micelles is rather complex because, unlike other polysaccharides added directly to milk to stabilize the fermented product, EPS are gradually produced during fermentation. The interaction of EPS with milk proteins is influenced by the protein charge, hydrophobicity, and other characteristics that also change during fermentation [22]. The effect of capsule-forming nonropy cultures on structure formation was also monitored using low-shear dynamic measurements [36]. The gelation point occurred at a higher pH value in milk fermented with the capsule-forming nonropy culture (gelation at pH 5.5) than in milk fermented with either a ropy or an EPS-nonproducing culture (gelation at 5.3 to 5.4). Yogurt made with EPS-producing cultures is less susceptible to syneresis, more viscous, and had more water holding capacity than that made with EPS-non-producing cultures [22].

14.4 Development of Cheese Texture

Cheese was probably the first food material to attract the serious attention of rheologists. Starting with liquid milk and finishing up with a product that may range from a smooth paste to a near solid call for the use of the whole gamut of rheological techniques [37]. The instrumental measurement of the texture properties of cheese is performed for two reasons; as a quality control method for cheese makers, and as a technique for scientists to study cheese structure. The texture properties of cheese can be as important as flavour, and are a large part of the total score awarded by the cheese grader [38]. The texture properties of cheese are a function of cheese composition, microstructure, macrostructure and the physicochemical state of its components [39].

There are two main cheese groups, which are usually designated as "hard" and "soft ". The hard cheeses are characterised by their use: they are eaten by the place, and the consumer requires something which can be readily handled, cut, and then chewed. The soft cheeses are usually spread on bread or a cracker and the ease of spreading and smoothness in the mouth are desirable properties [37].

The initial steps of cheese making involve the coagulation of casein micelles via three possible methods: enzymic coagulation (using calf rennet or other proteolytic enzymes from animals, plans, microorganisms), acid coagulation (using lactic acid bacteria or addition of acids) and heat, or combinations of acid and proteolytic enzymes or heat and acid coagulation. Various curd handling steps convert this weak gel network into a fresh "green" curd. The art or science of cheese making is all about managing five key factors; milk composition, rate, extent of acid development, moisture content, curd manipulation, and maturation conditions [11].

The physical properties of cheese (i.e., body/texture, viscoelasticity, shredding, melt/ stretch, and colour) are influenced by initial cheese milk composition, manufacturing procedures, and maturation conditions. Two of the most important factors influencing these properties are the condition of the casein particles in cheese (e.g., interactions between and within molecules, as well as the amount of Ca associated with these particles) and the extent of proteolysis. These are in turn influenced by various environmental conditions such as pH development, temperature, and ionic strength [11]. Therefore, how individual casein molecules, or aggregates of many casein molecules interact, is vital in understanding the functional properties of cheese. It has recognized that proteins to fat ratio, moisture in non fat substance, pH, temperature, Ca levels, and proteolysis play an important role in functional properties of cheese [2, 11, 37, 39]

14.4.1 Factors affecting the functionality of cheese

Texture and microstructure properties of cheese are affected by several factors. Some of these factors also have an effect on flavour, appearance, and other attributes important to consumers. Each individual parameter used during the cheese manufacture has some effect on the produced cheese. However, new technologies and new ingredients offer new directions in cheese manufacture. Figure 14.2 shows the tentative scheme for the development of cheese texture and the factors affecting functionality of cheese during making and ripening.

14.4.1.1 Milk

Milk from various animals, particularly cattle, buffaloes, goats, and sheep is used to make cheese. It is well recognized that the quality of the milk supply has a major impact on the quality of the resultant cheese. As the primary raw material, the quality and properties of milk have a direct effect on cheese functional properties. Such factors as the breed of cattle, stage of lactation, milking season, and feeding [2, 40].

Milk is normally standardized to minimize some of the variations e.g., due to composition. The standardization is performed with a target casein-to-fat ratio. When the casein-to-fat ratio is not properly controlled, the cheese may be either too soft or too hard, unless adjustments are made to change water content in the curd. Fat composition is directly related to cheese melting, stretching, and related properties of cheese at elevated temperature. Thus, seasonal variations may affect cheese properties even if the milk is standardized [2, 40].

Homogenization of milk is a process during which large fat globules are disintegrated into considerably smaller particles. Their total surface is up to 6-fold larger than was the total surface of the original fat globules. Since the original fat globule membranes were fragmented by homogenization, there is a large area of unprotected exposed fat surface. Consequently, the small fat particles react immediately with any available proteins in the medium until all bare fat is again well covered. Even entire casein micelles are used for this purpose [41].

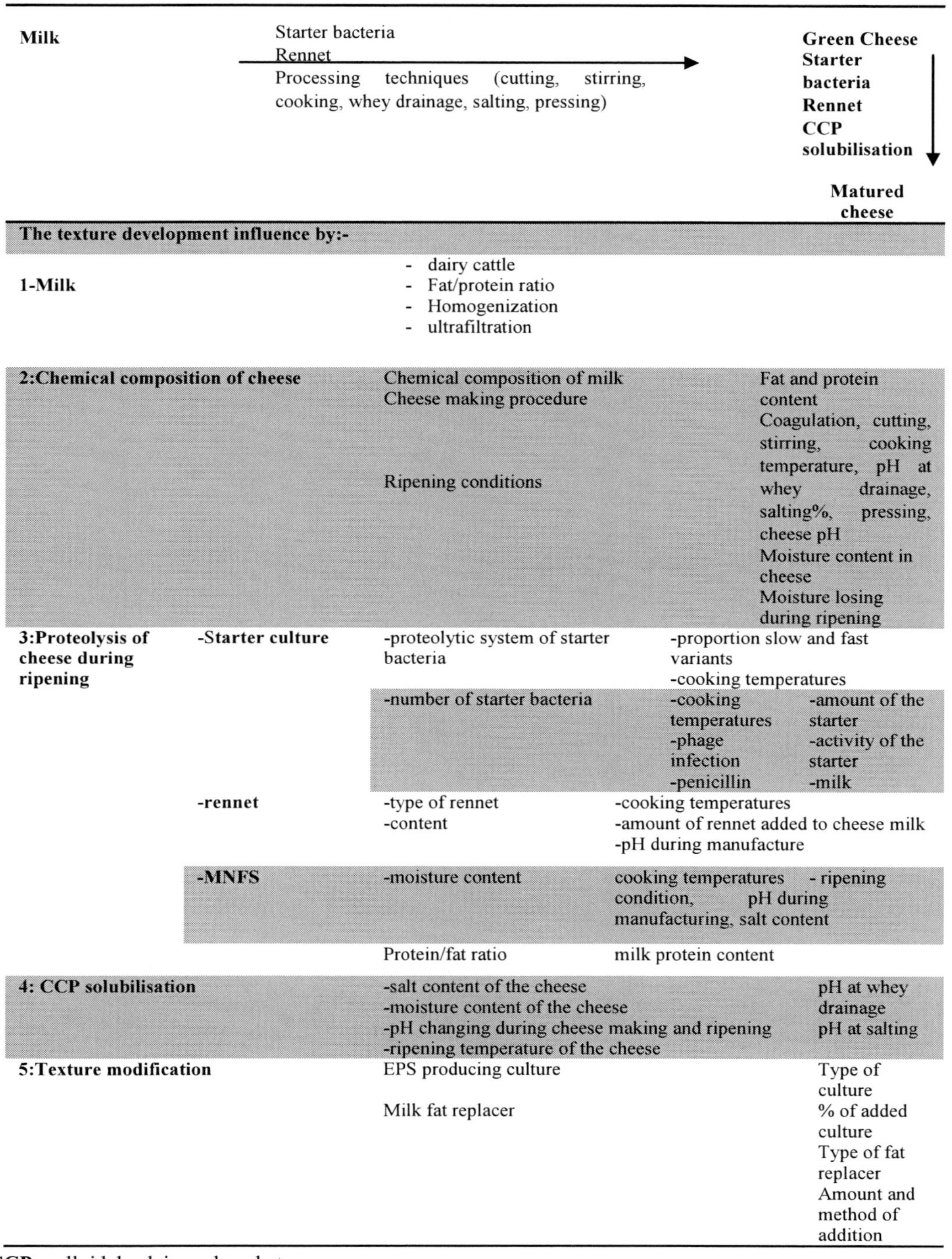

The texture development influence by:-				
1-Milk		- dairy cattle - Fat/protein ratio - Homogenization - ultrafiltration		
2:Chemical composition of cheese		Chemical composition of milk Cheese making procedure Ripening conditions	Fat and protein content Coagulation, cutting, stirring, cooking temperature, pH at whey drainage, salting%, pressing, cheese pH Moisture content in cheese Moisture losing during ripening	
3:Proteolysis of cheese during ripening	-Starter culture	-proteolytic system of starter bacteria	-proportion slow and fast variants -cooking temperatures	
		-number of starter bacteria	-cooking temperatures -phage infection -penicillin	-amount of the starter -activity of the starter -milk
	-rennet	-type of rennet -content	-cooking temperatures -amount of rennet added to cheese milk -pH during manufacture	
	-MNFS	-moisture content	cooking temperatures - ripening condition, pH during manufacturing, salt content	
		Protein/fat ratio	milk protein content	
4: CCP solubilisation		-salt content of the cheese -moisture content of the cheese -pH changing during cheese making and ripening -ripening temperature of the cheese	pH at whey drainage pH at salting	
5:Texture modification		EPS producing culture Milk fat replacer	Type of culture % of added culture Type of fat replacer Amount and method of addition	

CCP: colloidal calcium phosphate
MNFS: moisture in the non fat substance

Figure 14.2. Tentative scheme for the development of cheese texture and the factors affecting functionality of cheese during making and ripening.

Homogenized milk produces a different kind of curds. It is firmer than the curd made from non homogenized milk provided that all other parameters have been left unchanged (casein/fat ratio, moisture content, coagulant, pH). This phenomenon was known even before the use of electron microscopy, but now the reason for the difference is clear: The minute fat globules with casein micelles anchored on their surfaces have become part of the protein matrix. Fat is no more an inert inclusion but has become a structural constituent [2, 41].

Pasteurisation of milk markedly retained more whey into the curd, and produced softer cheeses comparing with cheese made from raw milk [42]. The addition of calcium salts to the pasteurised milks, improved the curd properties. Although, Pyne [43] claimed that Ca has only a minor influence on gel strength, other authors [44-46] reported that the addition of up to 10 mM Ca increased gel strength.

High pressure (600 MPa) treatment of skim milk results in a cheese with a dense network of casein strands, as observed by SEM, formed from partially disintegrated casein micelles. High pressure treatment of milk increases the rate of firming of renneted milk and gel strength [10].

The effects of vacuum-condensed (CM) and ultrafiltered (UF) milk on some compositional and functional properties of Cheddar cheese was studied by Acharya and Mistry [47]. Calcium content was higher in UF cheeses than in CM cheeses followed by control, and it increased with protein content in cheese milk. Condensed milk cheeses exhibited a higher level of proteolysis than UF cheeses. Extent as well as method of concentration influenced the melting characteristics of the cheeses. Melting was greatest in the control cheeses and least in cheese made from condensed milk and decreased with increasing level of milk protein concentration. Vacuum condensing and ultrafiltration resulted in Cheddar cheeses of distinctly different quality.

14.4.1.2 Starter culture

The main purpose of adding starter to milk cheese is for acid production. The rate of acid production is critical in carefully controlling cheese composition and meltability. The proteolytic activity of the starter culture affects rheological and texture properties of cheese during storage [2].

The microorganisms involved in cheese making and cheese ripening can be divided into two major groups: (1) microorganisms that are added to the cheese milk after being carefully selected by the starter manufacturer or the cheese-making company, and (2) non-starter lactic acid bacteria [48, 49]. Starter and non-starter bacteria are important not only for the acid development which hastens the milk coagulation and assists in expulsion of whey, but also from the standpoint of its influence on flavour, body and texture of finished cheese [42].

The role of the bacteria is to assist curdling by decreasing pH of the milk. This is achieved by converting lactose into lactic acid. After the whey had been removed and the curd salted and pressed, the next stage "ripening" takes places at a lower temperature for several weeks or months. This is the time when bacteria slowly degrade the milk proteins and produce substances which give the cheese its characteristic structure. The great variety of cheeses is made possible by the combinations of many varieties of specific bacteria. However, some cheeses are ripened by moulds (*fungi*) such as the *Penicillium camemberti, Penicillium roqueforti,* rather than bacteria. A small group of cheeses (Paneer, Queso Blanco, White cheeses) is made by coagulating milk while it is hot, with an acid, such as lactic acid. Such cheeses are not ripened [41].

Exopolysaccharides (EPS) produced by some strains of lactic acid bacteria (LAB) improved functional properties of cheeses and fermented dairy products. The structure-function relationship of EPS has been the subject of many studies [50-54], and recently reviewed by Hassan [22]. The selection criteria of such cultures depend on the physical characteristics desired in the product.

14.4.1.3. Coagulation

Most natural cheese types are made by the use of rennet to coagulate the casein micelles in milk and the addition of starter culture to produce acid [11]. The primary function of the milk clotting enzyme is to initiate the coagulation of the cheese milk to which it has been added. This includes the rapid and highly specific cleavage of the Phe_{105}-Met_{106} bond of k-casein resulting in the appearance of para-k-casein and caseinomacropeptide (CMP). The casein micelles are kept colloidally dispersed in the milk by steric and electrostatic repulsion involving the negatively charged CMP part of the k-casein molecules. When this repulsive barrier has been taken away by the enzymatic removal of the CMP parts, the micelles become unstable; then, at an appropriate temperature, the milk starts to coagulate under the influence of Ca^{2+} ions in the medium [55].

During the cheese-making process the visual clotting time is the key parameter for programming the apparatus operations. The too soft rennet gel loses the milk protein and fat with whey [56]. In case of the manufacture of Domiati cheese using salted milk, the pH value serves as indirect measure for the structure formation process and final properties. Solubilisation of calcium during cheese-making occurs as a function of NaCl added or pH reduction [11, 57-59]; as a result, the colloidal calcium phosphate (CCP) dissociates from the casein micelle, leaving calcium and phosphate at the terminals of casein. Awad [60] reported that the milk containing 5 and 10 g NaCl/100 g did not coagulate by rennet at pH 5.0. The gel firmness increased with decreasing unsalted milk pH to 6.2 and then the firmness decreased as the pH of milk decreased below 6.2. In case of using salted milk, the firmness decreased as the pH at renneting dropped below 6.4.

Electron microscopy shows the enzymatic coagulum form a thin matrix consisting of their clusters and short chains, encapsulating fat globules. Void spaces in the matrix are filled with the liquid milk serum, which is a solution of lactose, minerals, and vitamins, and a suspension of whey proteins [28, 30].

It is a great challenge for the cheese producer to keep sensory attributes of a particular product constant in a world where, for example, the original rennet has to be replaced with a similar product from a different source because there is a global shortage of calf stomachs. Enzymes with a high proteolytic activity may be efficient in quickly curdling the milk. The high proteolytic enzymes may also lead to a weakened casein matrix and alter the characteristic consistency of the cheese. Gel formation is greatly influenced by pH, Ca concentration, protein content, and temperature [11, 61, 62]. Calcium concentration varies in practice due to changes in milk composition, acid development, and the addition of $CaCl_2$. Gelation temperature is selected by the cheesemakers based on cheese type and experience.

When the gel has attained sufficient firmness, which is traditionally determined subjectively by the cheesemaker, it is cut with knives. If the curd is cut when it is very soft, the moisture content of the resulting cheese is lower [63, 64]. If the gel is left for a longer time before cutting, the moisture content of the cheese is higher.

Presumably, this change in the moisture content is a reflection of the extent of bonding between and within casein particles, which increases with time. Additional events, such as fusion of particles, rearrangement processes, and further incorporation of micelles into the network, all contribute to the growth in gel strength [11].

If milk is preacidified or ripened with starter culture, then the pH at renneting is lower than the natural pH of milk, gelation proceeds will be faster due to the reduction in pH primarily due to the reduction in charge repulsion between micelles and accelerated rennet activity [11]. Gel firmness increases with a reduction in pH up to a maximum at pH 6.0 to 6.2; at lower pH values the firmness decreases [11, 60].

In ricotta and queso blanco style cheeses, milk is heated to 80-85 °C. Acidulants (e.g., acetic, lactic, or citric) are added to the hot milk to bring the pH to 5.9 to 5.4. Flocculation of casein occurs rapidly under these conditions, before it is mechanically dewheyed. In these cheeses, both casein and β-lactoglobulin jointly precipitate and forming the curd structure [11].

In cottage cheese, milk is incubated with culture (22 to 35°C), and the length of time before cutting varies from 5 to 16 h. A low level of rennet may be added in cottage cheese, which helps to form a gel suitable for cutting, and aids in moisture expulsion of large curd cottage cheese [11]. Rennet is often added after some acidity has been developed by the starter culture (e.g., after 1 to 2 h). When rennet is added, the gel is ready to be cut at a higher pH (e.g., 4.8) than in its absence (e.g., 4.6). The presence of denatured whey proteins in acid induced casein gels made from heated milk results in gelation, occurring at higher pH values due to the higher isoelectric point of β-lactoglobulin [11, 62].

14.4.1.4. Curd handling

In rennet-induced gels, most of the serum is lost in whey after the coagulum is cut, but some remains between and within casein aggregates. Most cheeses are cooked, i.e., heated to temperatures higher than those used for gelation, mainly as a means of increasing the syneresis of the curd particles [11]. For cheeses in which low moisture content is required, the coagulum is cut into small curd particles using fine knives and a high cook temperature is also used. After cutting, the curd particles are continuously stirred during the cooking and holding stage in the vat, and the collisions between particles also increase syneresis. A higher cooking temperature normally results in lower moisture content in cheese due to curd shrinkage. Scalding temperature enhances the metabolic activity of bacteria in the curd, which increases lactic acid production and thus lowers pH, which further helps to contract the curd, expelling more whey. This renders cheese acidic, hard, crumbly, and dry [2]. High cooking temperature may also affect cheese properties by decreasing the residual proteolytic activity of the coagulant and starter culture.

After whey drainage, the curd particles start to fuse together unless fusion is prevented by stirring in the vat. Fusion of particles becomes more obvious at pH values <6.0 and corresponds to the "cheddaring" stage in traditional Cheddar cheese manufacture [11]. Cheddaring is a process, during which slabs of the warm curd are piled up in the cheese vat, subjecting the curd to a slow flow. It aligns the proteins and fat globules into a 'fibrous' structure reminiscent of a baked chicken breast. A section cut parallel with the 'fibres' shows the internal organization of the curd. Similar kinds of structuring ('stretching') may be found in Italian-style Mozzarella cheese and, in particular, in 'string cheeses'. In pasta-filata type cheeses, the curd is subjected to a stretching process in hot water (>70 °C). Cheesemakers subjectively decide when the curd is suitable for stretching [11]. This point depends on many factors including the composition of the cheese and method used for acidification. The pH can range from 5.6 in directly acidified Mozzarella (depends on type of acid used) to 5.1 to 5.2 in low-moisture part-skim Mozzarella to 5.0 in very low fat Mozzarella. The cooking and stretching process helps to confer a plastic appearance to the curd and promotes the formation of a fibrous structure. The heat treatment reduces the amount of residual rennet activity in the curd, which decreases the amount of primary proteolysis that occurs during ripening [11].

In Cheddar and other dry-salted cheeses, the curd is kept granular by stirring (stirred-curd) or matted and later milled into small pieces before the application of dry salt. The pieces of curd are thoroughly mixed before they are hooped and pressed. The pressure encourages the formation of bonds between these curd pieces, although junction zones are still partly visible in aged cheese. One consequence of forming curd pieces by milling, and then salt addition, is that the continuity of the cheese matrix is much reduced compared with the original gel network [11]. Most other cheeses are salted by immersion of the cheese block into saturated brine for a length of time that is dictated by the type of cheese, and the shape and size of the block. In brine salted cheeses there is diffusion of salt into the cheese and outward migration of water. Cheese composition and proteolysis varies from the inside to the outside of the block, and this is reflected in differences in textural and functional properties within the cheese block [11].

Solubilisation of calcium during cheese-making occurs as a function of NaCl added or pH reduction [11, 57-59, 65], as a result, the colloidal calcium phosphate (CCP) dissociates from the casein micelle, leaving calcium and phosphate at the terminals of casein. The decrease in calcium binding to casein is attributed to a decrease in hydrophobic binding sites of submicelles, which results in weakening of the extent of binding strength between submicelles (11, 66]. The swelling, hydration and solubility of casein micelles in renneted milk are greatly increased in the presence of NaCl [57, 60, 67].

14.4.1.5 Pressing and ripening

The increase in the density of the curd matrix as a result of whey removal and pressing has been followed in various cheeses by TEM [30]. The casein micelle clusters in impressed cheddar cheese are more compacted than in curd during cheddaring stage. The fat globules are in contact with each other, having retained their fat globule membranes.

The typical cheese flavour results from lipolysis, proteolysis and further degradation of amino acids by starter cultures and nonstarter lactic acid bacteria. Proteolysis is a major determinant of the intact casein which has a large impact on the texture of Cheddar cheese [66]. For the development of an acceptable Cheddar cheese flavour, a well-balanced breakdown of the curd protein (that is, casein) into small peptides and amino acids is necessary [68]. These products of proteolysis either contribute directly to flavour [55] or act as precursors of flavour compounds.

Major textural changes occur during curing. The texture of cheese depends upon the cheese composition and the extent of biochemical changes during ripening [4, 40, 69]. The casein network is greatly weakened when Phe_{23}-Phe_{24} bond in the α_{s1}-casein is hydrolyzed by the residual coagulant to give the peptide α_{s1}-I- casein [6, 66].

14.4.2 Development of cheese body and texture during cheese ripening

In cheese grading or judging, the physical properties are commonly identified in the terms "body" and "texture". By convention, in the dairy industry the term "body" denotes the consistency of the product (e.g., firmness, softness, cohesiveness, rubberiness, elasticity, plasticity, pastiness, brittleness, curdiness, crumbliness), whereas "texture" refers to the relative number, type, and size of openings that can be observed visually (e.g., close, open, gassy, slit-openings or mechanical openness) or by the sense of touch (as in mealy/grainy) to reveal internal particles [70, 71]. Confusingly, all the terms listed above for the "body" of cheese are widely used outside the dairy industry to describe the textural, rheological, and fracture properties of foods [72].

Many factors influence texture development of cheese. These include those factors that affect the curd moisture content (scalding temperature, fineness of the curd, duration of stirring, etc), cheese composition, pH, interactions between casein and serum proteins, proteolysis, Ca content, ionic strength, salt content, and manufacturing protocol, especially rate and extent of acid development, [39, 73-75].

Different cheese varieties have a wide range of textural characteristics, and these also greatly change with aging due to proteolysis, moisture loss, salt uptake, pH change, and the slow dissolution of residual Ca associated with casein particles. Cheese composition (i.e., moisture, protein, fat, NaCl, milk salts, and pH) has a major impact on the body and texture of cheeses [11]. Table 14.2 describes the texture properties of some dairy products.

Cheese composition is mainly controlled by the initial composition of the cheese milk, which is modified by the method used for milk standardization and the manufacturing protocols (e.g., pH at renneting and draining, size of curd particles, cooking temperature, method of salting) used for cheese making. Factors such as species and breed of animal, stage of lactation, and seasonality, can all affect the initial milk composition and alter the texture of cheese [2, 11, 40]. Traditional standardization of milk usually means maintaining particular casein to fat ratio by cream addition/removal or addition of skim milk. This does not mean that cheese is always made from totally uniform milk as the total contents of casein, minerals, fat, and lactose can still vary with this procedure [11].

Table 14.2. Description the texture properties of some dairy products.

Product	Description
Yogurts, whipped cream, and fermented cream	Supplied in containers due to unsupported structure. Soft products with weak gelled structure.
Cottage cheese and curds	Consumed or handled in bulk. Small, irregular and non-uniform particulate pieces with viscous solid structure.
Butter, process cheese, Ricotta, quark and cream cheese	Self-supporting at low temperature, becoming fluid as the temperature rises. Some of these products (Ricotta, process cheese spread and cream cheese) supplied in container due to temperature-related softening. Smooth viscous pastes with uniform structure.
Brie and Camembert cheeses	These products have a rind on the outside, which holds very soft creamy interior. The ripened products have a soft texture
Cheddar and Parmesan cheeses	Waxy and elastic solid materials with low moisture content. Generally matured to develop specific texture and flavours.

The moisture content in cheese is affected by various factors such as cooking temperature, salt content, etc. It is generally established that the greater the moisture content, the softer the cheese and the better its meltability. However, high-moisture cheese has poor shredability [2].

Fat content in the cheese is responsible for its many desirable functional, texture, and sensory properties [76]. Fat present in cheese curd acts as a plasticizer and inhibits the formation of cross links between the casein chains. The weaker and more porous the protein network, the more readily the fat is lost from the network. Higher fat content allows cheeses to melt better, but it may be more difficult to shred [2, 22, 43, 45].

The texture and fracture properties of a cheese are largely determined by the nature and arrangement of its structural network. In full-fat Cheddar cheese, structure is usually described as a continuous protein matrix in which fat globules and residual whey are dispersed [22, 54, 77]. When fat decreases, the protein matrix becomes closer with less fat globule dispersion leading to a more compact structure that affects cheese texture characteristics and overall acceptability [77]. This effect seems to be proportionally correlated with the amount of fat removed. Beal and Mittal [78] reported that hardness, gumminess, and chewiness increased linearly, and cohesiveness and springiness decreased nonlinearly with fat content decrease in Cheddar cheese.

Another important textural property is the proper eye development in Emmental or Gouda cheese (i.e., size, spacing, and number) [11]. For this to occur, the curd should be pliable enough so that as gas is being produced slowly at the appropriate time (e.g., in the hot room stage for Emmental) a smooth eye is formed rather than a crack or split, which can occur if excessive gas is produced too rapidly or if fracture occurs at a relatively small deformation ("short cheese") [79, 80]. The main factors influencing the deformation at fracture are pH and the amount of Ca associated with the casein with cracks more likely to occur in Gouda or Emmental cheese if the pH is below 5.15 and if gas is produced at a later stage of maturation due to the proteolysis reduces the fracture strain [11, 79, 81-83].

The structural matrix of cheese is a cross-linked casein–calcium phosphate network in which fat globules are physically entrapped [37], this protein matrix is elastic when the casein is largely intact [84]. After 1 mo of ripening, a sharp decrease in the hardness of Cheddar cheese was observed [53]. The decrease in hardness during the early stages of ripening is due to the rapid transformation of the rubbery texture of young cheese into a smoother and softer product [85]. This early change in the texture is attributed to a number of factors such as proteolysis of casein network by rennet [11], increasing protein hydration by absorbing serum from the fat-serum channels [6, 39], and solubilization of colloidal calcium phosphate [11, 86]. Water redistribution, which occurs mainly during the first few weeks of ripening [6], seemed to play a major role in cheese softening. After 2 and 4 mo of ripening, the hardness of Cheddar cheese started to increase again after the sharp decrease noticed in the first few weeks. The increase in hardness during ripening might result from the reduction in the level of free water, which increases cheese resistance to deformation [6, 53, 87].

Smaller cavities surrounding fat globules and more direct contact between fat and protein network were observed by Cryo-Scanning Electron Microscopy in the 6-mo-old full fat cheese compared with the young cheese, which indicated a reduction in the amount of the expressible water as cheese aged [54]. This observation is consistent with findings of other researchers [6, 88, 89]. The reduction in the expressible serum that resulted in more fat-protein interactions might have been responsible for the increase in firmness of full fat cheese as it aged [22, 53].

14.4.3 Soft cheese

Soft cheeses are classified as cheeses with moisture contents between 48% and 80%. They are sub-divided further into either:

- Surface ripened, ripening mould added e.g. Brie, Camembert
- Unripened, e.g. Cottage, Lactic, Ricotta

Lactic soft cheeses are generally manufactured from skimmed milk (although milk with a fat content as high as 25% can be used). Mesophilic starter cultures, usually containing *Lactococcus lactis* ssp. *lactis* or ssp. *cremoris* are used to form an acid coagulum. Whey is drained from the curd until it is sufficiently dry, the resulting cheese is salted and packaged. Texture analysis plays a vital role in the quantification of curd characteristics where it acts as a rheological predictor of finished product quality, as well as a potential indicator of moisture content [41].

Lactose naturally present in milk is fermented by the starter culture into lactic acid. The production of acid causes a decrease in product pH, which results in destabilisation of the casein. The milk begins to curdle at around pH 5.2 and the casein is precipitated in the form of flocculent curds entrapping the whey phase in a 3-dimensional network, forming a soft gel [22].

The elevated protein content (solids-not-fat) of the light and extra light soft cheese samples, form strong casein-casein bonds uncharacteristic in a full fat soft cheese, where homogenised fat globules are partially covered with casein, facilitating protein-protein interactions. Fat trapped within this protein network imparts the smooth creamy mouthfeel and viscosity characteristic of full fat soft cheese. Fat is therefore considered a critical component in defining food texture and mouthfeel and consequently eating quality [41].

14.4.4 Cheeses matured in brine

Brined cheeses are produced in East-Mediterranean countries under various names: Feta, Telemes (Greece); Telemea/Branza de Braila (Romania); Bjalo salamureno sirene/Bjalo sirene (Bulgaria); Bieno sirenje (FYROM); Mohant (Slovenia); Sjenicki, Homoljski, Zlatarski, Svrljiški (Serbia); Pljevaljski, Polimsko-Vasojevaski, Ulcinjski (Montenegro); Travnicki/Vlasicki (Bosnia-Herzegovina); Beyaz peynir, Edirne peyniri (Turkey); Liqvan, Iranian white (Iran); Akawi (Lebanon); Domiati, Mish (Egypt) [90].

Cheeses matured in brine share the characteristic of being preserved in brine and in most cases without the need for refrigeration. This method of cheese storage has a determinable effect on biochemical, texture and structural changes that occur in these cheeses and lead to the development of their characteristic flavour and texture. Basically, cheeses matured in brine are rindless and vary in moisture content from soft to semi-hard type [91]. Salt has a significant effect on the rheological properties of Feta cheese; a high percentage of salt develops a harder texture, a higher pH and lower moisture content.

Domiati is unique among cheese varieties in the addition of large quantities of NaCl to milk before renneting. The pH of pickled Domiati cheese is close to the isoelectric point of caseinate and partially solubilizes the colloidal calcium which causes shrinkage of the cheese matrix and exudation of cheese serum into the pickle [90, 91].

The ultra structure of Domiati cheese is composed of a framework of spherical casein aggregates held by bridges and enclosing fat. The internal structure of the cheese is affected by the changes that occur during ripening. Most of these changes occur in the protein matrix as casein aggregates dissociate into smaller spherical particles, forming a loose structure. The fat globules in the cheese undergo slight lypolysis; otherwise they are unlikely to change during storage [91].

14.4.5 Cheese made from ultrafiltered milk (UF-cheese)

Using milk concentrated by ultrafiltration (UF) for cheese making offers the advantages of increased plant efficiency, savings in energy and labor costs, lower rennet and starter costs, reduced cheese vat requirements, increase yields, and more flexibility in disposal of the unwanted milk components. Other advantage of UF is to reduce the transportation cost, if the milk is processed at the farm. No significant changes in the process appear to be required for satisfactory Cheddar cheese making from milk concentrated by UF up to 2-fold [92]. However, the use of a conventional making procedure with higher concentration of milk results in losses of fat, and in the production of cheeses of abnormal composition, flavour and texture [93, 94]. Calcium content was higher in UF cheeses than in control. Ultrafiltered milk produced cheese with higher protein content than normal milk. The breakdown of $\alpha s1$-casein and αs_1-I-casein fractions was highest in the control and decreased in Uf-cheeses [47, 95]. The higher calcium content and lower proteolysis in UF cheese increased the hardness [47].

The use of high level of rennet in the manufacture of UF-Feta leads to a firmer and less adhesive cheese. This is attributed to an increased aggregation rate of altered casein micelles, leading to a denser network structure and a firmer, grittier texture. Also, the rheological properties of UF-Feta are affected by the level of proteolysis during ripening [91]. The texture characteristic of UF and traditional Domiati cheese are significantly different. Fresh UF-Domiati cheese is firmer and more adhesive than conventional cheese which may be attributed to the high retention of calcium in curd and to greater curd firming in UF-cheese. On the other hand, conventional Domiati cheese is chewy and gummy. These differences can also be observed by sensory evaluation. The method of packaging also affects the textural attributes of UF-Domiati cheese [91].

14.4.6 Low fat cheese

The term low fat cheese generally refers to cheeses whose fat content is lower than its corresponding full fat variety [76]. As a result, there is a major shift in the compositional balance of the various components of cheese compared with its full fat counterpart.

Specifically, as the fat content of cheese is lowered, moisture content increases and protein plays a greater role in texture development [53, 96]. This change in the microenvironment is largely responsible for the shifts in the functional and rheological characteristics of the cheese [97, 98]. Fat reduction is associated with many textural and functional defects in cheese [76]. The high casein content in reduced-fat Cheddar cheese imparts a firm and rubbery body and texture [5, 76, 96, 99].

Texture development in cheese occurs due to the breakdown of αs_1-casein during ripening [4]. Furthermore, milk fat normally provides a typical smoothness to a full fat cheese by being evenly distributed within the casein matrix of cheese. In low fat variants there is inadequate breakdown of casein and, therefore, the cheese appears to have a relatively firm texture [76]. Low fat versions of unripened varieties such as Cream, Cottage, Mozzarella, and others also possess certain unique characteristics that are not desirable to the consumer.

14.4.6.1 Low fat cheese texture improvement

Cheddar cheese is made using a combination of carefully selected starter cultures. Such cultures should meet 4 main criteria: rapid acid production, bacteriophage resistance, salt sensitivity, and ripening activity [22]. Exopolysaccharide-producing cultures are good candidates for reduced-fat Cheddar cheese making for several reasons. They have the ability to bind water and increase the moisture in the non fat substance (MNFS) with no need to modify the cheese-making protocol [22, 53]. This is an important function because fat reduction results in lower MNFS. The MNFS plays a major role in texture development in cheese and it would be very desirable to increase it to levels similar to those in the full-fat counterpart [4, 100]. To increase the moisture in reduced-fat Cheddar cheese, manufacturers lower the cooking temperature and drain the whey at a higher pH value. Such modifications might have a negative effect on cheese flavour development [22, 76]. In addition, increasing moisture might result in increased levels of free moisture in cheese. During the first few weeks of ripening, redistribution of free water would produce pasty, difficult to shred reduced-fat cheeses [22, 53]. Exopolysaccharides increase moisture retention by water binding or entrapment within their 3-dimensional network. In addition, EPS seem to act as nuclei for the formation of large pores in cheese. Moisture bound or entrapped in the EPS might not be available for protein hydration and would not produce pasty cheese. Exopolysaccharides also increase the viscosity of the aqueous phase in cheese and modify its flow characteristics. In addition, EPS interfere with protein-protein interactions physically or through their interaction with proteins [22].

Several studies highlighted the positive effect of EPS producing cultures on the physical and functional properties of reduced-fat Cheddar cheese [53, 54, 101-104]. Unlike reduced-fat cheese made with no EPS, hardness of both full-fat cheese and reduced-fat cheese made with the EPS culture increased during the first month of ripening [53]. This is a significant improvement in reduced-fat cheese containing high moisture levels, and would have a direct impact on reduced fat cheese used as an ingredient. Scanning electron microscopy micrographs showed that fresh, EPS negative, reduced-fat cheese contained fewer and larger pores than the EPS-positive cheese [54].

Other methods to increase moisture retention include inclusion of whey proteins and sweet buttermilk in cheese [76]. Whey proteins denatured by high heat treatment (>80 °C), have increased water absorption capacity and have been used in the manufacture of reduced fat Havarti-type cheese [105] and low fat Edam cheese [106]. Excessive whey protein addition is likely to interfere with rennet curd formation and ultimately adversely affect cheese quality [107]. Inclusion of sweet buttermilk in low fat cheese also helps retain moisture and also improved the body and texture of cheeses perhaps because of the inclusion of the milk fat globule membrane in the buttermilk [76, 108].

Processes involving homogenization have also been developed with the specific goal of improving the body and texture of low fat cheeses. Tunick et al. [109] reported on the use of milk homogenized at 10,300 and 17,200 kPa for manufacturing low fat Mozzarella cheese. Improvements in textural and melting characteristics of cheeses were reported by such treatment. Homogenization of milk not only reduces the size of milk fat globules as the interfacial forces at the new fat globule surface may disrupt casein micelles [110] and leads to curd shattering and yield loss. Metzger and Mistry [5, 111] developed a procedure in which 40% fat cream is homogenized and blended with skim milk to the desired fat content for the manufacture of low fat Cheddar cheese. Homogenization in this manner has minimal effect on milk proteins but provides the needed reduction in fat globule size and consequently an increase in fat globule surface area and numbers. Cheeses had excellent body and texture, less free oil in melted cheese than in control cheeses, and improved yield due to increased fat and protein recovery.

14.4.7 Process Cheese

Process cheese has a relatively short history. First experiments started at the end of the 19th century but success was achieved only in 1912 in Switzerland, when citric acid was introduced as a melting salt. A few years later, sodium phosphates were added to sodium citrate and have been used since that time.

However, when cheese is melted without any additive, fat separates from protein and the result is terrible. The secret of 'processing' cheese is in keeping the fat in the protein matrix. Heating, however, decreases the ability of the cheese proteins to keep the fat globules in the dispersed state, which means that the emulsifying capability of the proteins has been reduced. Melting salts restore it by binding calcium which is present in the caseins. Melting salts with very strong calcium-binding ability lead to the production of hard process cheeses which contain fat in the form of very small globules. Process cheese has become a very important product in the cheese industry. Process cheese has many unique functional characteristics compared with natural cheese. These include meltability, plasticity, and shelf-stability [112].

Process cheese is a generic term used to describe 3 separate categories of cheese. These categories are pasteurized process cheese, pasteurized process cheese food and pasteurized process cheese spread (Code of Federal Regulations, [113]). These categories differ on the basis of the requirements for minimum fat content, dry matter basis and the maximum allowed moisture content as well as the quality and the number of optional ingredients that can be used.

14.4.7.1 Factors affecting the functionality of process cheese

Process cheese is manufactured by blending different natural cheeses and additional ingredients including emulsifying salts. The type, characteristic, and age of the natural cheese, type and amount of emulsifying salt, and processing conditions play a major role in controlling the textural, viscoelastic, functional, microstructural, and sensorial properties of process cheese [114-118]. The properties of process cheese also depend on its composition and the interaction between the protein network and the dispersed lipid phase [114, 119, 120].

14.4.7.1.1 Natural cheese used in process cheese making

The characteristics of natural cheese utilized to manufacture process cheese have a major influence on process cheese characteristics. Researchers have highlighted some of the important physicochemical characteristics of a natural cheese that influence the functional properties of process cheese. These include pH, Ca content, and age or amount of intact casein present in the natural cheese and natural cheese made with EPS producing culture [118, 121-127].

Natural cheese made from concentrated milk has been found to influence the chemical as well as functional properties of process cheese [117]. Appropriate selection of natural cheese is important to achieve a process cheese with the desired chemical and functional characteristics. The importance of natural cheese pH on process cheese properties has been highlighted in a study performed by Olson et al. [122] and Kapoor et al. [126]. Their results indicated that even after the final pH of the process cheese was adjusted to 5.4 to 5.5, the process cheese made using Cheddar cheese with the higher pH was harder and less meltable at all stages of ripening when compared with the process cheese made using Cheddar cheese with the normal pH.

Natural cheese containing more Ca, P, and salt/moisture (S/M), significantly increased total Ca and P, pH, and intact casein in the process cheese food. With the increase in natural cheese Ca and P and S/M, there was a significant increase in the texture profile analysis (TPA)-hardness and the viscous properties of process cheese food, whereas the meltability of the process cheese food significantly decreased [126].

A typical process cheese is usually made using young and aged cheeses. Young cheese is preferred for textural and performance issues, whereas aged cheese offers better flavour. As cheese ages, fats and proteins break down through lipolysis and proteolysis into smaller units [118]. The aging process increases flavor intensity in cheese and decreases its emulsifying ability [127].

Protein associations at a given level of emulsification determine process-cheese texture. Short proteins have fewer chances than long proteins to interact with each other. As a result, an aged cheese tends to produce a shorter, more crumbly texture. The proteins become more water-soluble as protein-protein interactions weaken. This situation can temporarily enhance emulsification, but as proteins continue to break down, the decrease in protein-protein interactions leads to a general loss of structure and poor emulsification. To solve this problem, young cheese or rennet casein could be added. Too much of either of these, could result in overly viscous product for processing [128, 129]. A cheese pH also affects the protein's configuration. Proteins roll up into spheres and reduce their interactions with other phases at specific pH values that correspond with the particular proteins and their isoelectric points [127].

14.4.7.1.2 Emulsifying salts

Emulsifying salts provide an effective way to control cheese properties. Some salts can modify pH, but as a group, they are primarily used for their calcium-binding ability. Phosphoric and citric acid salts are commonly used in process cheese including: sodium citrate, sodium aluminum phosphate (SALP), monosodium phosphate (MSP), disodium phosphate (DSP), trisodium phosphate (TSP), tetrasodium tripolyphosphate (TSTPP), sodium tripolyphosphate (STPP), sodium hexametaphosphate (SHMP), and insoluble metaphosphate (IMP). All of these salts have a natural tendency to bind calcium, including that in casein fragments [127-130].

Salts that bind weakly to calcium, yielding a weak emulsion, are most commonly used for soft, easily melted cheeses. Salts in this category include sodium citrate, SALP, DSP and TSP. They all bind calcium at a similar strength, but sodium citrate and SALP cannot bind as much calcium by weight as DSP or TSP. For this reason, sodium citrate and SALP have typical usage levels of 3%, while DSP and TSP are used at 2%. TSP is usually used in combination with other salts because it also raises the pH of the cheese. MSP does not bind strongly to calcium either, but it isn't used often because it decreases cheese pH to unacceptable levels [125, 127-129].

14.4.7.1.3 pH of process cheese

Casein's isoelectric point is about pH 5. Normally, cheese pH is higher than this, producing an excess of negative charge on the protein. As cheese pH is reduced to 5, a crumbly texture can develop due to weakening of protein-protein bonds and the fat can start to demulsify. Increasing the pH to less than 6.5 improves solubility and strengthens protein bonds, creating a more elastic and better-emulsified cheese [11, 27, 125].

The effects of pH make it important to consider the pH of a cheese's emulsifier. MSP, with a pH of 4.2, produces a dry, crumbly cheese. TSP has a pH of 13, and produces a moist, elastic cheese. To adjust pH in process cheese products, the FDA allows organic acids such as vinegar, lactic acid, citric acid, acetic acid and phosphoric acid [127].

The acidity of the process cheese has a most decisive influence on its consistency and structure. This acidity is expressed as pH dependent on the type of cheese, a pH below 5.4 causes a definite firming up of the cheese and, if carried further, can result in coagulation. As the pH rises, the consistency becomes thin and less viscous. The process cheese with a firm body should have a pH value below 5.7, whereas process cheese of soft and spreadable consistency should have a pH above 5.7 [124, 127-129].

14.4.7.1.4 Fat content

Cheese protein: fat ratio determines to what extent the cheese's texture can be modified. In process cheese, this ratio determines hardness and non-melting properties [118].

14.4.7.1.5 Water

The water dissolves the salt and disperses the casein. Even when the raw material for processing contains a certain quantity of moisture, this is firmly bound up in the molecule and would, in most cases, not be enough to dissolve the salts and to peptize the casein. Therefore, in every case, the addition of water is necessary. The quantity added must have a certain relation to the protein content, at content and emulsifying salt, if a good dispersion of the casein and thereby a homogeneous processed structure is to be attained [41, 119, 124, 127].

Process cheese foods and spreads often require additional ingredients to bind the extra water added to these products. Hydrocolloids and gums bind water, control viscosity during processing, and contribute to the finished texture of the cheese product. Other ingredients, such as whey proteins, provide some of the same functionalities, and are very cost effective.

These proteins add body and provide a smooth, creamy texture, but do not melt, stretch, spread or retain finished-cheese firmness, as caseins do [119, 125, 127, 130].

14.4.7.1.6 Heating

Heat is a decisive factor in processing. It is, however, possible to obtain with emulsifying salt a certain result in cold conditions when the cheese mass is mechanically treated very vigorously. A minimum temperature of 65-70 °C is desirable for processing. Viscosity and structure changes occur in a process cheese mass at a temperature above 100 °C [124, 128, 129].

14.4.7.1.7 Duration of processing

The length of time during which the cheese is subject to thermal and mechanical effects plays an important role. The duration of processing depends to a great extent on the consistency of the raw material and the type of product to be made. Temperature and duration of processing are, to a certain extent, dependent on each other. As the temperature is increased, the time of processing must be correspondingly shortened. Whereas a process cheese can be held at 75 °C for 15 min without any noteworthy change in structure and consistency, at 145 °C, the time must be reduced to only a few sec if a detrimental influence on the finished product is to be avoided [124, 125, 127, 129].

During the cooling of process cheese products, the homogeneous, molten, viscous mass sets to form a characteristic body, which, depending on the blend formulation, processing conditions, and cooling rate, may vary from firm and sliceable to semi-soft and spreadable. Factors that contribute to structure formation during cooling include solidification (crystallization) of fat and protein-protein interactions, which result in the formation of a new matrix. It is envisaged that newly formed emulsified fat globules become an integral part of the matrix owing to interaction of their para caseinate membrane with the para caseinate matrix [119, 125, 127].

14.5 Butter Texture Development

The rheological properties of milk fat and butter are an essential quality mark. They determine, e.g., spreadability, ease of cutting and stand-up. The most important textural property in butter is spreadability. Many factors may influence butter firmness, including the nature of the cream, the thermal and mechanical treatments of cream, the manufacturing techniques, such as conventional churning or continuous manufacturing methods, the post-manufacturing handling and storage, chemical composition of butter, and solid fat content [131]. This subject has been examined in comprehensive literature and has been reviewed by Walstra et al. [132]. It is well known that the composition of butterfat partially determines its consistency. A strong correlation between the concentrations of C14:0, C16:0 and C18:1 and the butter firmness was reported. The relationship between firmness and triglyceride composition is more significant than fatty acid content [132].

The processing conditions during manufacturing play an important role in the butter texture (e.g. cream ripening condition, churning temperature, cooling rate). The hard texture could be obtained by reducing the cooling rate [133].

With constant processing conditions, a relationship was found between solid fat content and texture properties. Fat composition will directly affect the solid fat content and thus consistency [134]. Milk and butter samples from cows with a more unsaturated milk fatty acid composition had a lower atherogenic index, and the butter samples were more spreadable, softer, and less adhesive [134].

Conclusion

Texture is a critical characteristic for both the consumer and the manufacturer perception of dairy products. The texture of dairy products could be analyzed by human sensory and/or instrumentals mechanical. The exact nature of the texture changes is difficult to determine using sensory approach. Texture Profile Analysis (TPA) is the key instrumental method used to correlate with sensorial textural parameters. Analyses of the force-time curve led to the extraction of seven textural parameters, five measured (fracturability, hardness, cohesiveness, adhesiveness, springiness) and two calculated (gumminess, chewiness) from the measured parameters. Some others modern complex instrumentations, such as transmission, scanning, cry-scanning and environmental electron microscopy, dynamic oscillatory low-strain rheology, confocal laser scanning microscopy, dynamic light scattering, and nuclear magnetic resonance have been used in studying the structure and function properties of dairy products. Much of the early work on structure of some dairy products has been conducted by various microscopic techniques. There is a need to confirm this information using these modern instruments with sensory properties, texture profile analysis, rheological measurements, and functional properties. Sensory and functional properties are important in developing new dairy products. Modern instrumentation is expensive, but it will provide a basis to develop new food ingredients to add value to dairy products.

References

[1] ISO, (1992). Sensory Analysis-Vocabulary. *International Organization for Standardization*, ISO5492.

[2] Sundaram, G., & Ak, M.M. (2003). *Cheese rheology and texture*. CRC Press. Boca raton, London

[3] Foegeding, E.A., & Drake, M.A. (2007). Sensory and Mechanical Properties of Cheese Texture. Review, *Journal of Dairy Science*, *90*, 611-1624.

[4] Lawrence, R.C., Creamer, L.K., & Gilles, J. (1987). Texture development during cheese ripening. *Journal of Dairy Science*, *70*, 1748-1760.

[5] Metzger, L.E., & Mistry, V.V. (1995). A new approach using homogenization of cream in the manufacture of reduced fat Cheddar cheese. 2. Microstructure, fat globule distribution, and free oil. *Journal of Dairy Science*, *78*, 1883-1895.

[6] McMahon, D.J., Fife, R.L., & Oberg, C.J. (1999). Water partitioning in Mozzarella cheese and its relationship to cheese meltability. *Journal of Dairy Science*, *82*, 1361-1369.

[7] Guerzoni, M.E., Vannini, L., Chaves Lopez, C., Lanciotti, R., Suzzi, G., & Gianotti, A. (1999). Effect of high pressure homogenization on microbial and chemico-physical characteristics of goat cheeses. *Journal of Dairy Science*, *82*, 851-862.

[8] Sainani, M.R., Vyas, H.K., & Tong, P. S. (2004). Characterization of particles in cream cheese. *Journal of Dairy Science, 87,* 2854-2863.

[9] Raval, D.M., & Mistry, V.V. (1999). Application of ultrafiltered sweet buttermilk in the manufacture of reduced fat process cheese. *Journal of Dairy Science*, *82*, 2334-2343.

[10] Everett, D.W., & Auty M.A.E. (2008). Cheese structure and current methods of analysis. *International Dairy Journal*, *18*, 759-773

[11] Lucey, J.A., Johnson, M.E., & Horne, D.S. (2003). Perspectives on the basis of the rheology and texture properties of cheese. *Journal of Dairy Science*, *86*, 2725-2743.

[12] Walstra, P., & van Vliet, T. (1982). Rheology of cheese. (pp. 22-27). IDF No. 153, *International Dairy Federation*, Brussels, Belgium.

[13] Tamime, A.Y., & Robinson, R.K. (2000). *Yogurt: Science and Technology*. (2nd edition). Published in the United Kingdom by Woodhead Publishing Limited Abington Hall, Abington, Cambridge, CB1 6AH, England.

[14] Lucey, J.A., Tamehana, M. Singh, H., & Munro, P.A. (2000). Rheological properties of milk gels formed by a combination of rennet and glucono-δ-lactone. *Journal of Dairy Research*, *67*, 415-427.

[15] Walstra, P., & van Vliet, T. (1986). The physical chemistry of curd making. *Netherlands Milk and Dairy Journal*, *40*, 241-259.

[16] Van Vliet, T. (1988). Rheological properties of filled gels. Influence of filler matrix interaction. *Colloid Polymers Science*, *266*, 518-524.

[17] Lawless, H.T., & H. Heymann. (1999). *Sensory Evaluation of Food*. Aspen Publishers, Gaithersburg, MD.

[18] Bourne, M. (1978). Texture profile analysis. *Food Technology*, *32*, 62-66.

[19] Brown, J.A., Foegeding E.A., Daubert, C.R., Drake, M.A., & Gumpertz, M. (2003). Relationships among rheological and sensorial properties young cheese. *Journal of Dairy Science*, *86*, 3054-3067.

[20] Frost, M.B., & Janhoj, T. (2007). Understanding creaminess. *International Dairy Journal*, *17*, 1298-1311

[21] Heertje, I., Visser, J., & Smits, P. (1985). Structure formation in acid milk gels. *Food Microstructure*, *4*, 267-277.

[22] Hassan, A.N. (2008). ADSA Foundation Scholar Award: Possibilities and Challenges of Exopolysaccharide-Producing Lactic Cultures in Dairy Foods. *Journal of Dairy Science*, *91*, 1282-1298.

[23] Parnell-Clunies E.M., Kakuda, Y., Deman, J.M., & Cazzola, F. (1988). Gelation profiles of yogurt as affected by heat treatment of milk. *Journal of Dairy Science*, *71*, 582-588.

[24] Pouliot, Y., Boulet, M., & Paquin, P. (1989). Experiments on the heat-induced salt balance changes in cow's milk. *Journal of Dairy Research*, *56*, 513-519.

[25] Kalab, M. (1979). Microstructure of dairy foods. 1. Milk products based on protein. *Journal of Dairy Science*, *62*, 1352-1364.

[26] Kalab, M. (1981). Scanning electron microscopy of dairy products; an overview. In D., Holcomb, & M., Kalab (Eds.), *Studies of Food Microstructure: Scanning electron microscopy*. AMF O'Hare, IL.

[27] Kalab, M. (1992). *Food structure and milk products*. In: Y.H. Hui, (Ed), Encyclopedia of Food Science and Technology (Vol. 2, pp. 1170-1196). John Wiley and Sons Inc., New York

[28] Kalab, M. (1993). Practical aspects of electron microscopy in dairy research. *Food Structure*, *12*, 95-114.

[29] Kalab, M., Emmons, D.B., & Sargant, A.G. (1976). Milk gel structure. IV. Microstructure of yogurts in relation to the presence of thickening agents. *Journal of Dairy Research*, *42*, 453-458.

[30] Kalab, M., Allan-Wojtas, P., & Miller, S.S. (1995). Microscopy and other imaging techniques in food structure analysis. *Trends in Food Science and Technology*, *6*, 177-186.

[31] Harwalkar, V.R., & Kalab, M. (1980) Milk gel structure. XI. Electron microscopy of glucono-Δ-lactone-induced skim milk gels. *Journal of Texture Studies*, *11*, 35-49.

[32] Renan, M., Arnoult-Delest, V., Paquet, D., Brulé, G., & Famelart, M.H. (2008). Changes in the rheological properties of stirred acid milk gels as induced by the acidification procedure, *Dairy Science and Technology*, *88*, 341-353.

[33] Renan, M., Guyomarch, F., Arnoult-Delest, V., Paquet, D., Brulé, G., & Famelart, M.H. (2009). Rheological properties of stirred yogurt as affected by gel pH on stirring, storage temperature and pH changes after stirring. *International dairy Journal*, *19*, 142-148

[34] Hassan, A.N., Frank, J.F., Farmer, M.A., Schmidt, K.A., & Shalabi, S.I. (1995). Formation of yogurt microstructure and threedimensional visualization as determined by confocal scanning laser microscopy. *Journal of Dairy Science*, *78*, 2629-2636.

[35] Hassan, A.N., Frank, J.F., Schmidt, K.A., & Shalabi, S.I. (1996). Rheological properties of yogurt made with encapsulated nonropy lactic cultures. *Journal of Dairy Science*, *79*, 2091-2097.

[36] Hassan, A.N., Corredig, M., & Frank, J.F. (2002). Capsule formation by nonropy starter cultures affects the viscoelastic properties of yogurt during structure formation. *Journal of Dairy Science*, *85*, 716-720.

[37] Prentice, J.H., Langley, K.R., & Marshall, R.J. (1993). Cheese rheology. In P.F. Fox, (ed.). (Vol. 1 pp 303-340). *Cheese: Chemistry, Physics and Microbiology*. Elsevier, London.

[38] Farkye, N.Y., & Fox, P.F. (1990). Observations on plasmin activity in cheese. *Journal of Dairy Research*, *57*, 413-418.

[39] Guinee, T.P. (2002). *Cheese rheology*. In H., Roginski, J.W., Fuquay, & P.F., Fox, (Eds.), *Encyclopedia of Dairy Science*, (pp 341-349). Academic Press, London, UK.

[40] Fox, P.F., Guinee, T.P., Cogan, T.M., & McSweeney, P.L.H. (2000). *Fundamental of cheese science. Cheese rheology and texture* (pp. 305-340). Jaithers Burg, MD: Aspen Publisher Inc.

[41] Kosikowski, F.V., & Mistry, V.V. (1997). *Cheese and fermented milk foods*. Vol 1. Origins and principles. Westport, CT: F.V. Kosikowski L.L.C.

[42] Awad, S. (2006). Texture and flavour development in Ras cheese made from raw and pasteurised milk. *Food Chemistry*, *97*, 394-400.

[43] Pyne, G.T. (1955). The chemistry of casein: a review of the literature. *Dairy Science Abstracts*, *17*, 531-554.

[44] Jen, J.J., & Ashworth, U.S. (1970). Factors influencing the curd tension of rennet-coagulated milk. Salt balance. *Journal of Dairy Science*, *53*, 1201-1206.

[45] Marshall, R.J., Hatfield, D.S., & Green, M.L. (1982). Assessment of two instruments for continuous measurement of the curd-firming of renneted milk. *Journal of Dairy Research*, *49*, 127-135.

[46] Storry, J.E., & Ford, G.D. (1982). Some factors affecting the post clotted development of coagulum strength in renneted milk. *Journal of Dairy Research*, *49*, 469-470.

[47] Acharya, M.R., & Mistry, V.V. (2004). Comparison of effect of vacuum-condensed and ultrafiltered milk on Cheddar cheese. *Journal of Dairy Science*, *87*, 4004-4012.

[48] Fox, P.F., McSweeney, P.L.H., & Lynch, C.M. (1998). Significance on non starter lactic acid bacteria in Cheddar cheese. *Australian Journal of Dairy Technology*, *53*, 83-89.

[49] Johnson, M.E. (1998). Cheese products. In E.H., Marth, & J.L., Steele (Eds.), *Applied Dairy Microbiology* (pp. 213-249). New York: Marcel Dekker Inc.

[50] Faber, E.J., Zoon, P., Kamerling, J.P., & Vliegenthart J.F. (1998). The exopolysaccharides produced by *Streptococcus thermophilus* Rs and Sts have the same repeating unit but differ in viscosity of their milk cultures. *Carbohydr. Res. 310*, 269276.

[51] Goh, K.T., Hemar, Y., & Singh, H. (2005). Viscometric and static light scattering studies on an exopolysaccharide produced by *Lactobacillus delbrueckii* subspecies *bulgaricus* NCFB 2483. *Biopolymers*, *77*, 98-106.

[52] Korakli, M., & Vogel, R.F. (2006). Structure/function relationship of homopolysaccharide producing glycansucrases and therapeutic potential of their synthesized glycans. Appllied. *Microbiology and Biotechnology*, *71*, 790-803.

[53] Awad, S., Hassan, A.N., & Muthukumarappan, K. (2005). Applications of exopolysaccharides producing cultures in reduced-fat cheddar cheese: Texture and melting properties. *Journal of Dairy Science*, *88*, 4204-4213.

[54] Hassan, A.N., & Awad, S. (2005). Application of exopolysaccharideproducing cultures in reduced-fat Cheddar cheese. Cryo-scanning electron microscopy observations. *Journal of Dairy Science*, *88*, 4214-4220.

[55] Visser, S. (1993). Proteolytic enzymes and their relation to cheese ripening and flavour. *Journal of Dairy Science*, *76*, 329-350.

[56] Robinson, R.K., & Wilbey, R.A. (1998). *Cheesemaking practice*. In R. Scoot (Ed.), (3rd ed.). Gaithersburg, MD: Aspen publishers, Inc.

[57] Creamer, L. K. (1985). Water absorption by renneted casein micelles. *Milchwissenschaft*, *40*, 589-591.

[58] Gatti, C., & Pires, M. (1995). Effect of monovalent cations on the kinetics of renneted milk coagulation. *Journal of Dairy Research*, *62*, 667-672.

[59] Gaucheron, F., Le Graet, Y., & Briard, V. (2000). Effect of NaCl addition on the mineral equilibrium of concentrated and acidified casein micelles. *Milchwissenschaft*, *55*, 82-86.

[60] Awad, S. (2007). Effect of sodium chloride and pH on the rennet coagulation and gel firmness. *Lebensmittel-Wissenschaft und Technology*, *40*, 220-224.

[61] Lomholt, S.B., & Qvist, K.B. (1999). *The formation of cheese curd*. In B.A. Law (Ed.), Technology of cheesemaking (pp. 66-98). Sheffield, UK: Sheffield Academic Press.

[62] Lucey, J.A. (2002). Formation and physical properties of milk protein gels. *Journal of Dairy Science*, *85*, 281-294.

[63] Johnson, M.E., & Chen, C.M. (1995). Technology of manufacturing reduced-fat Cheddar cheese. In E.L., Malin, & M.H., Tunick (Eds.), *Chemistry of Structure-Function Relationships in Cheese* (pp. 331-337). Plenum Press, New York.

[64] Johnson, M.E., Chen, C.M., & Jaeggi, J.J. (2001). Effect of Rennet Coagulation Time on Composition, Yield, and Quality of Reduced-Fat Cheddar Cheese. *Journal of Dairy Science*, *84*, 1027-1033.

[65] Metzger, L.E., Barbano, D.M., Kindstedt, P.S., & Guo, M.R. (2001). Effect of milk preacidification on low fat Mozzarella cheese: II Chemical and functional properties during storage. *Journal of Dairy Science*, *84*, 1348-1356.

[66] Creamer, L.K., & Olson, N.F. (1982). Rheological evaluation of maturing Cheddar cheese. *Journal of Food Science*, *47*, 632-636, 646.

[67] Pastorino, A.J., Hansen, C.L., & McMahon, D.J. (2003). Effect of salt on structure–function relationships of cheese. *Journal of Dairy Science*, *86*, 60-69.

[68] Singh, T.K., Drake, M.A., & Cadwallader, K.R. (2003). Flavour of Cheddar cheese: A chemical and sensory perspectives. *Comprehensive Reviews in Food Science and Food Safety*, *2*, 139-162.

[69] De Jong, L. (1976). Protein breakdown in soft cheese and its relation to consistency. 1. Proteolysis and consistency of 'Noordhollandse Meshanger' cheese. *Netherlands Milk and Dairy Journal*, *30*, 242-253.

[70] Van Slyke, L.L., & Price, W.V. (1979). *Cheese*. (2nd edition pp 239-253). Ridgeview Publ. Company, Reseda, CA.

[71] Bodyfelt, F.W., Tobias, J., & Trout, G.M. (1988). *The Sensory Evaluation of Dairy Products*. Van Nostrand Reinhold, New York.

[72] Bourne, M.C. (1982). *Food Texture and Viscosity; Concept and Measurement*. Academic Press, San Diego, CA.

[73] Kindstedt, P.S. (1991). Functional properties of Mozzarella cheese: A review. *Cultured Dairy Products Journal*, *26*, 27-31.

[74] Kindstedt, P.S. (1993). Effect of manufacturing factors, composition, and proteolysis on the functional characteristics of Mozzarella cheese. *Food Science and Nutrition*, *33*, 167-187.

[75] McMahon, D.J., Oberg, C.J., & McManus, W. (1993). Functionality of Mozzarella cheese. *Australian Journal of Dairy Technology*, *48*, 99-104.

[76] Mistry, V.V. (2001). *Low fat cheese technology. International of Dairy Journal*, *11*, 413-422.

[77] Ustunol, Z., Kawachi,, K., & Steffe, J. (1995). Rheological properties of Cheddar cheese as influenced by fat reduction and ripening time. *Journal of Food Science*, *60*, 1208-1210.

[78] Beal, P., & Mittal G.S. (2000). Vibration and compression responses of Cheddar cheese at different fat content and age. *Milchwissenschaft*, *55*,139-142.

[79] Zoon, P., & Allersma, D. (1996). Eye and crack formation in cheese by carbon dioxide from decarboxylation of glutamic acid. *Netherlands Milk and Dairy Journal*, *50*, 309-318.

[80] Noel, Y., Boyaval, P., Thierry, A., Gagnaire, V., & Grappin, R. (1999). Eye formation and Swiss-type cheeses. In B. A. Law, (ed.) (pp 222-250). *Technology of Cheesemaking*. Sheffield Academic Press, Sheffield, UK.

[81] Akkerman, J.C., Walstra, P., & van Dijk, H.J.M. (1989). Holes in Dutch-type cheese. 1. Conditions allowing eye formation. *Netherlands Milk and Dairy Journal*, *43*, 453-476.

[82] Luyten, H., van Vliet, T., & Walstra, P. (1991). Characterization of the consistency of Gouda cheese: Rheological properties. *Netherlands Milk and Dairy Journal*, *45*, 33-53.

[83] Luyten, H., & van Vliet, T. (1996). Effect of maturation on large deformation and fracture properties of (semi-) hard cheeses. *Netherlands Milk and Dairy Journal*, *50*, 295-307.

[84] Jameson, M.E. (1990). Cheese with less fat. *Australian Journal of Dairy Techno*logy, *11*, 93-98.

[85] Lawrence, R.C., & Gilles, J. (1987). *Cheddar cheese and related dry-salted cheese varieties.* In P. Fox, (ed.), (Vol. 1., pp 1-44). Cheese: Chemistry, Physics, and Microbiology. Elsevier Applied Science Publishers, New York, NY.

[86] Lawrence, R.C., & Gilles, J. (1982). Factors that determine the pH of young Cheddar cheese. *New Zealand Journal of Dairy Science and Technology*, *17*, 1-14.

[87] Beal, P., & Mittal, G.S. (2000). Vibration and compression responses of Cheddar cheese at different fat content and age. *Milchwissenschaft*, *55*, 139-142.

[88] Guo, M., & Kindstedt, P. (1995). Age-related changes in the water phase of Mozzarella cheese. *Journal of Dairy Science*, *78*, 2099-2107.

[89] Ramkumar, C., Creamer, L., Johnston, K., & Bennett, R. (1997). Effect of pH and time on the quantity of readily available water within fresh cheese curd. *Journal of Dairy Research*, *64*, 123-134.

[90] Alichanidis, E.P.A. (2008). Characteristics of major traditional regional cheese varieties of East-Mediterranean countries: a review. *Dairy Science and Technology*, *88*, 495-510.

[91] El Soda, M., & Abd El-Salam, M.H. (2002). *Cheeses matured in brine*. In: H. Roginski, J.W., Fuquay, & P.F., Fox (Eds.). Encyclopedia of Dairy Science (pp 404-411). Academic Press, London, UK.

[92] Chapman, H.R., Bines, V.E., Glover, F.A., & Skudder, P.J. (1974). Use of milk concentrated by ultrafihration for making hard cheese, soft cheese, and yogurt. *Journal of the Society of Dairy Technology*, *27*, 151-155.

[93] Green, M.L. (1985). Effect of pretreatment and making conditions on the properties of Cheddar cheese from milk concentrated by ultrafiltration. *Journal of Dairy Research*, *52*, 555-564.

[94] Sutherland, B.J., & Jameson, G.W. (1981). Composition of hard cheese manufactured by ultrafiltration. *Australian Journal of Dairy Technology*, *36*, 136-143.

[95] Nawar, M.A., Awad, S., Shamsia, S., & Ali, A.H. (2007). Effect of milk concentration by ultrafiltration on the proteolysis and rheological properties of low-fat soft white cheese. *Mansoura University Journal of Agricultural Science*, *32*, 277-289.

[96] Mistry, V.V., & Anderson, D.L. (1993). Composition and microstructure of commercial full-fat and low-fat cheeses. *Food Structure*, *12*, 259-266.

[97] Banks, J.M., Hunter, E.A., & Muir, D.D. (1993). Sensory properties of low fat Cheddar cheese: Effect of salt content and adjunct culture. *Journal of Society of Dairy Technology*, *46*, 119-123.

[98] Bryant, A., Ustunol, Z., & Steffe, J. (1995). Texture of Cheddar cheeses as influenced by fat reduction. *Journal of Food Science*, *60*, 1216-1219.

[99] Emmons, D.B., Kalab, M., Larmond, E., & Lowrie, R.J. (1980). Milk gel structure X. Texture and microstructure in Cheddar cheese made from whole milk and from homogenized low fat milk. *Journal of Texture Studies*, *11*, 15-34.

[100] Fenelon, M.A., Ryan, M.P., Rea, M.C., Guinee, T.P., Ross, R.P., Hill, C., & Harrington, D. (1999). Elevated temperature ripening of reduced fat Cheddar made with or without lacticin-producing starter culture. *Journal of Dairy Science*, *82*, 10-22.

[101] Awad, S., Hassan, A.N., & Halaweish, F. (2005). Applications of exopolysaccharides-producing cultures in reduced-fat cheddar cheese: Composition and proteolysis. *Journal of Dairy Science*, *88*, 4195-4203.

[102] Dabour, N., Kheadr, E.E., Fliss, I., & LaPointe G. (2005). Impact of ropy and capsular exopolysaccharide-producing strains of *Lactococcus lactis* subsp. *cremoris* on reduced-fat Cheddar cheese production and whey composition. *International Dairy Journal*, *15*, 459-471.

[103] Dabour, N., Kheadr, E., Benhamou, N., Fliss, I., & LaPointe, G. (2006). Improvement of texture and structure of reduced-fat Cheddar cheese by exopolysaccharide-producing lactococci. *Journal of Dairy Science*, *89*, 95-110.

[104] Hassan, A.N., Awad, S., & Muthukumarappan, K. (2005). Effects of exopolysaccharide- producing cultures on the viscoelastic properties of reduced-fat Cheddar cheese. *Journal of Dairy Science*, *88*, 4221-4227.

[105] Lo, C.G., & Bastian, E.D. (1998). Incorporation of native and denatured whey proteins into cheese curd for manufacture of reduced-fat Havarti-type cheese. *Journal of Dairy Science*, *81*, 16-24.

[106] Schreiber, R., Neuhauser, S., Schindler, S., & Kessler, H.G. (1998). Incorporation of whey protein aggregates in semi-hard cheese.Part 1: Optimizing processing parameters. *Deutsche Milchwirtschaft*, *49*, 958-962.

[107] Guinee, T.P., Fenelon, M.A., Mulholland, E.O., O'Kennedy, B.T., O'Brien, N., & Reville, W.J. (1998). The influence of milk pasteurization temperature and pH at curd milling on the composition, texture and maturation of reduced fat Cheddar cheese. *International Journal of Dairy Technology*, *1,* 1-10.

[108] Mistry, V.V., & Anderson, D.L. (1993). Composition and microstructure of commercial full-fat and low-fat cheeses. *Food Structure*, *12*, 259-266.

[109] Tunick, M.H., Mackey, K.L., Shieh, J.J., Smith, P.W., Cooke, P., & Malin, E.L. (1993). Rheology and microstructure of low-fat Mozzarella cheese. *International Dairy Journal*, *3*, 649-662.

[110] Darling, D.F., & Butcher, D.W. (1978). Milk fat globule membrane in homogenized cream. *Journal of Dairy Research*, *45*, 197-208.

[111] Metzger, L.E., & Mistry, V.V. (1994). A new approach using homogenization of cream in the manufacture of reduced fat Cheddar cheese. Manufacture, composition, and yield. *Journal of Dairy Science*, *77*, 3506-3516.

[112] Drake, M.A., Gerard, P.D., Truong, V.D., & Daubert, C.R. (1999). Relationship between instrumental and Sensory measurements of cheese texture. *Journal of Texture Studies*, *30*, 451-476.

[113] Code of Federal Regulations. (2003). Section 133.169 US Dept. Health Human Services, Washington, DC.

[114] Brown, J.A., Foegeding, E.A., Daubert, C.R., Drake, M.A., & Gumpertz, M. (2003). Relationships among rheological and sensorial properties young cheese. *Journal of. Dairy Science*, *86*, 3054-3067.

[115] French, S.J., Lee, K.M., Decastro, M., & Harper, W.J. (2002). Effects of different protein concentrates and emulsifying salt conditions on the characteristics of a processed cheese product. *Milchwissenshaft*, *57*, 79-82.

[116] Glenn, T.A., Daubert, C.R., Farkas, B.E., & Stefanski, L.A. (2003). Statistical analyses of creaming variables impacting processed cheese melt quality. *Journal of Food Quality*, *26*, 299-321.

[117] Acharya, M.R., & Mistry, V.V. (2005). Effect of vacuum-condensed or ultrafiltered milk on pasteurized process cheese. *Journal of Dairy Science*, *88*, 3037-3043.

[118] Hassan, A.N., Awad, S. & Mistry, V.V. (2007). Reduced fat process cheese made from young reduced fat Cheddar cheese manufactured with exopolysaccharide-producing cultures. *Journal of Dairy Science*, *90*, 3604-3612.

[119] Van Vliet, T. (1988). Rheological properties of filled gels. Influence of filler matrix interaction. *Colloid Polymers Science*, *266*, 518-524.

[120] Schar, W., & Bosset, J.O. (2002). Chemical and physico-chemical changes in processed cheese and ready-made fondue during storage. *Lebensmittel-Wissenschaft und Technology*, *35*, 15-20.

[121] Templeton, H.L., & Sommer, H.H. (1930). Some observations on processed cheese. *Journal of Dairy Science*, *13*, 203-220.

[122] Olson, N.F., Gunasekaran, S., & Bogenrief, D.D. (1996). Chemical and physical properties of cheese and their interactions. *Netherlands Milk and Dairy Journal*, *50*, 279-294.

[123] Meyer, A. (1973). *Processed Cheese Manufacture.* Food Trade Press Ltd., London, UK.

[124] Thomas, M.A. (1973). *The Manufacture of Processed Cheese, Scientific Principles*. (1st edition). New South Wales Department of Agriculture,

[125] Fox, P.F., Guinee, T.P., Cogan, T.M., & McSweeney, P.L.H. (2000). *Fundamental of cheese science. Processed cheese and substitute or imitation cheese products* (pp. 429-45). Jaithers Burg, MD: Aspen Publisher Inc.

[126] Kapoor, R., Metzger, L.E., Biswas, A.C., & Muthukummarappan, K. (2007). Effect of natural cheese characteristics on process cheese properties. *Journal of Dairy Science*, *90*, 1625-1634.

[127] Klostermeyer, H., & Uhlman, G. (1998). *Processed cheese manufacture.* A Joha Guide. BK Giulini Chemie & Co. OHG, Ladenburg.

[128] Shimp, L.A. (1985). Process cheese principles. *Food Technology*, *39*, 63-70.

[129] Caric, M., & Kaláb, M. (1987). Processed cheese products. In: P. F. Fox (Ed), *Cheese: chemistry, physics and microbiology*. Volume 2. Major cheese groups (pp. 339-383). London, New York: Elsevier Applied Science.

[130] Caric, M., Gantar, M., & Kalab, M. (1985). Effects of emulsifying agents on the microstructure and other characteristics of process cheese. *Food Microstructure*, *4*, 297-312.

[131] Bornaz, S., Fanni, J., & Parmentier, M. (1993). Butter Texture: The Prevalent Triglycerides. *Journal of the American Oil Chemists' Society*, *70*, 1075-1079.

[132] Walstra, P., van Vliet, T., & Kloek, W. (1995). Crystallization and rheological properties of milk fat spreads. In P. Fox (Ed.). *Advanced dairy chemistry-2 Lipids*. (pp. 179-212). Chapman & Hall., 2-6 Boundary Row, London, UK.

[133] Vanhoutte, B., & Huyghebaert, A. (2003). Physical properties of milk fat. In: B. Rossell. (Ed.). *Oils and fats. V3. Dairy fats*. (pp 99-138). Leatherhead International. Randalla Road, Leatherhead, Surrey KT22 7RY, UK.

[134] Bobe, G., Hammond, E.G., Freeman, A.E., Lindberg, G.L., & Beitz, D.C. (2003).Texture of butter from cows with different milk fatty acid compositions. *Journal of Dairy Science*, *86*, 3122-3127.

In: Practical Food and Research
Editor: Rui M. S. Cruz, pp. 393-406
ISBN: 978-1-61728-506-6

Chapter XV

VITAMINS

Pamela Manzi and Laura Pizzoferrato

Istituto Nazionale di Ricerca per gli Alimenti e la Nutrizione (INRAN), Food Science Department Via Ardeatina, 546 – 00178 Roma, Italia, manzi@inran.it and pizzoferrato@inran.it

ABSTRACT

European and Italian legislations act as a guide among the numerous milk technological treatments. Fat and water soluble vitamins are chosen as quality indicators to study the effect of processing on milks, fermented milks, and cheeses.

Actually vitamins either fat or water soluble are quite labile components: some are sensitive to heat, some to oxygen or light. These factors, with time of treatment, are important variables, responsible for product safety and quality and they should be optimized during a process development. Moreover, vitamins are sensitive indices of a processed food, useful to describe both the nutritional, and the functional food quality (e.g. some vitamins are natural antioxidant).

In this chapter the effect of treatments on fat soluble (vitamin A, carotene, vitamin E) and water soluble vitamins (folic acid, vitamin B_2 and B_{12}, vitamin C) are considered.

Two tracing parameters (DAP - Degree of Antioxidant Protection and DRI – Degree of Retinol Isomerization), obtained by easy mathematical calculations based on vitamin data, are also reported. They are useful as quality indicators, and have already been utilized to evaluate quality changes occurring in milk and cheeses.

Finally, the effects of storage and packaging conditions are also briefly addressed, much more work should be necessary to deepen these aspects with particular reference to the development of new packaging materials.

15.1 INTRODUCTION

European and Italian laws ratify different milk products - and consequently their industrial processes - that can be commercialized in the Community.

The European Regulation N. 1411/71 [1] defines different milk categories:

- *full cream milk* (fat content is either at least 3.50% naturally or has been brought to at least 3.50%);
- *semi-skimmed milk* (fat content has been brought to at least 1.50% and at most 1.80%);
- *skimmed milk* (fat content has been brought to not more than 0.30%).

The involved process is simply a skimming, but – as we shall discuss afterwards – even this mechanical process can affect vitamin equilibrium in milk.

A Council Directive in 1992 [2] provides the rules for the production and commercialization of heat treated milk. In particular:

Raw milk is produced by secretion of the mammary glands of one or more cows, ewes, goats or buffaloes, which has not been heated beyond 40 °C or undergone any treatment that has an equivalent effect. Raw milk has a short shelf-life but it can be processed by heat treatments in order to make milk safe for human consumption and to extend its shelf-life.

Pasteurized milk is a milk treated at high temperature for a short time (at least 71.7 °C for 15 s or any equivalent combination). The pasteurized milk shows a negative reaction to the phosphatase test and a positive reaction to the peroxidase test. However, the production of pasteurized milk which shows a negative reaction to the peroxidase test is authorized, provided that the milk is labelled as "*High-temperature Pasteurized*".

High Quality Pasteurized milk is defined by Italian legislation [3] as a selected milk treated at 72 °C for 15 s and with a percentage of soluble whey proteins not less than 15.5%.

UHT milk is obtained by applying to the raw milk a continuous flow of heat entailing the application of a high temperature for a short time (not less than 135 °C for not less than a 1 s) with the aim to destroy all residual spoilage microorganisms and their spores. Such thermal process can cause changes in the nutritional, organoleptic, or technological properties of milk, but the UHT treatment can now be considered a "mild" technology if compared with the traditional sterilized milk, heated and sterilized in hermetically sealed wrappings or containers.

Recently a "*micro-filtered pasteurized milk*" has been allowed in Italy [4], this is a long-life milk produced by a mechanical treatment (micro-filtration) and a mild heat treatment.

Fermented milks, obtained by lactic bacteria fermentation, nowadays represent an increasing market share of dairy products. In Italy, *Streptococcus thermophilus* and *Lactobacillus bulgaricus* are solely utilized to produce "yogurt" [5], but fermented milks with other lactic bacteria (*Bifidobacterium* or *Lactobacillus*) are commercialized. Among these lactic bacteria, "probiotics" are microorganisms able to adhere to intestinal epithelial tissues, colonize the gastrointestinal tract, produce antimicrobial substances, modulate the immune response, influence metabolism [6-8] and, in all, beneficially affect the host animal by improving its intestinal microbial balance [9].

Milk is also the raw material for cheese production. Most types of cheese are made involving a number of common steps: the milk (raw or thermally treated) is added of an appropriated bacteria culture and mixed with rennet that causes the coagulation of milk (curd). The curd is cut with special cutting tools into small cubes of the desired size to facilitate expulsion of whey. The finished curd is placed in cheese moulds of metal, wood or plastic, which determine the shape of the finished cheese, and pressed. Treatments during curd making and pressing determine the characteristics of cheeses. In particular, flavour characteristics are developed during the cheese ripening [10].

The cited treatments can cause modifications on product quality and, in particular, vitamins can be affected.

In this chapter the main modifications due to industrial milk processes on fat soluble (in particular A and E) and water soluble vitamins (B-group vitamins and vitamin C) are described.

15.2 Effects of Treatments on Fat-Soluble Vitamins

15.2.1 Vitamin A, functional to epithelial cells, vision, gene regulation and immune cell function.

In foods of animal origin, vitamin A is present as retinols (*trans* and *cis* isomers), as free alchool or in esterified form. In plant foods vitamin A is contained as pro-vitamin molecules, mainly alpha and beta carotene that can be converted in vitamin A after human absorption. In cow milk both the animal and plant vitamin A forms (retinols and beta carotene) are present.

An analytical complete separation and the quantification of all-*trans* retinol, *cis* retinol isomers and carotenes is necessary for the evaluation of the actual vitamin A quality of milk and dairy products. Actually the isomer all-*trans* retinol has the highest vitamin activity (100%), but the *cis* isomers are less active (13-*cis* isomer has 75% activity), while carotenes have different pro-vitamin activity (beta carotene has 16% of vitamin A activity).

Vitamin A is generally considered a heat-resistant compound [11], but during heat treatments or light exposure of milk a *trans* to *cis* isomerization can occur [12]. The vitamin A (all-*trans* retinol) content of milk after exposure to 2200 lx intensity fluorescent light decreases to 23.6% of the original content in semi-skimmed milk and to 4.2% in skimmed milk [13].

The content of vitamin A (all-*trans* retinol, 13-*cis* retinol, and beta carotene) shows large variability with feed, season, lactation period and direct comparisons among products can be affected by this variability. The Degree of Retinol Isomerization (DRI), a ratio between two forms of vitamin A (13-*cis* and all-*trans* retinol) overcomes this problem and acts as an indicator of milk processing free from other variables [14, 15].

In Table 15.1 the DRI, Degree of Retinol Isomerization in commercial heat treated milk is reported. In micro-filtered milk the DRI value is very low, actually microfiltration is a mild tecnique used to mechanically remove bacterial contaminants and suspended materials and milk is slightly heat treated.

Sterilized milk has the highest degree of isomerization due to the severe thermal treatment and also due to light permeability of the packing. Actually this milk is generally sterilized and commercialized in the same bottle, made of plastic (lower DRI) or glass (higher DRI). In any case, this index is well correlated with other well known heat treatment indices that describe the modifications of protein (furosine and whey proteins) and carbohydrate (lactulose) [16].

Table 15.1. Degree of Retinol Isomerization (DRI%) in heat treated cow milk.

Milk (cow)	Samples N°	DRI% (average)
Raw	10	0.2
High-quality pasteurized	12	2.8
Micro filtered pasteurized	10	1.5
Pasteurized	27	7.9
High temperature pasteurized	10	8.0
UHT	14	19.4
Sterilized in bottle	9	34.6

[personal data]

The *trans-cis* isomerization can also occur during milk fermentation and cheese production. In probiotic fermented milk and synbiotic fermented milk the DRI varies from 3.1% to 23.7% [17]. The presence of *cis* isomers in fermented milk suggests that fermentation, directly or indirectly, induces *trans-cis* isomerization, confirming the opportunity to assess any individual forms in evaluating the vitamin A content in dairy products [18] and, particularly during nutritional surveys, the all *trans* retinol content alone can not be assumed to be the total vitamin A level.

The variability of vitamin A content in cheese is due to the milk composition and to the cheesemaking process [19]. Actually, the DRI value in cheese varies from 16.5% in Groviera cheese to 35% in Crescenza cheese and these results suggest that there is no general relationship between ripening period and retinol isomerization and that the differences in the DRI value cannot be explain solely by heat treatments during manufacture. It is likely that other factors such as light and oxygen levels during production and storage play different, but probably synergistic roles in retinol isomerization. [14].

The level of DRI and retinol isomers is also studied in Mozzarella di Bufala Campana PDO cheeses during storage [20]. At the beginning of the experiment (1 day post-production) the DRI is meanly 15%, but during storage (traditional storage conditions, 10 days at room temperature) this value increases in all studied Mozzarella di Bufala Campana cheeses due to both *cis* isomers increase and all-*trans* retinol decrease. The same storage experiment carried out in refrigerated conditions (4 °C in the dark) shows a not significant increase in the DRI. Actually the kinetics of isomerization reactions, as mostly chemical reactions, is accelerated by temperature increase.

The presence of carotenes in milk is specie–dependent and in buffalo, goat, and sheep milk, carotenes are not analytically detectable. In human milk the main carotenes are lutein, cryptoxanthin, alpha carotene, beta carotene, lycopene, while in bovine milk beta carotene is preponderant [21].

Carotenes, besides their well known function as vitamin A precursor, have other valuable properties. They are anti-oxidant factors (radical and singlet oxygen quenching), act in cell differentiation, are precursors of nuclear receptor ligand, and inductors of cell-communication [22].

Retinol and carotenes in Dutch milk and dairy products are studied by Hulshof and coworkers [23]. According to these Authors preparation of yogurt or butter leads to retinol and beta carotene losses up to 20%. Their study confirms that winter milk contains a lower

(20%) amount of retinol and beta carotene than summer milk. Moreover, differences in retinol and carotene content between young and mature Gouda cheese are small, indicating that, in this case, ripening does not affect retinol and/or beta carotene levels.

Effects of package light transmittance on vitamin A in pasteurized whole milk [24], UHT whole milk [25] and fortified UHT low-fat milk [26] can be very different. Pasteurized whole milk (3% fat), stored under fluorescent light at 8 °C in clear polyethylene terephthalate (PET) bottles and in three pigmented PET bottles, shows a reduction of 22% of vitamin A in clear PET bottles, while in all pigmented PET bottles the vitamin loss is significantly lower. In UHT whole milk a vitamin A loss, ranging from 88 to 66%, is observed. Fortified UHT low-fat milk shows in clear PET bottles, a reduction of 93% in vitamin A and of 66% in vitamin D3. In all pigmented PET bottles, the vitamin retention is only slightly higher: 70-90% for vitamin A and 35-65% for vitamin D3 depending on the pigmentation level.

15.2.2 Vitamin E, functional as *in vivo* and *in vitro* antioxidant

Vitamin E is a family of eight naturally occurring homologues: 4 tocopherol homologues (alpha, beta, gamma, and delta) with a saturated 16-carbon phytyl side chain and 4 tocotrienols (alpha, beta, gamma, and delta) with three double bonds on the side chain.

Alpha tocopherol, the main form of vitamin E has the highest vitamin E activity (100%), followed by beta tocopherol (50%), alpha tocotrienol (30%) and gamma tocopherol (10%) [27].

In foods vitamin E is studied for its nutritional role and also for its properties to prevent oxidation of poliunsaturated fatty acids, in milk and dairy products these antioxidant characteristics can also be useful to avoid cholesterol oxidation.

Milk contains approximately 12 mg/100g of cholesterol and the probability to find out oxy-sterols in milk or fresh dairy products is very poor: the medium is a liquid, highly dispersed and the oxygen content is not sufficient to start the reaction. More severe treatments such as milk spray-drying, especially after long storage, or butter cooking cause the formation of measurable amounts of cholesterol oxidation products. Whole milk powder, stored for 2 years contains 7-alpha-hydroxycholesterol (3.9 mg/kg fresh product) and alpha-epoxycholesterol (4.1 mg/kg fresh product) [28]. The formation of oxy-sterols in dairy products can also occur during storage of grated or sliced cheeses when a large surface is exposed to light and oxygen.

Oxy-sterols, even in very small amounts, could be dangerous for the human health [29] and the presence in dairy products of antioxidant molecules such as alpha tocopherol and beta carotene with cholesterol is a favourable coincidence to avoid or slow down oxidation reactions.

The Degree of Antioxidant Protection (DAP), parameter calculated as the molar ratio between antioxidant compounds and an oxidation target, is useful just to estimate the potential oxidative stability of fat in foods [30].

In milk and dairy products, the antioxidant compounds considered are alpha tocopherol and beta carotene (when it is present), and, for lack of significant amounts of very easily oxidable poliunsaturated fatty acids, cholesterol is the oxidation target molecule [31].

In Table 15.2, the DAP, Degree of Antioxidant Protection in different milk products is reported. Generally the higher the DAP value, the higher the protection from oxidation. With the increase of milk treatment severity, DAP value decreases due to the decrease of alpha tocopherol and beta carotene and to the constancy of cholesterol.

Table 15.2. Degree of Antioxidant Protection (DAP) in cow milk.

Milk (cow)	Samples N°	DAP (average)
Raw	10	7.0
High quality pasteurized	12	6.1
Micro filtered pasteurized	10	6.0
Pasteurized	27	5.6
High temperature pasteurized	10	5.8
UHT	14	5.5
Sterilized in bottle	9	4.9
Whole milk	5	7.0
Semi-Skimmed milk	5	5.4
Skimmed milk	5	1-5

[personal data]

Surprisingly more severe treatments (milk dehydration and concentration), allow an increase of milk antioxidant properties [32]. This unexpected result is attributed to the formation of new antioxidant species induced by the early stage of the Maillard reaction occurring in milk during these industrial processes.

But why the DAP value decreases from all fat milk to skimmed milk?

Skimmed dairy products - obviously - contain less fat, less cholesterol and also less fat soluble vitamins than whole products and a good linear correlation between these compounds (alpha tocopherol, beta carotene, cholesterol) and fat content in different dairy samples can be observed. On the contrary – as confirmed by the DAP value - the composition of milk fat changes differently during skimming process: cholesterol concentration increases, while antioxidant compound concentration decreases or remains unchanged in the residual fat matter. The skimming process mainly induces a depletion of the largest fat globules (cholesterol-poor) and the remaining smallest globules are cholesterol-rich, while antioxidant compound concentration decreases or remains unchanged in the residual fat. As a matter of fact, the residual cholesterol is more susceptible to oxidation in skimmed than in whole products [33]. Actually this lower cholesterol protection in skimmed milk products is not a real risk for health's consumers but it is a good example of what can happen when a natural product is industrially changed. Even the skimming process, a very mild technology, can induce unexpected modifications – not always positive - on the natural characteristics of foods.

15.3 EFFECTS OF TREATMENTS ON WATER-SOLUBLE VITAMINS

15.3.1 Folates, functional against neural tube defects

This class of compounds has a chemical structure and nutritional activity similar to that of folic acid and represents important B-group vitamins, participating in one-carbon transfer reactions required in many metabolic pathways, purine and pyrimidine biosynthesis and amino acid interconversions. Recent studies on folates focus on their protective role against neural tube defects occurring during early pregnancy [34].

On average, milk and dairy products provide 10% to 15% of the daily folate intake in many Western countries, especially among the younger population. Milk and especially fermented dairy products like yogurt, buttermilk and different kind of cheeses are recognized as good dietary sources of folates and in particular the major form 5-methyl-THF [35].

The stability of folates is pH-dependent: they less stable between pH 4–6. For their chemical reactivity, folates are very vulnerable to food processing (light and heat). In a pioneering work [36] the effect of heat treatments and storage on human milk are reported: total folate concentration decreases similarly by all heat treatments (62.5 °C for 30 min; 72 °C for 15 s; 88 °C for 5 s and 100 °C for 5 min). These data have recently been confirmed in cow milk [35]: the folate content in milk (whole, semi-skimmed or skimmed) ranges between 4 to 10 μg/100 g. During pasteurization losses of folate are moderate (<10%), while during UHT process losses are more important (<43%). Moreover oxygen in the packaging headspace and packaging with oxygen permeable seals play an important role in the folate level decreasing.

In fermented milk, data on folates are variable from 3 to 18 μg/100g, depending on the different bacteria culture used.

In cheeses, folate contents depend on the cheese making procedures and on the ripening time. Generally during cheese manufacture folates are recovered in whey (50%).

In Brazil, dairy products (powdered milk, sterilized milk, dairy beverages and petit Suisse cheese) are often enriched with folic acid. The analyses of these products show that 16 samples have folic acid levels with only slight deviations from the values declared, 7 samples (mainly sterilized milks and ready to drink dairy beverages) have very low values, while no folic acid is detected in 2 out of 3 petit Suisse cheeses. These data confirm that folic acid can be destroyed during UHT process and during cheese manufacture [37].

15.3.2 Riboflavin, functional as a catalyst for redox reactions in numerous metabolic pathways and in energy production

Riboflavin, or Vitamin B_2, is a water-soluble, yellow, fluorescent compound. This vitamin is present as part of the coenzymes flavin mononucleotide (FMN) and flavin-adenine dinucleotide (FAD) [27].

Riboflavin is relatively stable during thermal food processing and storage, but it is very sensitive to light. It can accept or donate a pair of hydrogen atoms, it can act as a photosensitizer or a prooxidant for food components under light. Photosensitization of riboflavin causes production of reactive oxygen species that can accelerate the decomposition of proteins, lipids, carbohydrates, and vitamins, and could cause significant nutrient loss in foods [38].

Although milk is one of the primary sources of riboflavin the amounts of this vitamin and other flavin derivatives have not been accurately quantified. In a study of Roughead and McCormick [39] a comprehensive assessment of milk is performed in order to identify and quantify the flavins: riboflavin and flavin adenine dinucleotide (FAD) are the predominant flavins in all milk samples.

Haddad and Loewenstein [40] study in their work the effects on some water soluble vitamins of four time-temperature treatments of milk: high temperature short time pasteurization (72 °C for 16 s); commercial HTST pasteurization (80 °C for 16 s); commercial UHT-sterilization (110 °C for 3.5 s) and UHT-sterilization (140 °C for 3.5 s). These four different treatments do not cause significant riboflavin or thiamine losses. On the contrary, vitamin C content decreases as thermal treatment severity increases.

In pasteurized whole milk (3% fat), stored under fluorescent light at 8 °C in clear polyethylene terephthalate (PET) bottles and in three pigmented PET bottles a 33% vitamin B_2 reduction has observed, while the vitamin B_{12} content remains almost stable, higher vitamin recovery is observed in all pigmented PET bottles [24]. In the same experimental conditions, UHT whole milk and fortified UHT low-fat milk show a B_2 complete decomposition in the clear PET bottles while in the pigmented ones the vitamin retention is 63-95% for vitamin B_2 depending on the pigmentation level [25, 26].

In Italian Caciotta cheeses, produced with reduced salt or with semi skimmed milk, riboflavin levels are not modified by salt reduction while the low fat cheeses show riboflavin reduced contents [41]. This reduction in a water soluble vitamin is probably due to the skimming removing riboflavin bound to flavoproteins located on the fat globule membrane and it is another unexpected effect of the skimming procedure.

15.3.3 Cyanocobalamin, functional as grow factor and antianaemic molecule

Cyanocobalamin, or vitamin B_{12}, is synthesized by bacteria (e.g. rumen) and animal foods are considered the major dietary source.

Vitamin B_{12} content in milk is not high (0.36 μg/100 g cow milk; 0.07 μg/100 g goat milk; 0.36 μg/100 g buffalo milk; 0.71 μg/100 g sheep milk) [42], however dairy products can be considered significant contributors of vitamin B_{12} intake, because of their generally high presence in human diet.

In bovine milk, appreciable losses of vitamin B_{12} are reported during milk processing: boiling for 2-5 min and 30 min cause 30% and 50% loss, respectively. Pasteurization leads to 5%-10% loss and also fluorescent light induces a considerable decreasing [43].

In fermented milk vitamin B_{12} concentration decreases significantly: these data are well known, in a pioneering work Reddy and coworkers [44] determine the effect of various factors upon B-vitamin content of cultured yogurt and direct acidified yogurt. Folic acid and vitamin B_{12} contents decrease 29 and 60% respectively in cultured yogurt and 48 and 54% in acidified yogurt.

Data of reduction of vitamin B_{12} during fermentation are confirmed by Sato and coworkers [45]: the content of vitamin B_{12} in whey is reduced considerably during lactic acid fermentation.

In cheese, only from 20 to 60% of vitamin B_{12} is recovered in Cottage cheese, hard cheese, and blue cheese [43].

Data on Cottage cheese are in accordance with Reif and coauthors [46]. In their work Cottage cheese contains, on the average, 257 μg niacin, 24 μg vitamin B_6, 2.1 μg vitamin B_{12}, and 40.6 μg folic acid per 100 g. In general, the higher the vitamin content of milk, the higher the vitamin content of cheese curd. According to these authors considerable amounts of vitamins are lost in the whey during the manufacturing process of Cottage cheese, but niacin, vitamin B_6, and vitamin B_{12} are retained from 16.0 to 63.7%.

Papastoyiannidis and coworkers [47] study four fermented cow's milks fortified with B-group vitamins (B_1, B_2, pyridoxine, pyridoxal, pyridoxamine and folic acid) and inoculated with different mixed probiotic cultures. Their data show that some vitamins are partially lost during heating and fermentation with different microorganism species and strains, but during storage for 16 days at 4 °C, the content of all vitamins remains stable. Moreover fermented fortified products, combining beneficial effects of probiotics with vitamins, are reported to be a good alternative to dietary supplements.

15.3.4 Ascorbic acid, functional as antioxidant

Ascorbic acid and dehydroascorbic acid, its first oxidized product, are water soluble vitamins known as vitamin C.

Vitamin C is very heat-sensitive especially in the presence of light, oxygen, iron and copper. Milk and milk products are poor sources of vitamin C: cow milk, goat milk, sheep milk and buffalo milk contain 1.5 mg/100 g, 1.3 mg/100 g, 4.2 mg/100 g and 2.3 mg/100 g of vitamin C, respectively [42].

Some Authors [48] demonstrate in ewe's milk fermented by *Streptococcus thermophilus*, *Lactobacillus acidophilus* and *Bifidobacterium ssp.* starter culture that vitamin C can be synthesized by some lactic bacteria (*Lactococcus*).

Vitamin C retention is influenced by packaging, Gliguem and Birlouez-Aragon [49] analyze the vitamin C content on fortified milk samples (growth milks) for children (1 to 3 years) and find out a vitamin complete oxidation in a 3-layered opaque bottle after 1 month of storage. In a 6-layered opaque bottle, with an oxygen barrier, a 75% vitamin C retention is observed after a storage of 4 months.

Vitamin C decrease during fermentation ranges from 29 to 32% [50].

In a survey, Scott and Bishop [51] supply information on water-soluble vitamins in retailed milk. In UHT whole milk samples, the authors found negligible levels of vitamin C and folic acid, while vitamin B_6 and vitamin B_{12} levels were, respectively, 73 and 56% of their relevant levels in pasteurized milk.

Vitamin concentration (C, folic acid and B_{12}) in sterilized milk are lower than in pasteurized milk, while, with the exception of vitamin B_{12}, B-group vitamins and especially folic acid, are in higher concentration in yogurt than in pasteurized milk [51].

15.4 Analytical Methods for Vitamin Determinations in Milk and Dairy Products

The main procedures available for the determination of fat soluble vitamins involve mostly high performance liquid chromatographic (HPLC), and gas chromatographic separation techniques.

The vitamin extraction is usually performed either by a direct solvent treatment either by a basic hydrolysis of foods (saponification of the fat fraction) followed by a solvent extraction of the non hydrolyzed fraction (unsaponifiable). Saponification is used to hydrolyze esterified forms of vitamins A, E, D, and carotenoids and should be performed with addition of an antioxidant compound (generally ascorbic acid, pyrogallol, or butylated hydroxy toluene) to avoid vitamin oxidation [52].

In particular for vitamin D in some dairy products (milk and infant formulas) the low contents and the presence of interfering compounds require sample treatment and purification of the extract, before the determination by high performance liquid chromatography [53].

Recently the use of supercritical fluids in fat soluble vitamin analyses is considered an interesting tool to decrease the amount of organic solvents utilized. The main advantages, besides the minimal consumption of organic solvents, are the exclusion of oxygen, the reduction of heat and the shorter extraction time. Future possibilities for supercritical fluids will be the simultaneous determination of fat-soluble vitamins, triacylglycerols, diacylglycerols, monoacylglycerols, free fatty acids, cholesterol and other steroids and lipid esters, all in one chrimatographic run [54].

Traditional protocols for B-group vitamins are often based on microbiological assays, however, these methods are lengthy, manipulative and generally of poor precision. Nowadays, the determination of water soluble vitamins occurs with the simultaneous determination using HPLC with UV or fluorescence detection.

The chromatographic separation is generally carried out using reversed-phase high performance liquid chromatography (RP-HPLC) on a C_{18} column, and the vitamins are detected at different wavelengths by either fluorescence or UV-visible detection. The relevant advantages of these methods are the simultaneous determination in foods and dairy products with a reduction of the time required for quantitative extraction. [55].

Nevertheless, according to Woollard and Indyk [56] for an estimation of B-vitamin content, the extraction procedure is a critical factor and generally a combination of acid hydrolysis and enzymatic digestion is used to release endogenous protein-bound and phosphorylated B-group vitamins.

The study of Woollard and Indyk [56] shows a rapid, simultaneous determination of riboflavin, FMN, niacinamide, pyridoxal, and pyridoxine, with consecutive measurement of thiamine with use of dioctylsulfosuccinate as counter-ion under reversed-phase LC conditions, with a high degree of precision compared with single-vitamin reference methods.

A recent method [57] for the simultaneous determination of four water-soluble vitamins (B-group), using ultra-performance liquid chromatography electrospray ionization tandem triple quadrupole mass spectrometry, has been developed and validated for the determination of trace amounts of vitamins in fortified milk powers. A reversed phase C_{18} column and a binary gradient acetonitrile-water mobile phase are applied for the separation of the water-soluble vitamins with recoveries ranging from 86.0 to 101.5%.

Conclusion

Milk industrial treatments have a common purpose: to make milk and cheese safe for human consumption and to prolong their shelf-life.

The well known heat treatments are reliable and effective tools to hit this target, but the other side of the coin is a possible reduction of the nutritional and organoleptic quality of milk. Nowadays heat treated milks consumed in Europe are probably the best compromise among safety (the first requisite), nutritional and organoleptic quality and convenience. Actually many studies have been carried out to optimize the industrial conditions for each commercial category of milk and the European market is rich of optimal milk products.

The industrial treatment effects on vitamins strongly depend on the chemical structure of each vitamin. All-*trans* retinol, the main form of vitamin A in milk, isomerizes into the 13-*cis* retinol form and alpha tocopherol and beta carotene act as natural antioxidant and protects milk from oxidative reactions.

In the case of water soluble vitamins, heat treatments do not cause significant losses of thiamine contents, while vitamin C, folate and vitamin B_{12} contents decrease as thermal treatment severity increases. Riboflavin, relatively stable during thermal food processing and storage, is very sensitive to light.

Milk fermentation causes water soluble vitamin modifications: vitamin B_{12} concentration decreases while the content of folates depends on the different bacteria culture used.

Further research is necessary to better understand the effect of packaging on milk and cheese quality.

Storage condition is one of the most critical points in the dairy chain and new packaging materials, and above all, will play a key role in the future market of milk and cheese.

References

[1] Regulation EEC N. 1411/71 of the Council of 29 June 1971 laying down additional rules on the common organisation of the market in milk and milk products.

[2] Council Directive 92/46/EEC of 16 June 1992 laying down the health rules for the production and placing on the market of raw milk, heat-treated milk and milk-based products.

[3] Italian Law. (1989). Disciplina del trattamento e della commercializzazione del latte alimentare vaccino. Legge n. 169 del 03/05/1989, *Gazzetta Ufficiale* n. 108 del 11 Maggio 1989.

[4] Ministero della Salute. (2002). Trattamento di microfiltrazione nel processo di produzione del latte alimentare. Decreto del 17/06/2002, *Gazzetta Ufficiale* n.178 del 31 Luglio 2002.

[5] Ministro della Sanità. (1972). Produzione e commercio dello yogurt. *Circolare del Ministro della Sanit*à n.2 del 04/01/1972.

[6] Ziemer, C.J., & Gibson, G.R. (1998). An overview of probiotics, prebiotics and synbiotics in the functional food concept: perspectives and future strategies. *Intenational Dairy Journal*, *8*, 473-479.

[7] Saarela, M., Mogensen, G., Fondén, R., Mättö, J., & Mattila-Sandholm, T. (2000). Probiotic bacteria: safety, functional and technological properties. *Journal of Biotechnology*, *84*, 197-215.

[8] Teitelbaum, J.E., & Walker, W.A. (2002). Nutritional impact of pre and probiotics as protective gastrointestinal organisms. *Annual Review of Nutrition*, *22*, 107-138.

[9] Fuller, R. (1989). Probiotics in man and animals. *Journal of Applied Bacteriology*, *66*, 365-378.

[10] Ottogalli, G. (2001). *Atlante dei formaggi*. Milano, Hoepli Publishers.

[11] Burton, H. (1988). *Ultra-high temperature processing of milk and milk products*. New York, Elsevier Applied Science Publishers.

[12] Murpyh, P.A., Engelhardt, R., & Smith, S.E. (1988). Isomerization of retinyl palmitate in fortified skim milk under retail fluorescent lighting. *Journal of Agricultural and Food Chemistry*, *36*, 592-595.

[13] De Man, J.M. (1981). Light-induced destruction of vitamin A in milk. *Journal of Dairy Science*, *64*, 2031-2032.

[14] Panfili, G., Manzi, P., & Pizzoferrato, L. (1998). Influence of thermal and other manufacturing stresses on retinol isomerization in milk and dairy products. *Journal of Dairy Research*, *65*, 253-260.

[15] Panfili, G., Chimisso, E., Manzi, P., & Pizzoferrato, L. (1999). Il grado di isomerizzazione del retinolo può rappresentare un indice di processo in prodotti lattiero-caseari? *Scienza e Tecnica Lattiero Casearia*, *50*(2), 138-146.

[16] Pellegrino, L., De Noni, I., & Resmini, P. (1995). Coupling of lactulose and furosine indices for quality evaluation of sterilized milk. *International Dairy Journal*, *5*(7), 647-659.

[17] Manzi, P., Marconi, S., & Pizzoferrato, L. (2007). New functional milk-based products in the Italian market. *Food Chemistry*, *104*, 808-813.

[18] Panfili, G., Fratianni, A., Di Croscio, T., Gammariello, D., & Sorrentino E. (2006). Influenza dei processi di produzione e dello sviluppo microbico sul grado di isomerizzazione del retinolo nel latte e derivati. In Chiriotti Editore (Eds.), *Proceedings of the VII Congresso Italiano Scienza e Tecnologia degli Alimenti,* (Vol I, pp. 1126-1131). Pinerolo (Italy).

[19] Lucas, A., Rock, E., Chamba, J.F., Verdier-Metz, I., Brachet, P., & Coulon, J.B. (2006). Respective effects of milk composition and the cheese-making process on cheese compositional variability in components of nutritional interest. *Lait*, *86*, 21-41.

[20] Manzi, P., & Pizzoferrato, L. (2008). Kinetic study on unsaponifiable fraction changes and lactose hydrolysis during storage of Mozzarella di Bufala Campana PDO cheese. *International Journal of Food Sciences and Nutrition, iFirst article,* DOI: 10.1080/09637480802158176.

[21] Kim, Y., English, C., Reich, P., Gerber, L.E., & Simpson, K.L. (1990). Vitamin A and carotenoids in human milk. *Journal of Agricultural and Food Chemistry*, *38*, 1930-1933.

[22] Russel, R.M. (1998). Physiological and clinical significance of carotenoids. *International Journal of Vitamin and Nutrition Research*, *68*, 349-353.

[23] Hulshof, P.J.M., Van Roekel-Jansen, T., Van de Bovenkamp, P., & West, C.E. (2006). Variation in retinol and carotenoid content of milk and milk products in The Netherlands. *Journal of Food Composition and Analysis*, *19*, 67-75.

[24] Saffert, A., Pieper, G., & Jetten, J. (2006). Effect of package light transmittance on the vitamin content of pasteurized whole milk. *Packaging Technology and Science*, *19*, 211-218.

[25] Saffert, A., Pieper, G., & Jetten, J. (2008). Effect of package light transmittance on vitamin content of milk. Part 2: UHT whole milk. *Packaging Technology and Science*, *21*, 47-55.

[26] Saffert, A., Pieper, G., & Jetten, J. (2009). Effect of package light transmittance on the vitamin content of milk, part 3: Fortified UHT low-fat milk. *Packaging Technology and Science*, *22*, 31-37.

[27] FAO. (2001). *Human Vitamin and Mineral Requirements: Report of a joint FAO/WHO*. Food and Nutrition Division. Rome.

[28] Sieber, R. (2005). Oxidised cholesterol in milk and dairy products. *International Dairy Journal*, *15*, 191-206.

[29] Kumar, N., & Singhal, O.P. (1991). Cholesterol oxides and atherosclerosis: a review. *Journal of Science and Food Agriculture*, *55*, 497-510.

[30] Pizzoferrato, L., Manzi, P., & Marconi, S. (1999). Functional milk-based foods consumed in Italy. In R., Lasztity, W., Pfannhauser, L.S., Sarkadi, & S., Tomoskozi (Eds.), *Proceedings of Euro Food Chem X European conference on: Functional foods. A new challenge for the food chemist* (Vol.2, pp.195-201). Budapest (Hungary): Publishing Company of TUB.

[31] Pizzoferrato, L., Manzi, P., Marconi, S., Fedele, V., Claps, S., & Rubino, R. (2007). Degree of Antioxidant Protection: a parameter to trace the origin and quality of goat's milk and cheese. *Journal of Dairy Science*, *90*, 4569-4574.

[32] Barbano, D.M., Ma, Y., & Santos, M.V. (2006). Influence of raw milk quality on fluid milk shelf life. *Journal of Dairy Science*, *89*(supp. E), E15-E19.

[33] Manzi, P., & Pizzoferrato, L. (2008). Cholesterol and antioxidant vitamins in fat fraction of whole and skimmed dairy products. *Food and Bioprocess Technology*, on line first. DOI: 10.1007/s11947-008-0060-3.

[34] Lumley, J., Watson, L., Watson, M., & Bower, C. (2001). Periconceptional supplementation with folate and/or multivitamins for preventing neural tube defects. *Cochrane Database of Systematic,* 3. Art. No.: CD001056. DOI: 10.1002/14651858.

[35] Forssén, K. M., Jägerstad, M. I., Wigertz, K. & Witthöft, C. M. (2000). Folates and dairy products: a critical update. *Journal of the American College of Nutrition, 19(2),* 100S-110S.

[36] Goldsmith, S.J., Eitenmiller, R.R., Toledo, R.T., & Barnhart, H.M. (1983). Effects of processing and storage on the water-soluble vitamin content of human milk. *Journal of Food Science*, *48*, 994-997.

[37] Catharino, R.R., Lima, J.A., Godoy, H.T. (2007). Determination of folic acid in enriched dairy products. *Acta Alimentaria*, *36*, 139-147.

[38] Choe, E., Huan, R., & Min, D.B. (2005) Chemical reactions and stability of riboflavin in foods. *Journal of Food Science*, *70*, R28-R36.

[39] Roughead, Z.K., & McCormick, D.B. (1990). Qualitative and quantitative assessment of flavins in cow's milk. *Journal of Nutrition*, *120*, 382-388.

[40] Haddad, G.S., & Loewenstein, M. (1983). Effect of several heat treatments and frozen storage on thiamine, riboflavin, and ascorbic acid content of milk. *Journal of Dairy Science*, *66*, 1601-1606.

[41] Quattrucci, E., Bruschi, L., Manzi, P., Aromolo, R., & Panfili G. (1997). Nutritional evaluation of typical and reformulated Italian cheese. *Journal of Science and Food Agriculture*, *73*, 46-52.

[42] U.S. Department of Agriculture, Agricultural Research Service. (2007). USDA National Nutrient Database for Standard Reference, Release 21. Available from: http://www.ars.usda.gov/nutrientdata.

[43] Watanabe, F. (2007). Vitamin B_{12} sources and bioavailability. *Experimental Biology and Medicine*, *232*, 1266-1274.

[44] Reddy, K.P., Shahani, K.M., & Kulkarni, S.M. (1976). B-complex vitamins in cultured and acidified yogurt. *Journal of Dairy Science*, *59*, 191-195.

[45] Sato, K., Wang, X., & Mizoguchi, K. (1997). A modified form of a vitamin B-12 compound extracted from whey fermented by *Lactobacillus helveticus*. *Journal of Dairy Science*, *80*, 2701-2705.

[46] Reif D., Shahani K.M., Vakil J.R., & Crowe L.K. (1976). Factors affecting B-complex vitamin content of Cottage Cheese. *Journal of Dairy Science*, *59*, 410-415.

[47] Papastoyiannidis, G., Polychroniadou, A., Michaelidou, A.M., & Alichanidis, E. (2006). Fermented milks fortified with B-group vitamins: vitamin stability and effect on resulting products. *Food Science and Technology International*, *12*, 521-529.

[48] Bonczar, G., Wszolek, M., & Siuta, A. (2002). The effects of certain factors on the properties of yogurt made from ewe's milk. *Food Chemistry*, *79*, 85-91.

[49] Gliguem, H., & Birlouez-Aragon, I. (2005). Effects of sterilization, packaging, and storage on vitamin C degradation, protein denaturation, and glycation in fortified milks. *Journal of Dairy Science*, *88*, 891-899.

[50] Urbiene, S., & Mitkute, D. (2007). Changes of vitamin C during milk fermentation. *Milchwissenschaft*, *62*, 130-132.

[51] Scott K.J., & Bishop D.R. (1986). Nutrient content of milk and milk products: vitamins of the B complex and vitamin C in retail market milk and milk products. *International Journal of Dairy Technology*, *39*, 32-35.

[52] Panfili, G., Manzi, P., & Pizzoferrato, L. (1994). HPLC simultaneous determination of tocopherol, carotenes, retinol and its geometric isomers in Italian cheeses. *Analyst*, *119*, 1161-1165.

[53] Perales S., Alegría A., Barberá R., & Farré, R. (2005). Determination of vitamin D in dairy products by High Performance Liquid Chromatography. *Food Science and Technology International*, *11*, 451-462.

[54] Turner, C., King J.W., & Mathiasson, L. (2001). Supercritical fluid extraction and chromatography for fat-soluble vitamin analysis. *Journal of Chromatography*, *936*, 215-237.

[55] Blake, C.J. (2007). Committee on Food Nutrition. *Journal of AOAC International*, *90*, 18B-25B.

[56] Woollard, D.C., & Indyk, H.E. (2002). Rapid determination of thiamine, riboflavin, pyridoxine, and niacinamide in infant formulas by liquid chromatography. *Journal of AOAC International*, *85*, 945-951.

[57] Baiyi, L., Yiping, R., Baifen, H., Wenqun, L., Zengxuan, C., & Xiaowei, T. (2008). Simultaneous determination of four water-soluble vitamins in fortified infant foods by ultra-performance liquid chromatography coupled with triple quadrupole mass spectrometry. *Journal of Chromatographic Science*, *46*, 225-232.

PART IV

STATISTICAL ANALYSIS

In: Practical Food and Research
Editor: Rui M. S. Cruz, pp. 409-451
ISBN: 978-1-61728-506-6

Chapter XVI

STATISTICAL ANALYSIS IN FOOD SCIENCE

Eduardo Esteves
Instituto Superior de Engenharia, Universidade do Algarve, Campus da Penha, 8005-139 Faro and CCMAR- Centro de Ciências do Mar, Campus de Gambelas, 8005-139 Faro, Portugal, eesteves@ualg.pt

ABSTRACT

Several statistical techniques commonly employed in the area of food science and technology are presented as a function of to studies objectives: descriptive statistics, confidence intervals, one-sample tests, comparison of two or more samples, relationship between variables, and design of experiments. Although the procedures are usually performed using statistical software programs, equations are provided in order to facilitate those still using spreadsheet programs. Assumptions are emphasized and alternative methods or remedial actions are pointed out whenever possible. This chapter is more of a starting point than the finishing line.

16.1 INTRODUCTION

Why Study Statistics?

Two major areas of the food engineering practice are quality control, and research and development (R&D). Both areas require expertise in the use of statistics. Quality control (*sensu* statistical process control) is outside the scope of this chapter and dealt with elsewhere [1-3]. The statistical techniques abridged in this chapter are commonly used in different stages of the R&D process of developing (new) food products.

The aims are to present the techniques in a structured way, directed at studies' objectives, emphasizing tests assumptions, discussing some of the problems and pointing out remedial actions, and clarifying the interpretation of results. Moreover, the validity of parametric methods depends on knowing the population distribution function. Because sometimes this assumption does not hold, alternative nonparametric tests [vide 4-6] are suggested whenever suitable.

What are Statistics?

Statistics is about data. Definitions of the term abound. According to the Organization for Economic Co-operation and Development (OECD at http://stats.oecd.org/glossary/index.htm accessed on 3/10/2008) or the American Statistical Association (AmStat at http://www.amstat.org on 5/10/2008), it is the discipline concerned with the collection, summarization and analysis of data in order to make statements about the real world (Figure 16.1).

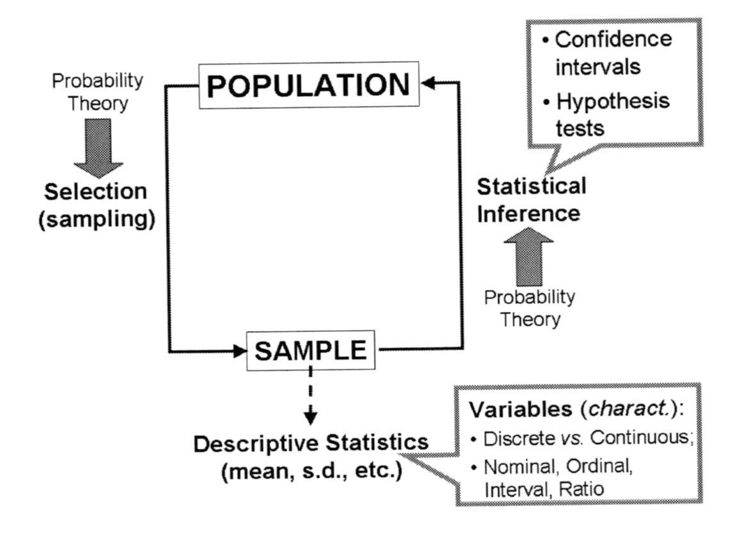

Figure 16.1. Diagrammatic interpretation of Statistics.

Data collection is steered by sampling and design of experiments which draw on probability theory and statistical inference. Sampling deals with strategies of obtaining information from finite populations "as they are" without concern with probability modeling. On the other hand, in experimental design one deliberately makes an alteration to a group (or groups) of experimental units in order to study the modifications observed thereof. The empirical information gathered can be summarized graphically or through statistical measures that often facilitate the communication of results. Probability theory sustains the induction (or statistical inference) from sample data to the population from which they were drawn: estimating population parameters or testing hypothesis. The ultimate aim of statistical analysis is to infer characteristics of a population from the study of a representative sample – i.e. a subset of data – and/or to compare populations and predict future events – specifically to evaluate their probabilities.

Why Do We Need Probabilities?

Application of statistical methods requires an understanding of data and its characteristics, namely average and variation. These characteristics are reflected either by empirical or theoretical distributions. Most statistical methods are based on theoretical distributions that approximate the actual distributions, thus facilitating the analysis of even relatively complex data. Therefore, the user of statistical methods must be familiar with both.

Everyday concepts of "likelihood" and "certainty" are formalized by the branch of mathematics called probability. The "mathematical" investigations of probability go back to the exchange of correspondence between two XVII-century French mathematicians, Blaise Pascal [1623-1662] and Pierre de Fermat [1601-1655]. A thorough discourse on probability is well beyond the scope and intent of this section, but some aspects and considerations of probability theory that underlie the procedures discussed in the following sections should be presented albeit briefly and informally. For an English-speaking audience, worthwhile presentations of probability are found in [7] and [8]. Notwithstanding, excellent textbooks are available in other languages, e.g. in Portuguese [9].

The dictionary definition of Probability is that "it is the quality or fact of being probable". There are several competing interpretations of the "actual meaning" of probabilities: the classical, *a priori* definition by Pierre-Simon de Laplace [1749-1827], that probability of an event is the ratio of favorable outcomes to the whole number of outcomes possible, if the outcomes are equally likely; the frequencist, *a posteriori* definition that the probability of an event is the limit of the percentage of times that the event occurs in repeated, independent trials under essentially the same circumstances; and the subjective definition, developed after the "revival" of the Bayes Theorem due to Thomas Bayes [1702-1761] in the 1930s, that probability measures the speaker's "degree of belief" that the event will occur. Each definition has situations in which it is most natural; notwithstanding they all seem to have shortcomings.

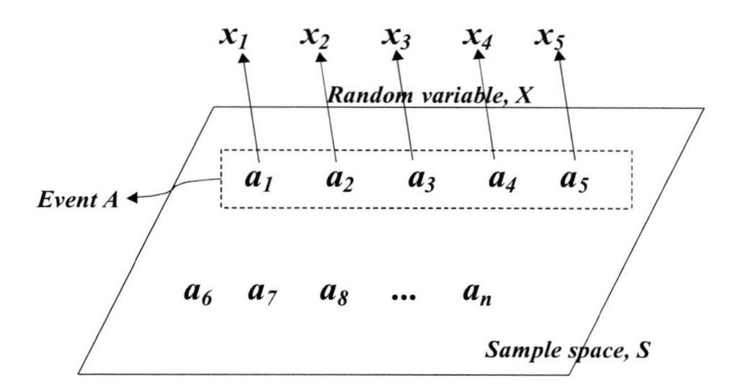

Figure 16.2. Illustration of probability theory concepts: sample space S, possible outcome a_i, event A, random variable (function) X, and result x_i.

Essentially, the link between probability theory and applied statistics is the notion of random variable. Intuitively, a random variable (usually noted X, Y or Z) can be regarded as a function mapping the sample space of a random process to real numbers (Figure 16.2). A random process or experiment is an activity, operation or procedure for which the outcome(s) can not be predicted with certainty. For example, rolling a die or sampling a package from the production line. The set of all possible outcomes of such an experiment constitutes the sample space. There are two types of random variable: discrete and continuous. Discrete random variables take on one of a set of specific values, for example the number of defective products in a lot. Continuous random variables can be realized with any of a range of values, e.g. the net weight of a food product. Every random variable has an associated probability distribution, which exhibits long-term regularities: the expected value (or mean) *E(X)* that represents the average amount one "expects" as the outcome of a random trial when identical odds are repeated many times, and the variance *V(X)* that captures the dispersion of probability distributions.

16.2 Some Important Probability Distributions

Some discrete and continuous probability distributions that arise frequently in practical problems of food engineering are briefly presented below. Their probability functions can be obtained with spreadsheet programs.

Consider a simple random experiment with only two mutually exclusive outcomes, called "success" and "failure" (these can be noted 1 and 0), with constant probabilities p and q, respectively. This is also called a Bernoulli trial. In a sequence of n such experiments, one can use the Binomial distribution (Figure 16.3), due to Jakob Bernoulli [1654-1705], to determine the probability of the number of "successes" i.e. the random variable X:

$$P(X=x)=C_x^n \cdot p^x \cdot (1-p)^{n-x}$$

where x=1,2,…,n, is the "actual number of successes" and C_x^n is the binomial coefficient. If X is a binomially distributed random variable, then the expected value of X is $E(X)=np$ and the variance is $V(X)=np(1-p)$.

Example 1. Evidence from population surveys suggests that obesity prevalence in EU countries reaches up to 38% in women. In a random sample of 90 European women, what is the probability of finding half of them classified as obese?

$$P(X=45)=C_{45}^{90}(0.38)^{45}(1-0.38)^{45}=0.0058$$, i.e. 0.58%.

The Poisson distribution (due to Simon Poisson [1781-1840]) can be thought of as a particular case of the Binomial distribution where p is small and/or n is quite large (some authors use the n>50 and np<5 rule-of-thumb to decide whether to use the Poisson distribution). Thus, this probability distribution is particularly useful when studying "rare events". The Poisson probability density with mean $\lambda = n \cdot p$ (where $\lambda > 0$) is given by:

$$P(X=x)=\frac{\lambda^x}{x!}\cdot e^{-\lambda}$$

The expected value equals the variance, $E(X)=\lambda=n \cdot p=V(X)$.

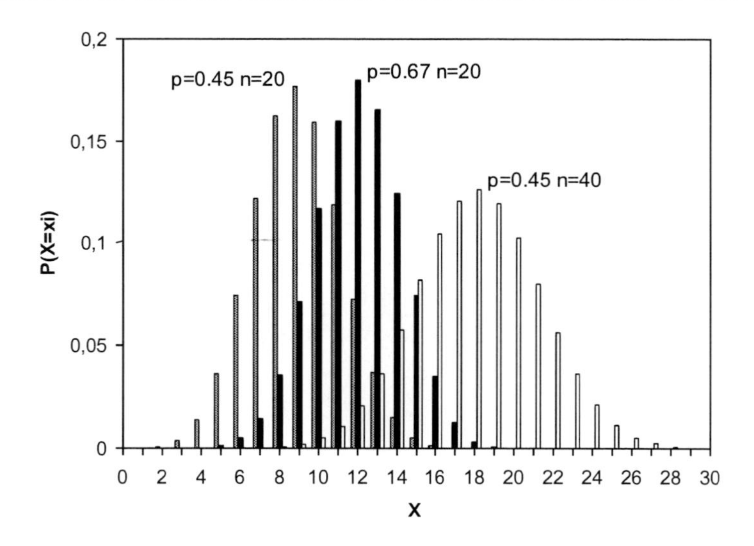

Figure 16.3. Binomial probability density for pairs of parameters p and n.

Example 2. In a factory, 1% of the products manufactured are defective. What is the probability of finding 5 non-conforming products in a lot of 150 items? The expected number of defective items is $E(X) = \lambda = 150(0.01) = 1.5$, thus

$P(X = 5) = (1.5^5/5!) \cdot e^{-1.5} = 0.0141$, i.e. 1.4%.

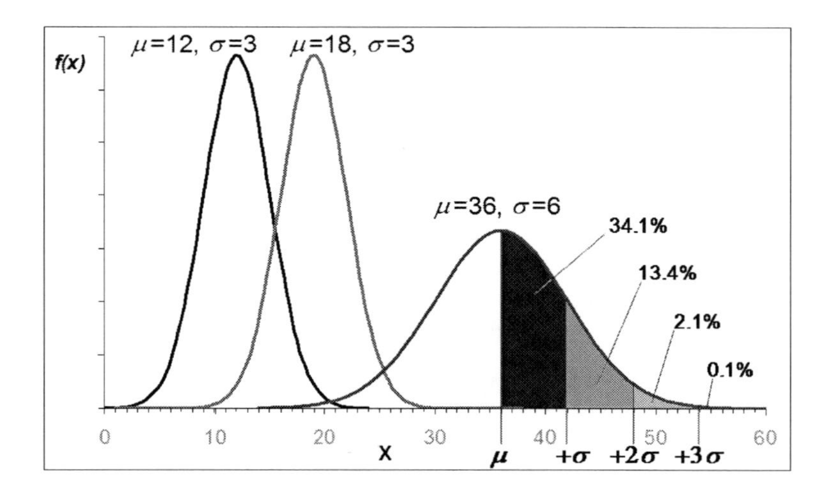

Figure 16.4. Probability density functions of the Normal distribution for different combinations of parameters, mean μ and standard-deviation σ. The rightmost normal distribution has the probabilities (areas) colored in shades of blue (e.g. the interval -3σ to $+3\sigma$ accounts for about 99.7% of the values).

In many problems, however, it is necessary or mathematically simpler to consider continuous random variables, e.g. the net weight of packages, the reaction rate of an enzyme or the dissolved sugar-to-water mass ratio (°Brix) of juice. Among the probability distributions of continuous random variables, the Normal distribution (a.k.a. the Normal curve, the Gaussian distribution, etc.) stands out because it is applicable in many fields, ranging from psychological to physical phenomena (due in part to the Central Limit theorem), and it is mathematically quite amenable. The Normal distribution is due to Abraham de Moivre [1667-1754] and Pierre-Simon Laplace [1749-1827] but Carl Friedrich Gauss [1777-1855] pioneered its usage (therefore the "Gaussian"). The normal distribution is defined by two parameters, of location and scale, the mean μ and variance σ^2, respectively (Figure 16.4). When $\sigma > 0$, about 68% of values drawn from a normal distribution are within one standard deviation away from the mean μ; about 95% of the values are within two standard deviations and about 99.7% lie within three standard deviations (Figure 16.4).

If population parameters μ and σ are known, it is possible to relate all normal random variables to the standard normal distribution since

$$Z = \frac{X - \mu}{\sigma}$$

is a standard normal random variable with mean of zero and standard deviation of one. However, it is uncommon to know the population parameters μ and σ. Moreover, in the case of small samples (i.e. n<30) it is not possible to assume that sample statistics (mean $\bar{x}$ and standard-deviation s) are unbiased estimators of μ and σ. In 1908, William S. Gosset [1876-1937] under the pseudonym Student published the t-distribution (thence the usual designation of Student-t distribution) which constitutes a great alternative to Z. The distribution of

$$t = \frac{X - \bar{x}}{s}$$

where the sample standard deviation $s = \sqrt{\sum (x_i - \bar{x})^2 / (n-1)}$, is normal and has one parameter only, the degrees of freedom (*df*) $\nu = n - 1$.

Example 3. The diameter of plums is normally distributed with an average of 5 cm and a standard-deviation of 1.2 cm. What is the proportion of plums with a diameter between 3.5 and 7 cm? $P(3.5<X<7)=P(-1.25<Z<1.67)=P(Z>1.67)-P(Z<-1.25)=0.952-0.106=0.847$.

Example 4. In a randomly chosen day, 11 measurements of the duration of a certain task were made: $\bar{x} =$ 4.1 s and $s = 1.69$. What is the maximum duration of 90% of the tasks carried out? For $P(X < x_i) \equiv P(t < t_i) = 0.90$, $t_i = 1{,}372$ with 10 *df*, thus $x_i = 1.372 \times 1.69 + 4.1 = 6{,}4$ s.

The chi-square(d) χ^2 distribution is another widely used theoretical probability distribution in inferential statistics, namely to test the differences among proportions or for goodness-of-fit between observed and theoretical distributions. The chi-square distribution (Figure 16.5) has one parameter, its *df* (or *k*). It has a positive skew; as the *df* increase, the chi-square distribution approaches a normal distribution. The mean of a chi-square distribution is $E(X) = k$ and the variance is $V(X) = 2k$.

Finally, the F-distribution (Figure 16.6), also known as Snedecor's F distribution or the Fisher-Snedecor distribution (after Sir Ronald A. Fisher [1890-1962] and George W. Snedecor [1881-1974]), is a continuous probability distribution, that arises frequently as the null distribution of test statistics, e.g. the F-test and perhaps most notably in the analysis of variance (ANOVA). Its mean and variance are $\mu = d_2/(d_2 - 2)$ for degrees of freedom $d_2 > 2$ and $\sigma^2 = 2d_2^2(d_1 + d_2 - 2)/d_1(d_2 - 2)^2(d_2 - 4)$ for $d_2 > 4$, respectively.

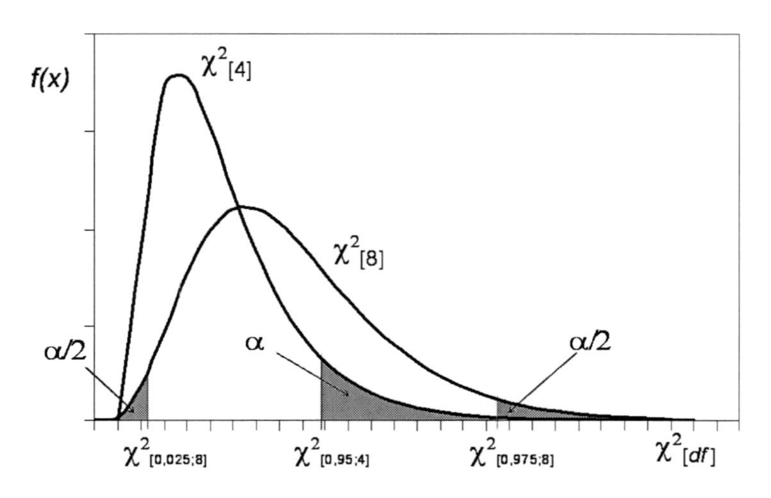

Figure 16.5. Probability density functions of the chi-square distribution, for 4 and 8 *df*.

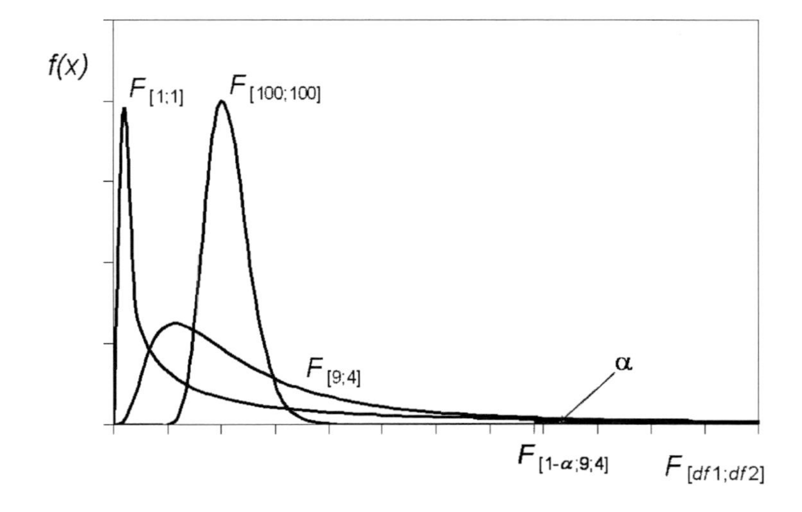

Figure 16.6. Probability density function of the F-distribution with different numerator/denominator *df*.

16.3 Sampling, Samples and Descriptive Statistica

Sampling without assuming a subjacent model is the traditional approach to select a representative subset of n observations from the population of N items. Random is the keyword in statistical sampling because it allows the assessment of observations' uncertainty. Randomness can be accomplished by using random-number tables or software-generated random numbers when it is possible to assign each member of a population a unique number (i.e. in the case of finite populations). Very often this is not possible, e.g. when sampling from an ongoing process that makes listing every element impossible. Thus, a random sample from an "infinite" population is a sample selected such that each element is selected independently and comes from the same population. Random sampling gives each item in the population an equal chance of being selected and measured. There are different variants of random sampling, e.g. (simple) random sampling with or without replacement, stratified sampling, cluster sampling, and systematic sampling.

Determining sample size *n*, where $1 \leq n \leq N$, is usually one of the first difficulties faced by the researcher and there is no simple or correct answer without additional information, *viz.* the risk of rejecting a true hypothesis α, the risk of accepting a false hypothesis β (see Section 16.5), and the value of the population standard deviation σ. The size of the sample is mainly related to the study's objective and the behavior of the variable(s) to be measured but also dependent on other factors, e.g. cost, ease of access to equipment, etc. Van Belle [10] provides and discusses a helpful set of rule-of-thumb formulas for the calculation of sample size. Lenth [11] gives suggestions for successful and meaningful sample-size determination. Several software programs, both commercial (e.g. Power and Precision, from BioStat Inc.) and freeware (e.g. R [12] or G*Power [13]), include algorithms for sample size determination for the most complicated or computationally intensive methods presented below. Moreover, various online (re)sources devote significant effort to this topic [14, 15].

In the context of simple random sampling, if one wants to estimate the average of a certain variable with 95% confidence (c=1-α), the most common and simple approach is to use

$$n = \left(z \frac{\sigma}{d} \right)^2$$

where *n* is the sample size (round up to the next integer), *z* is the standard normal (1-α/2) percentile $z_{[1-\alpha/2]}$ (e.g. z=1.96 if α=0.05), *d* is the sampling error (usually expressed in units of the standard deviation) and σ is the standard deviation of the variable in the population (this is usually not known beforehand and a good estimate is necessary – e.g. the standard deviation, *s*, of a preliminary sample of more than 30 items). To compare two means, assuming the most common choices of α=0.05 and β=0.20, use the basic formula

$$n = \frac{16}{d^2}$$

where $d = (\mu_1 - \mu_2)/\sigma = \delta/\sigma$ is the treatment difference to be detected in units of the standard deviation – the standardized difference. This equation can also be used in the one sample case, i.e. when a single sample is compared with a known population value, where the numerator is 8 instead of 16.

Example 5. If one wants to estimate the mean protein content of ham (in %) to within 0.5 of the population mean, with 95% confidence and the standard deviation is 2%, then should sample $n = (1.96 \times 2/0.5)^2 = 61.5 \approx 62$ items.

When comparing several means, the "standard" parametric approach is to use analysis of variance or ANOVA (see Section 16.8). Computing the required sample size for experiments to be analyzed by ANOVA is complicated, with lots of possibilities. A decision on the size of the experiment is largely a matter of judgment (and budget) and some of the more formal approaches to do it have spurious precision [16]. In the simplest case, if one wants to compare k means or groups (treatments/levels of one factor in the ANOVA jargon), Sokal and Rohlf [17] and Montgomery [18] propose an iterative, graphical method to evaluate ϕ (a quantity related to the F-distribution),

$$\phi = \sqrt{n\delta^2/2ks^2}$$

where δ is the assumed difference between the most different means. This method of computing the number of replicates per treatment/level/group n, depends on the question asked and needs a first guess of n and knowledge about the error s^2, for example from prior data or a judgment estimate. Alternatively, given the effect size f, significance level (α) and power ($1-\beta$), one can determine sample size per group using

$$f = \sqrt{\frac{1}{\sigma^2}\sum_{i=1}^{k} p_i \cdot (\mu_i - \mu)^2}$$

where $p_i = n_i/N$, n_i is the number of observations in group i, N is the total number of observations, μ_i is the mean of group i, μ is the grand mean and σ^2 is the error variance within groups. Cohen [19] suggests that f values of 0.1, 0.25 and 0.4 represent small, medium and large effect sizes, respectively. Maxwell and Delaney [20] present several ways of determining sample size, including tables, for designed experiments.

On the other hand, when studies involve data in the form of counts or proportions, the sample size should be "as big as one can afford!" because there is little information in such data [21]. Although the sampling distribution of proportions actually follows a binomial distribution, the normal approximation is used for this derivation. Thus, in order to estimate the "true" proportion one should sample

$$n \geq \left(\frac{z}{d}\right)^2 \hat{p}(1-\hat{p})$$

items, where n is the sample size (round up to the next integer), z is the standard normal ($1-\alpha/2$) percentile, d is the sampling margin of error and $\hat{p}$ is an "educated guess" about the proportion of "successes" (frequently unknown at the start of the study). Use $\hat{p} = 1/2$, unless it is known that p belongs to an interval a≤p≤b that does not include 0.5, in which case substitute the interval endpoint nearer to 0.5 for $\hat{p}$. In survey sampling, the formula above is usually simplified to $E = 1/\sqrt{n}$. If n = 1000 then the sampling error E = ~3%, or vice-versa. These figures are quoted often in news reports of opinion polls and other sample surveys. Moreover, to compare two proportions, p_1 and p_2, use the formula

$$n = \frac{16\bar{p}(1-\bar{p})}{(\hat{p}_1 - \hat{p}_2)^2}$$

to calculate the required sample size per group (where $\bar{p} = (\hat{p}_1 + \hat{p}_2)/2$ is used to estimate the pooled variance). For $\hat{p} < 0.05$, the Poisson distribution can be used to determine the sample size per group n:

$$n = \frac{4}{(\sqrt{\lambda_1} - \sqrt{\lambda_2})^2}$$

where λ_1 and λ_2 are the means of populations 1 and 2.

Example 6. The engineer wants to estimate, at 95% confidence, the output yield (good/bad) of a new process with an accuracy of d=0.10. The current process has been yielding 65% "good parts", i.e. p=0.65. Thus, she should draw $n = (1.96/0.1)^2(0.65)(0.35) = 87{,}3 \approx 88$ random parts from the output of the new process to estimate the yield.

Sample observations can be coded in several different ways or scales: a *nominal* (or categorical) – observations are classified by some quality it possesses and grouped into categories, thought of as labels, whose sequence is arbitrary, e.g. categories of food products – beverages, meat and meat products, fish and shellfish, dairy and egg products, etc.; *ordinal* (a.k.a. ranked) data, e.g. food products in a sensory test are ordered according to preference; the age classes of respondents to a survey 18-24, 25-31, 32-39 yr., etc.; *interval* – observations are measured in a scale composed of equal-sized intervals in which the origin is arbitrary (a great example is that of the two most common temperature scales: Celsius and Fahrenheit); and *ratio* – measurement scales having a constant interval size and a true zero point that correspond to the cancellation of the characteristics under study, e.g. lengths in cm, volumes in m^3, or duration in seconds. Ratio scales allow the interpretation of a score of $2x$ as twice the level of a score x. The type of variable recorded determines the kind of statistical techniques used afterwards.

The most elementary degree of data analysis is its description. Classically, data are thought to be exact, representative and adapted to/represented by the mathematical models being used and thus constitute a (statistical) sample. However, often this is not the case and one is confronted with a mere collection of data. Tukey [22] coined the term Exploratory Data Analysis (EDA) to encompass several techniques (mostly graphical in nature) that shed light on the structure of such data without having pre-conceived ideas about it. There is a large collection of statistical tools for descriptive statistics and EDA, which include e.g. frequency and contingency tables, histograms, box plots, scatter plots, probability plots. These help the researcher gain insight into a data set in terms of assumptions testing, model selection, relationship identification and/or outlier detection, etc. EDA facilitates the interactive and usually undirected search for structures and trends.

Dot plots can be used to present qualitative data or explore how small amounts of quantitative data (<15 observations) are distributed (Figure 16.7a). When collecting and summarizing larger amounts of data, it is often helpful to record the data in the form of a frequency table (one variable) or a contingency table (two or more variables). The distribution of the total number of observations among the various categories is termed a frequency distribution. The categories can be related to nominal/ordinal data (Tables 16.1 and 16.2) or derived from quantitative (interval or ratio) data (Table 16.3). In the later case, a set of n observations can be distributed among k categories, obtained using for example the Sturges Rule $k \approx I(\log_2 n)+1$ where $I(x)$ is the larger integer less than x. Frequency distributions are useful largely as a means of indicating which type of theoretical distribution best describes the statistical properties of the population under study [23]. Frequency table(s) can be presented graphically by means of a histogram (Figures 16.7b and 16.7c) that show, through (adjacent) bars, what proportion of cases fall into each of several categories (or bins). A box-and-whisker plot (Figure 16.7d) is another option to examine the distribution of data. In its simplest form, a box-and-whisker plot shows a box that contains the middle 50% of the data values, a heavy line that divides the data in half (the median) and two whiskers that extend from the box to the maximum and minimum values. Often extreme values (a.k.a. outliers) which strongly deviate from the remainder are also shown as asterisks or dots. Finally, scatter plots (Figure 16.7e) allow the detection of putative relationship(s) between (quantitative) variable(s).

Descriptive statistics is concerned with summarizing and presenting data concisely. There are several statistical measures that in a sense represent all data in a sample. Estimates of population parameters (which are rarely known) obtained from random samples are called statistics. In samples, one generally finds a preponderance of values somewhere around the middle of the range of observed values. The description of this concentration near the middle is a measure of central tendency (or location). Examples of this are the arithmetic mean $\bar{x}$ and the median $\tilde{x}$. In addition, it is also desirable to have a measure of dispersion of data that indicates the spread of the measurements around the center of the distribution. The standard-deviation s and the interquartile range IQR are relevant examples of those measures. Other measures relate location and scale: e.g. the coefficient of variation (a measure of relative variation), skewness (which characterizes the degree of asymmetry of a distribution around its mean) and kurtosis (that "measures" the flatness/peakedness of the distribution). These are less used but highly informative.

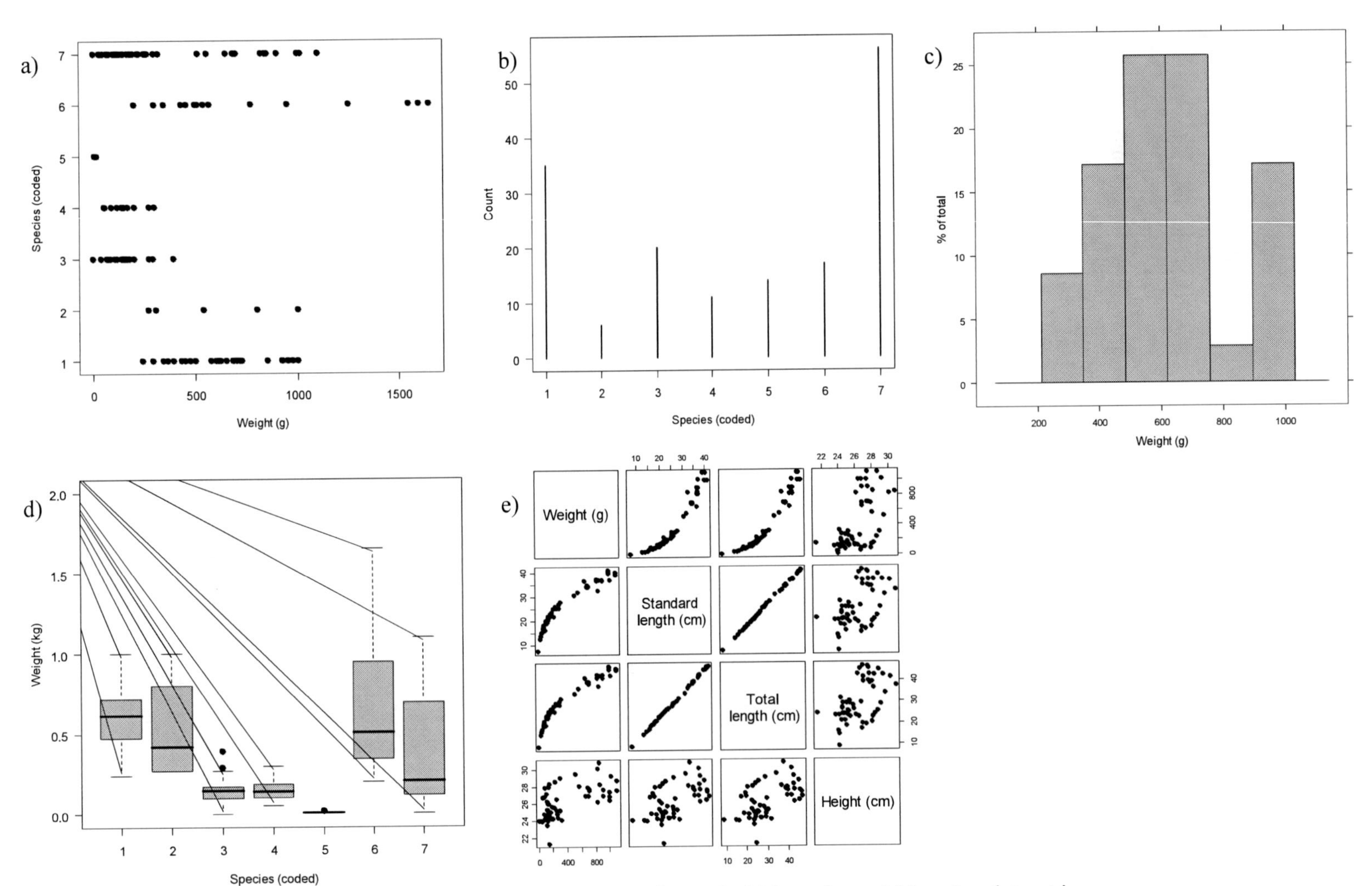

Figure 16.7. EDA graphical techniques: (a) dot-plots, (b and c) histograms, (d) box-and-whiskers plot, and (e) scatter plot matrix.

Table 16.1. Number of registered organic producers in the European Union (UE15), as of October, 2007. The other seven EU15 countries not shown have no registered producers. Source: EuroStat at http://ec.europa.eu/eurostat.

Country	Number
Belgium	1525
Germany	26820
Greece	24725
Latvia	4120
Lithuania	2881
Netherlands	2417
Norway	3082
United Kingdom	7631

Table 16.2. Contingency table of data related to product quality and supplier, obtained from lot inspection.

	Rejected products	Acceptable products	Excellent products
Supplier A	12	23	89
Supplier B	8	12	62
Supplier C	21	30	119

Table 16.3. Frequency table obtained from a sample of 21 simulated, random measurements using Sturges's rule to define the classes/categories.

Lower limit	Upper limit	Class center	Count (F)	Relative frequency (f)	Cumulative f
14.5	18.7	16.6	5	0.238	0.238
18.7	22.9	20.8	2	0.095	0.333
22.9	27.1	25.0	7	0.333	0.667
27.1	31.3	29.2	3	0.143	0.810
31.3	35.5	33.4	4	0.190	1.000

Because the mean and standard-deviation have the same units, the expression $\bar{x} \pm s$ is often used empirically and it is an abbreviated way of stating that about 2/3 of the values are within that interval (*cf.* Figure 16.4). Sometimes, the mean-and-standard deviation-approach to sample description is unsuitable. The median which is the middle measurement of an ordered set of n items is less sensitive to extreme scores (gross errors and outliers) than the mean and this makes it a better measure for highly skewed (non-symmetrical) distributions. For example, the median income is usually more informative than the mean income.

Moreover, the interquartile range, i.e. the difference between the 3rd and 1st quartiles (noted Q_3 and Q_1 respectively) corresponding to the middle 50% of data values, together with the median sensibly describes a sample (*vd.* Figure 16.7d).

16.4 More Than Sample Description

Provided a randomly drawn sample of independent data one can estimate the value of some parameter of the population θ. For example, the mean net weight of a sample of canned food products can be used to estimate the mean net weight of all the products manufactured – point estimation. However, a point (single) estimate gives no idea of its precision. It is possible to find an interval about the estimate, named a confidence interval, which has a given probability of including the true value of θ – interval estimation. Parameter estimation (sometimes named estimation theory) is one of the two closely types of problems in statistical inference. The other, testing of hypothesis (or decision theory), will be addressed in the next section.

A statistic is actually a random variable, because its value changes from sample to sample. The sampling distribution of a sample statistic, calculated from a sample of n measurements, is the probability distribution of the statistic. Sampling distributions of statistics are very important as they afford measures of precision for the estimates concerned. For practical applications, it is not crucial that the population be normally distributed. The reason for this is an important result concerning sampling distributions, called Central Limit Theorem (CLT): the sampling distribution of an average tends to be normal (with mean μ and standard-deviation $\sigma_{\bar{y}} = \sigma/\sqrt{n}$) as the sample size n increases, even when the distribution from which the average is computed is decidedly non-normal. For most populations, if $n \geq 30$ the CLT approximation is good.

The limits of the (1-α)100% confidence interval (CI) for the mean μ of large samples, with $n>30$, can be calculated on the basis of a range of z times the standard-deviation each side of the mean, i.e.

$$\bar{x} \pm z(\sigma/\sqrt{n})$$

where z is $z_{[1-\alpha/2]}$, i.e. $z=1.96$ if $\alpha=0.05$. More commonly, however, σ is unknown and the sample standard-deviation s can be used in place of σ to obtain the CI:

$$\bar{x} \pm t(s/\sqrt{n})$$

where t is $t_{[1-\alpha/2,n-1]}$ (e.g. $t=2.093$ if $\alpha=0.05$ and $n=20$). The formula above can be used as long as the population is normally distributed with mean μ (the proof of this is beyond the scope of this chapter). One of several normality tests (e.g. Shapiro-Wilk normality test or the Lillefors test) may be used to assess the normality assumption.

Example 7. The maturity index of a certain fruit is normally distributed (ranging from 1 to 8). A sample of seven fruits was selected to estimate the lot average maturity. Based on sample data ($\bar{x} = 6.86$ and s=0.69), the 95% confidence interval for fruit maturity is [6.22; 7.50]. Since the CI does not include (in fact it exceeds) a pre-determined value of 6, harvesting is advised.

Setting a CI for the variance of a single population σ^2 is done using the sample variance s^2 through

$$\frac{(n-1)s^2}{\chi^2_{[1-\alpha/2,n-1]}} < \sigma^2 < \frac{(n-1)s^2}{\chi^2_{[\alpha/2,n-1]}}$$

where, for example, χ^2=1.69 for $\alpha/2$=0,025 and df=7.

In general, the sample proportion $\hat{p} = x/n$ (where x is the number of "successes" and n is sample size) can be viewed as a sample average; hence we can invoke the CLT to determine a confidence interval for the population proportion π. If n is large enough to ensure that $\hat{p} \pm 3\sigma_{\hat{p}}$ does not include 0 or 1, then the distribution of $\hat{p}$ may be assumed to be normal (some authors propose the more conservative rule-of-thumb, $n\hat{p} > 10$ and $n(1-\hat{p}) > 10$). Under these conditions, the confidence interval for p is:

$$\hat{p} \pm z\sqrt{\hat{p}(1-\hat{p})/n}\,.$$

When $\hat{p}$ approximates 0 or 1, or situations with small samples sizes, other (competing) formulas are available that perform better, e.g. Wilson score interval or the Agresti-Coull interval. One of the forms of the Wilson Interval is given in [24]:

$$\frac{2n\hat{p} + z^2 \pm z\sqrt{4n\hat{p}(1-\hat{p})}}{2(n+z^2)}$$

where z is $z_{[1-\alpha/2]}$. Brown et al. [25] revisited the problem of interval estimation of a binomial proportion and showed that the chaotic coverage properties of the Normal approximation interval are far more persistent than is appreciated. Those authors recommend the Wilson interval for small n and the Agresti–Coull interval for larger n (i.e for practical use when $n \geq 40$), which is defined as

$$\tilde{p} \pm z\sqrt{\frac{\tilde{p}(1-\tilde{p})}{\tilde{n}}}$$

where $\tilde{p} = \tilde{X}/\tilde{n}$, $\tilde{X} = X + z^2/2$, $\tilde{n} = n + z^2$ and z is $z_{[1-\alpha/2]}$. The parameter $\tilde{p}$ is a better choice than $\hat{p} = x/n$. Even for small sample sizes, the easy-to-present Agresti–Coull interval is much preferable to the standard one.

Several confidence intervals for the population median have been proposed including, for example, the simple CI presented by Olive [26]:

$\tilde{x} \pm t \cdot se(\tilde{x})$ where t is $t_{[1-\alpha/2,p]}$ (with $p \approx \sqrt{n}$) and $se(\tilde{x}) = 0.5(x_{Un} - x_{Ln+1})$ with $Un = n - Ln$ and $Ln = n/2 - \left\lceil \sqrt{n/4} \right\rceil$ ($\lceil x \rceil$ denotes the smallest integer equal to or greater than x, e.g. $\lceil 3.3 \rceil = 4$). For large samples, the CI for a certain quantile q can be obtained using the Binomial distribution [4]. In fact, the j-th and k-th observations in the ordered data set (x'_j and x'_k), which constitute the lower and upper limits of the CI, can be found using $j = nq - z_{[1-\alpha/2]}\sqrt{nq(1-q)}$ and $k = nq + z_{[1-\alpha/2]}\sqrt{nq(1-q)}$ rounded up to the next integers and considering that q=0.5 for the median.

16.5 Comparing Samples

Besides parameter estimation, statistical inference deals with a closely related type of problem: the testing of hypothesis (sometimes referred to as decision theory). To test a hypothesis is to make a decision regarding the reasonableness of a statement, or hypothesis, taking into consideration newly and objectively collected facts (data). If these facts do not support the stated belief, the hypothesis is rejected. Nonetheless, one does not make an absolute rejection; instead we state our degree of confidence in our decision being correct.

The "classical" hypothesis testing procedure consists of three major steps. First, statement of the null and alternative hypothesis, respectively H_0 and H_1 (or H_A), and selection of the level of significance α (usually 5% or 1%; where confidence c=1–α). In general, the two kinds of alternative hypothesis considered are: "equal" *vs.* "different" (e.g. H_0:μ=μ_0 vs. H_1:$\mu \neq \mu_0$, two-tailed test of hypothesis); or "higher (lower) than" (i.e. H_0:$\mu \leq \mu_0$ vs. H_1:$\mu > \mu_0$, one-tailed (upper) test of hypothesis; H_0:$\mu \geq \mu_0$ vs. H_1:$\mu < \mu_0$, one-tail (lower) test of hypothesis).

Then, computation of the test statistic *e.t.*, which is based on the null hypothesis and calculated from the observed data, and determination of its probability (a.k.a. *p*-value[1]).

Finally, rejection or not of the null hypothesis is made after confronting the *e.t.* with a critical, theoretical value belonging to the appropriate probability distribution, or by comparing its *p*-value with α (if *p*-value≤α, reject H_0) and the statement of conclusion (e.g. "mean A is significantly greater than mean B at p<0.05"). Classically, in a two-tailed test if the |*e.t.*|>critical value then reject H_0 with (1-α)100% confidence, e.g. if $|z^*| > z_{[1-\alpha/2]}$ then reject H_0: $\mu_X = \mu_Y$. On the other hand, if one wants to test one-tailed hypothesis, e.g. H_0: $\mu \leq$

[1] The p-value is the probability of finding by chance alone a test statistic equal or greater than the observed value, assuming that H_0 is true. The *p*-value can be easily obtained using spreadsheet programs and is provided by statistical software packages.

μ_0 vs. H_1: $\mu > \mu_0$, reject the null hypothesis if the *e.t.*>critical value (for example, if $t^* > t_{[1-\alpha;n-1]}$, where t is now the (1-α) percentile of the probability distribution). Alternatively, if the p-value<α reject the null hypothesis (both one- and two-tail).

In carrying out this kind of decision process, one can incur two possible types of errors: state that the null hypothesis is false (i.e. reject H_0) when it is really correct; or decide in favor of the null hypothesis (not reject H_0) when it is in fact false. The first is commonly called a type I error and its probability denoted α, while the second is called type II error. The (conditional) probability of making the later type of error is designated β and depends on the true parameter value. The power of a test, i.e. the probability of detecting the specific alternative hypothesis, is 1–β.

16.6 One Sample *Vs.* Theory

On the basis of experimental results one might wish to test the theory that the mean μ has a particular value μ_0. Statistically, this can be re-written as H_0: $\mu = \mu_0$ vs. H_1: $\mu \neq \mu_0$ when dealing with observations that are normally distributed. The hypothesis can be tested using e.g. the test statistic

$$z^* = (\bar{x} - \mu_0)/(\sigma/\sqrt{n}).$$

This is sometimes called a z-test. If σ is unknown but the sample size is reasonably large ($n \geq 30$, preferably $n > 100$) or data is normally distributed, the sample standard deviation s can be used instead of σ in the above equations. Otherwise, a test statistic based on the Student-t distribution with (n–1) *df* which is usually more appropriate for real world problems can be used if data is normally distributed:

$$t^* = (\bar{x} - \mu_0)/(s/\sqrt{n}).$$

Example 8. The variance of protein content in a new source of cereal grain is 2.3% and it is hypothesized that protein content equals 12%. A random sample of 100 grains was analyzed and found to have a mean protein content of 11.7%. This is significantly different from the hypothesized content of 12%, at the α=0.05 level, since the modulus of $z^* = (11.7 - 12)/\sqrt{2.3/100} = -1.98$ is greater than z=1.96 (alternatively, because p=0.0479 < 0.05). Would one reach the same conclusion if α=0.01?

Example 9. Certain food product should contain no more than 10% (w/w) of lipids, i.e. H_0: $\mu \geq 10\%$ vs. H_1: $\mu < 10\%$. A sample of 13 products was analyzed and found to have 9% (w/w) of lipids (with s=1.3%). Because $t^* = (9 - 10)/(1.3/\sqrt{13}) = -2.77$ is less than $-t_{[1-\alpha,12]} = -1.78$ (or p=0.0084), one can conclude with 95% confidence that lipid content does not exceed 10%.

To test hypothesis concerning a population proportion π_0, e.g. H$_0$: $\pi=\pi_0$, and when n is large the test statistic

$$z^* = \frac{\hat{p} - \pi_0}{\sqrt{\pi_0(1-\pi_0)/n}}$$

is compared to the (1-α/2) percentile of the standard normal distribution $z_{[1-\alpha/2]}$. When $n \leq 20$, the number of "successes" x (used to calculate $\hat{p}$) can be used as the test statistic T. The null distribution of T is tabulated [4]. This later procedure is called Binomial test.

16.7 Comparing Two Samples

In R&D projects, an experiment will seldom involve only one sample. Therefore, the comparison of means, variances or proportions from two samples is considerably of greater interest. Usually, one wants to statistically test if two populations are identical based on random samples of those populations. Two sample data may arise from either independent or paired samples. In the former case, observations included in one sample are not associated with observations in the other sample whereas in the later each observation on one sample has a corresponding observation in the other, which is similar in one or several respects, except for the treatment effect being studied. Herein, both data sets (samples) are supposed normally distributed.

When comparing the means of two independent, random samples from two supposedly different populations, X and Y, the test statistics used to test H$_0$: $\mu_X = \mu_Y$ (or $\mu_X - \mu_Y = 0$) against H$_1$: $\mu_X \neq \mu_Y$ is

$$z^* = \frac{\bar{x} - \bar{y}}{\sqrt{\sigma_x^2/n_x + \sigma_y^2/n_y}}$$

where σ^2 are known population variances and n_x and n_y are sample sizes.

In the case the population variances are unknown but assumed equal (use the F-test to assess this assumption – see below), the common variance σ^2 is estimated by the pooled variance $S^2 = [(n_x - 1)s_x^2 + (n_y - 1)s_y^2]/(n_x + n_y - 2)$ and the test statistic, with $(n_x + n_y - 2)$ df, is

$$t^* = \frac{\bar{x} - \bar{y}}{S\sqrt{1/n_x + 1/n_y}}.$$

Example 10. Two sensory panels of experienced subjects evaluated independently the acidity of orange juices obtained from fruits of two different suppliers, A and B, in order to test if juice A is more acid than juice B, i.e. H$_1$: $\mu_A > \mu_B$. They used a 120-mm line scale limited by the terms "imperceptible" and "very pronounced". Sample data provided the following statistics: $\bar{x}_A = 74.1$, $s_A^2 = 95.54$, $\bar{x}_B = 63.9$, $s_B^2 = 85.49$, $n_A = 10$ and $n_B = 11$. The variances were judged not significantly different (see next example). Since $t^* = 2.45$ is greater than $t_{[0.95;19]} = 1.73$ (or because p=0.0121), one can conclude in favor of H$_1$ with 95% confidence.

If the assumption of equal variances is found to be untenable, an exact procedure to compare independent samples is unavailable but the following approximation is satisfactory for most cases. The test statistic is:

$$t' = \frac{\bar{x} - \bar{y}}{\sqrt{s_x^2/n_x + s_y^2/n_y}}$$

with (approximate) *df* given by $\nu' = (\upsilon_x + \upsilon_y)^2 / [\upsilon_x^2/(n_x - 1) + \upsilon_y^2/(n_y - 1)]$, where $\upsilon_x = s_x^2/n_x$, $\upsilon_y = s_y^2/n_y$ and ν' is rounded to nearest integer when using the Student-*t* distribution.

When comparing the means of paired samples, one focuses on the difference between the (paired) measurements, thus e.g. H$_0$: $\mu_D = D_0$ (where $\mu_D = \mu_x - \mu_y$ and D_0 is often zero) *vs*. H$_1$: $\mu_D \neq D_0$. Use the following test statistic

$$t_D = \frac{\bar{d} - D_0}{S_{\bar{d}}}$$

with $(n-1)$ *df*, where n is the number of paired observations, $\bar{d}$ is their average difference and $S_{\bar{d}} = \frac{1}{\sqrt{n}}\sqrt{[n\sum d_i^2 - (\sum d_i)^2]/(n-1)}$.

When the assumptions of normality and equal variances are not valid but the sample sizes are large, the results using Student-*t* test statistics are approximately correct. There are, however, alternative (nonparametric) test procedures that require less stringent conditions – the Mann-Whitney *U* test to compare two independent samples and the Wilcoxon signed rank test for differences between pairs of measurements.

One can test if two populations have the same variance (formally H$_0$: $\sigma_1^2 = \sigma_2^2$ *vs*. H$_1$: $\sigma_1^2 \neq \sigma_2^2$), for example prior to a *t*-test to compare two means, using the test statistic

$$F^* = \frac{s_1^2}{s_2^2}$$

where $s_1^2 > s_2^2$ and with (n_1–1) *df* in the numerator and (n_2–1) *df* in the denominator (e.g. F=5.19 for α=0.05, 4 and 5 *df*). Notice that this test, similarly to the one-sample approach, is severely and adversely affected by sampling non-normal populations. Unfortunately there seems to be no truly non-parametric alternative to it.

Example 11. In order to test if the variances of two independent samples of sensory scores are equal (see previous example), $s_A^2 = 95.54$ and $s_B^2 = 85.49$, a F-test was done. Because F^*=1.13 is less than $F_{[0.95,9,10]} = 3.02$ (or p-value=0.4231) there is no significant difference between the variances.

To test the hypothesis that two (binomial) proportions[2] are equal (H_0: $\pi_1 = \pi_2$), we can use the test statistic

$$z^* = \frac{\hat{p}_1 - \hat{p}_2}{\sqrt{\hat{p}(1-\hat{p})\left(1/n_1 + 1/n_2\right)}}$$

where n_1 and n_2 are the sample sizes, $\hat{p}_1 = x_1/n_1$ and $\hat{p}_2 = x_2/n_2$ are the estimated sample proportions and $\hat{p} = (x_1 + x_2)/(n_1 + n_2)$ is the pooled estimate of the common p. This procedure is appropriate if np_i and $n(1 - p_i)$ are 5 or more, for i=1,2. Otherwise, use the chi-square test (see example 13 below for an application of this test).

16.8 Comparing More Than Two Samples

A technique known as analysis of variance (ANOVA) employs tests based on variance ratios to determine whether or not significant differences exist among means of several groups of observations. These are assumed normally distributed with the same variance. The homogeneity of variances assumption can be tested using the Bartlett's test (see below). In a one-way ANOVA[3], the k groups differ along a single dimension or factor. The total, overall variation in the data (SS_T) can be divided into a part due to the variation (or differences)

[2] In most applications in food science, the normal approximation to binomial is used to deal with proportions (which are in fact binomial variables) since the proportion can be thought of as the mean number of observations per class (this makes use of the CLT).

[3] The "approach" described herein, considers that factor levels are assigned *a priori* by the researcher and constitutes his/her primary interest (a.k.a. fixed effects ANOVA). Otherwise, the levels can be randomly chosen from among varying levels and the researcher wishes to make inferences beyond the particular values of the factors used in the study (i.e. random effects ANOVA). Other more complex designs exist, e.g. two-way ANOVA, Latin-Square Designs, Factorial ANOVA.

between the groups (SS_G) and another part related to "natural, random experimental error" (SS_E): $SS_G + SS_E = SS_T$.

Using the sum of squares between the groups (SS_G) and the error sum of squares (SS_E), ANOVA provides a means of testing if "all means μ_j are equal" (H_0: $\mu_1 = \mu_2 = ... = \mu_j$) or "at least two means μ_j are not equal". The ANOVA F-statistic

$$F = \frac{SS_G/(k-1)}{SS_E/(n-k)}$$

is compared to the (1-α) percentile of the F distribution with (k–1) and (n–k) degrees of freedom, $F_{[1-\alpha,k-1,n-k]}$. If $F > F_{[1-\alpha,k-1,n-k]}$ then reject H_0 at a (1–α)100% confidence level.

Example 12. In order to assess the difference in resistance of four packaging materials (coded A – D), measurements (in standardized units) were randomly obtained from samples of those materials. The average standardized resistance varied between 2.43 for D and 3.30 for C ($\bar{x}_A = 2.90$ and $\bar{x}_B = 2.88$). The ANOVA table obtained using Microsoft Excel® Data Analysis Tools (and slightly edited) is presented below (Table 16.4). Since $F = 11.47 > F_{[0.95,3,12]} = 3.49$, one can conclude with 95% confidence that the (mean) resistance of, at least, one of the materials is different from the others.

Table 16.4. ANOVA table for the packaging resistance example obtained using Microsoft Excel® Data Analysis tools (edited).

Source of variation	*SS*	*df*	*MS*	*F*	*F*(critical)	*p-value*
Between groups	1.63	3	0.54	11.47	3.49	0.0008
Within (Error)	0.57	12	0.05			
Total	2.19	15				

It is quite common that nuisance factors affect the results (i.e. influence the response) but are not of primary interest. When these are known and controllable, e.g. day or batch[4], blocking can be used to eliminate their effects. Thence, Randomized (Complete) Block Designs (RCBD) are more efficient than simple one-way ANOVA (a.k.a. Completely Randomized Designs). In RCBD, experimental units are first divided into groups based on those "control variables", i.e. nuisance factors, and then are randomly assigned to the different categories of the primary factor – in fact, blocks represent a restriction in randomization. The overall variation in data is now subdivided accordingly: $SS_G + SS_B + SS_E = SS_T$. The F-statistic obtained thereof is now compared to $F_{[1-\alpha,k-1,(k-1)(b-1)]}$.

[4] For example, the runs of an experiment can not be performed in one day (or shorter period of time) or the experimental material comes from different batches or lots.

When the ANOVA assumptions are seriously compromised, the nonparametric alternative to one-way ANOVA – the Kruskal-Wallis test – can be used instead. If the observations are somehow related among groups, the Friedman test is a useful alternative.

Example 13. The washing effectiveness (in retarding bacterial growth in milk containers) of different solutions (1 to 3) was analyzed in a laboratory. Only three trials could be run per day, thus a RCBD was used to analyze the results for four days. There are significant differences between solutions (ANOVA $F = 40.72 > F_{[0.95,2,6]} = 5.14$) when allowing for the (known, expected) between-day differences (Table 16.5).

Table 16.5. ANOVA table obtained using Microsoft® Excel Data Analysis Tools (edited) for the washing effectiveness example.

Source of variation	*SS*	*df*	*MS*	*F*	*F (critical)*	*p-value*
Solution	703.5	2	351.8	40.72	5.14	0.0003
Block (Days)	1106.9	3	369.0	42.72	4.76	0.0002
Error	51.8	6	8.6			
Total	1862.3	11				

If the H_0 hypothesis is rejected at the end of ANOVA, we do not know which means differs from each other. Several multiple-comparisons methods have been proposed to answer that question. In the case of independent comparisons or for preplanned comparisons, the Fisher's Least Significant Difference (*LSD*) can be performed to determine which of the samples differ(s) significantly. For a specified α, use

$$LSD = t \cdot \sqrt{MS_E (1/n_i + 1/n_j)}$$

where t is $t_{[1-\alpha/2;n-k]}$, $MS_E = SS_E/(n-k)$, and n_i and n_j are the respective sample sizes from groups i and j being compared. Two means are declared to be significantly different at the α-level of significance if their values differ by more than the value of *LSD*. When comparisons are decided only after having observed some or all of the data, several alternative procedures for these *post-hoc* comparisons based on the Studentized range distribution q have been proposed, e.g. the Tukey Honestly Significant Difference (Tukey HSD) or its modification the Student-Newman-Keuls test, the Dunnett's test for comparisons with a control, and the more general procedure (although more conservative test) to make all possible comparisons among the k populations proposed by Scheffé. These procedures are easier to carry out *via* software programs.

Example 12. (continued) One can further compare the resistance of packaging materials using the Fisher's LSD test. For α=0.05, $LSD = 2.18 \cdot \sqrt{0.05(2/4)} = 0.34$. Thus, the average resistances can be ordered schematically as C > (A = B) > D.

The Bartlett's test can be used to compare three or more variances σ_k^2, e.g. to test the homogeneity of variances prior to ANOVA, using the test statistic

$$B = (\ln s_p^2)(\sum \nu_i) - \sum \nu_i \cdot \ln s_i^2$$

where $\nu_i = n_i - 1$, n_i is the sample size of group i and $s_p^2 = \sum SS_i / \sum \nu_i$ is the pooled variance. If the corrected test statistic $B_C = B/C$, where $C = 1 + 1/(3k-3)(\sum 1/\nu_i - 1/\sum \nu_i)$, is higher than $\chi^2_{[1-\alpha/2,k-1]}$, reject H_0. The chi-square approximation is less satisfactory if most of the ν_i are less than 5.

We may extend the chi-square method (briefly mentioned earlier) to test the difference among proportions from c independent populations (although we will have to back transform proportions into the actual counts). Data, arranged in a contingency table with two rows and c columns (Table 16.6), where the observed frequencies f_0 (i.e. number of "successes" and "failures" in the jargon of the Binomial distribution) per column are presented instead of the respective proportions, is used to determine the chi-square χ^2 test statistic

$$\chi^2 = \sum_{j=1}^{c} \frac{(f_o - f_e)^2}{f_e}$$

where $f_e = n_{row} n_{colum} / n$. If χ^2 is greater than the (1-α) percentile of the χ^2 distribution with (c–1) df, reject H_0: $p_1 = p_2 = \ldots = p_c$. This is sometimes called a chi-square test for independence.

Example 14. In 1994, the businesses in the municipalities of Albufeira and Faro (south Portugal) were distributed by the various areas of business (coded with numbers, e.g. 6201 "food and beverage", 6202 "chemicals and pharmaceuticals", 6203 "textiles and shoes",... , 6209 "non-specified") as shown in the Table 16.6. Because f_e <5 for the category labeled 6208, the two last categories are aggregated. Thus, since $\chi^2 = 74.58 > \chi^2_{[0.95,7]} = 14.07$, there seems to exist differences in the types of businesses between the two municipalities.

Table 16.6. Distribution of (retail) companies by the various areas of business (coded) in two municipalities, Faro and Albufeira (south Portugal), in 1994. Source: Marktest *Sales índex*. Análise do poder de compra regional – 1994.

Code Municipalities	6201	6202	6203	6204	6205	6206	6207	6208	6209
Albufeira	272	13	98	24	50	10	3	0	113
Faro	388	40	142	104	43	89	15	2	167

16.9 Relationships Between Variables

Scientists and engineers are frequently faced with the need to find mathematical models describing physical, chemical, biological, etc. phenomena. Sometimes, particularly in physics and chemistry, there is an exact (functional) relationship between the variables, mathematically $y = g(x)$. Given a particular value of *X*, the function *g* indicates the corresponding value of *Y*. In such a case there is very little experimental uncertainty and no statistical analysis is really required.

Commonly, however, one has to derive the (regression) models, e.g. $y = f(x) + \varepsilon$, from observations or experiments – empirical data – considering that a (unknown, unspecified) relationship is supposed to exist between variables and that data points do not exactly match that model (Figure 16.8). The better *f*(*x*) fits the data, the more rigorously it describes the relationship between the variables. Moreover, understanding the associations among variables can be useful in many ways, and one of the most important and common is prediction. The problem of finding the model that describes the relationship between variables and serves to predict the values of a response variable using one or more predictor variables is named regression analysis. Regression analysis is more eloquently and extensively presented in [27-30]. Related multivariate techniques, e.g. principal component analysis, are well presented elsewhere [31-33].

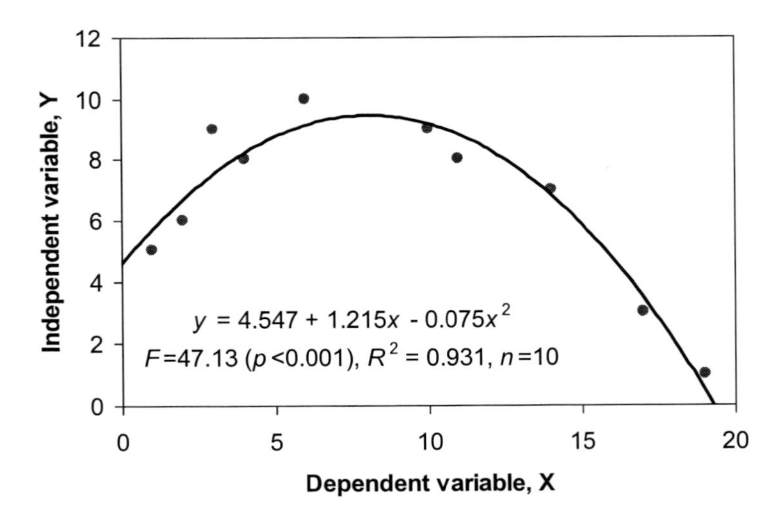

Figure 16.8. Illustration of a possible statistical relationship between two simulated variables – a 2nd degree polynomial (line) is plotted against empirical data (dots).

16.10 Simple Linear Regression

The first step in any regression analysis should be to plot the data in a scatter diagram. Although we may lose the exact values of the observations in graphing data, the plot(s) usually suggests the kind of relationship (if any) there exists between (the two) variables. The data illustrated in Figure 16.9 suggests that a straight line would adequately describe the relationship between an explanatory or independent variable X that is controlled by the researcher and is error-free (or measured with negligible error), and one response or dependent variable Y obtained experimentally or through observation – simple linear regression (OLR). Nonetheless, some dispersion is observed and data points do not exactly match that line – errors. Thus, regression analysis is used to find the line, mathematically $y = \beta_0 + \beta_1 x + \varepsilon$, that best fits the data, i.e. obtain the estimates of the parameters, b_0 (intercept) and b_1 (slope), which minimize the errors ε.

Regression theory assumes that errors are mutually independent and normally distributed, with a mean of zero and variance equal to the variance of Y. One method of estimating model parameters while minimizing the errors was originally proposed by Karl Gauss [1777-1855] and is named Least Squares Method. It is aimed at minimizing the sum of squared errors, SS_E,

$$SS_E = \sum \varepsilon^2 = \sum (y_i - \hat{y})^2$$

where the errors $\varepsilon = (y_i - \hat{y})$ are in fact the vertical distances between data (points) and the model (line) (Figure 16.9). For a sample of bivariate data, i.e. a set of n pairs of observations (x_1, y_1), (x_2, y_2), ..., (x_n, y_n), the least squares estimate of the slope b_1 is geometrically the inclination of the line but can also be understood as the variation in Y per unit change of X. The estimated intercept b_0 is the estimated value of Y when X equals zero. Similarly, it can be viewed as the estimated (average) value of the response if there is no effect of/relationship with X.

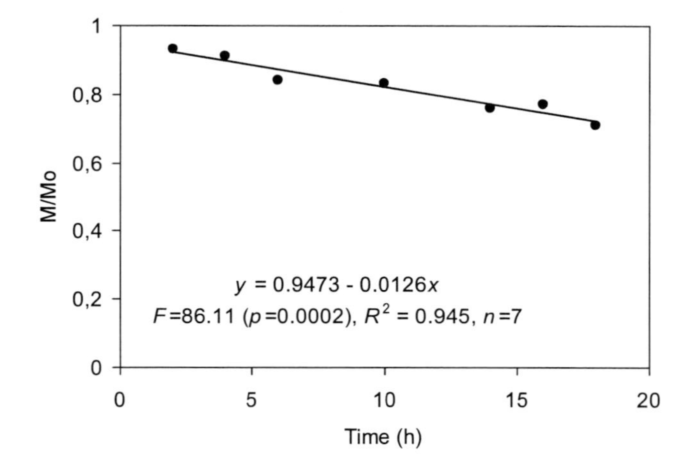

Figure 16.9. Plot of experimental data on moisture reduction (M/M_0) and time (h) of fruit dehydrated through osmosis at syrup concentration of *ca.* 60° Brix and temperature of about 50 °C. The fitted regression model and statistics are also shown.

The estimated b_0 and b_1 can (and should) be further examined by calculating their respective confidence intervals using respectively

$$b_0 \pm t \cdot s_{b_0}$$

and

$$b_1 \pm t \cdot s_{b_1},$$

where t is $t_{[1-\alpha/2;n-2]}$. The standard-errors of b_0 and b_1 are given by $s_{b_0} = \sqrt{\hat{\sigma}^2 \left[n^{-1} + \bar{x}^2 / S_{xx}\right]}$ and $s_{b_1} = \sqrt{\hat{\sigma}^2 / S_{xx}}$, where $\hat{\sigma}^2 = SS_E/(n-2)$, and $S_{xx} = \sum_{i=1}^{n} (x_i - \bar{x})^2$.

Example 15. The regression line obtained for the data plotted in Figure 16.9 is $y = 0.947 - 0.013x$. The 95% CI for b_0 and b_1 are [0.907, 0.988] and [–0,016, –0,009], respectively.

It is quite uncommon that the regression equation obtained exactly coincides with empirical data. Several tests and/or procedures can/should be used to assess the usefulness, evaluate the goodness-of-fit and check the assumptions of the regression model: ANOVA, R^2 and residuals' analysis.

ANOVA can be used to test the significance of regression models, i.e. answer the question "is the model useful?" In simple terms, the total variation of the data in a regression problem is divided into two complementary parts: one is "contained in"/"explained by" the model; the other is natural, random variation of Y that is not described by the regression equation. This idea, the ANOVA identity (see also Section 16.8), is algebraically summarized as $SS_T = SS_R + SS_E$ (where SS stands for "sum of squares"). If the F test-statistic obtained thereafter is significant then the regression model is judged to significantly describe the relationship between the variables.

If a regression is found to be significant for the data at hand, one can determine "how well the regression line approximates the real data points" by using the coefficient of determination, R^2. This coefficient varies between 0 (no fit) and 1 (perfect fit) and it is obtained from the SS mentioned above through two equivalent forms

$$R^2 = \frac{SS_R}{SS_T} = 1 - \frac{SS_E}{SS_T}.$$

The value obtained can be interpreted as the proportion (or percentage) of the variation of the dependent variable Y that can be explained, i.e. that is correctly predicted, by the independent variable X through the regression model obtained. Nonetheless, the R^2 does not tell whether the independent variable is the true cause of the changes in the dependent variable (other variables(s) could be involved) or that the correct regression model has been found (other models could fit the data better). Furthermore, in simple linear regression, the coefficient of determination is, actually, the square of the correlation coefficient (i.e. the Pearson product-moment correlation coefficient, usually noted *r*). This fact contributed to the generalized confusion involving the two coefficients and their interpretation. While the R^2 is derived from regression ANOVA, *r* is calculated by dividing the covariance of two variables by the product of their standard deviations. In statistical usage, correlation refers to the departure of two variables from independence (i.e. measures association) and can not be validly used to infer a causal relationship between the variables – a distinct interpretation from R^2.

Notwithstanding, even the smallest, barely detectable effect (i.e. slope) not explaining much of the variance (i.e. with a low R^2) will finally become highly significant if the sample size is large enough. Most of the times, these results are examples of the difference between "statistical significance" and "practical or biological relevance". Draper and Smith [27] present a simplified way of checking regression model's usefulness by finding the ratio

$$(\max \hat{Y}_i - \min \hat{Y}_i) / \sqrt{p\hat{\sigma}^2/n}$$

where p is the number of parameters in the equation (two in OLR), $\hat{\sigma}^2 = SS_E/(n-p)$ and n is the number observations used. Provided that this is equal or greater than four and no other defect is observed in the fit, a worthwhile regression interpretation is likely to be possible.

Example 15. (continued) The regression model obtained earlier for the moisture reduction data was judged as significant (since ANOVA $F_0 = 86.11 > F_{[0.95,1,5]} = 6.61$) and found to explain a large fraction of the variation in the data (R^2=0.945). The ratio proposed by [27] is 18.26, several times greater than four, confirming the usefulness of the model for moisture reduction prediction.

A common objective in regression analysis is to estimate *Y*, $\hat{y}$, for a certain value of *X* (within the scope of the model) or to predict a new observation of *Y*, $\hat{y}_0$, corresponding to a given level *X* of the predictor variable x_0. The confidence interval for the estimated *Y* given a particular value of *X*, x_0, is calculated by

$$\hat{y} \pm t \cdot \sqrt{\hat{\sigma}^2[n^{-1} + (x_0 - \bar{x})^2 / S_{xx}]}$$

where $\hat{y} = b_0 + b_1 x_0$, *t* is $t_{[1-\alpha/2;n-2]}$, $\hat{\sigma}^2 = SS_E/(n-2)$, and $S_{xx} = \sum_{i=1}^{n}(x_i - \bar{x})^2$. On the other hand, one can construct a $100(1-\alpha)\%$ confidence interval for a "future" realization of *Y* using

$$\hat{y}_0 \pm t \cdot \sqrt{\hat{\sigma}^2[1+n^{-1}+(x_0-\bar{x})^2/S_{xx}]}$$

where $\hat{y}_0$ is again obtained directly from the regression model. This prediction is viewed as the result of a new trial, independent of the trials used to estimate the regression model although assuming that the significant and relevant regression model is applicable.

Frequently, the fitted regression is used to estimate X, $\hat{x}$, from a given value of Y, y_0 – inverse prediction or calibration problem. These problems have aroused controversy among statisticians and are still being researched. Nonetheless, by re-arranging the regression equation into $\hat{x}=(y_0-b_0)/b_1$ (where $b_1 \neq 0$), one can derive a point estimate of X. Furthermore, the corresponding CI can be calculated using

$$\hat{x} \pm t \cdot \sqrt{(\hat{\sigma}^2/b_1^2)[1+n^{-1}+(\hat{x}-\bar{x})^2/S_{xx}]}$$

if the quantity $t^2\hat{\sigma}^2/(b_1^2 S_{xx})$ is small (say less than 0.1). Slight differences exist when one wants to estimate the CI of $\hat{x}$ not from a single observation of Y but from an average of q observations. Readers are encouraged to consult e.g. [29]

Example 15. (continued) What is the expected (average) moisture reduction if the fruit sample is treated for 10 h? Using $0.796 \pm 2.57\sqrt{0.0004[7^{-1}+(12-10)^2/232]}$, the following 95% confidence interval is obtained [0.775, 0.818].

Inference procedures for regression analysis are strictly valid only when the model assumptions on which the procedures are based are satisfied. However, models are approximations to reality, and so model assumptions will never hold exactly. Nevertheless, if a model is a reasonable approximation of reality, then inferences based on the model may be adequate for real applications [30]. If the model is appropriate then the residuals e_i should be close to the random errors ε_i and contain little or no information, thus representing only natural sampling variation (that can not be attributed to any specific source) [28].

Residual analysis is one of the most common approaches to check the fitted regression model; lack-of-fit test and the Durbin-Watson test for serial correlation being the other basic methods (not discussed herein). Plots of the residuals e_i, or better still of the standardized (or studentized) residuals, $r_i = \hat{e}_i/(\hat{\sigma}\sqrt{1-h_{i,i}})$, where the hat values $h_{i,i}=n^{-1}+(x_i-\bar{x})^2/S_{xx}$, can be used to check the assumptions of regression analysis. A satisfactory plot is one that shows a (more or less) horizontal band of points (Figure 16.10a). There are many possible

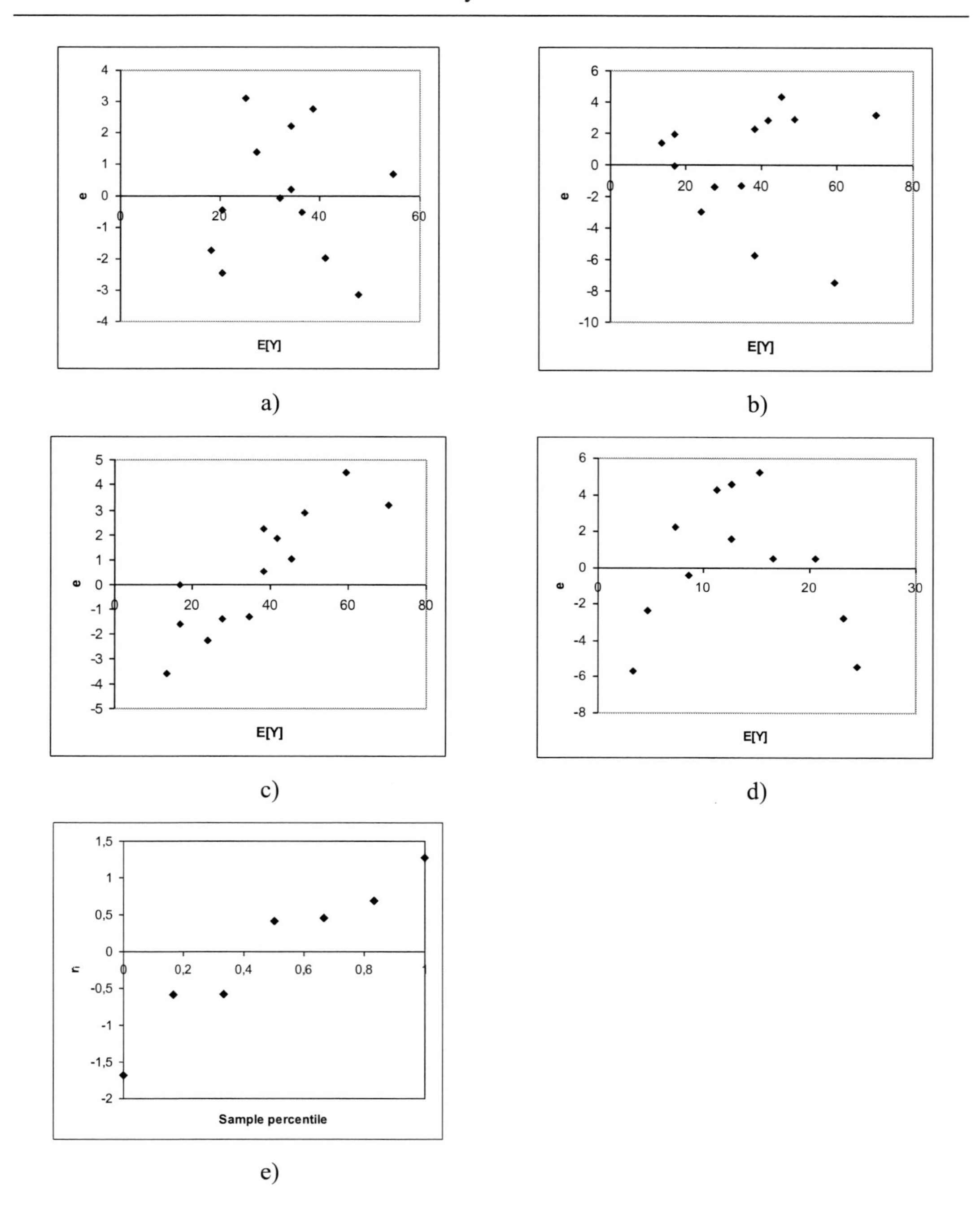

Figure 16.10. (a-d) Plots of regression residuals against the expected value of Y illustrating an adequate model and several different unsatisfactory patterns (see text for further details); (e) Normal probability plot of residuals.

unsatisfactory plots. A funnel pattern reveals non-constant variance or lack-of-fit of the model (Figure 16.10b). A log-transformation of response variable Y could be used to remedy this problem. An upward (or downward) trend is usually related to the wrongful omission of β_0 from the model (Figure 16.10c). Finally, a curvature (Figure 16.10d) is possibly due to an inadequate model. Adding extra terms to the model (including second-order terms) or transforming the response variable should be considered. Furthermore, a normal probability plot of r_i can be used to evaluate the validity of the assumption that each subpopulation of *Y* values, determined by *X*, is a Gaussian population (Figure 16.10e). Checking assumptions involves a great deal of computing, plotting, etc. which are easily done using one of several statistical computing packages (or even with popular, general-purpose spreadsheet programs). Plotting r_i against time order, helps to check for non-independence of observations.

16.11 Multiple Linear Regression

Intuitively, one can improve the predictive power of simple regression if other predictor or independent variables, up to a reasonable number, are added to the (simple) regression model. The concepts presented earlier can be extended to the multiple linear regression (MLR) case. MLR can even accommodate independent variables which are nominal (using "dummy" variables) or study case(s) where the dependent variable is itself nominal (logistic regression). The computational power available to almost every researcher (e.g. spreadsheet programs) facilitates the task of regression analysis immensely.

A linear relationship between the response variable *Y* (independently sampled from a normally distributed population) and *k* explanatory, independent variables x_j $(j = 1,...,k)$, which are not-random, can be written as

$$y_i = \beta_0 + \beta_1 x_{i1} + \beta_2 x_{i2} + \ldots + \beta_k x_{ik} + \varepsilon_i = \beta_0 + \sum_{j=1}^{k} \beta_j x_{ij} + \varepsilon_i$$

where β_j, $j = 0,1,\ldots,k$, are the p=1+k parameters (or regression coefficients) and ε is the random errors (normally distributed with mean of zero). Equivalently, the equation above can be written in matrix notation as $\mathrm{Y} = \mathrm{X}\beta + \varepsilon$ where Y is the vector of observations for the response variable, X is matrix of values for the *k* independent variables, β is the vector of regression coefficients and **ε** is the vector of errors. In this form, the equation is mathematically more amenable. As before the objective is to find the vector $\hat{\beta}$ of regression coefficients that minimizes the sum of squared errors – least squares method, using $\hat{\beta} = (\mathrm{X}^{\mathrm{T}}\mathrm{X})^{-1}\mathrm{X}^{\mathrm{T}}\mathrm{Y}$.

The estimated regression coefficients can (should) again be presented as confidence intervals. The limits of the $100(1-\alpha)\%$ confidence interval for the *j*-th coefficient $\hat{\beta}_j$ (or b_j) can be obtained using

$$b_j \pm t \cdot s_{b_j}$$

where t is $t_{[1-\alpha/2;n-p]}$ and the standard error of b_j is $s_{b_j} = \sqrt{\hat{\sigma}^2 C_{jj}}$ (where $\hat{\sigma}^2 = SS_E/(n-p)$ and C_{jj} is the value in the j-th line, j-th column of the matrix $C = (X^T X)^{-1}$).

Example 16. Colour changes (ΔE) during osmotic dehydration of seafood samples were analyzed at different times (t, 20 to 240 min.), temperatures (T, 30 to 38 ºC) and brine concentrations (C, 0.2 and 0.3 g NaCl/g). Results are presented in Table 16.7. The regression model is ΔE=8.366+1.252C–0.073T–0.018t. The 95% CI for b_0, b_{Conc} b_{Temp} b_{Time} are [6.442, 10.289], [–3.206 , 5.710], [–0,120, –0,026], and [0,016, 0,020], respectively.

Table 16.7. Results of an experiment performed to study the colour changes (ΔE) during osmotic dehydration of seafood samples at different times, temperatures and brine concentrations.

Concentration (g NaCl/g sample)	Temperature (ºC)	Time (min.)	ΔE
0.21	30	20	6.68
0.21	36	40	7.11
0.27	32	90	7.86
0.21	32	60	7.04
0.27	32	180	9.74
0.27	36	20	6.01
0.27	38	240	9.91
0.21	38	20	6.68
0.27	30	40	7.11
0.21	36	60	7.22
0.21	38	60	7.05
0.21	32	180	9.74
0.27	36	240	10.09
0.27	38	120	8.23

The significance of the fitted regression model can be assessed using ANOVA as before. The null hypothesis is now $H_0 : \beta_1 = \beta_2 = \ldots = \beta_k = 0$; if $f_0 > f_{[1-\alpha,k,n-p]}$ then reject H_0 with $100(1-\alpha)\%$ confidence and conclude that at least one regressor included in the fitted regression equation contributes significantly to the variation of the response. Residual analysis as described above for OLR is used to check the adequacy of the fitted regression model.

In MLR, the coefficient of determination R^2 can again be used to "measure" the usefulness of the regression equation. However, because R^2 is inflated by adding explanatory variables to a regression equation regardless of their significance, several authors propose the use of the adjusted-R^2 instead. This is given by

$$R^2_{adj.} = 1 - \frac{SS_E/(n-p)}{SS_T/(n-1)} = 1 - \left(\frac{n-1}{n-p}\right)(1-R^2)$$

where R^2 is obtained as before for OLR, n is the number of data points and p is the number of parameters in the regression model. Again, R^2 (or its adjusted version) does not tell whether the independent variables are the true causes of the change in the response, the correct regression was used and/or the most appropriate set of independent, explanatory variables has been chosen.

Example 16. (continued). The regression model obtained earlier for the colour changes of seafood was judged as significant (since ANOVA $F_0 = 165.78 > F_{[0.95,3,10]} = 3.71$) and found to explain a large fraction of the variation in the data ($R^2_{Adj} = 0.974$).

Regression analysis is a (virtually) never-ending procedure: one can tentatively exclude or include explanatory variables to the regression model in pursuit of the "best model"! The meaning of best is controversial. A good model selection technique should balance goodness-of-fit and complexity (ultimately the number of parameters/variables). Many methods have been proposed for selecting a suitable set of predictors in multiple regressions. If only one knew the magnitude of σ^2 for any single well-defined problem it would be much easier to choose the best regression equation [27].

Classical methods of variable selection include backward elimination, forward selection, and stepwise regression. They sequentially delete and/or add predictors by means of (modified) mean squared error criteria. In backward elimination, one should first look at the correlations of each of the explanatory variables with the response variable. Then, the starting model contains only explanatory variables significantly correlated with the response. Examine the significance of each of the explanatory variables in your model using t-tests, i.e. the test statistic

$$T_0 = \frac{b_j}{s_{b_j}}$$

If the null hypothesis H$_0$: $\beta_j = 0$ is rejected, that explanatory variable x_j has an effect on the outcome variable while controlling for other X's and it should therefore be retained in the model. One should remove the remaining variable with the highest p-value from the model and examine what happens to R^2 and $s^2 = \sum e_i^2/(n-p-1)$. Basically, R^2 should not decrease too rapidly and s^2 should not increase too rapidly when a variable is dropped, or else that variable should be added back into the model. This process should be continued until all nonsignificant variables have been removed from the model. Only one variable, however, should be removed from the model at a time.

On the other hand, stepwise regression is a sequential process for fitting the least squares model, where at each step a single explanatory variable is either added to or removed from the model in the next fit. The most commonly used criterion for the addition or deletion of a variable in stepwise regression is based on partial F-statistic:

$$F_{partial} = \frac{(R_2^2 - R_1^2)/(k_2 - k_1)}{(1 - R_2^2)/(n - k_2 - 1)}$$

The suffix 2 refers to the larger model, whereas the suffix 1 refers to the reduced model (with $k_1 < k_2$). The test values for the partial F-statistic are often called "F to enter" and "F to remove" (some programs use a fixed F-test value, e.g. F=4, rather than individual percentage points of the F distribution, e.g. α=0.05, as the *df* change).

Selecting a model of the right form to fit a set of data is always an iterative process and usually requires the use of empirical evidence in the data, knowledge of the process and some trial-and-error experimentation. When in doubt between models remember the "law of parsimony", or Ockham's razor: "an explanation should use no more variables than are necessary" (a principle proposed by William of Ockham in the fourteenth century).

Example 16. (continued). The t-tests statistics for the variables included in regression model were t_0=0.626, t_0=–3.46 and t_0=21.98, respectively for concentration, temperature and time. Thus, brine concentration is considered non-significant and could be dropped from the model. The "reduced" model $y = 8.672 - 0.073T + 0.018t$ is highly significant (ANOVA $F_0 = 165.78 > F_{[0.95,3,10]} = 3.71$) and useful ($R^2_{Adj} = 0.976$).

In MLR, when explanatory variables are correlated there is difficulty in interpreting the effect of those variables on the outcome. This problem of multicollinearity can have significant impact on the quality and stability of the fitted regression model. The simplest method for detecting multicollinearity is the correlation matrix among explanatory variables. A common approach to remedy a multicollinearity problem is to omit explanatory variables. For example, if x_1 and x_2 are highly correlated (e.g. r>0.9), the simplest approach would be to use only one of them, since one variable conveys essentially all the information in the other variable.

Again, one of the main objectives of regression analysis is to estimate or predict the value of the response variable Y given the values of explanatory variables x_j. The $100(1-\alpha)\%$ confidence interval of the expected value of Y can be obtained using

$$\hat{y} \pm t \cdot \sqrt{\hat{\sigma}^2 x_0^T C x_0}\ .$$

where $\hat{y} = x_0^T B$ (or alternatively from $\hat{y} = b_0 + b_1 x_1 + b_2 x_2 + \ldots$) and $x_0^T = [1 \quad x_{01} \quad x_{02} \quad \ldots \quad x_{0k}]$ is a line vector of values of x_j. Moreover, one can construct a CI for a "future" realization of Y (noted $\hat{y}_0$) using

$$\hat{y}_0 \pm t \cdot \sqrt{\hat{\sigma}^2[1 + x_0^T (X^T X)^{-1} x_0]}.$$

Predicting "new" realizations of the response value should be done with care, as it is easy to extrapolate outside the data range used in the analysis (Figure 16.11). As in simple linear regression analysis, the final model can be used to estimate CI of *X's* from a given (or average) value of *Y*. The method extends the procedure presented earlier for the case of just one *X*.

Example 16. (continued). What is the expected colour change of a seafood sample to be dehydrated for 3 h at 34 °C in 0.21 g NaCl/g brine? Using $9.398 \pm 2.20\sqrt{0.0474(1+8.526)}$ the 95% confidence interval of ΔE is [7.92, 10.88].

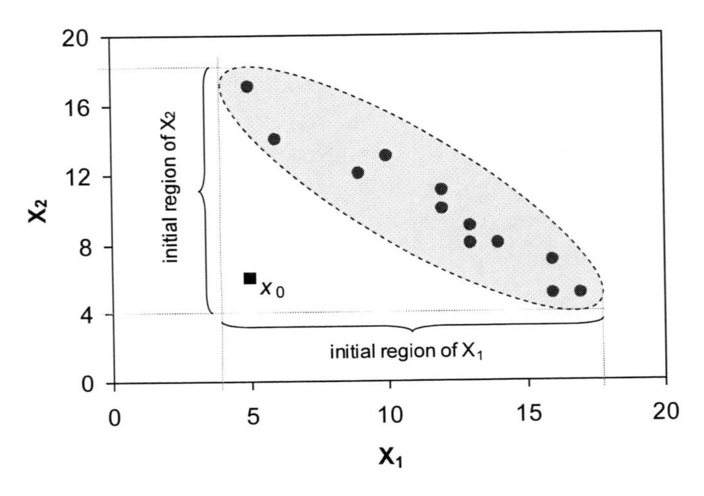

Figure 16.11. Illustration of extrapolation (x_o) beyond the "region" of initial observations (gray ellipse) of a regression model with two independent variables, X_1 and X_2.

16.12 Non-Linear Regression

In contrast to the "simpler" regression analysis of linear models, it is more difficult to fit non-linear[5] functions to data, *e.g.* $y = a\exp(bx)$. Commonly, variables are transformed in order to linearize the relationship and allow its study through ordinary least squares regression. This is possible for intrinsically linear relationships. Take for example the function $y = a\exp(bx)$ and obtain the natural logarithms of both terms to get $\ln y = \ln a + bx$, i.e. linear model of the form $y' = a' + bx$. However, problems with the

[5] Polynomials, such as $y = a + bx + cx^2 + dx^3$, are in fact linear models and can therefore be studied using the OLR: fit the model $\hat{y} = b_0 + b_1x_1 + b_2x_2 + b_3x_3$, where $x_1 = x$, $x_2 = x^2$ and $x_3 = x^3$. The likely problems of multicollinearity among x_j of this approach can be circumvented through the scalling of explanatory variables.

necessary assumption of residuals normality and/or orthogonality among explanatory variables may arise. Nowadays, it is possible to perform non-linear regression analysis even with common spreadsheet programs like Microsoft Excel® (through the Solver® add-in), although "dedicated" software, *e.g.* SPSS® (SPSS Inc., Chicago) or R [12] should be preferred. Alternatively, online resources, e.g. [34], can be used with confidence. In most situations, all these should provide similar estimates for a certain dataset.

In this chapter the most simple and relatively common situation is presented. A dependent, response variable Y obtained experimentally and normally distributed is supposedly related to an independent, explanatory variable X (controlled by the researcher and error-free) through $y = f(x) + \varepsilon$ (where $f(x)$ is a function with one or more parameters θ and ε are independent and normally distributed errors). The reader is referred to e.g. [35] for a more thorough discussion of this topic. Often, the equation that relates Y e X is chosen by the researcher according to theory or a working hypothesis – deterministic approach. Conversely, there exist statistical methods to describe the relationship between variables without prior knowledge or an underlying model (for example, polynomial or *spline* functions) – empirical perspective – which for extension and complexity reasons are not discussed herein.

The Least Squares Method can be used to fit $f(x)$ to empirical data. However, in the case of non-linear models it is not possible to obtain the estimates $\hat{\theta}$ in one step as in linear regression analysis. Therefore, SS_E is minimized through an iterative process using an appropriate algorithm that requires initial estimates of parameters θ_0 and stops when the changes in parameter values produce a negligible reduction in SS_E.

Initial estimates of function parameters should be specified taking into consideration the researcher's experience, preliminary experiments or an "educated guess". Knowledge about the meaning of model parameters facilitates the choice of initial values. A poor selection can increase computation time, prevent algorithm convergence or originate an "inappropriate" solution, because the algorithm converged at a local minimum. The choice of initial parameter estimates is more influential in the case of models with many parameters [35].

According to Draper and Smith [27], the usual tests used in the linear model case are, in general, not appropriate when the model is nonlinear. Notwithstanding, an alternative approach is presented below.

The goodness-of-fit of the fitted model can be assessed visually, by simultaneously plotting the data and the curve. This is a simple yet powerful way of assessing the final model's adequacy. Some authors suggest the additional representation of the CI for $\hat{y}$,

$$\hat{y} \pm t \cdot se(Y)$$

where $\hat{y}$ is the estimated value of Y for a certain value of X, x_0, t is $t_{[1-\alpha/2;n-p]}$, and $se(Y) = \sqrt{SS_E/(n-p)}$ is the standard error of Y (n is the sample size i.e. number of data points, and p is the number of model parameters).

Best-fit parameter values reported by nonlinear regression should be checked for scientific plausibility (constraining the parameters to a sensible range can remedy this) and precision [35]. Results of statistical analysis software include the standard error (SE) and confidence intervals (CI) of the estimated parameters. When the error terms are independent and normally distributed and the sample size is reasonably large, SE and CI for parameters can be obtained from the estimated covariance matrix $\hat{\sigma}^2(W^T W)^{-1}$, where W is a $n \times p$ gradient matrix of first derivatives of the function *f* evaluated at $\hat{\beta}$, and $\hat{\sigma}^2 = SS_E/(n-p)$. If the CI is really wide, there is a problem!

In linear regression analysis, the coefficient of determination R^2 is commonly used to assess goodness-of-fit. Although it can be used in non-linear regression problems, several authors caution its use because it is difficult to meet some of its assumptions.

Model adequacy can also be assessed using: a plot of residuals vs. observed values of X, the Lillefors test for normality of residuals and the (Wald-Wolfowitz) runs test of residuals. The plot of residuals for an adequate model does not shown tendencies or patterns. The Lillefors test is a modification of the "standard" Kolmogorov-Smirnov (K-S) test and is used to test if residuals have a normal distribution. Finally, the runs test is useful to examine the hypothesis that the residuals are mutually independent (i.e. if there are systematic deviations between data and the model). However, Conover [4] alerts to the very little power of runs tests.

The goodness-of-fit of two different models fitted to the same data can be compared using their respective SS_E. One can use the F-test statistic,

$$F_0 = \frac{(SS_1 - SS_2)/(df_1 - df_2)}{SS_2/df_2}$$

where SS is the SS_E, df are the degrees of freedom (*i.e.* number of points minus the number of model parameters) and indices 1 and 2 denote the model with less and more parameters, respectively. If *p-value* $< \alpha$, then the more complex model (with higher number of parameters) fits better the data at hand than the simpler model. Notwithstanding, model comparison should not be entirely statistical. More important are the physical, chemical or biological plausibility of the model and its consistency with the data.

Example 17. To study the kinetics of an enzyme, reaction rates V_o for several concentrations of substrate [s] were obtained. The fitted Michaelis-Menten kinetics model is $V_o = 212.68 \cdot [s]/(0.064 + [s])$. Parameters and regression model statistics obtained with Microsoft® Excel Solver® add-in are presented in Figure 16.12. Using R [12], the standard-errors of V_{max} and K_m are 3.33 and 0.004, respectively.

16.13 "EFFICIENT" DESIGN OF EXPERIMENTS

The essence of an experiment is to make purposeful changes to a system (through the independent variable(s) – a.k.a. factors) and to study the effect of these changes (upon the dependent variable, i.e. output response). The word experiment is used in a quite precise sense to mean an investigation where the system being studied is under the control of the investigator. By contrast, in an observational study some aspects of the experimental material, factors/treatments and/or measurements and in particular the allocation of individuals to treatment groups, are outside the investigator control. In practice the distinction between experiments and observational studies can sometimes become blurred [16]. Sir Ronald A. Fisher [1890-1962] originally developed experimental design in the 1920s. Complete and helpful references on this subject are [16, 18, 36]. Hu [37] provides a hands-on, computer-aided statistical approach to food product design, namely for food process and recipe modeling and optimization.

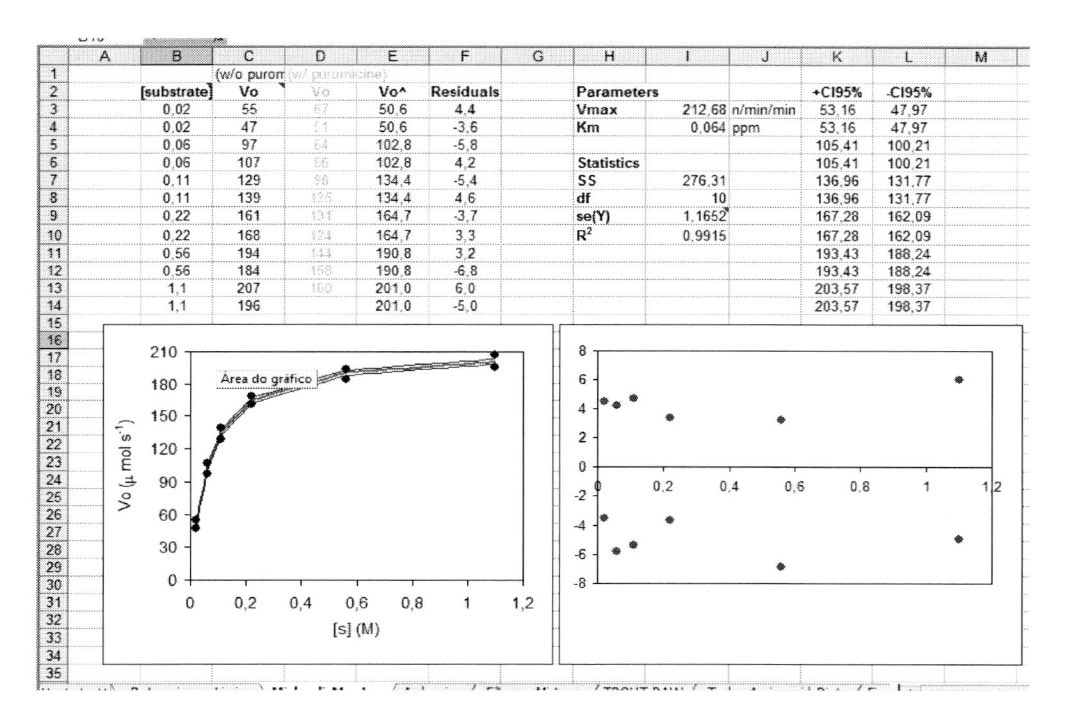

[substrate]	Vo (w/o puron	Vo^	Residuals	Parameters			+CI95%	-CI95%
0,02	55	50,6	4,4	Vmax	212,68	n/min/min	53,16	47,97
0,02	47	50,6	-3,6	Km	0,064	ppm	53,16	47,97
0,06	97	102,8	-5,8				105,41	100,21
0,06	107	102,8	4,2	Statistics			105,41	100,21
0,11	129	134,4	-5,4	SS	276,31		136,96	131,77
0,11	139	134,4	4,6	df	10		136,96	131,77
0,22	161	164,7	-3,7	se(Y)	1,1652		167,28	162,09
0,22	168	164,7	3,3	R^2	0,9915		167,28	162,09
0,56	194	190,8	3,2				193,43	188,24
0,56	184	190,8	-6,8				193,43	188,24
1,1	207	201,0	6,0				203,57	198,37
1,1	196	201,0	-5,0				203,57	198,37

Figure 16.12. Parameters and regression model statistics of the Michaelis-Menten kinetics model fitted to data of reaction rates V_o for several concentrations of substrate [s] using the Solver® tool of Microsoft® Excel: $V_o = 212.68 \cdot [s]/(0.064 + [s])$. The 95% confidence intervals of predicted V_o (rightmost columns noted +CI95% and –CI95%) are plotted together with experimental data (left plot) and residuals are plotted against the expected values of V_o (right plot).

Experimental design is a critically important tool in the field of (industrial) food production and processing, for characterizing a process (usually the interest is determining which process variables affect the response and rank the important through unimportant factors – screening experiments) and optimizing a process (i.e. to model the process and then determine the optimal settings of the process factors). In food product design there are two different kinds of system problems – process and mixture problems – that should be dealt with by different statistical methodologies. Notwithstanding, it is possible to simultaneously study k process and q mixture factors (sometimes named crossed mixture-process designs). In process problems, all the independent variables are not related to each other i.e. the change in one variable is not restricted by another variable – orthogonality. Recipe (or mixture) problems are obviously important for successful food products. A recipe usually includes several ingredients (i.e. independent variables) which can not vary without changing at least one of the other ingredients in the mixture, because all the ingredients will be part of a constant sum of 100%. Some valuable albeit more complex methodologies, e.g. response surface methods and analysis of mixture experiments, are covered in elsewhere [18, 37, 38],

Montgomery [18] points out three basic principles of statistical design of experiments: replication, randomization and blocking. Replication is the repetition of the basic experiment. Randomization, i.e. the random allocation of experimental units/subjects to treatments/factor levels, is a cornerstone underlying the use of statistical methods. Blocking is a design technique used to improve the precision by reducing or eliminating the variation transmitted from known and controllable nuisance factors i.e. process variables that may influence the response but in which one is not directly interested (e.g. day or batch). Sometimes a nuisance factor is unknown and uncontrollable. Randomization is used to guard against such factor(s).

Moreover, the choice of appropriate response variable(s) for measurement, the avoidance of systematic errors (or biases), the assessment of the magnitude of experimental-wise errors, and the advantageous use of any special structure in the treatments (for example, when these are specified by combinations of factors) are key aspects to consider in planning and conducting an experiment. There is a close connection between design of experiment(s) and (data) analysis in that an objective of design is to make both analysis and interpretation of results as simple and clear as possible.

Many experiments involve the study of two or more factors. Factorial experiments, in which factors (or process variables) are varied together, is almost unanimously considered the correct and most efficient approach to experimental design when compared to the best-guess approach or to the one-factor-at-a-time strategy. From the several factorial designs available, the one with k factors each at only two levels that may quantitative, e.g. 110/130 °C, 200/250 psi, or qualitative, such as two operators, presence/absence of factor – 2^k factorial designs – is highlighted herein since it useful both for the researcher and "plant engineer". There are several statistics software packages that will set up and analyze two-level factorial designs (e.g. Design-Expert from StatEase Inc., or R [12]).

One can distinguish complete (or full) and incomplete designs. In the former, every possible combination of (the levels of) factors is tested whereas in the later, only a fraction of the experiments is actually tried (thence the designation fractional designs). For the remaining of this section we will assume that conclusions apply only to the factor levels considered in the experiment (fixed factors effects), designs are completely randomized and observations are independent and normally distributed.

The simplest case of two-level factorial designs involves only two factors (A and B), each at two levels (low and high). For each of the 2^k combination of factors' levels (also called runs or trials), replicate observations can (should) be measured in order to assess experimental error (Figure 16.13).

Using ANOVA, one can test a set of hypothesis concerning the effects of the factors per se and the effect of their interaction respectively:

H_0: $\alpha_1 = \alpha_2 = 0$ *vs*. H_1: Some $\alpha_i \neq 0$

H_0: $\beta_1 = \beta_2 = 0$ *vs*. H_1: Some $\beta_j \neq 0$

H_0: $\alpha\beta_{ij} = 0$, for every i, j *vs*. H_1: $(\alpha\beta)_{ij} \neq 0$.

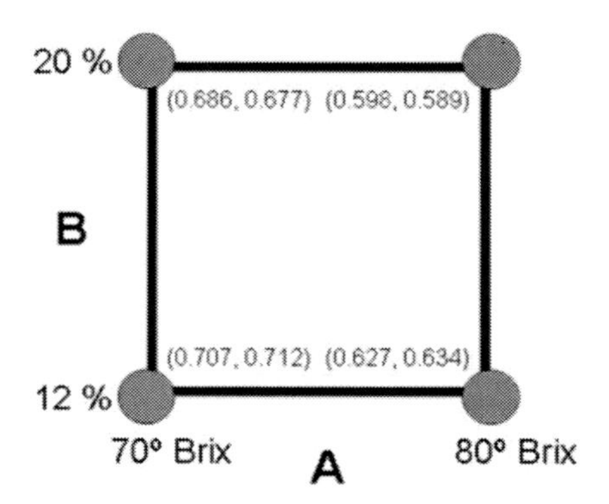

Figure 16.13. Illustration of a 2-level factorial design with two factors, A (syrup concentration, 70 and 80° Brix) and B (fiber content, 12 and 20%). Numbers within parenthesis are duplicate measurements of the response variable (a_W of cereal bar after 60 days at room temperature) per combination of factors.

For each null hypothesis, the derived F_0 statistic is compared with the (1–α) percentile of the corresponding F-distribution to decide whether to reject or not the stated hypothesis. Alternatively, the respective p-values could be used to make a decision. The first hypothesis to be tested is the one concerning the interaction effect. Its rejection means that factors are non-additive, i.e. they interact, thus is ineffectual to study the mains effects of single factors per se.

Significant results can be expressed in terms of a regression model, $y=\beta_0+\beta_1x_1+\beta_2x_2+\beta_3x_1x_2+\varepsilon$, where x_1 and x_2 are the explanatory variables (i.e. the factors A and B) and x_1x_2 is their interaction (AB). The coefficients are obtained after coding the results (e.g. using the –1 and +1 notation for the low and high levels). This model can be used to confidently predict the response variable. Model goodness-of-fit can be ascertained and compared among fitted models through several statistics, namely the adjusted-R^2, the predicted-R^2 ($R^2_{pred.} = 1 - PRESS/S_{yy}$ where $S_{yy} = \sum_{i=1}^{n}(y_i - \bar{y})^2$) and the $PRESS$. The closer the values of adjusted-R^2 and predicted-R^2, the better the fit – differences below 0.1 correspond to an adequate model. The predicted residual sum of squares $PRESS = \sum_{i=1}^{n}(y_i - \hat{y}_{[i-1]})^2$, which measures the fit of the model to each observation not considered in the model (noted [i–1] above), provides a useful scaling of residuals. A small value of $PRESS$ indicates that the model is likely to be a good predictor. The ratio proposed by [27] (this chapter, page 43), sometimes named "Adequate Precision", can be used as a measure of model usefulness. Finally, the adequacy of the fitted model can be assessed through residual analysis (*cf.* section on regression analysis).

In two-level factorial designs involving 5 or more factors, the number of experiments (runs or trials) needed to study every combination of factor levels is most of the times excessive (e.g. a design with five factors represents 2^5=32 different combinations, or runs). If one can dismiss the higher order, more complex interactions, because these are usually not significant and have no physical meaning, or when one is searching for the important, critical factors out of an initial set of potential variables, i.e screening experiments, fractional designs are very useful. In fact, systems or processes are more strongly conditioned by a small number of factors and low-order interactions than more complex interactions – sparsity-of-effects principle or hierarchical ordering principle [sensu 39]. Moreover, a fractional design can be projected into stronger, larger designs and allow sequential and complementary experimentation to confirm conclusion(s).

Response: **aW**
ANOVA for Selected Factorial Model
Analysis of variance table [Partial sum of squares]

Source	Sum of Squares	DF	Mean Square	F Value	Prob > F
Model	0.016	2	8.029E-03	253.26	< 0.0001
A	0.014	1	0.014	439.89	< 0.0001
B	2.112E-03	1	2.112E-03	66.64	0.0004
Residual	1.585E-04	5	3.170E-05		
Lack of Fit	4.050E-05	1	4.050E-05	1.37	0.3063
Pure Error	1.180E-04	4	2.950E-05		
Cor Total	0.016	7			

Std. Dev.	5.630E-03	R-Squared	0.9902
Mean	0.65	Adj R-Squared	0.9863
C.V.	0.86	Pred R-Squared	0.9750
PRESS	4.058E-04	Adeq Precision	33.644

Factor	Coefficient Estimate	DF	Standard Error	95% CI Low	95% CI High
Intercept	0.65	1	1.991E-03	0.65	0.66
A-Brix	-0.042	1	1.991E-03	-0.047	-0.037
B-Fiber	-0.016	1	1.991E-03	-0.021	-0.011

Final Equation in Terms of Coded Factors:

aW	=
+0.65	
-0.042	* A
-0.016	* B

Final Equation in Terms of Actual Factors:

aW	=
+1.34500	
-8.35000E-003	* Brix
-4.06250E-003	* Fiber

Diagnostics Case Statistics

Standard Order	Actual Value	Predicted Value	Residual	Leverage	Student Residual	Cook's Distance	Outlier t
1	0.71	0.71	-4.750E-03	0.375	-1.067	0.228	-1.086
2	0.71	0.71	2.500E-04	0.375	0.056	0.001	0.058
3	0.63	0.63	5.750E-03	0.375	1.292	0.334	1.416
4	0.63	0.63	-1.250E-03	0.375	-0.281	0.016	-0.253
5	0.69	0.68	6.750E-03	0.375	1.516	0.460	1.846
6	0.68	0.68	-2.250E-03	0.375	-0.505	0.051	-0.464
7	0.60	0.60	2.250E-03	0.375	0.505	0.051	0.464
8	0.59	0.60	-6.750E-03	0.375	-1.516	0.460	-1.846

Figure 16.14. Output from DX®6 analysis of the data illustrated in Figure 16.13 (see text for further explanation).

Example 18. The statistical analysis (i.e. DX®6 output) of the experiment illustrated in Figure 16.13 is presented in Figure 16.14. The Brix×Fiber interaction effect was found to be non-significant in the full model (F=1.37, *p-value*=0.3063) and thus removed from the final (reduced) model actually shown.

In a two-level fractional design, sometimes noted 2^{k-p}, only $1/2^{p}$ experiments are run and the significance of $(2^{k-p}-1)$ of the factors is tested. Rules and relatively complex algorithms exist to construct such designs. Several textbooks [e.g. 18, 36], provide tables that help plan fractional factorial experiments those not using "dedicated" software programs.

A fractional factorial design requires the selection of p independent generators. These determine the alias structure of the design through the defining relation. The generators should be chosen to obtain the highest possible resolution. For example, in a half-fraction five-factor design one decides to confound the fourth-order interaction effect of ABCD with the effect of factor E (i.e. the generator ABCD=E). The corresponding defining relation is ABCDE=I, which produces the following alias structure: A=BCDE, B=ACDE, ..., AB=CDE, AC=BDE, etc. This means, for example, that the estimated effect of A is in fact the effect of A together with the negligible effect of the fourth-order interaction BCDE or the interaction effect of AB is confounded (i.e. estimated together) with that of the interaction CDE. The design just presented is a resolution V design. However, the smaller the fraction of runs, the lower the resolution of the design. The interpretation of results is similar to full factorial designs although taking into consideration the confounding of effects.

CONCLUSION

In this chapter, the most commonly used statistical techniques of the numerous available to the researcher in the field of food science and technology are presented. Whenever, empirical data does not meet the assumptions of "classical", parametric tests, other methodologies are suggested as alternatives. For obvious reasons of space and scope, a few statistical tests useful in food science and technology R&D are not addressed, e.g. response surface methods and analysis of mixture experiments, or not detailed, e.g. two-level fractional factorial designs, in this chapter albeit being important. These are well presented elsewhere, e.g. [18]. This chapter is more of a starting point than the finishing line.

REFERENCES

[1] Juran, J.M., Gryna, F.M., & Bingham Jr., R.S. (1959). *Quality control handbook* (3rd ed.). New York, McGraw-Hill Book Co.

[2] Mitra, A. (1993). *Fundamentals of quality control and improvement*. U.S.A., Macmillan Publishing Company.

[3] Montgomery, D.C. (2005). *Introduction to statistical quality control* (5th ed.). New York, John Wiley & Sons.

[4] Conover, W.J. (1999). *Practical nonparametric statistics*. New York, John Wiley & Sons. Inc.

[5] Siegel, S. (1975). *Estatística não-paramétrica (para as ciências do comportamento)*. São Paulo, McGraw-Hill.

[6] Vieira, S. (2003). *Bioestatística. Tópicos avançados*. Rio de Janeiro, Editora Campus Ltda (Elsevier Science).

[7] Ross, S.M. (2003). *Introduction to probability models* (8th ed.). Amsterdam, Academic Press.
[8] Papoulis, A. (1990). *Probability and statistics*. New Jersy, Prentice-Hall International Editors.
[9] Pestana, D., & Velosa, S. (2006). *Introdução à probabilidade e estatística* (2ª ed.). Lisboa, Fundação Calouste Gulbenkian.
[10] Van Belle, G. (2002). *Statistical rules of thumb*. U.S.A., John Wiley & Sons.
[11] Lenth, R.V. (2001). Some practical guidelines for effective sample size determination. *The American Statistician*, *55*, 187-193.
[12] R Development Core Team (2007). *R: A language and environment for statistical computing*. Vienna, Austria, R Foundation for Statistical Computing.
[13] Faul, F., Erdfelder, E., Lang, A.-G., & Buchner, A. (2007). G*Power 3: A flexible statistical power analysis program for the social, behavioral, and biomedical sciences. *Behavior Research Methods*, *39(2)*, 175-191.
[14] Lenth, R.V. Java applets for power and sample size [online]. 2006-9 [4 September 2009]. Available from http://stat.uiowa.edu/~rlenth/Power/.
[15] Harsham, H. Sample size determination [online]. 1994-2009 [4.9.2009]. Available from http://home.ubalt.edu/ntsbarsh/Business-stat/otherapplets/SampleSize.
htm#rabsol.
[16] Cox, D.R., & Reid, N. (2000). *The theory of the design of experiments*. Boca Raton, Chapman & Hall/CRC Press.
[17] Sokal, R.R., & Rohlf, F.J. (1995). *Biometry. The principles and practices of statistics in biological research* (3rd ed.). New York, W.H. Freeman and Company.
[18] Montgomery, D.C. (2001). *Design and analysis of experiments* (5th ed.). New York, John Wiley & Sons.
[19] Cohen, J. (1988). *Statistical power analysis for the behavioral sciences* (2nd ed.). New York, Academic Press.
[20] Maxwell, S.E., & Delaney, H.D. (2004). *Designing experiments and analyzing data. A model comparison perspective* (2nd ed.). London, Lawrence Erlbaum Associates, Publishers.
[21] Wild, C., & Seber, G. (2000). *Chance encounters: A first course in data analysis and inference*. New York, John Wiley & Sons.
[22] Tukey, J.W. (1977). *Exploratory data analysis*. Massachusetts, Addison-Wesley.
[23] Bethea, R.M., Duran, B.S., & Boullion, T.L. (1995). *Statistical methods for engineers and scientists*. New York, Marcel Dekker Inc.
[24] Newcombe, R.G. (1998). Interval estimation for the difference between independent proportions: comparison of eleven methods. *Statistics in Medicine*, *17*(8), 873-890.
[25] Brown, L.D., Cai, T.T., & DasGupta, A. (2001). Interval Estimation for a Binomial Proportion. *Statistical Science*, *16(2)*, 101-133.
[26] Olive, D.J. (2005). A Simple Confidence Interval for the Median [online]. [2/3/2009]. Available from www.math.siu.edu/olive/ppmedci.pdf.
[27] Draper, N.R., & Smith, H. (1998). *Applied Regression Analysis* (3rd ed.). New York, John Wiley & Sons Inc.
[28] Frees, E.W. (1996). *Data analysis using regression models. The business perspective*. New Jersey, Prentice-Hall.

[29] Neter, J., Kutner, M.H., Nachtsheim, C.J., & Wasserman, W. (1996). *Applied linear regression models* (3rd ed.). Chicago, McGraw-Hill Inc. and Irwin Inc.

[30] Graybill, F.A., & Iyer, H.K. (1994). *Regression analysis: concepts and applications.* Belmont, Duxbury Press.

[31] Everitt, B.S., & Dunn, G. (2001). *Applied multivariate data analysis* (2nd ed.). London, Arnold Publishers.

[32] Manly, B.F.J. (1994). *Multivariate statistical methods. A primer* (3rd ed.). New York, CRC Press.

[33] Sharma, S. (1996). *Applied multivariate techniques.* New York, John Wiley & Sons.

[34] Phillips, J.R. ZunZun.com Online Curve Fitting and Surface Fitting Web Site [online]. 2009 [2.3.2009]. Available from http://zunzun.com/.

[35] Motulsky, H., & Christopoulos, A. (2004). *Fitting models to biological data using linear and nonlinear regression: a practical guide to curve fitting.* U.S.A., Oxford University Press.

[36] Box, G.E.P., Hunter, J.S., & Hunter, W.G. (1978). *Statistics for experimenters. An introduction to design, data analysis, and model building.* New York, John Wiley & Sons.

[37] Hu, R. (1999). *Food product design. A computer-aided statistical approach.* Pennsylvania, Technomic Publishing Company Inc.

[38] Wadsworth, H.M. (1990). *Handbook of statistical methods for engineers and scientists.* N.Y., McGraw-Hill Publishing Company.

[39] Wu, C.F.J., & Hamada, M. (2000). *Experiments: Planning, analysis, and parameter design optimization.* New York, John Wiley & Sons.

CONTRIBUTORS

Adriana Descalzo
Instituto Tecnología de Alimentos
Centro de Investigación de Agroindustria, Instituto Nacional de Tecnología Agropecuaria
INTA, Morón, Buenos Aires
Argentina
adescalzo@cnia.inta.gov.ar

Amadeu F. Brigas
CIQA- Centro de Investigação em Química do Algarve and Departamento de Química e Farmácia, Faculdade de Ciências e Tecnologia
Universidade do Algarve
Campus de Gambelas 8005-139 Faro
Portugal
abrigas@ualg.pt

Amparo Andrés-Bello
Instituto de Ingeniería de Alimentos
Universidad Politécnica de Valencia
Camino de Vera s/n
46022 Valencia
Spain

Ana C. Figueira
CIEO- Centro de Investigação sobre Espaço e Organizações and Departamento de Engenharia Alimentar, Instituto Superior de Engenharia
Universidade do Algarve
Campus da Penha, 8005-139 Faro
Portugal
afiguei@ualg.pt

Ana L. Barbosa
Escola Superior de Turismo e Tecnologia do Mar
Instituto Politécnico de Leiria, Santuário N.ª Sra. dos Remédios, Peniche
Portugal
ac_barbosa@estm.ipleiria.pt

Andrea Biolatto
EEA Concepción del Uruguay
Instituto Nacional de Tecnología Agropecuaria INTA
Concepción del Uruguay,
Entre Ríos
Argentina
abiolatto@concepcion.inta.gov.ar

Beyza Ersoy
Department of Fishing and Fish
Processing Technology
Faculty of Fisheries
Mustafa Kemal University, Hatay
Turkey
beyzaersoy@gmail.com

B.K. Tiwari
Biosystems Engineering
UCD School of Agriculture
Food Science and Veterinary Medicine
University College Dublin
Belfield, Dublin 4
Ireland
brijesh.tiwari@ucd.ie

Célia Quintas
Departamento de Engenharia Alimentar
Instituto Superior de Engenharia
Universidade do Algarve
Campus da Penha, 8005-139 Faro
Portugal
cquintas@ualg.pt

Colm P. O'Donnell
Biosystems Engineering
UCD School of Agriculture
Food Science and Veterinary Medicine
University College Dublin
Belfield, Dublin 4
Ireland
colm.odonnell@ucd.ie

Cornelis Versteeg
Food Science Australia
A joint venture of CSIRO and the Victorian Government
671 Sneydes Road, Werribee, VIC 3030
Australia
Kees.versteeg@foodscience.afisc.csiro.au

Cristina L. M. Silva
CBQF- Centro de Biotecnologia e Química Fina and Escola Superior de Biotecnologia, Universidade Católica Portuguesa
Rua Dr. António Bernardino de Almeida, 4200-072 Porto
Portugal
clsilva@esb.ucp.pt

Eduardo Esteves
CCMAR- Centro de Ciências do Mar and Departamento de Engenharia Alimentar, Instituto Superior de Engenharia, Universidade do Algarve
Campus da Penha, 8005-139 Faro
Portugal
eesteves@ualg.pt

Fátima A. Miller
CBQF- Centro de Biotecnologia e Química Fina and Escola Superior de Biotecnologia, Universidade Católica Portuguesa
Rua Dr. António Bernardino de Almeida, 4200-072 Porto
Portugal
fmiller@esb.ucp.pt

Gabriela Grigioni
Instituto Tecnología de Alimentos
Centro de Investigación de Agroindustria. Instituto Nacional de Tecnología Agropecuaria and Facultad de Agronomía y Ciencias Agroalimentarias, Universidad de Morón Cabildo 134
Morón
Argentina
ggrigioni@cnia.inta.gov.ar

Golfo Moatsou
Laboratory of Dairy Research, Department of Food Science and Technology, Agricultural University of Athens
Iera Odos 75, 11855, Athens
Greece
mg@aua.gr

Igor V. Khmelinskii
CIQA- Centro de Investigação em Química do Algarve and Departamento de Química e Farmácia, Faculdade de Ciências e Tecnologia, Universidade do Algarve
Campus de Gambelas, 8005-139 Faro
Portugal
ikhmelin@ualg.pt

Jaime M. C. Aníbal
CIMA- Centro de Investigação Marinha e Ambiental and Departamento de Engenharia Alimentar, Instituto Superior de Engenharia, Universidade do Algarve
Campus da Penha, 8005-139 Faro
Portugal
janibal@ualg.pt

Janka Porubská
Department of Risk Assessment and Food Composition Data Bank
VUP Food Research Institute
Bratislava
Slovak Republic
janka.porubska@mail.t-com.sk

Javiera F. Rubilar
CIQA- Centro de Investigação em Química do Algarve and Departamento de Química e Farmácia, Faculdade de Ciências e Tecnologia, Universidade do Algarve
Campus de Gambelas, 8005-139 Faro
Portugal
javiera.rubilar@gmail.com

Javier Martínez-Monzó
Instituto de Ingeniería de Alimentos, Ciudad Politécnica de la Innovación
Universidad Politécnica de Valencia, Camino de Vera s/n
46022 Valencia
Spain
xmartine@tal.upv.es

K. Muthukumarappan
Agricultural and Biosystems Engineering Department
South Dakota State University
Brookings, SD
USA
muthukum@sdstate.edu

Laura Pizzoferrato
Istituto Nazionale di Ricerca per gli Alimenti e la Nutrizione (INRAN)
Food Science Department
Via Ardeatina, Roma
Italia
pizzoferrato@inran.it

Leandro Langman
Instituto Tecnología de Alimentos, Centro de Investigación de Agroindustria
Instituto Nacional de Tecnología Agropecuaria and Facultad de Agronomía y Ciencias Agroalimentarias, Universidad de Morón Cabildo 134
Morón, Buenos Aires
Argentina
llangman@cnia.inta.gov.ar

Margarida C. Vieira
CIQA- Centro de Investigação em Química do Algarve and Departamento de Engenharia Alimentar, Instituto Superior de Engenharia, Universidade do Algarve
Campus da Penha, 8005-139 Faro
Portugal
mvieira@ualg.pt

Maria M. Gil
Escola Superior de Turismo e Tecnologia do Mar - Instituto Politécnico de Leiria, Santuário N.ª Sra. dos Remédios, Peniche
Portugal
mariamanuel@estm.ipleiria.pt

Martín Irurueta
Instituto Tecnología de Alimentos, Centro de Investigación de Agroindustria
Instituto Nacional de Tecnología Agropecuaria
Morón, Buenos Aires
Argentina
mirurueta@cnia.inta.gov.ar

Martin Polovka
Department of Chemistry and
Food Analysis
VUP Food Research Institute
Bratislava
Slovakia Republic
polovka@vup.sk

Miguel Taverna
EEA Rafaela
Instituto Nacional de Tecnología Agropecuaria INTA
Ruta Nacional 34 Km. 27, CC 22 (2300), Rafaela, Santa Fé
Argentina
mtaverna@rafaela.inta.gov.ar

Netsanet S. Terefe
Food Science Australia
A joint venture of CSIRO and the Victorian Government
671 Sneydes Road, Werribee, VIC 3030
Australia
Netsanet.Shiferawterefe@csiro.au

Pablo A. Ulloa
CIEO- Centro de Investigação sobre Espaço e Organizações and Departamento de Química e Farmácia, Faculdade de Ciências e Tecnologia Universidade do Algarve
Campus de Gambelas, 8005-139 Faro
Portugal
pablo.ulloa.f@gmail.com

Pamela Manzi
Istituto Nazionale di Ricerca per gli Alimenti e la Nutrizione (INRAN)
Food Science Department
Via Ardeatina, 546 – 00178, Roma
Italia
manzi@inran.it

P. J. Cullen
School of Food Science and Environmental Health
Dublin Institute of Technology
Dublin 1
Ireland
pjcullen@dit.ie

Purificación García Segovia
Instituto de Ingeniería de Alimentos
Ciudad Politécnica de la Innovación
Universidad Politécnica de Valencia
Camino de Vera s/n
46022 Valencia
Spain
pugarse@tal.upv.es

Roxana Páez
EEA Rafaela
Instituto Nacional de Tecnología Agropecuaria INTA
Ruta Nacional 34-Km. 27, CC 22 (2300), Rafaela, Santa Fé
Argentina
rpaez@rafaela.inta.gov.ar

Rui M. S. Cruz
CIQA- Centro de Investigação em Química do Algarve and Departamento de Engenharia Alimentar, Instituto Superior de Engenharia, Universidade do Algarve
Campus da Penha, 8005-139 Faro
Portugal
rcruz@ualg.pt

Sameh Awad
Department of Dairy Science and Technology
Alexandria University
Egypt
Sameh111eg@yahoo.com

Teresa R. S. Brandão
CBQF- Centro de Biotecnologia e Química Fina and Escola Superior de Biotecnologia
Universidade Católica Portuguesa
Rua Dr. António Bernardino de Almeida, 4200-072 Porto
Portugal
tsbrandao@mail.esb.ucp.pt

INDEX

A

B

C

E

F

G

H

I

J

K

L

M

N

O

P

Q

R

S

T

W

X

Y

Z